# Master Handbook
# of Electronic Tables
# and Formulas

### 5th Edition

# Master Handbook of Electronic Tables and Formulas

## 5th Edition

Martin Clifford

**TAB Books**
Division of McGraw-Hill, Inc.
Blue Ridge Summit, PA 17294-0850

**TRADEMARKS**
**Amplifilm**™        AMP, Inc.
**Kapton**®, **Mylar**®, and **Teflon**®    E.I. DuPont de Nemours & Co., Inc.

FIFTH EDITION
THIRD PRINTING

© 1992 by **TAB Books**.
TAB Books is a division of McGraw-Hill, Inc.

Printed in the United States of America. All rights reserved. The publisher takes no
responsibility for the use of any of the materials or methods described in this book,
nor for the products thereof.

**Library of Congress Cataloging-in-Publication Data**

Clifford, Martin, 1910-
    Master handbook of electronic tables and formulas / by Martin
Clifford.—5th ed.
      p.  cm.
    Includes index.
    ISBN 0-8306-2192-X   ISBN 0-8306-2191-1 (pbk.)
    1. Electronics—Handbooks, manuals, etc.  2. Electronics—Tables.
I. Title.
TK7825.C56  1991
621.381′0212—dc20                       91-19752
                                            CIP

Acquisitions Editor: Roland S. Phelps
Managing Editor: Sandra L. Johnson
Book Editors: Christopher M. Cooke and B.J. Peterson
Director of Production: Katherine G. Brown
Series Design: Jaclyn J. Boone
Cover photo: Brent Blair Photography, Harrisburg, Pa.       WT1
Cover design: Lance Bush, Hagerstown, Md.           3739

To Kenneth, Paul, and Jerrold

**Other TAB Books by the Author**

No. 2675 *Microphones—3rd Edition*

# Contents

# Acknowledgments

I wish to thank the Digital Equipment Corporation, Maynard, Massachusetts for their kind permission to use their Powers of Two table. My thanks also go to my friend Marcus G. Scroggie and to Iliffe Books Ltd. for granting permission to use the decibel table that originally appeared in his *Radio Laboratory Handbook*, and to Dr. Bernhard Fischer and the Macmillan Company for the use of their Vector Conversion table.

I would also like to express my appreciation to the following for their considerable help:

AKG Acoustics, Inc.
American Radio Relay League, Inc.
Channel Master, Div. of Avnet, Inc.
GC Electronics
General Electric
*Television Cable and Fact Book* (Warren Publishing Inc.)
Zenith Video Technical Corp.

# Introduction

Problems in electronics can be solved in a number of ways. Possibly the most common method is to use a formula and to plug in or substitute numerical values. This technique calls for some arithmetic dexterity, and, quite often, a good working knowledge of algebraic and trigonometric functions. Aside from the work involved, the use of a formula has the disadvantage in that it supplies a single solution.

This book represents still another way of solving electronics problems. It consists of electronics data arranged in tabular form. In a few instances some arithmetic may be needed, but for the most part, if the elements of a problem are known, the answer is supplied immediately by a table.

The tables in this book are based on formulas commonly used in electronics. Many of the tables supply answers with a much higher order of accuracy than is generally needed in the solution of problems in electronics. Also, the tables supply a number of possible solutions, allowing the user a choice of practical component values that may be needed for a circuit.

There is a limit to the number of electronics tables that can be prepared. Tables are easily developed when only two variables are involved. Thus, it is simple enough to set up a table for capacitors in series or for resistance versus conductance. For involved formulas it is better to use the formulas directly.

What is the purpose of having a book of tables? Its function is to save time and work. Actually, there is no single best method for problem solving. Those who must solve problems in electronics as part of their educational training or work will find it helpful to be able to have a variety of techniques at their command—solving problems by formulas, solving problems by nomographs, by using a calculator and, with the help of this book, solving problems by tables and formulas.

This book has gone through a number of printings and is now in its fifth edition. It contains a substantial amount of new material, plus revisions to make the book more thorough. These revisions were necessary because the electronics field keeps expanding in many directions. In many instances, explanatory material has been added because a formula by itself is sometimes inadequate.

One of the advantages of a formula is that it lends itself to manipulation by the simple expedient of putting all the known values together, usually on the right-hand side of an equals sign, and the single unknown to its left.

# 1

CHAPTER

# Resistance

## Equivalent resistance—resistors in parallel

Whenever two resistors are connected in parallel, the total value of the shunt combination is always less than the value of the smaller unit. From a practical viewpoint, if one of the two parallel resistors has a value that is 10 or more times that of the other resistor, the equivalent value can be taken as being approximately equal to that of the smaller resistor. Use Table 1-1 to find the equivalent resistance of two resistors in parallel.

*Table 1-1. Equivalent resistance of two resistors in parallel.*

| | $R_1$ → | | | | | | | |
|---|---|---|---|---|---|---|---|---|
| **$R_2$** ↓ | **2.7** | **3.0** | **3.3** | **3.6** | **3.9** | **4.3** | **4.7** | **5.1** |
| **2.7** | 1.350 | 1.421 | 1.485 | 1.543 | 1.595 | 1.659 | 1.715 | 1.766 |
| **3.0** | 1.421 | 1.500 | 1.571 | 1.636 | 1.695 | 1.767 | 1.831 | 1.889 |
| **3.3** | 1.485 | 1.571 | 1.650 | 1.722 | 1.788 | 1.867 | 1.939 | 2.004 |
| **3.6** | 1.543 | 1.636 | 1.722 | 1.800 | 1.872 | 1.959 | 2.039 | 2.110 |
| **3.9** | 1.595 | 1.695 | 1.788 | 1.872 | 1.950 | 2.045 | 2.131 | 2.210 |
| **4.3** | 1.659 | 1.767 | 1.867 | 1.959 | 2.045 | 2.150 | 2.246 | 2.333 |
| **4.7** | 1.715 | 1.831 | 1.939 | 2.039 | 2.131 | 2.246 | 2.350 | 2.446 |
| **5.1** | 1.766 | 1.889 | 2.004 | 2.110 | 2.210 | 2.333 | 2.446 | 2.550 |
| **5.6** | 1.822 | 1.953 | 2.076 | 2.191 | 2.299 | 2.432 | 2.555 | 2.669 |
| **6.2** | 1.881 | 2.022 | 2.154 | 2.278 | 2.394 | 2.539 | 2.673 | 2.798 |
| **6.8** | 1.933 | 2.082 | 2.222 | 2.354 | 2.479 | 2.634 | 2.779 | 2.914 |
| **7.5** | 1.986 | 2.143 | 2.292 | 2.432 | 2.566 | 2.733 | 2.889 | 3.036 |
| **8.2** | 2.031 | 2.196 | 2.353 | 2.502 | 2.643 | 2.821 | 2.988 | 3.144 |
| **9.1** | 2.082 | 2.256 | 2.422 | 2.580 | 2.730 | 2.920 | 3.099 | 3.268 |
| **10** | 2.126 | 2.308 | 2.481 | 2.647 | 2.806 | 3.007 | 3.197 | 3.377 |

*Table 1-1. Continued.*

| $R_2$ | $R_1 \rightarrow$ 2.7 | 3.0 | 3.3 | 3.6 | 3.9 | 4.3 | 4.7 | 5.1 |
|---|---|---|---|---|---|---|---|---|
| ↓ | | | | | | | | |
| 11 | 2.168 | 2.351 | 2.538 | 2.712 | 2.879 | 3.092 | 3.293 | 3.484 |
| 12 | 2.204 | 2.400 | 2.588 | 2.769 | 2.943 | 3.166 | 3.377 | 3.579 |
| 13 | 2.236 | 2.438 | 2.632 | 2.819 | 3.000 | 3.231 | 3.452 | 3.663 |
| 15 | 2.288 | 2.500 | 2.705 | 2.903 | 3.095 | 3.342 | 3.579 | 3.806 |
| 16 | 2.310 | 2.526 | 2.736 | 2.939 | 3.136 | 3.389 | 3.633 | 3.867 |
| 18 | 2.348 | 2.571 | 2.789 | 3.000 | 3.205 | 3.471 | 3.727 | 3.974 |
| 20 | 2.379 | 2.609 | 2.833 | 3.051 | 3.264 | 3.539 | 3.806 | 4.064 |
| 22 | 2.405 | 2.640 | 2.870 | 3.094 | 3.313 | 3.597 | 3.873 | 4.140 |
| 24 | 2.427 | 2.667 | 2.901 | 3.130 | 3.355 | 3.647 | 3.930 | 4.206 |
| 27 | 2.455 | 2.700 | 2.941 | 3.176 | 3.408 | 3.709 | 4.003 | 4.290 |
| 30 | 2.477 | 2.737 | 2.973 | 3.214 | 3.451 | 3.761 | 4.063 | 4.359 |
| 33 | 2.496 | 2.750 | 3.000 | 3.246 | 3.488 | 3.804 | 4.114 | 4.417 |
| 36 | 2.512 | 2.769 | 3.023 | 3.273 | 3.519 | 3.841 | 4.157 | 4.467 |
| 39 | 2.525 | 2.786 | 3.043 | 3.296 | 3.545 | 3.873 | 4.195 | 4.510 |
| 43 | 2.540 | 2.804 | 3.065 | 3.322 | 3.576 | 3.909 | 4.237 | 4.559 |
| 47 | 2.553 | 2.820 | 3.083 | 3.344 | 3.601 | 3.939 | 4.273 | 4.601 |
| 51 | 2.564 | 2.833 | 3.099 | 3.363 | 3.623 | 3.966 | 4.303 | 4.636 |
| 56 | 2.576 | 2.847 | 3.116 | 3.383 | 3.646 | 3.993 | 4.336 | 4.674 |
| 62 | 2.587 | 2.862 | 3.133 | 3.402 | 3.669 | 4.021 | 4.369 | 4.712 |
| 68 | 2.597 | 2.873 | 3.147 | 3.419 | 3.688 | 4.044 | 4.396 | 4.744 |
| 75 | 2.606 | 2.885 | 3.161 | 3.435 | 3.707 | 4.067 | 4.423 | 4.775 |
| 82 | 2.614 | 2.894 | 3.172 | 3.449 | 3.723 | 4.086 | 4.445 | 4.801 |
| 91 | 2.622 | 2.904 | 3.185 | 3.463 | 3.740 | 4.106 | 4.469 | 4.829 |
| 100 | 2.629 | 2.913 | 3.195 | 3.475 | 3.754 | 4.123 | 4.489 | 4.853 |

| $R_2$ | $R_1 \rightarrow$ 5.6 | 6.2 | 6.8 | 7.5 | 8.2 | 9.1 | 10 |
|---|---|---|---|---|---|---|---|
| ↓ | | | | | | | |
| 2.7 | 1.822 | 1.881 | 1.933 | 1.986 | 2.031 | 2.082 | 2.126 |
| 3.0 | 1.953 | 2.022 | 2.082 | 2.143 | 2.196 | 2.256 | 2.308 |
| 3.3 | 2.076 | 2.154 | 2.222 | 2.292 | 2.353 | 2.422 | 2.481 |
| 3.6 | 2.191 | 2.278 | 2.354 | 2.432 | 2.502 | 2.580 | 2.647 |
| 3.9 | 2.299 | 2.394 | 2.479 | 2.566 | 2.643 | 2.730 | 2.806 |
| 4.3 | 2.432 | 2.539 | 2.634 | 2.733 | 2.821 | 2.920 | 3.007 |
| 4.7 | 2.555 | 2.673 | 2.779 | 2.889 | 2.988 | 3.099 | 3.197 |
| 5.1 | 2.669 | 2.798 | 2.914 | 3.036 | 3.144 | 3.268 | 3.377 |
| 5.6 | 2.800 | 2.942 | 3.071 | 3.206 | 3.328 | 3.467 | 3.590 |
| 6.2 | 2.942 | 3.100 | 3.243 | 3.394 | 3.531 | 3.688 | 3.827 |
| 6.8 | 3.071 | 3.243 | 3.400 | 3.566 | 3.717 | 3.892 | 4.048 |
| 7.5 | 3.206 | 3.394 | 3.566 | 3.750 | 3.917 | 4.111 | 4.286 |

*Table 1-1. Continued.*

| | $R_1 \rightarrow$ | | | | | | |
|---|---|---|---|---|---|---|---|
| $R_2$ ↓ | 5.6 | 6.2 | 6.8 | 7.5 | 8.2 | 9.1 | 10 |
| 8.2 | 3.328 | 3.531 | 3.717 | 3.917 | 4.100 | 4.313 | 4.505 |
| 9.1 | 3.467 | 3.688 | 3.892 | 4.111 | 4.313 | 4.550 | 4.764 |
| 10 | 3.590 | 3.827 | 4.048 | 4.286 | 4.505 | 4.764 | 5.000 |
| 11 | 3.711 | 3.965 | 4.202 | 4.459 | 4.698 | 4.980 | 5.238 |
| 12 | 3.818 | 4.088 | 4.340 | 4.615 | 4.871 | 5.175 | 5.455 |
| 13 | 3.914 | 4.198 | 4.465 | 4.756 | 5.028 | 5.353 | 5.652 |
| 15 | 4.078 | 4.387 | 4.679 | 5.000 | 5.302 | 5.664 | 6.000 |
| 16 | 4.148 | 4.468 | 4.772 | 5.106 | 5.421 | 5.801 | 6.154 |
| 18 | 4.271 | 4.612 | 4.935 | 5.294 | 5.634 | 6.044 | 6.429 |
| 20 | 4.375 | 4.733 | 5.075 | 5.455 | 5.816 | 6.254 | 6.667 |
| 22 | 4.464 | 4.837 | 5.194 | 5.593 | 5.974 | 6.437 | 6.875 |
| 24 | 4.541 | 4.927 | 5.299 | 5.714 | 6.112 | 6.598 | 7.059 |
| 27 | 4.638 | 5.042 | 5.432 | 5.870 | 6.290 | 6.806 | 7.297 |
| 30 | 4.719 | 5.138 | 5.543 | 6.000 | 6.440 | 6.982 | 7.500 |
| 33 | 4.788 | 5.219 | 5.638 | 6.111 | 6.568 | 7.133 | 7.674 |
| 36 | 4.846 | 5.289 | 5.720 | 6.207 | 6.679 | 7.264 | 7.826 |
| 39 | 4.897 | 5.350 | 5.790 | 6.290 | 6.775 | 7.378 | 7.959 |
| 43 | 4.955 | 5.419 | 5.871 | 6.386 | 6.887 | 7.511 | 8.113 |
| 47 | 5.004 | 5.477 | 5.941 | 6.468 | 6.982 | 7.624 | 8.246 |
| 51 | 5.046 | 5.528 | 6.000 | 6.538 | 7.064 | 7.722 | 8.361 |
| 56 | 5.091 | 5.582 | 6.064 | 6.614 | 7.153 | 7.828 | 8.485 |
| 62 | 5.136 | 5.636 | 6.128 | 6.691 | 7.242 | 7.935 | 8.611 |
| 68 | 5.174 | 5.682 | 6.182 | 6.755 | 7.318 | 8.026 | 8.718 |
| 75 | 5.211 | 5.727 | 6.235 | 6.818 | 7.392 | 8.115 | 8.824 |
| 82 | 5.242 | 5.764 | 6.279 | 6.872 | 7.455 | 8.191 | 8.913 |
| 91 | 5.275 | 5.805 | 6.327 | 6.929 | 7.522 | 8.273 | 9.010 |
| 100 | 5.303 | 5.838 | 6.367 | 6.977 | 7.579 | 8.341 | 9.091 |

| | $R_1 \rightarrow$ | | | | | | |
|---|---|---|---|---|---|---|---|
| $R_2$ ↓ | 11 | 12 | 13 | 15 | 16 | 18 | 20 |
| 2.7 | 2.168 | 2.204 | 2.236 | 2.288 | 2.310 | 2.348 | 2.379 |
| 3.0 | 2.357 | 2.400 | 2.438 | 2.500 | 2.526 | 2.571 | 2.609 |
| 3.3 | 2.538 | 2.588 | 2.632 | 2.705 | 2.736 | 2.789 | 2.833 |
| 3.6 | 2.712 | 2.769 | 2.819 | 2.903 | 2.939 | 3.000 | 3.051 |
| 3.9 | 2.879 | 2.943 | 3.000 | 3.095 | 3.136 | 3.205 | 3.264 |
| 4.3 | 3.092 | 3.166 | 3.231 | 3.342 | 3.389 | 3.471 | 3.539 |
| 4.7 | 3.293 | 3.377 | 3.452 | 3.579 | 3.633 | 3.727 | 3.806 |
| 5.1 | 3.484 | 3.579 | 3.663 | 3.806 | 3.867 | 3.974 | 4.064 |
| 5.6 | 3.711 | 3.818 | 3.914 | 4.078 | 4.148 | 4.271 | 4.375 |
| 6.2 | 3.965 | 4.088 | 4.198 | 4.387 | 4.468 | 4.612 | 4.733 |

*Table 1-1. Continued.*

| $R_2$ | 11 | 12 | 13 | 15 | 16 | 18 | 20 |
|---|---|---|---|---|---|---|---|
| 6.8 | 4.202 | 4.340 | 4.465 | 4.679 | 4.772 | 4.935 | 5.075 |
| 7.5 | 4.459 | 4.615 | 4.756 | 5.000 | 5.106 | 5.294 | 5.455 |
| 8.2 | 4.698 | 4.871 | 5.028 | 5.302 | 5.421 | 5.634 | 5.816 |
| 9.1 | 4.980 | 5.175 | 5.353 | 5.664 | 5.801 | 6.044 | 6.254 |
| 10 | 5.238 | 5.455 | 5.652 | 6.000 | 6.154 | 6.429 | 6.667 |
| 11 | 5.500 | 5.739 | 5.958 | 6.346 | 6.519 | 6.828 | 7.097 |
| 12 | 5.739 | 6.000 | 6.240 | 6.667 | 6.857 | 7.200 | 7.500 |
| 13 | 5.958 | 6.240 | 6.500 | 6.964 | 7.172 | 7.548 | 7.879 |
| 15 | 6.346 | 6.667 | 6.964 | 7.500 | 7.742 | 8.182 | 8.571 |
| 16 | 6.519 | 6.857 | 7.172 | 7.742 | 8.000 | 8.471 | 8.889 |
| 18 | 6.828 | 7.200 | 7.548 | 8.182 | 8.471 | 9.000 | 9.474 |
| 20 | 7.097 | 7.500 | 7.879 | 8.571 | 8.889 | 9.474 | 10.000 |
| 22 | 7.333 | 7.765 | 8.171 | 8.919 | 9.263 | 9.900 | 10.476 |
| 24 | 7.543 | 8.000 | 8.432 | 9.231 | 9.600 | 10.286 | 10.909 |
| 27 | 7.816 | 8.308 | 8.775 | 9.643 | 10.047 | 10.800 | 11.489 |
| 30 | 8.049 | 8.571 | 9.070 | 10.000 | 10.435 | 11.250 | 12.000 |
| 33 | 8.250 | 8.800 | 9.326 | 10.313 | 10.776 | 11.657 | 12.543 |
| 36 | 8.426 | 9.000 | 9.551 | 10.588 | 11.077 | 12.000 | 12.857 |
| 39 | 8.580 | 9.176 | 9.750 | 10.833 | 11.345 | 12.316 | 13.220 |
| 43 | 8.759 | 9.382 | 9.982 | 11.121 | 11.661 | 12.689 | 13.651 |
| 47 | 8.914 | 9.559 | 10.183 | 11.371 | 11.937 | 13.015 | 14.030 |
| 51 | 9.048 | 9.714 | 10.359 | 11.591 | 12.179 | 13.304 | 14.366 |
| 56 | 9.194 | 9.882 | 10.551 | 11.831 | 12.444 | 13.622 | 14.737 |
| 62 | 9.342 | 10.054 | 10.747 | 12.078 | 12.718 | 13.950 | 15.122 |
| 68 | 9.468 | 10.200 | 10.914 | 12.289 | 12.952 | 14.233 | 15.455 |
| 75 | 9.593 | 10.345 | 11.080 | 12.500 | 13.187 | 14.516 | 15.789 |
| 82 | 9.699 | 10.468 | 11.221 | 12.680 | 13.388 | 14.760 | 16.078 |
| 91 | 9.814 | 10.602 | 11.375 | 12.877 | 13.607 | 15.028 | 16.396 |
| 100 | 9.910 | 10.714 | 11.504 | 13.043 | 13.793 | 15.254 | 16.667 |

$R_1$

| $R_2$ | 22 | 24 | 27 | 30 | 33 | 36 | 39 | 43 |
|---|---|---|---|---|---|---|---|---|
| 2.7 | 2.405 | 2.427 | 2.455 | 2.477 | 2.496 | 2.512 | 2.525 | 2.540 |
| 3.0 | 2.640 | 2.667 | 2.700 | 2.727 | 2.750 | 2.769 | 2.786 | 2.804 |
| 3.3 | 2.870 | 2.901 | 2.941 | 2.973 | 3.000 | 3.023 | 3.043 | 3.065 |
| 3.6 | 3.094 | 3.130 | 3.176 | 3.214 | 3.246 | 3.273 | 3.296 | 3.322 |
| 3.9 | 3.313 | 3.355 | 3.408 | 3.451 | 3.488 | 3.519 | 3.545 | 3.576 |
| 4.3 | 3.597 | 3.647 | 3.709 | 3.761 | 3.804 | 3.841 | 3.873 | 3.909 |
| 4.7 | 3.873 | 3.930 | 4.003 | 4.063 | 4.114 | 4.157 | 4.195 | 4.237 |

*Table 1-1. Continued.*

| | | | | $R_1 \rightarrow$ | | | | |
|---|---|---|---|---|---|---|---|---|
| $R_2$ | 22 | 24 | 27 | 30 | 33 | 36 | 39 | 43 |
| 5.1 | 4.140 | 4.206 | 4.290 | 4.359 | 4.417 | 4.467 | 4.510 | 4.559 |
| 5.6 | 4.464 | 4.541 | 4.638 | 4.719 | 4.788 | 4.846 | 4.897 | 4.955 |
| 6.2 | 4.837 | 4.927 | 5.042 | 5.138 | 5.219 | 5.289 | 5.350 | 5.419 |
| 6.8 | 5.194 | 5.299 | 5.432 | 5.543 | 5.638 | 5.720 | 5.790 | 5.871 |
| 7.5 | 5.593 | 5.714 | 5.870 | 6.000 | 6.111 | 6.207 | 6.290 | 6.386 |
| 8.2 | 5.974 | 6.112 | 6.290 | 6.440 | 6.568 | 6.679 | 6.775 | 6.887 |
| 9.1 | 6.437 | 6.598 | 6.806 | 6.982 | 7.133 | 7.264 | 7.378 | 7.511 |
| 10 | 6.875 | 7.059 | 7.297 | 7.500 | 7.674 | 7.826 | 7.959 | 8.113 |
| 11 | 7.333 | 7.543 | 7.816 | 8.049 | 8.250 | 8.426 | 8.580 | 8.759 |
| 12 | 7.765 | 8.000 | 8.308 | 8.571 | 8.800 | 9.000 | 9.176 | 9.382 |
| 13 | 8.171 | 8.432 | 8.775 | 9.070 | 9.326 | 9.551 | 9.750 | 9.982 |
| 15 | 8.919 | 9.231 | 9.643 | 10.000 | 10.313 | 10.588 | 10.833 | 11.121 |
| 16 | 9.263 | 9.600 | 10.047 | 10.435 | 10.776 | 11.077 | 11.345 | 11.661 |
| 18 | 9.900 | 10.286 | 10.800 | 11.250 | 11.647 | 12.000 | 12.316 | 12.689 |
| 20 | 10.476 | 10.909 | 11.489 | 12.000 | 12.543 | 12.857 | 13.220 | 13.651 |
| 22 | 11.000 | 11.478 | 12.122 | 12.692 | 13.200 | 13.655 | 14.066 | 14.554 |
| 24 | 11.478 | 12.000 | 12.706 | 13.333 | 13.895 | 14.400 | 14.857 | 15.404 |
| 27 | 12.122 | 12.706 | 13.500 | 14.211 | 14.850 | 15.429 | 15.955 | 16.586 |
| 30 | 12.692 | 13.333 | 14.211 | 15.000 | 15.714 | 16.364 | 16.957 | 17.671 |
| 33 | 13.200 | 13.895 | 14.850 | 15.714 | 16.500 | 17.217 | 17.875 | 18.671 |
| 36 | 13.655 | 14.400 | 15.429 | 16.364 | 17.217 | 18.000 | 18.200 | 19.595 |
| 39 | 14.066 | 14.857 | 15.955 | 16.957 | 17.875 | 18.720 | 19.500 | 20.451 |
| 43 | 14.554 | 15.403 | 16.586 | 17.671 | 18.671 | 19.595 | 20.451 | 21.500 |
| 47 | 14.986 | 15.887 | 17.149 | 18.312 | 19.388 | 20.386 | 21.314 | 22.456 |
| 51 | 15.370 | 16.320 | 17.654 | 18.889 | 20.036 | 21.103 | 22.100 | 23.330 |
| 56 | 15.795 | 16.800 | 18.217 | 19.535 | 20.764 | 21.913 | 22.989 | 24.323 |
| 62 | 16.238 | 17.302 | 18.809 | 20.217 | 21.537 | 22.776 | 23.941 | 25.390 |
| 68 | 16.622 | 17.739 | 19.326 | 20.816 | 22.218 | 23.538 | 24.785 | 26.342 |
| 75 | 17.010 | 18.182 | 19.853 | 21.423 | 22.917 | 24.324 | 25.658 | 27.331 |
| 82 | 17.346 | 18.566 | 20.312 | 21.964 | 23.530 | 25.017 | 26.430 | 28.208 |
| 91 | 17.717 | 18.991 | 20.822 | 22.562 | 24.218 | 25.795 | 27.300 | 29.201 |
| 100 | 18.033 | 19.355 | 21.260 | 23.077 | 24.812 | 26.471 | 28.058 | 30.070 |

| | | | | $R_1 \rightarrow$ | | | | |
|---|---|---|---|---|---|---|---|---|
| $R_2$ | 47 | 51 | 56 | 62 | 68 | 75 | 82 | 91 | 100 |
| 2.7 | 2.553 | 2.564 | 2.576 | 2.587 | 2.597 | 2.606 | 2.614 | 2.622 | 2.629 |
| 3.0 | 2.820 | 2.833 | 2.847 | 2.862 | 2.873 | 2.885 | 2.894 | 2.904 | 2.913 |
| 3.3 | 3.083 | 3.099 | 3.116 | 3.133 | 3.147 | 3.161 | 3.172 | 3.185 | 3.195 |
| 3.6 | 3.344 | 3.363 | 3.383 | 3.402 | 3.419 | 3.435 | 3.449 | 3.463 | 3.475 |
| 3.9 | 3.601 | 3.623 | 3.646 | 3.669 | 3.688 | 3.707 | 3.723 | 3.740 | 3.754 |

*Table 1-1. Continued.*

| $R_2$ | $R_1 \longrightarrow$ 47 | 51 | 56 | 62 | 68 | 75 | 82 | 91 | 100 |
|---|---|---|---|---|---|---|---|---|---|
| 4.3 | 3.939 | 3.966 | 3.993 | 4.021 | 4.044 | 4.067 | 4.086 | 4.106 | 4.123 |
| 4.7 | 4.273 | 4.303 | 4.336 | 4.369 | 4.396 | 4.423 | 4.445 | 4.469 | 4.489 |
| 5.1 | 4.601 | 4.636 | 4.674 | 4.712 | 4.744 | 4.775 | 4.801 | 4.829 | 4.853 |
| 5.6 | 5.004 | 5.046 | 5.091 | 5.136 | 5.174 | 5.211 | 5.242 | 5.275 | 5.303 |
| 6.2 | 5.477 | 5.528 | 5.582 | 5.636 | 5.682 | 5.727 | 5.764 | 5.805 | 5.838 |
| 6.8 | 5.941 | 6.000 | 6.064 | 6.128 | 6.182 | 6.235 | 6.279 | 6.327 | 6.367 |
| 7.5 | 6.468 | 6.538 | 6.614 | 6.691 | 6.755 | 6.818 | 6.872 | 6.929 | 6.977 |
| 8.2 | 6.982 | 7.064 | 7.153 | 7.242 | 7.318 | 7.392 | 7.455 | 7.522 | 7.579 |
| 9.1 | 7.624 | 7.722 | 7.828 | 7.935 | 8.026 | 8.115 | 8.191 | 8.273 | 8.341 |
| 10.0 | 8.246 | 8.361 | 8.485 | 8.611 | 8.718 | 8.824 | 8.913 | 9.010 | 9.091 |
| 11 | 8.914 | 9.048 | 9.194 | 9.342 | 9.468 | 9.593 | 9.699 | 9.814 | 9.910 |
| 12 | 9.559 | 9.714 | 9.882 | 10.054 | 10.200 | 10.345 | 10.468 | 10.602 | 10.714 |
| 13 | 10.183 | 10.359 | 10.551 | 10.747 | 10.914 | 11.080 | 11.221 | 11.375 | 11.504 |
| 15 | 11.371 | 11.591 | 11.831 | 12.078 | 12.289 | 12.500 | 12.680 | 12.877 | 13.043 |
| 16 | 11.937 | 12.179 | 12.444 | 12.718 | 12.952 | 13.187 | 13.388 | 13.607 | 13.793 |
| 18 | 13.015 | 13.304 | 13.622 | 13.950 | 14.233 | 14.516 | 14.760 | 15.028 | 15.254 |
| 20 | 14.030 | 14.366 | 14.737 | 15.122 | 15.455 | 15.789 | 16.078 | 16.396 | 16.667 |
| 22 | 14.986 | 15.370 | 15.795 | 16.238 | 16.622 | 17.010 | 17.346 | 17.717 | 18.033 |
| 24 | 15.887 | 16.320 | 16.800 | 17.302 | 17.739 | 18.182 | 18.566 | 18.991 | 19.355 |
| 27 | 17.149 | 17.654 | 18.217 | 18.809 | 19.326 | 19.853 | 20.312 | 20.822 | 21.260 |
| 30 | 18.312 | 18.889 | 19.535 | 20.217 | 20.816 | 21.423 | 21.964 | 22.562 | 23.077 |
| 33 | 19.388 | 20.036 | 20.764 | 21.537 | 22.218 | 22.917 | 23.530 | 24.218 | 24.812 |
| 36 | 20.836 | 21.103 | 21.913 | 22.776 | 23.538 | 24.324 | 25.017 | 25.795 | 26.471 |
| 39 | 21.314 | 22.100 | 22.989 | 23.941 | 24.785 | 25.658 | 26.430 | 27.300 | 28.058 |
| 43 | 22.456 | 23.330 | 24.323 | 25.390 | 26.342 | 27.331 | 28.208 | 29.201 | 30.070 |
| 47 | 23.500 | 24.459 | 25.553 | 26.734 | 27.791 | 28.893 | 29.876 | 30.993 | 31.973 |
| 51 | 24.459 | 25.500 | 26.692 | 27.982 | 29.143 | 30.357 | 31.444 | 32.683 | 33.775 |
| 56 | 25.553 | 26.692 | 28.000 | 29.424 | 30.710 | 32.061 | 33.275 | 34.667 | 35.897 |
| 62 | 26.734 | 27.982 | 29.424 | 31.000 | 32.431 | 33.942 | 35.306 | 36.876 | 38.272 |
| 68 | 27.791 | 29.143 | 30.710 | 32.431 | 34.000 | 35.664 | 37.173 | 38.918 | 40.476 |
| 75 | 28.893 | 30.357 | 32.061 | 33.942 | 35.664 | 37.500 | 39.172 | 41.114 | 42.857 |
| 82 | 29.876 | 31.444 | 33.275 | 35.306 | 37.173 | 39.172 | 41.000 | 43.133 | 45.055 |
| 91 | 30.993 | 32.683 | 34.667 | 36.876 | 38.918 | 41.114 | 43.133 | 45.500 | 47.644 |
| 100 | 31.973 | 33.775 | 35.897 | 38.272 | 40.476 | 42.857 | 45.055 | 47.644 | 50.000 |

| $R_2$ | $R_1 \longrightarrow$ 10 | 11 | 12 | 13 | 15 | 16 | 18 | 20 | 22 |
|---|---|---|---|---|---|---|---|---|---|
| 100 | 9.091 | 9.910 | 10.714 | 11.504 | 13.043 | 13.793 | 15.254 | 16.667 | 18.033 |
| 110 | 9.167 | 10.000 | 10.820 | 11.626 | 13.200 | 13.968 | 15.469 | 16.923 | 18.333 |

*Table 1-1. Continued.*

| $R_2$ | $R_1 \rightarrow$ 10 | 11 | 12 | 13 | 15 | 16 | 18 | 20 | 22 |
|---|---|---|---|---|---|---|---|---|---|
| ↓ | | | | | | | | | |
| 120 | 9.231 | 10.076 | 10.909 | 11.729 | 13.333 | 14.118 | 15.652 | 17.143 | 18.592 |
| 130 | 9.286 | 10.142 | 10.986 | 11.818 | 13.448 | 14.247 | 15.811 | 17.333 | 18.816 |
| 150 | 9.375 | 10.248 | 11.111 | 11.963 | 13.636 | 14.458 | 16.071 | 17.647 | 19.186 |
| 160 | 9.412 | 10.292 | 11.163 | 12.023 | 13.714 | 14.545 | 16.180 | 17.778 | 19.341 |
| 180 | 9.474 | 10.366 | 11.250 | 12.124 | 13.846 | 14.694 | 16.364 | 18.000 | 19.604 |
| 200 | 9.524 | 10.427 | 11.321 | 12.207 | 13.953 | 14.815 | 16.514 | 18.182 | 19.820 |
| 220 | 9.565 | 10.476 | 11.380 | 12.275 | 14.043 | 14.915 | 16.639 | 18.333 | 20.000 |
| 240 | 9.600 | 10.518 | 11.423 | 12.332 | 14.118 | 15.000 | 16.744 | 18.462 | 20.153 |
| 270 | 9.643 | 10.569 | 11.489 | 12.403 | 14.211 | 15.105 | 16.875 | 18.621 | 20.342 |
| 300 | 9.677 | 10.611 | 11.538 | 12.460 | 14.286 | 15.190 | 16.981 | 18.750 | 20.497 |
| 330 | 9.706 | 10.645 | 11.579 | 12.507 | 14.348 | 15.260 | 17.069 | 18.857 | 20.625 |
| 360 | 9.730 | 10.674 | 11.613 | 12.547 | 14.400 | 15.319 | 17.143 | 18.947 | 20.733 |
| 390 | 9.750 | 10.698 | 11.642 | 12.581 | 14.444 | 15.369 | 17.206 | 19.024 | 20.825 |
| 430 | 9.773 | 10.726 | 11.674 | 12.619 | 14.494 | 15.426 | 17.277 | 19.111 | 20.929 |
| 470 | 9.792 | 10.748 | 11.701 | 12.650 | 14.536 | 15.473 | 17.336 | 19.184 | 21.016 |
| 510 | 9.808 | 10.768 | 11.724 | 12.677 | 14.571 | 15.513 | 17.386 | 19.245 | 21.090 |
| 560 | 9.825 | 10.788 | 11.748 | 12.705 | 14.609 | 15.556 | 17.439 | 19.310 | 21.168 |
| 620 | 9.841 | 10.808 | 11.772 | 12.733 | 14.646 | 15.597 | 17.492 | 19.375 | 21.246 |
| 680 | 9.855 | 10.825 | 11.792 | 12.756 | 14.676 | 15.632 | 17.536 | 19.429 | 21.311 |
| 750 | 9.868 | 10.841 | 11.811 | 12.779 | 14.706 | 15.666 | 17.578 | 19.481 | 21.373 |
| 820 | 9.880 | 10.854 | 11.827 | 12.797 | 14.731 | 15.694 | 17.613 | 19.524 | 21.425 |
| 910 | 9.891 | 10.869 | 11.844 | 12.817 | 14.747 | 15.724 | 17.651 | 19.570 | 21.481 |
| 1000 | 9.901 | 10.880 | 11.858 | 12.833 | 14.778 | 15.748 | 17.682 | 19.608 | 21.526 |

| $R_2$ | $R_1 \rightarrow$ 24 | 27 | 30 | 33 | 36 | 39 | 43 | 47 | 51 |
|---|---|---|---|---|---|---|---|---|---|
| ↓ | | | | | | | | | |
| 100 | 19.355 | 21.260 | 23.077 | 24.812 | 26.471 | 28.058 | 30.070 | 31.973 | 33.775 |
| 110 | 19.701 | 21.679 | 23.571 | 25.385 | 27.123 | 28.792 | 30.915 | 32.930 | 34.845 |
| 120 | 20.000 | 22.041 | 24.000 | 25.882 | 27.692 | 29.434 | 31.656 | 33.772 | 35.789 |
| 130 | 20.260 | 22.357 | 24.375 | 26.319 | 28.193 | 30.000 | 32.312 | 34.520 | 36.630 |
| 150 | 20.690 | 22.881 | 25.000 | 27.049 | 29.032 | 30.952 | 33.420 | 35.787 | 38.060 |
| 160 | 20.870 | 23.102 | 25.263 | 27.358 | 29.388 | 31.357 | 33.892 | 36.329 | 38.673 |
| 180 | 21.176 | 23.478 | 25.714 | 27.887 | 30.000 | 32.055 | 34.709 | 37.269 | 39.740 |
| 200 | 21.429 | 23.789 | 26.087 | 28.326 | 30.508 | 32.636 | 35.391 | 38.057 | 40.637 |
| 220 | 21.639 | 24.049 | 26.400 | 28.696 | 30.938 | 33.127 | 35.970 | 38.727 | 41.402 |
| 240 | 21.818 | 24.270 | 26.667 | 29.011 | 31.304 | 33.548 | 36.466 | 39.303 | 42.062 |
| 270 | 22.041 | 24.545 | 27.000 | 29.406 | 31.765 | 34.078 | 37.093 | 40.032 | 42.897 |
| 300 | 22.222 | 24.771 | 27.273 | 29.730 | 32.143 | 34.513 | 37.609 | 40.634 | 43.590 |
| 330 | 22.373 | 24.958 | 27.500 | 30.000 | 32.459 | 34.878 | 38.043 | 41.141 | 44.173 |

Table 1-1. Continued.

| $R_2$ ↓ | 24 | 27 | 30 | 33 | 36 | 39 | 43 | 47 | 51 |
|---|---|---|---|---|---|---|---|---|---|
| 360 | 22.500 | 25.116 | 27.692 | 30.229 | 32.727 | 35.188 | 38.412 | 41.572 | 44.672 |
| 390 | 22.609 | 25.252 | 27.857 | 30.426 | 32.958 | 35.455 | 38.730 | 41.945 | 45.102 |
| 430 | 22.731 | 25.405 | 28.043 | 30.648 | 33.219 | 35.757 | 39.091 | 42.369 | 45.593 |
| 470 | 22.834 | 25.533 | 28.200 | 30.835 | 33.439 | 36.012 | 39.396 | 42.727 | 46.008 |
| 510 | 22.921 | 25.642 | 28.333 | 30.994 | 33.626 | 36.230 | 39.656 | 43.034 | 46.364 |
| 560 | 23.014 | 25.758 | 28.475 | 31.164 | 33.826 | 36.461 | 39.934 | 43.361 | 46.743 |
| 620 | 23.106 | 25.873 | 28.615 | 31.332 | 34.024 | 36.692 | 40.211 | 43.688 | 47.124 |
| 680 | 23.182 | 25.969 | 28.732 | 31.473 | 34.190 | 36.885 | 40.443 | 43.961 | 47.442 |
| 750 | 23.256 | 26.062 | 28.846 | 31.609 | 34.351 | 37.072 | 40.668 | 44.228 | 47.753 |
| 820 | 23.318 | 26.139 | 28.941 | 31.723 | 34.486 | 37.229 | 40.857 | 44.452 | 48.014 |
| 910 | 23.383 | 26.222 | 29.043 | 31.845 | 34.630 | 37.397 | 41.060 | 44.692 | 48.293 |
| 1000 | 23.438 | 26.290 | 29.126 | 31.946 | 34.749 | 37.536 | 41.227 | 44.890 | 48.525 |

$R_1 \longrightarrow$

| $R_2$ ↓ | 56 | 62 | 68 | 75 | 82 | 91 | 100 |
|---|---|---|---|---|---|---|---|
| 100 | 35.897 | 38.272 | 40.476 | 42.857 | 45.055 | 47.644 | 50.000 |
| 110 | 37.108 | 39.651 | 42.022 | 44.595 | 46.979 | 49.801 | 52.381 |
| 120 | 38.182 | 40.879 | 43.404 | 46.154 | 48.713 | 51.754 | 55.545 |
| 130 | 39.140 | 41.979 | 44.646 | 47.461 | 50.283 | 53.529 | 56.522 |
| 150 | 40.777 | 43.868 | 46.789 | 50.000 | 53.017 | 56.639 | 60.000 |
| 160 | 41.481 | 44.685 | 47.719 | 51.064 | 54.215 | 58.008 | 61.538 |
| 180 | 42.712 | 46.116 | 49.355 | 52.941 | 56.336 | 60.443 | 64.286 |
| 200 | 43.750 | 47.328 | 50.746 | 54.545 | 58.156 | 62.543 | 66.667 |
| 220 | 44.638 | 48.369 | 51.944 | 55.932 | 59.735 | 64.373 | 68.750 |
| 240 | 45.405 | 49.272 | 52.987 | 57.143 | 61.118 | 65.982 | 70.588 |
| 270 | 46.380 | 50.422 | 54.320 | 58.696 | 62.898 | 68.061 | 72.973 |
| 300 | 47.191 | 51.381 | 55.435 | 60.000 | 64.398 | 69.821 | 75.000 |
| 330 | 47.876 | 52.194 | 56.382 | 61.111 | 65.680 | 71.330 | 76.744 |
| 360 | 48.462 | 52.891 | 57.196 | 62.069 | 66.787 | 72.639 | 78.261 |
| 390 | 48.969 | 53.496 | 57.904 | 62.903 | 67.754 | 73.784 | 79.592 |
| 430 | 49.547 | 54.187 | 58.715 | 63.861 | 68.867 | 75.106 | 81.132 |
| 470 | 50.038 | 54.774 | 59.405 | 64.679 | 69.819 | 76.239 | 82.456 |
| 510 | 50.459 | 55.280 | 60.000 | 65.385 | 70.642 | 77.221 | 83.607 |
| 560 | 50.909 | 55.820 | 60.637 | 66.142 | 71.526 | 78.280 | 84.848 |
| 620 | 51.361 | 56.364 | 61.279 | 66.906 | 72.422 | 79.353 | 86.111 |

| $R_1 \rightarrow$ | | | | | | | |
| $R_2 \downarrow$ | 56 | 62 | 68 | 75 | 82 | 91 | 100 |
| --- | --- | --- | --- | --- | --- | --- | --- |
| **680** | 51.739 | 56.819 | 61.818 | 67.550 | 73.176 | 80.259 | 87.179 |
| **750** | 52.109 | 57.266 | 62.347 | 68.182 | 73.918 | 81.153 | 88.235 |
| **820** | 52.420 | 57.642 | 62.793 | 68.715 | 74.545 | 81.910 | 89.130 |
| **910** | 52.754 | 58.045 | 63.272 | 69.289 | 75.222 | 82.727 | 90.099 |
| **1000** | 53.030 | 58.380 | 63.670 | 69.767 | 75.786 | 83.410 | 90.909 |

The tables shown on the following pages can also be used to find the equivalent resistance of three or more resistors in parallel (Fig. 1-1) if the problem is handled on a step-by-step basis. First, take any two of the resistors, and, using the tables, find the equivalent resistance. Consider this equivalent resistance just as

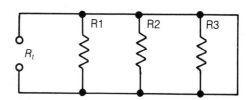

*Fig. 1-1. Resistors in parallel.*

though it were a physical unit and combine its value with the remaining resistor, again using the tables.

Sometimes a design problem involving resistors will yield a value that is not practical—not practical in the sense that a resistor having such a value will be unavailable. In this instance, the tables can again be used to advantage. Simply locate the nearest value in the tables and then move left to get the value of $R_2$ and upward to get the value of $R_1$. R1 and R2 will be standard, available resistors, that can be connected in parallel to supply the required resistance.

*Example*  What is the equivalent resistance of two resistors in parallel, one having a value of 5.6 Ω and the other a value of 9.1 Ω?

*Solution*  Locate 5.6 at the left, in the column marked $R_2$ in Table 1-1. Move across and locate the column headed by 9.1. The equivalent resistance is shown to be 3.467 Ω.

*Example*  What two resistors in parallel will give an equivalent value of 3 Ω?

*Solution*  The table shows possible combinations. You could use 3.3 and 33 Ω, or 3.6 and 18 Ω, or 3.9 and 13 Ω.

*Example*  What is the equivalent shunt resistance of a 68-Ω resistor and a 27-Ω resistor in parallel?

*Solution*   Locate 6.8 Ω in $R_2$ column. Move the decimal point one place to the right so that 6.8 becomes 68. Locate 2.7 Ω in the $R_1$ row and consider it now as 27 Ω. These two values meet at 1.933 in the table. However, this is now 19.33 Ω because you must again move the decimal point one place to the right.

This problem can also be solved directly in two ways. Locate 68 in the $R_1$ row and then move down to find the answer in the 27 row under the general heading of $R_2$. Or, locate 27 in the $R_1$ row and move down to find the answer in the 68 row under the general heading of $R_2$.

The values shown in Table 1-1 are median values and do not take tolerances into consideration. Resistor tolerances are usually 20% or less, and they can be plus or minus. However, Table 1-1 does supply a practical guide for the quick determination of two resistors in parallel, or for finding parallel resistor combinations which will be equivalent to a desired resistance value.

The assumption is made in Table 1-1 that the resistors to be wired in parallel are those having similar units, that is, in ohms, kilohms, or megohms. If one resistor has its value specified in ohms, and another in kilohms, for example, it would be necessary to change kilohms to ohms or ohms to kilohms.

*Example*   What is the equivalent resistance of a 1-kΩ resistor shunted by a 100-Ω resistor?

*Solution*   1 kΩ is equal to 1 000 Ω. Locate 1 000 in the $R_2$ column in Table 1-1 and then 100 in the last column to the right. The equivalent resistance is given as 90.909 Ω.

The range of ohmic values is wide and extends from as little as 1 Ω to multi-megohms. When solving formulas, it is helpful to move back and forth between these extremes. Table 1-2 shows the possible conversions between ohms,

*Table 1-2. Conversion factors for resistances.*

| Unit | | Equivalents | |
|------|------|------|------|
| 1 Ω | $= 10^{-3}$ kΩ | $= 0.001$ kΩ | $= 1/1\ 000$ kΩ |
| 1 Ω | $= 10^{-6}$ MΩ | $= 0.000\ 001$ MΩ | $= 1/1\ 000\ 000$ MΩ |
| 1 kΩ | $= 10^3$ Ω | $= 1\ 000$ Ω | |
| 1 kΩ | $= 10^{-3}$ MΩ | $= 0.001$ MΩ | $= 1/1\ 000$ MΩ |
| 1 MΩ | $= 10^3$ kΩ | $= 1\ 000$ kΩ | |
| 1 MΩ | $= 10^6$ Ω | $= 1\ 000\ 000$ Ω | |

kilohms, and megohms. Table 1-3 is similar to Table 1-2 but presents the conversion factors in a different way.

# Formulas for resistors in parallel

Although Table 1-1 is suitable for finding the equivalent resistance of parallel resistors, there are components other than physical resistors that have resistance.

*Table 1-3. Alternative arrangement of conversion factors for resistances.*

| | $\Omega$ | $k\Omega$ | $M\Omega$ |
|---|---|---|---|
| $\Omega$ | — | $\div$ 1 000 or $\times$ 10$^{-3}$ | $\div$ 1 000 000 or $\times$ 10$^{-6}$ |
| $k\Omega$ | $\times$ 1 000 or $\times$ 10$^3$ | — | $\div$ 1 000 or $\times$ 10$^{-3}$ |
| $M\Omega$ | $\times$ 1 000 000 or $\times$ 10$^6$ | $\times$ 1 000 or $\times$ 10$^3$ | — |

$\Omega$=ohms; $k\Omega$=kilohms; $M\Omega$=megohms.

A length of wire has resistance and so does a coil made of that wire. A metal bar can have resistance and so can a normally nonconductive material. Heating elements used for electric irons, ovens, and soldering irons have resistance. And, of course, there are also resistors that do not fit into the values shown in Table 1-1.

You can calculate the value of any pair of resistors using the following formula:

$$R_t = \frac{R_1 \times R_2}{R_1 + R_2}$$

$R_t$ is the total resistance. R1 is either of the resistors; R2 is the other. $R_1$ and $R_2$ must be in similar units: ohms, kilohms, or megohms.

If you have three resistors in parallel, you can still use this formula but in a two-step process. Find the equivalent value of any two of the resistors. Then, using the formula again, combine the answer with the remaining resistor. Alternatively, for three resistors you can use this formula:

$$R_t = \frac{1}{\dfrac{1}{R_1} + \dfrac{1}{R_2} + \dfrac{1}{R_3}}$$

**Example** What is the equivalent resistance of two resistors with values of 45 $\Omega$ and 95 $\Omega$? Note that neither of these values appears in Table 1-1.
*Solution*

$$R_t = \frac{45 \times 95}{45 + 95} = \frac{4\,275}{140} = 30.536$$

You can check on the answer by finding the values that are nearest to those given in the problem and using Table 1-1. The closest to 95 $\Omega$ is 91 $\Omega$. The closest to 45 $\Omega$ is 47 $\Omega$. Table 1-1 shows the equivalent resistance to be 30.993 $\Omega$.

**Example** What is the effective resistance of three parallel resistors with values of 75 $\Omega$, 95 $\Omega$, and 120 $\Omega$?
*Solution*

$$R_t = \frac{1}{\dfrac{1}{R_1} + \dfrac{1}{R_2} + \dfrac{1}{R_3}} = \frac{1}{\dfrac{1}{75} + \dfrac{1}{95} + \dfrac{1}{120}}$$

$$1/75 = 0.013\ 3 \qquad 1/95 = 0.010\ 5 \qquad 1/120 = 0.008\ 3$$

$$R_t = \frac{1}{0.013\ 3 + 0.010\ 5 + 0.008\ 3} = \frac{1}{0.032\ 1} = 31.15\ \Omega$$

The formula for finding three resistors in parallel can be extended to find the equivalent resistance of four or more shunt (parallel) resistors.

Still another formula for finding the equivalent resistance of a number of resistors in parallel is as follows:

$$\frac{1}{R_t} = \frac{1}{R_1} + \frac{1}{R_2} + \frac{1}{R_3}$$

Although the formula indicates just three resistors, it can be expanded for any number of resistors.

*Example* What is the equivalent resistance of 46 $\Omega$, 53 $\Omega$, and 75 $\Omega$ wired in parallel?

*Solution*

$$\frac{1}{R_t} = \frac{1}{R_1} + \frac{1}{R_2} + \frac{1}{R_3}$$
$$= \frac{1}{46} + \frac{1}{53} + \frac{1}{75}$$
$$= 0.021\ 7 + 0.019 + 0.013$$
$$= 0.053\ 7$$

The answer thus far isn't the resistance, but actually the conductance. To find the resistance, divide the answer into 1.

$$\frac{1}{R_t} = 0.053\ 7$$

$$R_t = \frac{1}{0.053\ 7} = 18.62\ \Omega$$

This answer is an approximation because the answers for the division of 46 into 1, 53 into 1 and 75 into 1 were carried out to only two or three decimal places. Thus, 1/46 is actually 0.021 739; 1/53 is 0.018 867 9 and 1/75 is 0.013 333. If these numbers are used the shunt resistance is 18.539 8 $\Omega$. This degree of accuracy might or might not be needed, depending on the work that is being done.

# Equal-value resistors in parallel

When two resistors in parallel, R1 and R2, have equal values of resistance, the total resistance, $R_t$, is equal to one-half of the value of either resistor.

$$R_t = \frac{1}{2} R_1 \text{ or } \frac{1}{2} R_2$$

When three resistors, R1, R2, and R3, are in parallel, the total resistance, $R_t$, is one third the value of any of the resistors.

$$R_t = \frac{1}{3} R_1 \text{ or } \frac{1}{3} R_2 \text{ or } \frac{1}{3} R_3$$

# Selecting a shunting resistor

A common problem is having a resistor on hand whose value is too high for a particular application. The resistor can be shunted with another resistor so that the two resistors in parallel will have the required resistance value.

$$R_1 = \frac{R_t \times R_2}{R_2 - R_t}$$

R1 and R2 are the two parallel resistors. $R_t$ is the equivalent value.

*Example*   What value resistor should be shunted across a 30-ohm resistor so that the equivalent value of the two parallel resistors will be 20 $\Omega$?
   *Solution*

$$R_1 = \frac{R_t \times R_2}{R_2 - R_t} = \frac{20 \times 30}{30 - 20} = \frac{600}{10} = 60 \ \Omega$$

A 30-$\Omega$ resistor when shunted by a 60-$\Omega$ resistor will have an equivalent resistance of 20 $\Omega$.

# Nomograms

The nomogram in N 1-1 is based on the formula:

$$R_t = \frac{R_1 \times R_2}{R_1 + R_2}$$

used for finding the equivalent resistance of two resistors in parallel. The two outer scales represent the known values of the two resistors. The center scale supplies the solution to the problem. The straight line in the nomogram shows how to find the equivalent value of a 180-$\Omega$ resistor in parallel with a 150-$\Omega$ resistor. The straightedge crosses the center scale at 82 $\Omega$. (The nomogram could also be used to find the equivalent of larger values, provided they are in identical powers of 10.)

Capacitors in series use the same formula as resistors in parallel, so this nomogram can also be used to solve for series capacitors. All three scales must be in the same capacitance units: microfarads or picofarads.

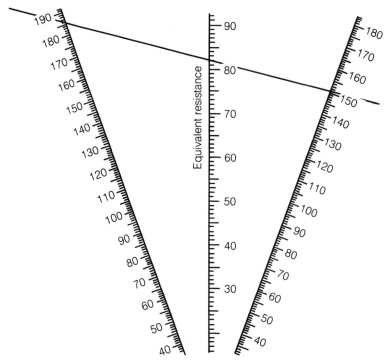

N 1-1. *Nomogram for determining the equivalent resistance of two resistors in parallel.*

An advantage of this nomogram is that you can use it to determine what resistor values or capacitor values you need to obtain a desired amount of resistance or capacitance. If, for example, you have a supply of miscellaneous resistors, but do not have a 60-Ω unit, put the straightedge on 60 and use this as a pivot point. The straightedge, in cutting across the left and right hand scales, will supply you with a large number of possible combinations. This technique can also be used for capacitors.

# Resistors in series

For resistors in series, the formula is as follows:

$$R_t = R_1 + R_2 + R_3$$

All that you need to do is to add the values of the resistors in the series circuit (Fig. 1-2). The resistance value must be in the same units: ohms, kilohms, or megohms.

# Resistors in series-parallel

Finding the total value of resistance for resistors connected in series-parallel can be obtained by using the formulas for series resistance and parallel resistance and

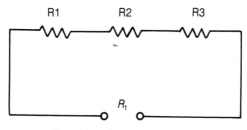

*Fig. 1-2.   Resistors in series.*

then adding the answers. In terms of a formula:

$$R_t = (R_1 + R_2) + \left( \frac{R_3 \times R_4}{R_3 + R_4} \right)$$

The other formulas for parallel resistances can be used instead.

## Resistance versus conductance

The opposition to the movement of an electrical current can be expressed in terms of resistance (measured in ohms) or in terms of conductance (expressed in siemens). It is often very convenient to be able to move back and forth quickly and easily between resistance and conductance in the solution of electronics problems. This can be readily done because resistance and conductance are reciprocals.

Sometimes, when working with resistances you will find the values are not covered by the tables and that using a formula to solve the problem will involve some laborious arithmetic. In that case, it might be easier and quicker to work with conductances. To find the total conductance of resistors in parallel, simply consider them as conductors and add the values of the individual units. Thus, if you have a number of resistors in parallel, use Table 1-4 to find the equivalent

*Table 1-4.*
*Resistance (ohms) versus conductance (siemens).*

| Ω | S | Ω | S | Ω | S |
|---|---|---|---|---|---|
| 0.1 | 10.000 0 | 2 | 0.500 0 | 12 | 0.083 3 |
| 0.2 | 5.000 0 | 3 | 0.333 3 | 13 | 0.076 9 |
| 0.3 | 3.333 3 | 4 | 0.250 0 | 14 | 0.071 4 |
| 0.4 | 2.500 0 | 5 | 0.200 0 | 15 | 0.066 7 |
| 0.5 | 2.000 0 | 6 | 0.166 7 | 16 | 0.062 5 |
| 0.6 | 1.666 7 | 7 | 0.142 9 | 17 | 0.058 8 |
| 0.7 | 1.428 6 | 8 | 0.125 0 | 18 | 0.055 6 |
| 0.8 | 1.250 0 | 9 | 0.111 1 | 19 | 0.052 6 |
| 0.9 | 1.111 1 | 10 | 0.100 0 | 20 | 0.050 0 |
| 1.0 | 1.000 0 | 11 | 0.090 9 | 21 | 0.047 6 |

*Table 1-4. Continued.*

| Ω | S | Ω | S | Ω | S |
|---|---|---|---|---|---|
| 22 | 0.045 5 | 49 | 0.020 4 | 76 | 0.013 2 |
| 23 | 0.043 5 | 50 | 0.020 0 | 77 | 0.013 0 |
| 24 | 0.041 7 | 51 | 0.019 6 | 78 | 0.012 8 |
| 25 | 0.040 0 | 52 | 0.019 2 | 79 | 0.012 7 |
| 26 | 0.038 5 | 53 | 0.018 9 | 80 | 0.012 5 |
| 27 | 0.037 0 | 54 | 0.018 5 | 81 | 0.012 3 |
| 28 | 0.035 7 | 55 | 0.018 2 | 82 | 0.012 2 |
| 29 | 0.034 5 | 56 | 0.017 9 | 83 | 0.012 0 |
| 30 | 0.033 3 | 57 | 0.017 5 | 84 | 0.011 9 |
| 31 | 0.032 3 | 58 | 0.017 2 | 85 | 0.011 8 |
| 32 | 0.031 3 | 59 | 0.016 9 | 86 | 0.011 6 |
| 33 | 0.030 3 | 60 | 0.016 7 | 87 | 0.011 5 |
| 34 | 0.029 4 | 61 | 0.016 4 | 88 | 0.011 4 |
| 35 | 0.028 6 | 62 | 0.016 1 | 89 | 0.011 2 |
| 36 | 0.027 8 | 63 | 0.015 9 | 90 | 0.011 1 |
| 37 | 0.027 0 | 64 | 0.015 6 | 91 | 0.011 0 |
| 38 | 0.026 3 | 65 | 0.015 4 | 92 | 0.010 9 |
| 39 | 0.025 6 | 66 | 0.015 2 | 93 | 0.010 8 |
| 40 | 0.025 0 | 67 | 0.014 9 | 94 | 0.010 6 |
| 41 | 0.024 4 | 68 | 0.014 7 | 95 | 0.010 5 |
| 42 | 0.023 8 | 69 | 0.014 5 | 96 | 0.010 4 |
| 43 | 0.023 3 | 70 | 0.014 3 | 97 | 0.010 3 |
| 44 | 0.022 7 | 71 | 0.014 1 | 98 | 0.010 2 |
| 45 | 0.022 2 | 72 | 0.013 9 | 99 | 0.010 1 |
| 46 | 0.021 7 | 73 | 0.013 7 | 100 | 0.010 0 |
| 47 | 0.021 3 | 74 | 0.013 5 | | |
| 48 | 0.020 8 | 75 | 0.013 3 | | |

conductance of each resistor. Add the conductances and then use the table once again to find the equivalent resistance.

The symbol for resistance is $R$; the symbol for conductance is $G$. The relationship between the two is expressed as $R=1/G$ or $G=1/R$. If you are considering a complete circuit, that is, a circuit consisting of a number of resistors in parallel, then the total resistance of the circuit is the reciprocal of the total conductance.

Conductance can be substituted into the different forms of Ohm's law. Thus, for resistance, you would have $R=E/I$. For conductance we would have $G=I/E$. The following formula regards the components as conductors.

$$G_t=G_1+G_2+G_3 \ldots \ldots$$

$G_t$ is the total conductance in siemens. Conversion can be made to the more familiar resistance form by dividing each term into the number 1.

$$G_t = G_1 + G_2 + G_3 = \frac{1}{R_t} = \frac{1}{R_1} + \frac{1}{R_2} + \frac{1}{R_3}$$

Where more than three resistors in parallel are involved, it is often more convenient to work with conductances rather than resistance. The formula for four resistors in parallel becomes fairly involved:

$$R_t = \frac{R_1 R_2 R_3 R_4}{R_1 R_2 R_3 + R_2 R_3 R_4 + R_1 R_2 R_4 + R_1 R_3 R_4}$$

Using conductances, the formula for four parallel resistors is a much simpler arrangement:

$$G_T = G_1 + G_2 + G_3 + G_4$$

*Example* What is the conductance of a resistor whose value, as measured, is 64 Ω?

*Solution* In Table 1-4, locate 64 in the column marked ohms. The value of conductance, as shown in the column (siemens) to the right, is 0.015 6 S.

*Example* What is the resistance of a component whose conductance is 0.055 6 S?

*Solution* The value of 0.055 6, in the siemens column, corresponds to 18 Ω, as indicated in the ohms column.

*Example* The values of four resistors, measured on a bridge, are 90, 83, 79, and 71 Ω, respectively. What is the equivalent resistance when these units are connected in parallel?

*Solution* Using Table 1-4 you will find that the corresponding conductance values are 0.011 1, 0.012 0, 0.012 7, and 0.014 1 S, respectively. Adding these results in a total conductance of 0.049 9 S. Using Table 1-4 once again, the closest conductance value is 0.050 0 S, and, as shown by the table, corresponds to 20 Ω.

# Standard EIA values for composition resistors

Composition resistors are available in values based on the recommendations of the EIA (Electronics Industries Association). These values are shown in Table 1-5.

The lower-case letter $k$ means kilo, or multiply by 1 000—120 k equals $120 \times 1\ 000$, or 120 000 Ω. The upper-case letter $M$ means mega, or multiply by 1 000 000—1.8 M equals $1.8 \times 1\ 000\ 000$, or 1 800 000 Ω.

Table 1-6 shows conductance values for standard composition resistors.

## Table 1-5. Standard Electronics Industry Association (EIA) values for composition resistors.

| Ω | Ω | Ω | Ω | Ω | Ω | Ω | Ω |
|---|---|---|---|---|---|---|---|
| ... | 10 | 100 | 1 k | 10 k | 100 k | 1 M | 10 M |
|  | 11 | 110 | 1.1 k | 11 k | 110 k | 1.1 M | 11 M |
|  | 12 | 120 | 1.2 k | 12 k | 120 k | 1.2 M | 12 M |
|  | 13 | 130 | 1.3 k | 13 k | 130 k | 1.3 M | 13 M |
|  | 15 | 150 | 1.5 k | 15 k | 150 k | 1.5 M | 15 M |
|  | 16 | 160 | 1.6 k | 16 k | 160 k | 1.6 M | 16 M |
|  | 18 | 180 | 1.8 k | 18 k | 180 k | 1.8 M | 18 M |
|  | 20 | 200 | 2 k | 20 k | 200 k | 2.0 M | 20 M |
|  | 22 | 220 | 2.2 k | 22 k | 220 k | 2.2 M | 22 M |
|  | 24 | 240 | 2.4 k | 24 k | 240 k | 2.4 M | . . . . . . |
| 2.7 | 27 | 270 | 2.7 k | 27 k | 270 k | 2.7 M |  |
| 3.0 | 30 | 300 | 3 k | 30 k | 300 k | 3.0 M |  |
| 3.3 | 33 | 330 | 3.3 k | 33 k | 330 k | 3.3 M |  |
| 3.6 | 36 | 360 | 3.6 k | 36 k | 360 k | 3.6 M |  |
| 3.9 | 39 | 390 | 3.9 k | 39 k | 390 k | 3.9 M |  |
| 4.3 | 43 | 430 | 4.3 k | 43 k | 430 k | 4.3 M |  |
| 4.7 | 47 | 470 | 4.7 k | 47 k | 470 k | 4.7 M |  |
| 5.1 | 51 | 510 | 5.1 k | 51 k | 510 k | 5.1 M |  |
| 5.6 | 56 | 560 | 5.6 k | 56 k | 560 k | 5.6 M |  |
| 6.2 | 62 | 620 | 6.2 k | 62 k | 620 k | 6.2 M |  |
| 6.8 | 68 | 680 | 6.8 k | 68 k | 680 k | 6.8 M |  |
| 7.5 | 75 | 750 | 7.5 k | 75 k | 750 k | 7.5 M |  |
| 8.2 | 82 | 820 | 8.2 k | 82 k | 820 k | 8.2 M |  |
| 9.1 | 91 | 910 | 9.1 k | 91 k | 910 k | 9.1 M |  |

## Table 1-6. Conductance of standard EIA values for composition resistors. (R, resistance in ohms; G, conductance in siemens).

| R | G | R | G | R | G | R | G | R | G |
|---|---|---|---|---|---|---|---|---|---|
|  |  | 11 | 0.090 91 | 110 | 0.009 09 | 1.1 k | 0.000 909 | 11 k | 0.000 090 |
|  |  | 12 | 0.833 33 | 120 | 0.008 33 | 1.2 k | 0.000 833 | 12 k | 0.000 083 |
|  |  | 13 | 0.076 92 | 130 | 0.007 69 | 1.3 k | 0.000 769 | 13 k | 0.000 077 |
|  |  | 15 | 0.066 67 | 150 | 0.006 67 | 1.5 k | 0.000 666 | 15 k | 0.000 066 |
|  |  | 16 | 0.062 50 | 160 | 0.006 25 | 1.6 k | 0.000 625 | 16 k | 0.000 062 |
|  |  | 18 | 0.055 56 | 180 | 0.005 56 | 1.8 k | 0.000 555 | 18 k | 0.000 055 |
|  |  | 20 | 0.050 00 | 200 | 0.005 00 | 2 k | 0.000 500 | 20 k | 0.000 050 |
|  |  | 22 | 0.045 45 | 220 | 0.004 55 | 2.2 k | 0.000 454 | 22 k | 0.000 045 |
|  |  | 24 | 0.041 67 | 240 | 0.004 17 | 2.4 k | 0.000 416 | 24 k | 0.000 041 |
| 2.7 | 0.370 37 | 27 | 0.037 04 | 270 | 0.003 70 | 2.7 k | 0.000 370 | 27 k | 0.000 037 |
| 3.0 | 0.333 33 | 30 | 0.033 33 | 300 | 0.003 33 | 3 k | 0.000 333 | 30 k | 0.000 033 |
| 3.3 | 0.303 03 | 33 | 0.030 30 | 330 | 0.003 03 | 3.3 k | 0.000 303 | 33 k | 0.000 030 |

*Table 1-6. Conductance of standard EIA values for composition resistors. (R, resistance in ohms; G, conductance in siemens).*

| R | G | R | G | R | G | R | G | R | G |
|---|---|---|---|---|---|---|---|---|---|
| 3.6 | 0.277 78 | 36 | 0.027 78 | 360 | 0.002 78 | 3.6 k | 0.000 277 | 36 k | 0.000 027 |
| 3.9 | 0.256 41 | 39 | 0.025 64 | 390 | 0.002 56 | 3.9 k | 0.000 256 | 39 k | 0.000 025 |
| 4.3 | 0.232 56 | 43 | 0.023 26 | 430 | 0.002 33 | 4.3 k | 0.000 232 | 43 k | 0.000 023 |
| 4.7 | 0.212 77 | 47 | 0.021 28 | 470 | 0.002 13 | 4.7 k | 0.000 212 | 47 k | 0.000 021 |
| 5.1 | 0.196 08 | 51 | 0.019 61 | 510 | 0.001 96 | 5.1 k | 0.000 196 | 51 k | 0.000 019 |
| 5.6 | 0.178 57 | 56 | 0.017 86 | 560 | 0.001 79 | 5.6 k | 0.000 178 | 56 k | 0.000 017 |
| 6.2 | 0.161 29 | 62 | 0.016 13 | 620 | 0.001 61 | 6.2 k | 0.000 161 | 62 k | 0.000 016 |
| 6.8 | 0.147 06 | 68 | 0.014 71 | 680 | 0.001 47 | 6.8 k | 0.000 147 | 68 k | 0.000 014 |
| 7.5 | 0.133 33 | 75 | 0.013 33 | 750 | 0.001 33 | 7.5 k | 0.000 133 | 75 k | 0.000 013 |
| 8.2 | 0.121 95 | 82 | 0.012 20 | 820 | 0.001 220 | 8.2 k | 0.000 121 | 82 k | 0.000 012 |
| 9.1 | 0.109 89 | 91 | 0.010 99 | 910 | 0.001 100 | 9.1 k | 0.000 109 | 91 k | 0.000 011 |
| 10 | 0.100 00 | 100 | 0.010 00 | 1 k | 0.001 000 | 10 k | 0.000 100 | 100 k | 0.000 010 |

Column *R* indicates the resistance values in ohms; column *G* the corresponding conductance values in siemens.

## Resistor types

There is a large variety of fixed resistors. These usually are named after some physical characteristic. Table 1-7 lists some that are more widely used.

*Table 1-7. Summary of resistor types.*

| | | |
|---|---|---|
| Axial | Fixed tapped | Precision |
| Carbon composition | Hybrid | Radial |
| Carbon film | Metal film | Rheostat |
| Ceramic | Metal oxide | Sliding tapped |
| Cermet | Miniature | Surface mount |
| Chip | Noninductive | Trimmer |
| Conductive plastic | PC board | Variable |
| Deposited film | Potentiometer | Wirewound |
| Fixed | Power | |

Resistors can be grouped into these basic types:

fixed

fixed tap

variable

Changes in electronics technology constantly presents a demand for new resistor types. Table 1-7 lists those in common use, but the table should be considered dynamic. As technology advances, some resistor types fall out of favor and new resistor types are introduced.

## Fixed resistors

Resistors are rated in terms of their power-handling capabilities in addition to their ohmic value. Standard resistors are available with different power ratings. Common power handling capabilities range from as little as 0.1 W to as much as 20 W. Usually, the larger the physical size of a resistor, the greater its power handling rating. Tolerance refers to how much a resistor can vary from its ohmic rating. Resistor tolerance ratings are usually 5%, 10%, or 20%. Resistors using carbon as the conductive element have a negative temperature coefficient (the resistance varies inversely with temperature). Most fixed resistors are usually either carbon or wirewound.

**Fixed tap**   Some wirewound resistors have a number of fixed taps. Each tap provides a different amount of resistance. Fixed-tap resistors often are used as voltage dividers.

## Variable resistors

Variable resistors are available as:

> potentiometers
> rheostats
> sliding tap

**Potentiometers**   The potentiometer is an electromechanical device having two terminals connected to opposite ends of a resistive element. It also has a sliding arm that makes wiping contact with the resistive element and is connected to a third terminal. The sliding arm is used to obtain a desired fraction of the input voltage. Hence, potentiometers are voltage dividing devices and have very limited current carrying ability.

**Rheostats**   The construction of a rheostat allows them to carry much more current than potentiometers. Rheostats are often wirewound. Rheostats often are used to control stage lights.

**Sliding tap**   This is a wirewound resistor with a movable tap. The tap can be moved to obtain a desired amount of resistance. The tap usually is tightened into position with a screw.

# Resistor color code

The ohmic value of a fixed resistor is indicated by bands of color around one end of the body of the resistor. Each color corresponds to a specific number, as indicated in Table 1-8.

The percentage of resistance variation, known as the tolerance of the resistor, is also color marked, as shown in Table 1-9. To read the color code, hold the

| Number | Color |
|--------|--------|
| 0 | Black |
| 1 | Brown |
| 2 | Red |
| 3 | Orange |
| 4 | Yellow |
| 5 | Green |
| 6 | Blue |
| 7 | Violet |
| 8 | Gray |
| 9 | White |

*Table 1-8.*
*Resistor color code.*

| Tolerance | Color |
|-----------|--------|
| 1% | Brown |
| 2% | Red |
| 3% | Orange |
| 4% | Yellow |
| 5% | Gold |
| 10% | Silver |
| 20% | No color |

*Table 1-9.*
*Tolerance color code.*

resistor (Fig. 1-3) so that the colors are at one end. The first and second colors represent the first and second digits of the resistor's value. The third color indicates the number of zeros that follow the first two numbers. The tolerance band will be the fourth color from the end.

*Example* A fixed resistor has the following color bands: brown, green, yellow, gold. What is its resistance and tolerance?

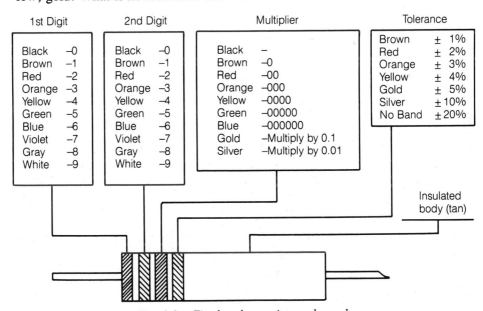

Fig. 1-3. Fixed-carbon resistor color code.

*Solution*  The colors have these values:

brown   green   yellow   gold

1          5          0000      5%

The third color, yellow in this example, indicates the number of zeros. Yellow is 4, so four zeros are used. The resistor has a value of 150 000 Ω and a tolerance of 5%.

Table 1-10 is a listing of fixed and variable resistor types and their characteristics.

*Table 1-10. Resistor types.* Radio Electronics.

| Types | Characteristics |
|---|---|
| Carbon composition | **Resistance range:** 2.7 Ω to 100 MΩ<br>**Power rating:** to 2 W<br>**Tolerance:** 1% to 20%<br>**Temperature coefficient:** −200 to −8.000 ppm/°C<br>General purpose. Excellent transient and surge handling capabilities. Resistance increases by 20% during storage under humid conditions. Made from a mixture of graphite and nonconductive binder such as clay to hold the particles in place. Resistance determined by the ratio of carbon to binder material, with the resistance varying from less than one ohm to several megohms. |
| Carbon composition potentiometer | **Resistance range:** 50 Ω to 10 MΩ<br>**Power rating:** to 5 W<br>**Temperature coefficient:** 1 000 ppm/°C<br>**Life expectancy:** 5 000 000 rotations<br>High shaft torque causes poor adjustability.<br>The resistive element is carbon composition or metallic film for potentiometers and resistance wire for rheostats. In some instances a conductive ceramic element is used for potentiometers. |
| Concentric potentiometer | Dual variable resistor equipped with two shafts, an inner and an outer. Each can be operated independently. |
| Fixed-film | Made of a core of a nonconductive material on which is sprayed a thin layer or film of a resistive substance. They have a higher power rating than carbon-composition resistors and are sometimes used as a substitute for wirewound types. The film can be applied in a continuous layer with the resistance determined by the thickness of the film. This film can be made of carbon whose particles are micro-crystalline, and is sometimes mixed with boron or various metal oxides. |
| Carbon film | **Resistance range:** 10 Ω to 25 MΩ<br>**Power rating:** 0.1 to 10 W<br>**Tolerance:** 2% to 10% |

*Table 1-10. Continued.*

| Types | Characteristics |
|---|---|
| | **Temperature coefficient:** $-200$ to $-1\ 000$ ppm/°C |
| | General purpose, cost less than carbon-composition units. |
| Metal film | **Resistance range:** 10 Ω to 3 MΩ (high voltage types: 1 kΩ to 30 GΩ) |
| | **Power rating:** to 10 W (high-voltage types: to 6 W) |
| | **Tolerance:** 0.1% to 2% |
| | **Temperature coefficient:** ±25 to ±175 ppm/°C |
| | **Life expectancy (potentiometers):** 100 000 rotations |
| | **Failure mode:** resistance change or catastrophic failure |
| | Fair degree of precision in lower value units. High stability, long life. |
| Film networks | **Resistance range:** 10 Ω to 33 MΩ |
| | **Power rating:** to 0.2 W per element, to 1.6 W per network |
| | **Tolerance:** 0.1% to 5% |
| | **Operating temperature range:** $-55$ to $+125$°C |
| | **Temperature coefficient:** ±25 to ±300 ppm/°C |
| Chip resistors | **Resistance range:** 1 Ω to 100 MΩ |
| | **Power rating:** to 2 W |
| | **Tolerance:** 1% to 20% |
| | **Operating temperature range:** $-55$ to $+125$°C |
| Power wirewound | **Resistance range:** 0.1 Ω to 180 kΩ |
| | **Power rating:** to greater than 225 W |
| | **Tolerance:** 5% to 10% |
| | **Temperature coefficient:** less than ±260 ppm/°C |
| | Wirewound resistors consist of resistance wire wound around a nonconductive core made of a heat resistive material such as ceramic, with the resistance dependent on the type of wire used. Their advantage is their higher power dissipation capability over resistors such as the carbon type. |
| Precision wirewound | **Resistance range:** 0.1 Ω to 800 kΩ |
| | **Power rating:** to 15 W |
| | **Tolerance:** 0.01% to 1% |
| | **Life expectancy (potentiometers):** (2 000 000 to 10 000 000 rotations |
| | Used in low-tolerance, high-power dissipation applications where ac performance is not critical. Power dissipation depends on heatsink or air flow around the device. When mounting on a PC (printed circuit) board, use standoffs to prevent charring the board. Wirewound potentiometers do not suffer from contact resistance variations. The units can be manufactured with low temperature coefficients and tight tolerances. Applications include motor speed controls. Precision types used in servo mechanisms. |

*Table 1-10. Continued.*

| Types | Characteristics |
|---|---|
| Cermet | **Resistance range:** 50 Ω to 5 MΩ<br>**Power rating:** to 2 W<br>**Life expectancy (potentiometers):** 50 to 500 000 rotations<br>Very stable under humid conditions. Low temperature coefficients. Low end resistance (2 Ω). Short life expectancy. Cermet is also the thick film used in resistor networks and chip resistors. |
| Ceramic | **Life expectancy:** Long in-service life<br>Made of carborundum, a carbon-silicon compound used as a diode demodulator in the early days of radio. Affected by the voltage across it with its resistance varying inversely with the applied emf. Made having either a positive or negative temperature coefficient. Available in fixed or variable form. Also known as a voltage-dependent resistor. |
| Conductive plastic potentiometers | **Resistance range:** 150 Ω to 5 MΩ<br>**Power rating:** to 1 W<br>**Temperature coefficient:** $-600$ to $-300$ ppm/°C<br>**Life expectancy:** 100 000 to 4 000 000 rotations |
| General purpose conductive plastic potentiometers | **Resistance range:** 1 Ω to 15 kΩ, depending on power rating<br>**Power rating:** to 1 000 W |
| Precision conductive plastic potentiometer | **Resistance range:** 100 Ω to 500 kΩ<br>**Power rating:** to 7 W<br>**Tolerance:** 3%<br>**Life expectancy:** Greater than 2 000 000 rotations |
| Conductive plastic trimmers | **Resistance range:** 10 to 100 000 Ω<br>**Power rating:** to 1 W<br>Plastic potentiometers have a long life expectancy. Resistance will shift if exposed to humidity. |
| Hybrid potentiometers | **Resistance range:** 200 to 250 000 Ω<br>**Power rating:** to 7 W<br>**Tolerance:** 5%<br>**Life expectancy:** 10 000 000 rotations |

## Tolerance

Electrical components have physical and electrical tolerances. Both are expressed in percent. To change percent to a decimal, divide percent by 100. To change a decimal to percent, multiply by 100.

There are three types of tolerances: the plus tolerance, the minus tolerance, and the plus-minus tolerance. Plus tolerances are indicated by a plus sign. Example: A wire has a diameter of 0.065 mil +2. This wire can have a diameter ranging between these limits: 0.065 mil and 0.065+2=0.067 mil. A minus tolerance is shown by a minus sign. Example: The chassis cutout for a transformer is 3 in −0.025 in. The cutout can have any dimension ranging from 3 in to 3−0.025 in 3 in to 2.975 in. A plus-minus tolerance is used for resistors. A resistor has a value of 1 000 Ω, ±2%. 20%=20/100=0.2. 1 000×0.2=200 Ω. 1 000+200 Ω=1 200 Ω. 1 000−200 Ω=800 Ω. A 1 000-Ω ±20% resistor can have any value between 800 and 1 200 Ω.

## Meter multipliers

Not only actual resistors, but components or components plus resistors can be put in parallel. Thus, a resistor can be mounted in shunt with a current meter, often an ammeter. The purpose of the shunt is to extend the range of the meter.

A formula for finding the value of shunt resistance is as follows:

$$R = \frac{R_i \times A}{I - A}$$

$R_i$ is the internal resistance of the meter. This value can be obtained from the manufacturer of the meter. $A$ is the maximum reading on the existing scale of the meter and $I$ is the current to be measured.

*Example*  A meter scale indicates a maximum possible reading of 1 A on a meter whose internal resistance is 0.01 Ω. What value of shunt resistance should be used to extend the scale to 10 A? See Fig. 1-4.

*Solution*  Based on the information supplied $R_i$ is 0.01 Ω, $A$ is 1 A and $I$ is 10 A.

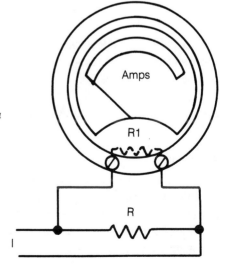

*Fig. 1-4.  Shunt used to extend the range of a current-reading meter.*

$$R = \frac{R_i \times A}{I - A} = \frac{0.01 \times 1}{10 - 1} = 0.001\ 1\ \Omega$$

Unlike ordinary resistors, such as those used in radio and television receivers, shunts must have a high order of accuracy and are often supplied by the manufacturers of the meters. To determine the multiplying effect of the shunt, use this formula:

$$MP = \frac{R_i}{R} + 1$$

$MP$ is the multiplying value, $R_i$ is the internal resistance, and $R$ is the value of the shunt.

In the example just given, $R_i$ is 0.01 Ω, $R$ is 0.001 1 Ω.

$$MP = \frac{0.01}{0.001\ 1} + 1 = 9.0 + 1 = 10$$

Sometimes more than one formula can be used to solve a problem in electronics. Although such formulas may look different, they are generally the same. They just use other symbols or have a different arrangement. When two resistive elements are in parallel, the same voltage appears across both. Thus:

$$I_m R_m = I_s R_s$$

$I_m$ is the current through the meter, $R_m$ is the meter resistance, $I_s$ is the shunt current, and $R_s$ is the value of shunt resistance. The formula can be transposed to solve for any one of the unknowns:

$$R_s = \frac{I_m R_m}{I_s}$$

*Example*  Assume you have a milliammeter having a full scale deflection of 1 mA and an internal resistance of 50 Ω. Suppose you want to extend the full scale reading by a factor of 20.

$$R_s = \frac{I_m R_m}{I_s} = \frac{(0.001)\ (50)}{0.020} = \frac{0.05}{0.02} = 2.5\ \Omega$$

Because the shunt must carry 20 times as much current as the meter, it must have 1/20th of the resistance of the meter. The internal resistance of the meter is 50 Ω; 1/20th of 50 Ω is 2.5 Ω.

A resistor wired in series with a voltmeter will extend its range based on the following formula:

$$R = \frac{E}{I} - R_i$$

$R_i$ is the internal resistance of the meter, $I$ is the current needed for full-scale deflection and $E$ is the voltage to be measured.

**Example**  A meter has an internal resistance of 1 000 Ω. It requires 1 mA for full-scale deflection. How can this instrument be made to measure 200 V?

**Solution**  The maximum voltage the instrument can measure, based on its electrical characteristics, is 1 V; 1 mA flowing through the internal resistance of 1 000 Ω is an electrical pressure of 1 V. $(E = I \times R = 0.001 \times 1\,000 = 1\ V)$.

$$R = \frac{E}{I} - R_i$$
$$= \frac{200}{0.001} - 1\,000$$
$$= 200\,000 - 1\,000 = 199\,000\ \Omega$$

By using a resistor in series with the meter, as shown in Fig. 1-5, the meter can be used to measure up to 200 V. (This is dc voltage, not ac.) In this example, the

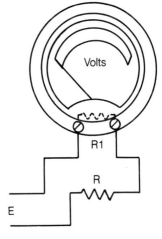

*Fig. 1-5.  Series resistor used for extending voltmeter range.*

multiplying power is 200. It can also be calculated by a formula:

$$MP = \frac{R_i + R}{R_i}$$
$$= \frac{1\,000 + 199\,000}{1\,000}$$
$$= \frac{200\,000}{1\,000} = 200$$

$R_i$ is the internal resistance and $R$ is the value of the multiplier resistance.

Table 1-11 supplies resistor values for commonly used voltmeters. Thus, a meter having a full-scale deflection of 1 V (first column) and requiring a current

*Table 1-11. Multiplier resistor values.*

| Range full scale (V) | Meter sensitivity (μA) | | | | mA | |
|---|---|---|---|---|---|---|
| | 50 | 100 | 200 | 500 | 1 | 5 |
| 1 | 20 000 | 10 000 | 5 000* | 2 000* | 1 000* | 200* |
| 1.5 | 30 000 | 15 000 | 7 500 | 3 000* | 1 500* | 300* |
| 2.5 | 50 000 | 25 000 | 12 500 | 5 000* | 2 500* | 500* |
| 3 | 60 000 | 30 000 | 15 000 | 6 000 | 3 000* | 600* |
| 5 | 100 000 | 50 000 | 25 000 | 10 000 | 5 000* | 1 000* |
| 6 | 120 000 | 60 000 | 30 000 | 12 000 | 6 000 | 1 200 |
| 10 | 200 000 | 100 000 | 50 000 | 20 000 | 10 000 | 2 000 |
| 12 | 240 000 | 120 000 | 60 000 | 24 000 | 12 000 | 2 400 |
| 15 | 300 000 | 150 000 | 75 000 | 30 000 | 15 000 | 3 000 |
| 25 | 500 000 | 250 000 | 125 000 | 50 000 | 25 000 | 5 000 |
| 30 | 600 000 | 300 000 | 150 000 | 60 000 | 30 000 | 6 000 |
| 50 | 1 000 000 | 500 000 | 250 000 | 100 000 | 50 000 | 10 000 |
| 60 | 1 200 000 | 600 000 | 300 000 | 120 000 | 60 000 | 12 000 |
| 100 | 2 000 000 | 1 000 000 | 500 000 | 200 000 | 100 000 | 20 000 |
| 150 | 3 000 000 | 1 500 000 | 750 000 | 300 000 | 150 000 | 30 000 |
| 250 | 5 000 000 | 2 500 000 | 1 250 000 | 500 000 | 250 000 | 50 000 |
| 300 | 6 000 000 | 3 000 000 | 1 500 000 | 600 000 | 300 000 | 60 000 |
| 500 | 10 000 000 | 5 000 000 | 2 500 000 | 1 000 000 | 500 000 | 100 000 |
| 1 000 | 20 000 000 | 10 000 000 | 5 000 000 | 2 000 000 | 1 000 000 | 200 000 |
| 1 500 | 30 000 000 | 15 000 000 | 7 500 000 | 3 000 000 | 1 500 000 | 300 000 |

* Meter internal resistance should be subtracted.

of 50 μA (second column) has a resistance of 20 000 Ω/V. In the left-hand column, move down until you reach the number 100. The column at the immediate right shows a resistance value of 2 000 000 Ω. To extend the range to 100 V would require a multiplier resistance of 20 MΩ.

# Conductance of metals

One of the distinguishing characteristics of metals is their ability to conduct electricity. As metals differ in hardness, ductility, density, tensile strength, malleability and melting point, they also differ in inherent conductance of electrical flow. See Table 1-12.

*Table 1-12. Relative*
*conductance of various metals.*

| Substance | Relative conductance (Silver = 100%) |
|---|---|
| Silver | 100 |
| Copper | 98 |
| Gold | 78 |
| Aluminum | 61 |
| Tungsten | 32 |
| Zinc | 30 |
| Platinum | 17 |
| Iron | 16 |
| Lead | 15 |
| Tin | 9 |
| Nickel | 7 |
| Mercury | 1 |
| Carbon | 0.05 |

# Resistivities of conductors

Metals conduct electricity. Some do it better than others. Table 1-13 shows how much resistance various conductors exhibit. Silver is the best electrical conductor and presents the least resistance to electrical flow. Although gold is the third best conductor, its cost makes gold impractical to use.

Table 1-13 supplies specific data on the resistance of various conductive materials measured at a temperature of 0 °C. Higher temperatures increase a material's resistance. Note that carbon has a higher resistivity per unit volume than any of the other substances.

# Codes and values

Usually, an electronic circuit will have a number of resistors. So you can identify them and be able to pinpoint their location, they are often assigned codes and values. Codes are alphanumeric with the letter R followed by a number used for resistors. R1 is resistor number 1, R2 is resistor number 2, etc. In a circuit diagram, resistors are numbered from left to right and from the upper part of the diagram to the lower. Figure 1-6 shows four different ways in which codes and values are used for resistor identification.

## Table 1-13. Resistivities of conductors at 0°C.

| Substance | μΩ Centimeter cube | Inch cube | Ω—Circular mils per foot round wires |
|---|---|---|---|
| Aluminum | 3.21 | 1.26 | 19.3 |
| Carbon | 4 000 to 10 000 | 1 600 to 2 800 | 24 000 to 42 000 |
| Constantan (Cu 60%, Ni 40%) | 49 | 19.3 | 295 |
| Copper | 1.72 | 0.68 | 10.4 |
| Iron | 12 to 14 | 4.7 to 5.5 | 72 to 84 |
| Lead | 20.8 | 8.2 | 125 |
| Manganin (Cu 84%, Ni 4%, Mn 12%) | 43 | 16.9 | 258 |
| Mercury | 95.76 | 37.6 | 575 |
| Nichrome (Ni 60%, Cr 12%, Fe 26%, Mn 2%) | 110 | 43 | 660 |
| Platinum | 11.0 | 4.3 | 66 |
| Silver | 1.65 | 0.65 | 9.9 |
| Tungsten | 5.5 | 2.15 | 33 |
| Zinc | 6.1 | 2.4 | 36.7 |

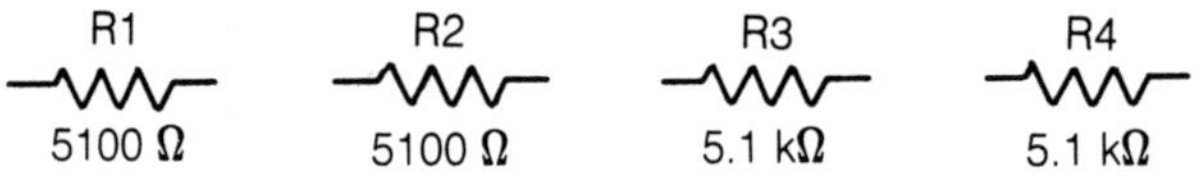

| R1 | R2 | R3 | R4 |
|---|---|---|---|
| 5100 Ω | 5100 Ω | 5.1 kΩ | 5.1 kΩ |

Fig. 1-6. Alphanumeric coding and values for fixed resistors.

# 2
## CHAPTER

# dc voltage, current, and power

## Types of dc voltages

The phrase *dc voltage* is a general heading for any voltage that does not reverse polarity, but the term supplies no further information about the characteristics of that voltage. A dc voltage can vary without changing its polarity. The output of an unfiltered rectifier is varying dc.

When a dc voltage has a constant amplitude it is referred to as *pure*. A pure dc voltage, when shunted across a load, might decrease its level but will remain constant. The current that is supplied will have zero frequency.

## Ohm's law for dc

Ohm's law is probably the best known and most widely used electronic formula. In its basic form it appears as:

$$E = I \times R$$

*E* is the voltage, *I* is the current in amperes, and *R* is the resistance in ohms. By transposing the terms, two additional formulas are obtained.

$$I = \frac{E}{R}$$

$$R = \frac{E}{I}$$

In these three formulas, the basic units of voltage, current, and resistance are used. The data must be converted to basic units. The voltage is pure dc; that is, voltage that does not contain any variations. In working with Ohm's law, it is customary to put the unknown on the left side of the equation.

### dc conductance

*Conductance* is the reciprocal of resistance. Consequently, Ohm's law can be expressed in terms of conductance:

$$E = \frac{I}{G}$$

$$I = E \times G$$

$$G = \frac{I}{E}$$

$E$ is the voltage in volts, $I$ is the current in amperes, and $G$ is the conductance in siemens.

# dc current flow through parallel resistors

The equation to determine current flow through two resistors connected in parallel and across a single dc voltage source is:

$$I_{R1} = \frac{I_t \times G_1}{G_1 + G_2}$$

$I_{R1}$ is the current flow in amperes through R1, $I_t$ is the total current, sometimes referred to as the line current, in amperes, $G_1$ is the conductance of R1 in siemens, and $G_2$ is the conductance of R2 in siemens. The equation to determine the current flow through R2 is:

$$I_{R2} = \frac{I_t \times G_2}{G_1 + G_2}$$

# dc power

The amount of dc power expended in a resistive component, whether that component is a resistor, a length of wire, or anything else, can be calculated from:

$$E = \frac{P}{I}$$

$$R = \frac{P}{I^2}$$

$$R = \frac{E^2}{P}$$

$P$ is the power in watts, $E$ is the dc source voltage in volts, $I$ is the current in amperes, and $R$ is the resistance in ohms. As in Ohm's law formulas, all of the values must be in basic units. If not, they must first be converted.

The three variations of Ohm's law and the three variations of the dc power laws use elements in common. This is indicated in the chart shown in Table 2-1.

*Table 2-1. Ohm's law and power law formulas.*

| Known values | $I =$ | $R =$ | $E =$ | $P =$ |
|---|---|---|---|---|
| $I\ \&\ R$ | | | $IR$ | $I^2 R$ |
| $I\ \&\ E$ | | $\dfrac{E}{I}$ | | $EI$ |
| $I\ \&\ P$ | | $\dfrac{P}{I^2}$ | $\dfrac{P}{I}$ | |
| $R\ \&\ E$ | $\dfrac{E}{R}$ | | | $\dfrac{E^2}{R}$ |
| $R\ \&\ P$ | $\sqrt{\dfrac{P}{R}}$ | | $\sqrt{PR}$ | |
| $E\ \&\ P$ | $\dfrac{P}{E}$ | $\dfrac{E^2}{P}$ | | |

Use the circular chart in Table 2-2 for calculating power, current, voltage, or resistance.

## Units and symbols

Various letters are used to identify basic units such as volts, ohms and amperes. The letter $E$ is an abbreviation for electromotive force, synonymous with voltage. (Sometimes the letter $V$ is used instead.) The letter $R$ represents resistance, and $I$ is used for current. The chart in Table 2-3 also shows multiples and submultiples of the basic units.

Large numbers are frequently used when solving Ohm's law problems. A megohm is a million ohms; a microampere is a millionth of an ampere. Solving

*Table 2-2. Memory device for calculating power, voltage, current and resistance.*

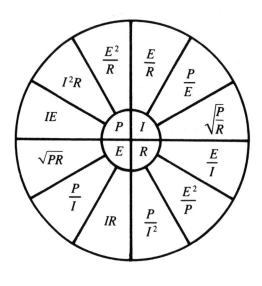

$$P = I\,E \qquad\qquad\qquad R = P/I^2$$

$$P = I^2 R \qquad\qquad\qquad R = E^2 P$$

$$P = E^2/R \qquad\qquad\qquad R = E/I$$

$$I = E/R \qquad\qquad\qquad E = \sqrt{P\,R}$$

$$I = P/E \qquad\qquad\qquad E = P/I$$

$$I = \sqrt{P/R} \qquad\qquad\qquad E = I\,R$$

problems involving these units can lead to arithmetic with large decimal numbers. A more convenient method is to use exponents. The exponents associated with resistance, current and voltage are shown in Table 2-4.

# Power

The basic unit of power is the watt. Power values range from a microwatt, or a millionth of a watt, to a megawatt, or a million watts. The units, their multiples, and the symbol for power, $P$, is presented in Table 2-5 and is supplied in terms of decimal numbers and exponents.

*Table 2-3. Units and symbols.*

| Unit | Symbol | Multiple | Value |
|---|---|---|---|
| volt | E | kilovolt (kV) | 1 000 V |
| volt | E | millivolt (mV) | 1/1 000 V |
| volt | E | microvolt ($\mu$V) | 1/1 000 000 V |
| ohm | R | kilohm (k$\Omega$) | 1 000 $\Omega$ |
| ohm | R | megohm (M$\Omega$) | 1 000 000 $\Omega$ |
| ampere | I | milliampere (mA) | 1/1 000 A |
| ampere | I | microampere ($\mu$A) | 1/1 000 000 A |

*Table 2-4. Units in exponential notation.*

| Unit | Equivalent unit | |
|---|---|---|
| 1 V | $10^3$ mV | $10^6$ $\mu$V |
| 1 mV | $10^{-3}$ V | $10^3$ $\mu$V |
| 1 $\mu$V | $10^{-6}$ V | $10^{-3}$ mV |
| 1 $\Omega$ | $10^{-3}$ kV | $10^{-6}$ M$\Omega$ |
| 1 k$\Omega$ | $10^3$ $\Omega$ | $10^{-3}$ M$\Omega$ |
| 1 m$\Omega$ | $10^6$ $\Omega$ | $10^3$ k$\Omega$ |
| 1 A | $10^3$ mA | $10^6$ $\mu$A |
| 1 mA | $10^{-3}$ A | $10^3$ $\mu$A |
| 1 $\mu$A | $10^{-6}$ A | $10^{-3}$ mA |

# Work and power

The unit of work is the foot-pound, but this unit does not involve any reference to time. When a 50-pound object is moved a distance of 20 ft, the work done is the weight multiplied by the distance. In this case it is $50\times20=1\ 000$ foot-pounds. The answer is the same whether the work was done in 10 minutes or an hour.

Power is work divided by time. If the work in the example above was completed in 40 s, the power expended would have been 1 000 foot-pounds in 40 s or 25 foot-pounds per second.

Work is power used per unit of time. This can be expressed as a formula:

$$\text{work}=P\times t$$

*Table 2-5. Symbols, multiple*
*and exponential values of dc power.*

| Unit | Symbol | Multiple | Value |
|------|--------|----------|-------|
| watt | P | μW | 1/1 000 000 W |
| watt | P | mW | 1/1 000 W |
| watt | P | MW | 1 000 000 W |

**Exponential notation**

| Unit | Equivalent units | |
|------|------------------|---|
| 1 W | $10^3$ mW | $10^6$ μW |
| 1 mW | $10^{-3}$ W | $10^3$ μW |
| 1 μW | $10^{-6}$ W | $10^{-3}$ mW |

The basic unit of work is the *watt-hour*. When large amounts of power are involved, power used per unit of time is expressed in kilowatt-hours, abbreviated as *kWh*.

$$kWh = \frac{P \times t}{1\ 000}$$

where *P* is the power in watts and *t* is the time in hours.

There are three systems used in measuring the amount of work done.

English: The units are the foot, the pound, and the second.
CGS: The units are the centimeter, gram, and the second.
MKS: The units are the meter, kilogram, and the second.

A detailed explanation of these three systems is supplied in chapter 7.

## Electrical power versus mechanical power

Horsepower (hp) is a unit of mechanical power and is the capacity of a machine, such as a motor, to do work.

1 hp = 746 W (actually 745.7 W)
1 hp = 550 ft•lbs/s
1 hp = 33 000 ft•lbs/min
1 hp = 550×60 ft•lbs/min

Although they look somewhat different, the last three are identical. When multiplied by 60 (because there are 60 s in 1 min) 1 hp=550 ft•lbs/s is exactly the same as 33 000 ft•lbs/min. Horsepower can be supplied in terms of kilowatts as well as in watts, as indicated by:

$$hp = \frac{kW \times 746}{1\ 000}$$

or by:

$$1 \text{ kW} = \frac{\text{hp} \times 1\,000}{746}$$

$$1 \text{ W} = 1/746 \text{ hp} = 0.001\,34 \text{ hp}$$

Motors are generally rated in terms of horsepower. But because a motor is an electromechanical device, it also has a rating in watts. Table 2-6 and Table 2-7

*Table 2-6. Watts versus horsepower.*

| W | hp | W | hp | W | hp | W | hp |
|---|---|---|---|---|---|---|---|
| 1 | 0.001 341 | 26 | 0.034 866 | 51 | 0.068 391 | 76 | 0.101 916 |
| 2 | 0.002 682 | 27 | 0.036 207 | 52 | 0.069 732 | 77 | 0.103 257 |
| 3 | 0.004 023 | 28 | 0.037 548 | 53 | 0.071 073 | 78 | 0.104 598 |
| 4 | 0.005 364 | 29 | 0.038 889 | 54 | 0.072 414 | 79 | 0.105 939 |
| 5 | 0.006 705 | 30 | 0.040 230 | 55 | 0.073 755 | 80 | 0.107 280 |
| 6 | 0.008 046 | 31 | 0.041 571 | 56 | 0.075 096 | 81 | 0.108 261 |
| 7 | 0.009 387 | 32 | 0.042 912 | 57 | 0.076 437 | 82 | 0.109 962 |
| 8 | 0.010 728 | 33 | 0.044 253 | 58 | 0.077 778 | 83 | 0.111 303 |
| 9 | 0.012 069 | 34 | 0.045 594 | 59 | 0.079 119 | 84 | 0.112 644 |
| 10 | 0.013 410 | 35 | 0.046 935 | 60 | 0.080 460 | 85 | 0.113 985 |
| 11 | 0.014 751 | 36 | 0.048 278 | 61 | 0.081 801 | 86 | 0.115 326 |
| 12 | 0.016 092 | 37 | 0.049 617 | 62 | 0.083 142 | 87 | 0.116 667 |
| 13 | 0.017 433 | 38 | 0.050 958 | 63 | 0.084 483 | 88 | 0.118 008 |
| 14 | 0.018 774 | 39 | 0.052 299 | 64 | 0.085 824 | 89 | 0.119 349 |
| 15 | 0.020 115 | 40 | 0.053 640 | 65 | 0.087 165 | 90 | 0.120 690 |
| 16 | 0.021 456 | 41 | 0.054 981 | 66 | 0.088 506 | 91 | 0.122 031 |
| 17 | 0.022 797 | 42 | 0.056 322 | 67 | 0.089 847 | 92 | 0.123 372 |
| 18 | 0.024 138 | 43 | 0.057 663 | 68 | 0.091 188 | 93 | 0.124 713 |
| 19 | 0.025 479 | 44 | 0.059 004 | 69 | 0.092 529 | 94 | 0.126 054 |
| 20 | 0.026 820 | 45 | 0.060 345 | 70 | 0.093 870 | 95 | 0.127 395 |
| 21 | 0.028 161 | 46 | 0.061 686 | 71 | 0.095 211 | 96 | 0.128 736 |
| 22 | 0.029 502 | 47 | 0.063 027 | 72 | 0.096 552 | 97 | 0.130 077 |
| 23 | 0.030 843 | 48 | 0.064 368 | 73 | 0.097 893 | 98 | 0.131 418 |
| 24 | 0.032 184 | 49 | 0.065 709 | 74 | 0.099 234 | 99 | 0.132 759 |
| 25 | 0.033 525 | 50 | 0.067 050 | 75 | 0.100 575 | 100 | 0.134 100 |

give the conversion between horsepower and watts. The tables can be extended by moving the decimal point in the same direction, for the same number of places in both columns. For horsepower and electrical power equivalents, see Table 2-7.

**Example**  A small motor is rated at 0.1 hp. What is its rating in watts?
**Solution**  The closest value to 0.1 hp in Table 2-6 is 0.100 575 shown in the

*Table 2-7. Horsepower versus watts.*

| hp | W | hp | W | hp | W | hp | W |
|------|---------|------|---------|------|---------|------|---------|
| 0.01 | 7.457 | 0.26 | 193.882 | 0.51 | 380.307 | 0.76 | 556.732 |
| 0.02 | 14.914 | 0.27 | 201.339 | 0.52 | 387.764 | 0.77 | 574.189 |
| 0.03 | 22.371 | 0.28 | 208.796 | 0.53 | 395.221 | 0.78 | 581.646 |
| 0.04 | 29.828 | 0.29 | 216.253 | 0.54 | 402.678 | 0.79 | 589.103 |
| 0.05 | 37.285 | 0.30 | 233.710 | 0.55 | 410.135 | 0.80 | 596.560 |
| 0.06 | 44.742 | 0.31 | 231.167 | 0.56 | 417.592 | 0.81 | 604.017 |
| 0.07 | 52.199 | 0.32 | 238.624 | 0.57 | 425.049 | 0.82 | 611.474 |
| 0.08 | 59.656 | 0.33 | 246.081 | 0.58 | 432.506 | 0.83 | 618.931 |
| 0.09 | 67.113 | 0.34 | 253.538 | 0.59 | 439.963 | 0.84 | 626.388 |
| 0.10 | 74.570 | 0.35 | 260.995 | 0.60 | 447.420 | 0.85 | 633.845 |
| 0.11 | 82.027 | 0.36 | 268.452 | 0.61 | 454.877 | 0.86 | 641.302 |
| 0.12 | 89.484 | 0.37 | 275.909 | 0.62 | 462.334 | 0.87 | 648.759 |
| 0.13 | 96.941 | 0.38 | 283.366 | 0.63 | 469.791 | 0.88 | 656.216 |
| 0.14 | 104.398 | 0.39 | 290.823 | 0.64 | 477.248 | 0.89 | 663.673 |
| 0.15 | 111.855 | 0.40 | 298.280 | 0.65 | 484.705 | 0.90 | 671.130 |
| 0.16 | 119.312 | 0.41 | 305.737 | 0.66 | 492.162 | 0.91 | 678.587 |
| 0.17 | 126.769 | 0.42 | 313.194 | 0.67 | 499.619 | 0.92 | 686.044 |
| 0.18 | 134.226 | 0.43 | 320.651 | 0.68 | 507.076 | 0.93 | 693.501 |
| 0.19 | 141.683 | 0.44 | 328.108 | 0.69 | 514.533 | 0.94 | 700.958 |
| 0.20 | 149.140 | 0.45 | 335.565 | 0.70 | 521.990 | 0.95 | 708.415 |
| 0.21 | 156.597 | 0.46 | 343.022 | 0.71 | 529.447 | 0.96 | 715.872 |
| 0.22 | 164.054 | 0.47 | 350.479 | 0.72 | 536.904 | 0.97 | 723.329 |
| 0.23 | 171.511 | 0.48 | 357.936 | 0.73 | 544.361 | 0.98 | 730.786 |
| 0.24 | 178.968 | 0.49 | 365.393 | 0.74 | 551.818 | 0.99 | 738.243 |
| 0.25 | 186.425 | 0.50 | 372.850 | 0.75 | 559.275 | 1.00 | 745.700 |

right-hand column under the heading hp. The power corresponding to this value is 75 W.

**Example**   A motor-generator is rated at 0.5 kW. What is its equivalent horsepower rating?

**Solution**   0.5 kW is 500 W. Table 2-6 does not list such a value but you can use the number 50 in place of 500. Locate the number 50 in the column headed by watts. Move the decimal point one place to the right and 50 becomes 500. The corresponding value of horsepower is 0.067 050 hp. Moving the decimal point of this number by one decimal place (to the right) supplies 0.670 5 hp. In practice this could be rounded off to 0.7 hp.

**Example**   What is the horsepower rating of a 1-kW generator?

**Solution**   1 kW is equal to 1 000 W. Use Table 2-6 by selecting the number 100 and moving its decimal point one place to the right, thus changing 100 to

1 000. The corresponding value of horsepower is 0.134 100, but remember to move the decimal point here as well. The answer is 1.341 hp.

**Example** A fractional horsepower motor is rated at 0.07 hp. What is its power rating in watts?

**Solution** There is no value in Table 2-6 that corresponds exactly to 0.07. There are two values, though, which are very close. One of these is 0.069 732 and the other 0.071 073. Thus, this motor has a rating between 52 and 53 W. However, if you want a more precise value than this, consult Table 2-7. There, you will see that 0.07 hp corresponds to 52.199 W.

# dc voltages in series aiding

When dc voltages are connected in series aiding, as shown in Fig. 2-1, the total voltage is equal to the sum of the individual voltages, represented by the batteries

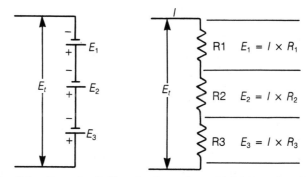

*Fig. 2-1. Batteries (left) and voltage drops (right) in series aiding.*

in the illustration. This is indicated as:

$$E_t = E_1 + E_2 + E_3$$

The battery connections in the drawing are referred to as *series aiding* because each battery contributes to the total voltage, $E_t$.

It might be necessary to move between low values of voltage, such as the microvolt (millionth of a volt), and the megavolt (millions of volts). There are two possible techniques—using decimal numbers, as in Table 2-8, or exponents, as in Table 2-9. Of the two, solving voltage conversion problems with exponents is easier, once experience has been obtained with this mathematical tool.

# dc voltages in series opposing

If one (or more) of the batteries in a circuit is transposed, it subtracts from the total voltage, with the circuit arrangement referred to as *series opposing*. It is possible to have the series-opposing voltage equal to the series-aiding voltage and in

*Table 2-8. Voltage conversions using decimal numbers.*

| From | μV | mV | V | kV | MV |
|---|---|---|---|---|---|
| | | | **To** | | |
| μV | — | ÷1 000 | ÷1 000 000 | ÷1 000 000 000 | ÷1 000 000 000 000 |
| mV | ×1 000 | — | ÷1 000 | ÷1 000 000 | ÷1 000 000 000 |
| V | ×1 000 000 | ×1 000 | — | ÷1 000 | ÷1 000 000 |
| kV | ×1 000 000 000 | ×1 000 000 | ×1 000 | — | ÷1 000 |
| MV | ×1 000 000 000 000 | ×1 000 000 000 | ×1 000 000 | ×1 000 | — |

*Table 2-9. Voltage conversions using exponents.*

| Unit | | Equivalent units | | | |
|---|---|---|---|---|---|
| 1 V | = | 1 000 mV | = | $10^3$ mV | = $10^6$ μV |
| 1 V | = | 1 000 000 μV | = | $10^6$ μV | |
| 1 V | = | 0.001 kV | = | $10^{-3}$ kV | |
| 1 V | = | 0.000 001 MV | = | $10^{-6}$ MV | |
| 1 kV | = | 1 000 V | = | $10^3$ V | |
| 1 kV | = | 0.001 MV | = | $10^{-3}$ MV | |
| 1 MV | = | 1 000 000 V | = | $10^6$ V | |
| 1 MV | = | 1 000 kV | = | $10^3$ kV | |
| 1 mV | = | 0.001 V | = | $10^{-3}$ V | |
| 1 mV | = | 1 000 μV | = | $10^3$ μV | |
| 1 μV | = | 0.000 001 V | = | $10^{-6}$ V | |
| 1 μV | = | 0.001 mV | = | $10^{-3}$ mV | |

that case no current will flow to an external circuit. In terms of a formula for the condition shown in Fig. 2-2:

$$E_t = (E_1 + E_2) - E_3$$

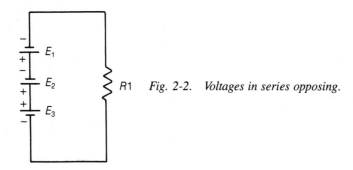

R1    Fig. 2-2.    Voltages in series opposing.

# IR drops

The flow of current through a resistor produces a voltage across it. This sometimes is referred to as a voltage drop or an *IR drop*. The voltage across the resistor is equal to the amount of the current (in amperes) multiplied by the ohmic value of the resistor.

# Resistors as voltage and current dividers

Resistors in dc circuits can be used conveniently as either voltage or current dividers. Further, the direction of current flow can also be used to determine the polarity of the voltage across current resistors (Fig. 2-3).

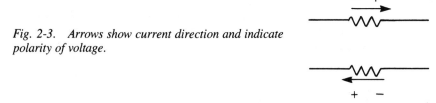

*Fig. 2-3.  Arrows show current direction and indicate polarity of voltage.*

When resistors are connected in series across a dc source voltage, the same amount of current flows through each. A voltage, commonly referred to as an IR drop, will appear across each of the resistors. For voltage drops, the total voltage is:

$$E_t = E_1 + E_2 + E_3 = IR_1 + IR_2 + IR_3$$

Since the resistors are in series their voltage drops will also be in series and will be additive.

# dc currents in parallel

The total current, or line current, flowing through parallel resistors (Fig. 2-4) is:

$$I_t = I_1 + I_2 + I_3$$

The currents through these resistors can be in microamperes, milliamperes, amperes, or any combination of these. The flow of current through all the resistors is in the same direction. The same voltage appears across each resistor and is equal to the source voltage.

In the case of two parallel resistors:

$$I_1 \times R_1 = I_2 \times R_2$$

By transposing terms it is possible to obtain:

$$\frac{I_1}{I_2} = \frac{R_2}{R_1}$$

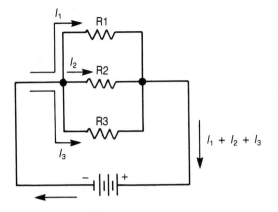

Fig. 2-4. *Division of current in a parallel circuit.*

This indicates that the ratio of the currents is inversely proportional to the ratio of the resistances. By transposing terms it is also possible to determine the values of $I_1$, $I_2$, $R_1$ or $R_2$. The formulas will be:

$$I_1 = \frac{I_2 \times R_2}{R_1}$$

$$I_2 = \frac{I_1 \times R_1}{R_2}$$

$$R_1 = \frac{I_2 \times R_2}{I_1}$$

$$R_2 = \frac{I_1 \times R_1}{I_2}$$

# The voltage divider

A fixed voltage, such as that supplied by a battery or power supply, is often not suitable for a particular circuit or component. Resistors can be used as voltage dividers to obtain any value of voltage **less** than that of the source. See Fig. 2-5.

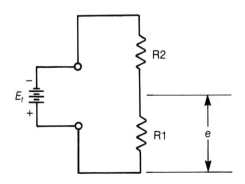

Fig. 2-5. *Resistors used as voltage dividers.*

The smaller voltage, $e$, is equal to the source voltage multiplied by the ratio of the resistances:

$$e = E_t \times \frac{R_1}{R_1 + R_2}$$

As $R_2$ is made smaller, the output voltage $e$ becomes larger, reaching the value of $E_t$ as a limit. The input voltage $E_t$ remains constant. Because the resistors, R1 and R2 are in series, the same current flows through them.

Suppose a load of nearly infinite resistance is placed across R1. If the load does not draw current, then the formula can be used as is to find the value of $e$. If the load does draw current, and its resistance is known, then R1 and the load may be considered as a pair of parallel resistors. Calculate the equivalent value of this parallel pair, substitute this value for $R_1$, and the formula can then be used to determine the amount of $e$.

## Voltage reference points

Ground, a common bus, or a neutral wire are commonly used as voltage reference points, but any voltage point can also act as a reference. In Fig. 2-6 circuit **A**,

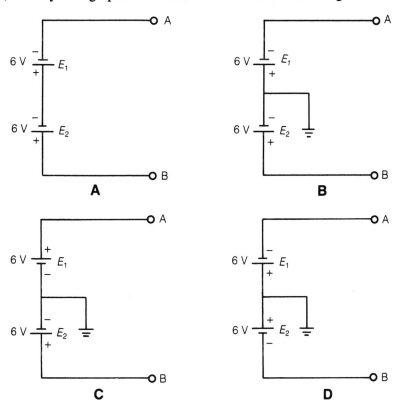

Fig. 2-6.    *Voltage is always measured between two points, one of which is the reference.*

point A is 12 V negative with respect to point B. Point B is 12 V positive with respect to point A. In drawing **B**, point A is 6 V negative with respect to ground. Point B is 6 V positive with respect to ground. The voltage from A to B is 12 V. In drawing **C**, both points, A and B, are 6 V positive with respect to ground. The voltage between points A and B is zero. In drawing **D** points A and B are both 6 V negative with respect to ground. The voltage between points A and B is zero.

# Batteries

Dc voltages can be obtained in four ways:

> electronic power supplies
> dc motor generators
> resistor voltage dividers
> batteries

Electronic power supplies depend on an ac input voltage and must be reasonably close to an ac receptacle. Dc motor generators are used in industry and in transportation to run large boats and trains. Resistive voltage dividers are used in electronic equipment when specific dc voltages are required. Batteries are used in portable equipment and in electronic and electrical equipment in the home.

Batteries can be categorized by voltage availability, current capability, weight, and in other ways. They can be categorized into the two main groups of rechargeable and nonrechargeable.

## Cells and batteries

The terms, *cells* and *batteries*, although often used interchangeably, are not synonymous. A cell is a single unit. A number of them may be connected in series or in parallel to form a battery. Batteries are used to supply a higher voltage or a greater current capability than a single cell.

## Identification of cells

There are a number of ways cells can be identified. Cells commonly are recognized by an assigned letter or groups of letters, such as AA, AAA, or D. Sometimes, as in the case of mercury cells, there is such a large variety that they only can be identified by function, such as hearing aid, paging, or watch. Another method of identification is a chemical characteristic such as zinc-carbon, alkaline, or mercury. A combination of electrical values (such as voltage) and a physical attribute (zinc) sometimes is used.

There are some cells that only can be identified by a manufacturer's part number. There is no standardization of part numbers, and each manufacturer supplies its own. Sometimes these numbers are cross-referenced in a manufacturer's literature.

Still another way of identifying is through standardized alphanumeric designations supplied by various agencies. One of these is the ANSI (American National Standards Institute). Another is the IEC (International Electrotechnical Commission). The IEC is represented in the U.S. by a committee that cooperates with ANSI for the establishment of component standards. Cell manufacturers usually use an ANSI recommended identification, or use both an ANSI and an IEC identification.

Batteries that cannot be recharged are sometimes referred to as primary types. Those that can be recharged, such as the lead-acid, are called secondary types. Another rechargeable type is the nickel-cadmium battery.

## Lead-acid batteries

Figure 2-7 is a cross-sectional view of a lead-acid battery. The electrical characteristics are listed in Table 2-10. The physical characteristics of lead-acid batteries are listed in Table 2-11.

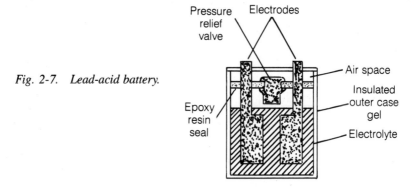

*Fig. 2-7. Lead-acid battery.*

*Table 2-10. Electrical characteristics of lead-acid batteries.*

| Characteristic | Description |
|---|---|
| Terminal voltage per cell | 2.15 V |
| Terminal voltage | 12.9 V |
| Number of cells | 6 |
| Current rating | 60 to 120 Ah with smaller and larger ratings available |

The four possible charging rates for these batteries are rapid, quick, standard, and trickle. See Table 2-12.

## Nickel-cadmium batteries

Nickel-cadmium batteries are more popularly known as *nicads*. These batteries are available in relatively small sizes and use the same letter designations as carbon-zinc dry cells. They can be purchased in sizes AAA, AA, C, Sub-C, and D.

*Table 2-11. Physical characteristics of lead-acid batteries.*

| Characteristic | Description |
|---|---|
| Negative plate | Pure sponge lead |
| Positive plate | Lead peroxide |
| Electrolyte | Sulfuric acid gel |
| Electrolyte seal | Epoxy resin |
| Electrodes | Two per cell |
| Number of cells | Six for 12-V battery |
| Battery position | Usually vertical but can be any position if electrolyte is gel |
| Cross section | See Fig. 2-7 |

*Table 2-12. Charging rate of lead-acid batteries.*

| Type of charge | Charging voltage (V dc per cell) | Charging current (% of cell capacity) | Charging time (h) |
|---|---|---|---|
| Rapid | 2.55−2.65 | 100 | 1−3 |
| Quick | 2.50−2.55 | 20−50 | 12−20 |
| Standard | 2.45−2.50 | 10−40 | 10−18 |
| Trickle | 2.28−2.32 | 10−20 | Continuous |

Like lead-acid batteries nicads are rechargeable and so are secondary types. Nicads, when compared to lead-acid types, are about one-third lighter for a given power output. Table 2-13 lists the electrical and physical characteristics of this type of battery. Like lead-acid cells the nicads have four charging rates. They are detailed in Table 2-14.

## Lead-acid and nicad discharge rates

Under identical operating conditions, the lead-acid and nicad batteries have fairly similar discharge rates. For about the first four hours the discharge is fairly flat, but following that, both batteries discharge quite rapidly and are completely discharged in about six hours. The graph in Fig. 2-8 shows this behavior.

## Zinc-carbon batteries

This battery is disposable and cannot be recharged. It is available in various sizes, designated as D, C, AA, AAA, and so on. A single cell has an output voltage of 1.5 V, but wired in series and combined in a package they can supply 6.0, 9.0, 12.0, 15.0, 225, 240, and 510 V. The deterioration of any single cell in the group can seriously affect the operation of the battery.

Table 2-15 lists the electrical characteristics of zinc-carbon batteries. Figure 2-9 is a cross-sectional view of a single battery, sometimes referred to as a cell.

*Table 2-13. Characteristics of nickel-cadmium cells.*

| Characteristic | Description |
|---|---|
| Basic type | Rechargeable |
| Current capacity | 80 mAh (9-V battery) |
| | 4 Ah (size D) |
| Shape | Cylindrical and rectangular |
| Compatibility | Interchangeable with comparable dry cells |
| Negative terminal | Case (nickel-plated steel) |
| Positive terminal | Circular metal plate on top |
| Electrode materials | Positive electrode: nickel hydroxide |
| | Negative electrode: cadmium hydroxide |
| Electrolyte | Nonreplaceable potassium hydroxide (alkaline) |
| Venting | Two types: vented and hermetically sealed |
| Open-circuit emf | 1.2 V |
| Load voltage | Essentially flat over current discharge range |
| Internal resistance | Very low, less than 100 m$\Omega$ |
| Internal capacitance | High |
| Series connection | Yes |
| Parallel connection | Not recommended |
| Discharge curve | Illustrated in Fig. 2-8 |
| Recharging time | Typically 14 to 16 h |
| Applications | Toys, consumer electronic products, flashlights, cameras, and watches |

*Table 2-14. Charging rates of nickel-cadmium batteries.*

| Type of charge | Charging voltage (volts per cell) | Charging current (% of cell capacity) | Charging time (h) |
|---|---|---|---|
| Rapid | Depends on cell | Depends on cell | 1−3 |
| Quick | 2.50−3.00 | 20 | 3−5 |
| Standard | 2.50−3.00 | 10 | 14 |
| Trickle | 2.50−3.00 | 5 | Continuous |

The size of a zinc-carbon battery is an important characteristic because the battery is used in applications where it makes a fairly snug fit. Table 2-16 supplies the measurements.

## Alkaline batteries

The alkaline battery types (Fig. 2-10) belong to the family of disposable batteries. Its single cell output voltage is the same as that of the zinc-carbon and is 1.5 V, except for the 9-V size. The alkaline advantages over the zinc carbon are that it has a greater energy storage capacity and it works better at cold and hot tempera-

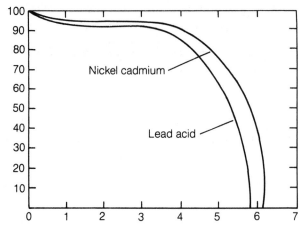

Fig. 2-8. *Discharge characteristics of lead-acid and nickel-cadmium cells.*

**Table 2-15. Electrical characteristics of zinc-carbon cells.**

| Characteristic | Description |
|---|---|
| Voltage | 1.5 V |
| Positive electrode | Carbon |
| Negative electrode | Zinc |
| Rechargeability | Not rechargeable |
| Electrolyte | Granulated carbon and manganese dioxide |
| 10-h current | 60 mA (approx.) |
| 100-h current | 10 mA (approx.) |

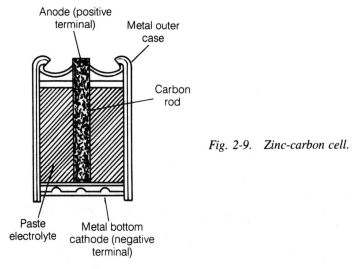

Fig. 2-9. *Zinc-carbon cell.*

*Table 2-16. Zinc-carbon dimensions.*

| Designation | D | C | AA | 9 V |
|---|---|---|---|---|
| Voltage | 1.5 | 1.5 | 1.5 | 9.0 |
| Diameter (in) | 1.344 | 1.0 | 0.560 | — |
| Height (in) | 2.406 | 1.951 | 1.969 | 1.906 |
| Length (in) | — | — | — | 1.031 |
| Width (in) | — | — | — | 0.656 |
| Primary application purpose | General purpose | General purpose | General purpose | Radios |

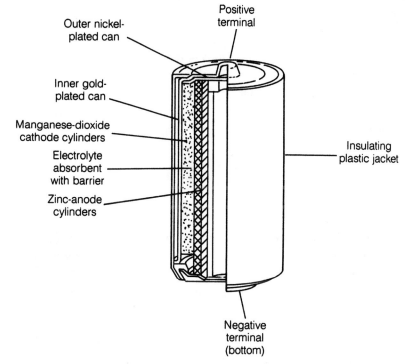

*Fig. 2-10. Alkaline cell.*

ture extremes. Table 2-17 lists some of the physical and electrical characteristics of alkaline batteries.

Alkaline batteries have the same letter designations as the zinc-carbon types, including sizes N, D, C, AA, and AAA. There is also a 9-V size commonly used in transistor radios. Table 2-18 is a listing of the different sizes.

## Mercury batteries

Mercury batteries are very small batteries used where space is at a premium as in hearing aids and watches. The unit is a dry type and is not rechargeable. Table 2-19 lists its physical and electrical characteristics.

*Table 2-17. Physical and electrical characteristics of alkaline batteries.*

| Characteristic | Description |
|---|---|
| Weight | About 50% more than a comparable zinc-carbon cell |
| Voltage· | 1.5 V (except 9 V transistor type) |
| Negative terminal | Manganese dioxide |
| Positive terminal | Powdered zinc |
| Electrolyte | Potassium hydroxide gel |
| Rechargeability | Units are primary cells and are not rechargeable |
| Internal resistance | Very low |
| Impedance | Very low, less than carbon-zinc |
| Shape | Available as cylinder or button type |

*Table 2-18. Dimensions of alkaline batteries.*

| Size | Voltage | Diameter (in) | Height (in) |
|---|---|---|---|
| D | 1.5 | 1.32 | 2.39 |
| C | 1.5 | 1.0 | 1.951 |
| AA | 1.5 | 0.56 | 1.969 |
| AAA | 1.5 | 0.410 | 1.75 |
| | **Width (in)** | **Length (in)** | |
| 9-V transistor | 0.65 | 1.031 | 1.906 |

*Table 2-19. Physical and electrical characteristics of mercury batteries.*

| Characteristic | Description |
|---|---|
| Voltage | 1.4 V, open circuit, slightly lower than carbon-zinc or alkaline; voltage does not decrease until cell is discharged |
| Rechargeability | Not rechargeable; mercury cells are primary types |
| Positive electrode | Amalgamated zinc pellets |
| Negative electrode | Mercuric oxide pellets |
| Electrolyte | Solution of potassium hydroxide and zinc oxide |
| Shelf life | Better than zinc carbon |
| Short-circuit current (AA cell) | 10.3 A |
| 9-V cell | 1.7 A |
| Case | Steel |
| Applications | Watches, calculators, hearing aids, smoke alarms, CB radio, microphones |

Mercury batteries are available in a variety of sizes as indicated in Table 2-20. They can be series connected to supply a wide range of voltages. Those shown in the table form a partial listing; they also supply 5.4, 6.75, 8.4, 9.8, 12.6 V, and other voltages. Figure 2-11 is an exposed view.

*Table 2-20. Sizes and current ratings of mercury batteries.*

| Voltage (V) | Diameter (in) | Height (in) | mAh |
|---|---|---|---|
| 2.7 | 0.665 | 0.607 | 250 |
| 5.6 | 0.600 | 0.787 | 110 |
| 5.6 | 0.500 | 0.787 | 150 |
| 1.35 | 0.625 | 0.440 | 500 |
| 1.35 | 0.455 | 0.135 | 80 |
| 1.35 | 0.625 | 0.645 | 1 000 |
| 5.4 | 0.675 | 1.767 | 500 |
| 1.35 | 0.455 | 0.210 | 190 |
| 1.35 | 0.615 | 0.238 | 250 |
| 4.05 | 0.660 | 0.845 | 250 |

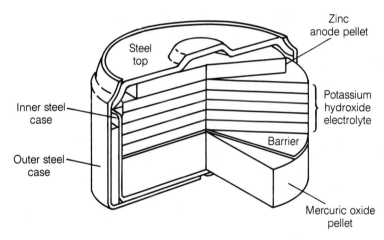

*Fig. 2-11.  Cross section of a mercury cell.*

## Lithium batteries

Lithium batteries can be made using a wide variety of chemicals as indicated in Table 2-21. The voltage measured across their output terminals with the battery unloaded is also shown in this table. Table 2-22 lists the lithium battery characteristics. The dimensions of various lithium batteries are indicated in Table 2-23.

## Silver-oxide batteries

Also known as silver-oxide alkaline-zinc batteries, these are primary types (non-rechargeable). The electrolyte is either potassium or sodium hydroxide. The useful operating life of the silver-oxide battery using sodium hydroxide ranges from 2 to 3 years. These batteries are available in a wide variety of dimensions and cur-

### Table 2-21. Chemical composition of lithium batteries.

| Electrolyte | Voltage |
|---|---|
| Lithium-thionyl chloride | 3.7 |
| Lithium-vanadium pentoxide | 3.4 |
| Lithium-silver chromate | 3.3 |
| Lithium-manganese dioxide | 3.0 |
| Lithium-sulfur dioxide | 2.9 |
| Lithium-carbon monofluoride | 2.8 |
| Lithium-iodine | 2.8 |
| Lithium-lead copper sulfide | 2.2 |
| Lithium-copper sulfide | 2.1 |
| Lithium-iron sulfide | 1.8 |
| Lithium-copper oxide | 1.8 |

### Table 2-22. Characteristics of lithium cells.

| Characteristic | Description |
|---|---|
| Voltage discharge curve | Flat |
| Rechargeability | Does not recharge |
| Internal resistance | Very low |
| Impedance | Low |
| Cathode (negative element) | Manganese dioxide and carbon black |
| Anode (positive element) | Lithium foil |
| Electrolyte | Lithium perchlorate in propylene carbonate |
| Reaction to temperature | Little or no effect |
| Voltage per cell | 3.0 V |
| Type | Dry cell button |
| Applications | Cardiac pacemakers, cameras, CMOS (complementary metal-oxide semiconductor) memory storage, power for LCDs in watches and calculators |

### Table 2-23. Dimensions of lithium cells.

| Voltage (V) | Diameter (in) | Height (in) | mAh |
|---|---|---|---|
| 3.0 | 0.460 | 0.420 | 160 |
| 6.0 | 0.510 | 0.990 | 160 |
| 3.0 | 0.630 | 0.079 | 60 |
| 3.0 | 0.787 | 0.063 | 75 |
| 3.0 | 0.787 | 0.098 | 145 |
| 3.0 | 0.787 | 0.126 | 190 |
| 3.0 | 0.965 | 0.079 | 120 |
| 3.0 | 0.965 | 0.118 | 250 |

rent outputs as listed in Table 2-24. The current ratings vary from a low of 15 mAh to as much as 250 mAh.

*Table 2-24. Dimensions and current ratings of silver-oxide batteries.*

| Diameter (in) | Height (in) | mAh |
|---|---|---|
| 0.455 | 0.110 | 70 |
| 0.455 | 0.122 | 70 |
| 0.455 | 0.081 | 35 |
| 0.305 | 0.143 | 38 |
| 0.305 | 0.210 | 70 |
| 0.374 | 0.141 | 45 |
| 0.375 | 0.102 | 67 |
| 0.311 | 0.102 | 24 |
| 0.311 | 0.102 | 24 |
| 0.455 | 0.135 | 83 |
| 0.455 | 0.137 | 90 |
| 0.610 | 0.190 | 250 |
| 0.455 | 0.210 | 180 |
| 0.311 | 0.082 | 16 |
| 0.267 | 0.084 | 15 |
| 0.311 | 0.141 | 38 |
| 0.455 | 0.165 | 120 |

## Zinc-chloride batteries

Zinc-chloride batteries are a variation of the zinc-carbon types and are sometimes designated as heavy duty. Physically, the two batteries, zinc-carbon and zinc-chloride, use a centrally located carbon electrode, functioning as the positive electrode.

## The ampere-hour

An *ampere* is a coulomb per second, but the second is a small time unit. In addition to the coulomb there is the ampere-hour, abbreviated as Ah. For smaller quantities of current, there is the milliampere-hour, abbreviated as mAh. There are 60 s in 1 min and 60 min in 1 h, $60 \times 60 = 3\ 600$, hence:

$$1 \text{ Ah} = 3\ 600 \text{ C}$$

The *ampere-hour* is a term used in connection with high battery current capability. The milliampere-hour is used for batteries having a lower output. A 100 Ah battery cannot deliver a current of 100 A for 1 h, but it can supply its approximate equivalent, consisting of 1 A for 100 h.

As the current increases, the current-delivering capability of the battery decreases. Excessive heat generated inside the battery and efficiency losses because of higher temperatures prevent large current outputs for extended time periods. Actually, the ampere-hour rating of a battery is not intended as a discharge factor, but is meant to specify the requirement for bringing a wholly or partially discharged battery to a fully charged state. At low current levels, and without the presence of heat, the discharge and charge current and time values are similar.

### Internal voltage drop of a battery

Any load put across a battery is effectively in series with the internal resistance of that battery. Unlike the load, which is generally a fixed value, the internal resistance of a battery can vary depending on the state of charge. A battery has minimum internal resistance when fully charged, with that resistance increasing when the battery discharges.

The current flowing through the internal resistance produces a voltage drop, thus supplying less voltage to the load. If the terminal voltage of a battery is 12 V and the internal voltage drop is 0.3 V, then the actual voltage that can be delivered to the load is $12.0 - 0.3 = 11.7$ V. Measuring the voltage across a battery without its load is of no significance, because under such circumstances the high internal resistance of the voltmeter means that the voltage drop inside the battery will be very small. Consequently, the measured terminal voltage of the battery will be maximum or close to it. The true voltage of the battery is obtained only when it is connected to its load. (See Fig. 2-12.) Table 2-25 is a summary of battery types and their voltages.

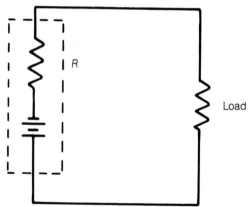

*Fig. 2-12. Internal resistance of battery, R, increases as battery is used, or gets older, and decreases when battery is charged.*

# Summary of Ohm's law and power formulas for dc

Table 2-26 is a summary of electrical and horsepower formulas.

### Table 2-25. Summary of battery types and voltages.

| Cell name | Type | Nominal open-circuit* voltage ($V_{dc}$) |
|---|---|---|
| Carbon-zinc | Primary | 1.5 |
| Zinc chloride | Primary | 1.5 |
| Manganese dioxide (alkaline) | Primary or secondary | 1.5 |
| Mercuric oxide | Primary | 1.35 |
| Silver oxide | Primary | 1.5 |
| Lead-acid | Primary | 2.1 |
| Nickel-cadmium | Secondary | 1.25 |
| Nickel-iron (Edison cell) | Secondary | 1.2 |
| Silver-zinc | Secondary | 1.5 |
| Silver-cadmium | Secondary | 1.1 |

*Open-circuit voltage is the terminal voltage without a load.

### Table 2-26. Summary of electrical and horsepower formulas.

| Factor | Formulas |
|---|---|
| Power | $P = I^2 \times R$ <br> $P = E^2/R$ <br> $P = E \times I$ <br> $kWh = \dfrac{E \times I}{1\ 000}$ or $\dfrac{P}{1\ 000}$ <br> $1\ kWh = \dfrac{hp \times 1\ 000}{746}$ <br> $1\ W = 1/746\ hp = 0.001\ 34\ hp$ |
| Current | $I = \sqrt{P/R}$ <br> $I = P/E$ <br> $I^2 = P/R$ |
| Voltage | $E = \sqrt{P \times R}$ <br> $E^2 = P \times R$ <br> $E = P/I$ |
| Resistance | $R = P/I \times I$ <br> $R = P/I^2$ <br> $R = E^2/P$ |
| Horsepower | $hp = \dfrac{kWh \times 746}{1\ 000}$ <br> $1\ hp = 746\ W$ <br> $= 550$ foot-pounds per second <br> $= 33\ 000$ foot-pounds per minute <br> $= 550 \times 60$ foot-pounds per minute |

# 3
## CHAPTER

# ac voltages and currents

## Sine waves

A single sine wave consists of a complete cycle, as shown in Fig. 3-1. Each sine wave has two peaks; a maximum positive and a maximum negative. One peak is at 90°, the other at 270°, with 360° required for a complete waveform. 180° of a cycle is one alternation or a half wave.

## Peak, peak-to-peak, average, and rms values of currents or voltages

The peak value of a sine wave of voltage or current is measured at either 90° or 270°. For this reason, peak (or peak-to-peak) values can be considered as instantaneous values. The average of all the instantaneous values over a complete cycle is zero; hence, average is generally understood to be the average of the instantaneous values over a half cycle. The average value is also equal to 2 divided by $\pi$. Taking the value of $\pi$ as 3.141 592 65, the average value of a sine wave of voltage or current is 0.636 619. This generally is rounded off to 0.637, the value used in this book. In some texts you will find the average value given as 0.636. Average value in Table 3-1 is 0.637 times the peak value.

The effective or root-mean-square (rms) value is also a form of instantaneous value averaging. Arithmetically, the effective value is obtained by dividing 1 by the square root of 2. Taking the square root of 2 as equal to 1.414 213, the effective or rms value of a sine wave of voltage or current is equal to 1/1.414 213, or

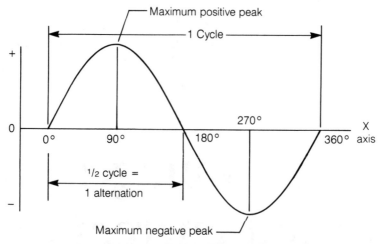

*Fig. 3-1. Details of a sine wave.*

0.707 107. In this text, as in other books on electricity and electronics, the rms value is rounded off to 0.707 times the peak value.

The data in Table 3-1 allows rapid movement among peak, peak-to-peak, average, and rms values of currents or voltages of sine waves. Also see Fig. 3-2.

*Table 3-1. Peak, peak-to-peak, average and rms (effective) values of sine wave currents or voltages.*

| Peak | Peak-to-peak | Average | rms | Peak | Peak-to-peak | Average | rms |
|---|---|---|---|---|---|---|---|
| 1 | 2 | 0.637 | 0.707 | 21 | 42 | 13.377 | 14.847 |
| 2 | 4 | 1.274 | 1.414 | 22 | 44 | 14.014 | 15.554 |
| 3 | 6 | 1.911 | 2.121 | 23 | 46 | 14.651 | 16.261 |
| 4 | 8 | 2.548 | 2.828 | 24 | 48 | 15.288 | 16.968 |
| 5 | 10 | 3.185 | 3.535 | 25 | 50 | 15.925 | 17.675 |
| 6 | 12 | 3.822 | 4.242 | 26 | 52 | 16.562 | 18.382 |
| 7 | 14 | 4.459 | 4.949 | 27 | 54 | 17.199 | 19.089 |
| 8 | 16 | 5.096 | 5.656 | 28 | 56 | 17.836 | 19.796 |
| 9 | 18 | 5.733 | 6.363 | 29 | 58 | 18.473 | 20.503 |
| 10 | 20 | 6.370 | 7.070 | 30 | 60 | 19.110 | 21.210 |
| 11 | 22 | 7.007 | 7.777 | 31 | 62 | 19.747 | 21.917 |
| 12 | 24 | 7.644 | 8.484 | 32 | 64 | 20.384 | 22.624 |
| 13 | 26 | 8.281 | 9.191 | 33 | 66 | 21.021 | 23.331 |
| 14 | 28 | 8.918 | 9.898 | 34 | 68 | 21.658 | 24.038 |
| 15 | 30 | 9.555 | 10.605 | 35 | 70 | 22.295 | 24.745 |
| 16 | 32 | 10.192 | 11.312 | 36 | 72 | 22.932 | 25.452 |
| 17 | 34 | 10.829 | 12.019 | 37 | 74 | 23.569 | 26.159 |
| 18 | 36 | 11.466 | 12.726 | 38 | 76 | 24.206 | 26.866 |
| 19 | 38 | 12.103 | 13.433 | 39 | 78 | 24.843 | 27.573 |
| 20 | 40 | 12.740 | 14.140 | 40 | 80 | 25.480 | 28.280 |

Table 3-1. Continued.

| Peak | Peak-to-peak | Average | rms | Peak | Peak-to-peak | Average | rms |
|------|------|------|------|------|------|------|------|
| 41 | 82 | 26.117 | 28.987 | 81 | 162 | 51.597 | 57.267 |
| 42 | 84 | 26.754 | 29.694 | 82 | 164 | 52.234 | 57.974 |
| 43 | 86 | 27.391 | 30.401 | 83 | 166 | 52.871 | 58.681 |
| 44 | 88 | 28.028 | 31.108 | 84 | 168 | 53.508 | 59.388 |
| 45 | 90 | 28.665 | 31.815 | 85 | 170 | 54.145 | 60.095 |
| 46 | 92 | 29.302 | 32.522 | 86 | 172 | 54.782 | 60.802 |
| 47 | 94 | 29.939 | 33.229 | 87 | 174 | 55.419 | 61.509 |
| 48 | 96 | 30.576 | 33.936 | 88 | 176 | 56.056 | 62.216 |
| 49 | 98 | 31.213 | 34.643 | 89 | 178 | 56.693 | 62.923 |
| 50 | 100 | 31.850 | 35.350 | 90 | 180 | 57.330 | 63.630 |
| 51 | 102 | 32.487 | 36.057 | 91 | 182 | 57.967 | 64.337 |
| 52 | 104 | 33.124 | 36.764 | 92 | 184 | 58.604 | 65.044 |
| 53 | 106 | 33.761 | 37.471 | 93 | 186 | 59.241 | 65.751 |
| 54 | 108 | 34.398 | 38.178 | 94 | 188 | 59.878 | 66.458 |
| 55 | 110 | 35.035 | 38.885 | 95 | 190 | 60.515 | 67.165 |
| 56 | 112 | 35.672 | 39.592 | 96 | 192 | 61.152 | 67.872 |
| 57 | 114 | 36.309 | 40.299 | 97 | 194 | 61.789 | 68.579 |
| 58 | 116 | 36.946 | 41.006 | 98 | 196 | 62.426 | 69.286 |
| 59 | 118 | 37.583 | 41.713 | 99 | 198 | 63.063 | 69.993 |
| 60 | 120 | 38.220 | 42.420 | 100 | 200 | 63.700 | 70.700 |
| 61 | 122 | 38.857 | 43.127 | 101 | 202 | 64.337 | 71.407 |
| 62 | 124 | 39.494 | 43.834 | 102 | 204 | 64.974 | 72.114 |
| 63 | 126 | 40.131 | 44.541 | 103 | 206 | 65.611 | 72.821 |
| 64 | 128 | 40.768 | 45.248 | 104 | 208 | 66.248 | 73.528 |
| 65 | 130 | 41.405 | 45.955 | 105 | 210 | 66.885 | 74.235 |
| 66 | 132 | 42.042 | 46.662 | 106 | 212 | 67.522 | 74.942 |
| 67 | 134 | 42.679 | 47.369 | 107 | 214 | 68.159 | 75.649 |
| 68 | 136 | 43.316 | 48.076 | 108 | 216 | 68.796 | 76.356 |
| 69 | 138 | 43.953 | 48.783 | 109 | 218 | 69.433 | 77.063 |
| 70 | 140 | 44.590 | 49.490 | 110 | 220 | 70.070 | 77.770 |
| 71 | 142 | 45.227 | 50.197 | 111 | 222 | 70.707 | 78.477 |
| 72 | 144 | 45.864 | 50.904 | 112 | 224 | 71.344 | 79.184 |
| 73 | 146 | 46.501 | 51.611 | 113 | 226 | 71.981 | 79.891 |
| 74 | 148 | 47.138 | 52.318 | 114 | 228 | 72.618 | 80.598 |
| 75 | 150 | 47.775 | 53.025 | 115 | 230 | 73.255 | 81.305 |
| 76 | 152 | 48.412 | 53.732 | 116 | 232 | 73.892 | 82.012 |
| 77 | 154 | 49.049 | 54.439 | 117 | 234 | 74.529 | 82.719 |
| 78 | 156 | 49.686 | 55.146 | 118 | 236 | 75.166 | 83.426 |
| 79 | 158 | 50.323 | 55.853 | 119 | 238 | 75.803 | 84.133 |
| 80 | 160 | 50.960 | 56.560 | 120 | 240 | 76.440 | 84.840 |

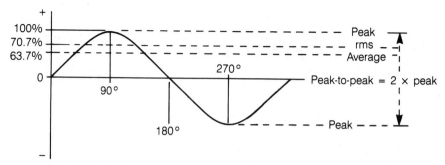

Fig. 3-2. Peak, rms, average, and peak-to-peak values of a sine wave.

*Example*  What is the peak value of a sine wave current whose effective (rms) value is measured as $17^1/_2$ V?

*Solution*  Locate the nearest value in the rms (effective) column. This is 17.675. Move to the left along the same line and locate 25 as the answer in the column marked peak.

*Example*  What is the average value of a voltage sine wave whose peak value is 160 V?

*Solution*  The maximum peak value shown in Table 3-1 is 120. You can extend the table, however, by multiplying each value by 10. Do this by adding a zero to the right of each whole number. Thus, in the peak column, 16 becomes 160. Move to the right and locate 10.192 in the average column. Multiply this value by 10 by moving the decimal point one place to the right. The average value is 101.92 V.

# Values of voltage or current of sine waves

Peak-to-peak, peak, rms and average values of voltage or current can be calculated from data in Table 3-2.

Table 3-2. Average, rms (effective), peak,
and peak-to-peak (p-p) values of sine waves of voltage or current.

| Given this value | Multiply by this value to get | | | |
| --- | --- | --- | --- | --- |
| | Average | rms (effective) | Peak | Peak-to-peak |
| Average | — | 1.11 | 1.57 | 3.14 ($\pi$) |
| rms (effective) | 0.9 | — | 1.414 | 2.828 |
| Peak | 0.637 | 0.707 | — | 2.0 |
| Peak-to-peak | 0.318 5 | 0.353 5 | 0.50 | — |

*Example*  The rms value of a sine wave is 3.14 V. What is its peak value?

*Solution*  Locate rms in the left column of Table 3-2. Move across to the peak column. The multiplication factor is 1.414. 1.414×3.14=4.439 9 V.

# Instantaneous values of voltage or current of sine waves

The instantaneous value of a wave is a function of the phase angle. At 0°, 180°, and 360° the instantaneous value of a sine wave is zero. It is a peak at 90° and 270°. See Fig. 3-3. These are the only values which may be known without the use of a table or formula. The instantaneous value of a sine voltage is:

$$e_i = E_{MAX} \sin \theta$$

Lowercase $e_i$ is instantaneous voltage, $E_{MAX}$ is peak voltage and $\theta$ is the degrees. For current substitute $I$ for $E$.

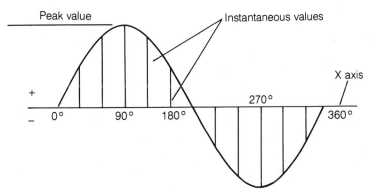

*Fig. 3-3.  Instantaneous values of voltage or current of a sine wave.*

Table 3-3 gives the instantaneous values of either voltage or current through a complete sine wave cycle of 360° for values of voltage ranging from 1 to 10. Other ranges can be obtained by moving the decimal point.

The values given in Table 3-3 under the heading of Peak voltage or current are those obtained from a table of natural trigonometric functions, and represent sine values. Thus, the sine of 10°=0.173 6 as shown by locating 10 in the extreme left-hand column and moving to the right and stopping in column 1. Greater accuracy can be obtained by consulting tables of natural trigonometric functions that supply a larger number of decimal places. Thus, a five-place table would show the sine of 10° as 0.173 65. For example, if a sine wave of voltage or current has a peak value of 1 V, its instantaneous value when the wave reaches 10° is 0.173 6 or 0.173 65 V depending on the accuracy you want. For a 2-V peak, the value would be 2×0.173 6 or 0.347 2. For a 3-V peak, the value would be 3×0.173 6 or

*Table 3-3. Instantaneous values of voltage or current of sine waves.*

| | | | | | Peak voltage or current | | | |
|---|---|---|---|---|---|---|---|---|
| **Phase angle (degrees)** | | | | **1** | **2** | **3** | **4** | **5** |
| 0 | 180 | 180 | 360 | 0.000 0 | 0.000 0 | 0.000 0 | 0.000 0 | 0.000 0 |
| 1 | 179 | 181 | 359 | 0.017 5 | 0.035 0 | 0.052 5 | 0.070 0 | 0.087 5 |
| 2 | 178 | 182 | 358 | 0.034 9 | 0.069 8 | 0.104 7 | 0.139 6 | 0.174 5 |
| 3 | 177 | 183 | 357 | 0.052 3 | 0.104 6 | 0.156 9 | 0.209 2 | 0.261 5 |
| 4 | 176 | 184 | 356 | 0.069 8 | 0.139 6 | 0.209 4 | 0.279 2 | 0.349 0 |
| 5 | 175 | 185 | 355 | 0.087 2 | 0.174 4 | 0.261 6 | 0.348 8 | 0.436 0 |
| 6 | 174 | 186 | 354 | 0.104 5 | 0.209 0 | 0.313 5 | 0.418 0 | 0.522 5 |
| 7 | 173 | 187 | 353 | 0.121 9 | 0.243 8 | 0.365 7 | 0.487 6 | 0.609 5 |
| 8 | 172 | 188 | 352 | 0.139 2 | 0.278 4 | 0.417 6 | 0.556 8 | 0.696 0 |
| 9 | 171 | 189 | 351 | 0.156 4 | 0.312 8 | 0.469 2 | 0.625 6 | 0.782 0 |
| 10 | 170 | 190 | 350 | 0.173 6 | 0.347 2 | 0.520 8 | 0.694 4 | 0.868 0 |
| 11 | 169 | 191 | 349 | 0.190 8 | 0.381 6 | 0.572 4 | 0.763 2 | 0.954 0 |
| 12 | 168 | 192 | 348 | 0.207 9 | 0.415 8 | 0.623 7 | 0.831 6 | 1.039 5 |
| 13 | 167 | 193 | 347 | 0.225 0 | 0.450 0 | 0.675 0 | 0.900 0 | 1.125 0 |
| 14 | 166 | 194 | 346 | 0.241 9 | 0.483 8 | 0.725 7 | 0.967 6 | 1.009 5 |
| 15 | 165 | 195 | 345 | 0.258 8 | 0.517 6 | 0.776 4 | 1.035 2 | 1.294 0 |
| 16 | 164 | 196 | 344 | 0.275 6 | 0.551 2 | 0.826 8 | 1.102 4 | 1.378 0 |
| 17 | 163 | 197 | 343 | 0.292 4 | 0.584 8 | 0.877 2 | 1.169 6 | 1.462 0 |
| 18 | 162 | 198 | 342 | 0.309 0 | 0.618 0 | 0.927 0 | 1.236 0 | 1.545 0 |
| 19 | 161 | 199 | 341 | 0.325 6 | 0.651 2 | 0.976 8 | 1.302 4 | 1.628 0 |
| 20 | 160 | 200 | 340 | 0.342 0 | 0.684 0 | 1.026 0 | 1.368 0 | 1.710 0 |
| 21 | 159 | 201 | 339 | 0.358 4 | 0.716 8 | 1.075 2 | 1.433 6 | 1.792 0 |
| 22 | 158 | 202 | 338 | 0.374 6 | 0.749 2 | 1.123 8 | 1.498 4 | 1.873 0 |
| 23 | 157 | 203 | 337 | 0.390 7 | 0.781 4 | 1.172 1 | 1.562 8 | 1.953 5 |
| 24 | 156 | 204 | 336 | 0.406 7 | 0.813 4 | 1.220 1 | 1.626 8 | 2.033 5 |
| 25 | 155 | 205 | 335 | 0.422 6 | 0.845 2 | 1.267 8 | 1.690 4 | 2.113 0 |
| 26 | 154 | 206 | 334 | 0.438 4 | 0.876 8 | 1.315 2 | 1.753 6 | 2.192 0 |
| 27 | 153 | 207 | 333 | 0.454 0 | 0.908 0 | 1.362 0 | 1.816 0 | 2.270 0 |
| 28 | 152 | 208 | 332 | 0.469 5 | 0.939 0 | 1.408 5 | 1.878 0 | 2.347 5 |
| 29 | 151 | 209 | 331 | 0.484 8 | 0.969 6 | 1.454 4 | 1.939 2 | 2.424 0 |
| 30 | 150 | 210 | 330 | 0.500 0 | 1.000 0 | 1.500 0 | 2.000 0 | 2.500 0 |
| 31 | 149 | 211 | 329 | 0.515 0 | 1.030 0 | 1.545 0 | 2.060 0 | 2.575 0 |
| 32 | 148 | 212 | 328 | 0.529 9 | 1.059 8 | 1.589 7 | 2.119 6 | 2.649 5 |
| 33 | 147 | 213 | 327 | 0.544 6 | 1.089 2 | 1.633 8 | 2.178 4 | 2.732 0 |
| 34 | 146 | 214 | 326 | 0.559 2 | 1.118 4 | 1.677 6 | 2.236 8 | 2.796 0 |
| 35 | 145 | 215 | 325 | 0.573 6 | 1.147 2 | 1.720 8 | 2.294 4 | 2.868 0 |
| 36 | 144 | 216 | 324 | 0.587 8 | 1.175 6 | 1.763 4 | 2.351 2 | 2.939 0 |
| 37 | 143 | 217 | 323 | 0.601 8 | 1.203 6 | 1.805 4 | 2.407 2 | 3.009 0 |
| 38 | 142 | 218 | 322 | 0.615 7 | 1.231 4 | 1.847 1 | 2.462 8 | 3.078 5 |
| 39 | 141 | 219 | 321 | 0.629 3 | 1.258 6 | 1.887 9 | 2.517 2 | 3.146 5 |
| 40 | 140 | 220 | 320 | 0.642 8 | 1.285 6 | 1.928 4 | 2.571 2 | 3.214 0 |
| 41 | 139 | 221 | 319 | 0.656 1 | 1.312 2 | 1.968 3 | 2.624 4 | 3.280 5 |
| 42 | 138 | 222 | 318 | 0.669 1 | 1.338 2 | 2.007 3 | 2.676 4 | 3.345 5 |

*Table 3-3. Continued.*

| Phase angle (degrees) | | | | Peak voltage or current | | | | |
|---|---|---|---|---|---|---|---|---|
| | | | | 1 | 2 | 3 | 4 | 5 |
| 43 | 137 | 223 | 317 | 0.682 0 | 1.364 0 | 2.046 0 | 2.728 0 | 3.410 0 |
| 44 | 136 | 224 | 316 | 0.694 7 | 1.389 4 | 2.084 1 | 2.778 8 | 3.473 5 |
| 45 | 135 | 225 | 315 | 0.707 1 | 1.414 2 | 2.121 3 | 2.828 4 | 3.535 5 |
| 46 | 134 | 226 | 314 | 0.719 3 | 1.438 6 | 2.157 9 | 2.877 2 | 3.596 5 |
| 47 | 133 | 227 | 313 | 0.731 4 | 1.462 8 | 2.194 2 | 2.925 6 | 3.657 0 |
| 48 | 132 | 228 | 312 | 0.743 1 | 1.486 2 | 2.229 3 | 2.972 4 | 3.715 5 |
| 49 | 131 | 229 | 311 | 0.754 7 | 1.509 4 | 2.264 1 | 3.018 8 | 3.773 5 |
| 50 | 130 | 230 | 310 | 0.766 0 | 1.532 0 | 2.298 0 | 3.064 0 | 3.830 0 |
| 51 | 129 | 231 | 309 | 0.777 1 | 1.554 2 | 2.331 3 | 3.108 4 | 3.885 5 |
| 52 | 128 | 232 | 308 | 0.788 0 | 1.576 0 | 2.364 0 | 3.152 0 | 3.940 0 |
| 53 | 127 | 233 | 307 | 0.798 6 | 1.597 2 | 2.395 8 | 3.195 4 | 3.993 0 |
| 54 | 126 | 234 | 306 | 0.809 0 | 1.618 0 | 2.427 0 | 3.236 0 | 4.045 0 |
| 55 | 125 | 235 | 305 | 0.819 2 | 1.638 4 | 2.457 6 | 3.276 8 | 4.096 0 |
| 56 | 124 | 236 | 304 | 0.829 0 | 1.658 0 | 2.487 0 | 3.316 0 | 4.145 0 |
| 57 | 123 | 237 | 303 | 0.838 7 | 1.677 4 | 2.516 1 | 3.354 8 | 4.193 5 |
| 58 | 122 | 238 | 302 | 0.848 0 | 1.696 0 | 2.544 0 | 3.392 0 | 4.240 0 |
| 59 | 121 | 239 | 301 | 0.857 2 | 1.714 4 | 2.571 6 | 3.428 8 | 4.286 0 |
| 60 | 120 | 240 | 300 | 0.866 0 | 1.732 0 | 2.598 0 | 3.464 0 | 4.330 0 |
| 61 | 119 | 241 | 299 | 0.874 6 | 1.749 2 | 2.623 8 | 3.498 4 | 4.373 0 |
| 62 | 118 | 242 | 298 | 0.882 9 | 1.765 8 | 2.648 7 | 3.531 6 | 4.414 5 |
| 63 | 117 | 243 | 297 | 0.891 0 | 1.782 0 | 2.673 0 | 3.564 0 | 4.455 0 |
| 64 | 116 | 244 | 296 | 0.898 8 | 1.797 6 | 2.696 4 | 3.595 2 | 4.494 0 |
| 65 | 115 | 245 | 295 | 0.906 3 | 1.812 6 | 2.718 9 | 3.625 2 | 4.531 5 |
| 66 | 114 | 246 | 294 | 0.913 5 | 1.827 0 | 2.740 5 | 3.654 0 | 4.567 5 |
| 67 | 113 | 247 | 293 | 0.920 5 | 1.841 0 | 2.761 5 | 3.682 0 | 4.602 5 |
| 68 | 112 | 248 | 292 | 0.927 2 | 1.854 4 | 2.781 6 | 3.708 8 | 4.636 0 |
| 69 | 111 | 249 | 291 | 0.933 6 | 1.867 2 | 2.800 8 | 3.734 4 | 4.668 0 |
| 70 | 110 | 250 | 290 | 0.939 7 | 1.879 4 | 2.819 1 | 3.758 8 | 4.698 5 |
| 71 | 109 | 251 | 289 | 0.945 5 | 1.891 0 | 2.836 5 | 3.782 0 | 4.727 5 |
| 72 | 108 | 252 | 288 | 0.951 1 | 1.902 2 | 2.853 3 | 3.804 4 | 4.755 5 |
| 73 | 107 | 253 | 287 | 0.956 3 | 1.912 6 | 2.868 9 | 3.825 2 | 4.781 5 |
| 74 | 106 | 254 | 286 | 0.961 3 | 1.922 6 | 2.883 9 | 3.845 2 | 4.806 5 |
| 75 | 105 | 255 | 285 | 0.965 9 | 1.931 8 | 2.897 7 | 3.863 6 | 4.829 5 |
| 76 | 104 | 256 | 284 | 0.970 3 | 1.940 6 | 2.910 9 | 3.881 2 | 4.851 5 |
| 77 | 103 | 257 | 283 | 0.974 4 | 1.948 8 | 2.923 2 | 3.897 6 | 4.872 0 |
| 78 | 102 | 258 | 282 | 0.978 1 | 1.956 2 | 2.934 3 | 3.912 4 | 4.890 5 |
| 79 | 101 | 259 | 281 | 0.981 6 | 1.963 2 | 2.944 8 | 3.926 4 | 4.908 0 |
| 80 | 100 | 260 | 280 | 0.984 8 | 1.969 6 | 2.954 4 | 3.939 2 | 4.924 0 |
| 81 | 99 | 261 | 279 | 0.987 7 | 1.975 4 | 2.963 1 | 3.950 8 | 4.938 5 |
| 82 | 98 | 262 | 278 | 0.990 3 | 1.980 6 | 2.970 9 | 3.961 2 | 4.951 5 |
| 83 | 97 | 263 | 277 | 0.992 5 | 1.985 0 | 2.977 5 | 3.970 0 | 4.962 5 |
| 84 | 96 | 264 | 276 | 0.994 5 | 1.989 0 | 2.983 5 | 3.978 0 | 4.972 5 |

*Table 3-3. Continued.*

|  |  |  |  | Peak voltage or current | | | | |
|---|---|---|---|---|---|---|---|---|
| Phase angle (degrees) | | | | **1** | **2** | **3** | **4** | **5** |
| 85 | 95 | 265 | 275 | 0.996 2 | 1.992 4 | 2.988 6 | 3.984 8 | 4.981 0 |
| 86 | 94 | 266 | 274 | 0.997 6 | 1.995 2 | 2.992 8 | 3.990 4 | 4.988 0 |
| 87 | 93 | 267 | 273 | 0.998 6 | 1.997 2 | 2.995 8 | 3.994 4 | 4.993 0 |
| 88 | 92 | 268 | 272 | 0.999 4 | 1.998 8 | 2.998 2 | 3.997 6 | 4.997 0 |
| 89 | 91 | 269 | 271 | 0.999 8 | 1.999 6 | 2.999 4 | 3.999 2 | 4.999 0 |
| 90 | 90 | 270 | 270 | 1.000 0 | 2.000 0 | 3.000 0 | 4.000 0 | 5.000 0 |

|  |  |  |  | Peak voltage or current | | | | |
|---|---|---|---|---|---|---|---|---|
| Phase angle (degrees) | | | | **6** | **7** | **8** | **9** | **10** |
| 0 | 180 | 180 | 360 | 0.000 0 | 0.000 0 | 0.000 0 | 0.000 0 | 0.000 0 |
| 1 | 179 | 181 | 359 | 0.105 0 | 0.122 5 | 0.140 0 | 0.157 5 | 0.175 0 |
| 2 | 178 | 182 | 358 | 0.209 4 | 0.244 3 | 0.279 2 | 0.314 1 | 0.349 0 |
| 3 | 177 | 183 | 357 | 0.313 8 | 0.366 1 | 0.418 4 | 0.470 7 | 0.523 0 |
| 4 | 176 | 184 | 356 | 0.418 8 | 0.488 6 | 0.558 4 | 0.628 2 | 0.698 0 |
| 5 | 175 | 185 | 355 | 0.523 2 | 0.610 4 | 0.697 6 | 0.784 8 | 0.872 0 |
| 6 | 174 | 186 | 354 | 0.627 0 | 0.731 5 | 0.836 0 | 0.940 5 | 1.045 0 |
| 7 | 173 | 187 | 353 | 0.731 4 | 0.853 3 | 0.975 2 | 1.097 1 | 1.219 0 |
| 8 | 172 | 188 | 352 | 0.835 2 | 0.974 4 | 1.113 6 | 1.252 8 | 1.392 0 |
| 9 | 171 | 189 | 351 | 0.938 4 | 1.094 8 | 1.251 2 | 1.407 6 | 1.564 0 |
| 10 | 170 | 190 | 350 | 1.041 6 | 1.215 2 | 1.388 8 | 1.562 4 | 1.736 0 |
| 11 | 169 | 191 | 349 | 1.444 8 | 1.335 6 | 1.526 4 | 1.717 2 | 1.908 0 |
| 12 | 168 | 192 | 348 | 1.247 4 | 1.455 3 | 1.663 2 | 1.871 1 | 2.079 0 |
| 13 | 167 | 193 | 347 | 1.350 0 | 1.575 0 | 1.800 0 | 2.025 0 | 2.250 0 |
| 14 | 166 | 194 | 346 | 1.451 4 | 1.693 3 | 1.935 2 | 2.177 1 | 2.419 0 |
| 15 | 165 | 195 | 345 | 1.552 8 | 1.811 6 | 2.070 4 | 2.329 2 | 2.588 0 |
| 16 | 164 | 196 | 344 | 1.653 6 | 1.929 2 | 2.204 8 | 2.480 4 | 2.756 0 |
| 17 | 163 | 197 | 343 | 1.754 4 | 2.046 8 | 2.339 2 | 2.631 6 | 2.924 0 |
| 18 | 162 | 198 | 342 | 1.854 0 | 2.163 0 | 2.472 0 | 2.781 0 | 3.090 0 |
| 19 | 161 | 199 | 341 | 1.953 6 | 2.279 2 | 2.604 8 | 2.930 4 | 3.256 0 |
| 20 | 160 | 200 | 340 | 2.052 0 | 2.394 0 | 2.736 0 | 3.078 0 | 3.420 0 |
| 21 | 159 | 201 | 339 | 2.150 4 | 2.508 8 | 2.867 2 | 3.225 6 | 3.584 0 |
| 22 | 158 | 202 | 338 | 2.247 6 | 2.622 2 | 2.996 8 | 3.371 4 | 3.746 0 |
| 23 | 157 | 203 | 337 | 2.344 2 | 2.734 9 | 3.125 6 | 3.516 3 | 3.907 0 |
| 24 | 156 | 204 | 336 | 2.440 2 | 2.846 9 | 3.253 6 | 3.660 3 | 4.067 0 |
| 25 | 155 | 205 | 335 | 2.535 6 | 2.958 2 | 3.380 8 | 3.803 4 | 4.226 0 |
| 26 | 154 | 206 | 334 | 2.630 4 | 3.068 8 | 3.507 2 | 3.945 6 | 4.384 0 |
| 27 | 153 | 207 | 333 | 2.724 0 | 3.178 0 | 3.632 0 | 4.086 0 | 4.540 0 |
| 28 | 152 | 208 | 332 | 2.817 0 | 3.286 5 | 3.756 0 | 4.225 5 | 4.695 0 |
| 29 | 151 | 209 | 331 | 2.908 8 | 3.393 6 | 3.878 4 | 4.363 2 | 4.848 0 |
| 30 | 150 | 210 | 330 | 3.000 0 | 3.500 0 | 4.000 0 | 4.500 0 | 5.000 0 |
| 31 | 149 | 211 | 329 | 3.090 0 | 3.605 0 | 4.120 0 | 4.635 0 | 5.150 0 |
| 32 | 148 | 212 | 328 | 3.179 4 | 3.709 3 | 4.239 2 | 4.769 1 | 5.299 0 |

*Table 3-3. Continued.*

| | Phase angle (degrees) | | | Peak voltage or current | | | | |
|---|---|---|---|---|---|---|---|---|
| | | | | 6 | 7 | 8 | 9 | 10 |
| 33 | 147 | 213 | 327 | 3.267 6 | 3.812 2 | 4.356 8 | 4.901 4 | 5.446 0 |
| 34 | 146 | 214 | 326 | 3.355 2 | 3.914 4 | 4.473 6 | 5.032 8 | 5.592 0 |
| 35 | 145 | 215 | 325 | 3.441 6 | 4.015 2 | 4.588 8 | 5.163 4 | 5.736 0 |
| 36 | 144 | 216 | 324 | 3.526 8 | 4.114 6 | 4.700 4 | 5.290 2 | 5.878 0 |
| 37 | 143 | 217 | 323 | 3.610 8 | 4.212 6 | 4.814 4 | 5.416 2 | 6.018 0 |
| 38 | 142 | 218 | 322 | 3.694 2 | 4.309 9 | 4.925 6 | 5.541 3 | 6.157 0 |
| 39 | 141 | 219 | 321 | 3.775 8 | 4.405 1 | 5.034 4 | 5.663 7 | 6.293 0 |
| 40 | 140 | 220 | 320 | 3.856 8 | 4.499 6 | 5.142 4 | 5.785 2 | 6.428 0 |
| 41 | 139 | 221 | 319 | 3.936 6 | 4.592 7 | 5.248 8 | 5.904 9 | 6.561 0 |
| 42 | 138 | 222 | 318 | 4.014 6 | 4.683 7 | 5.352 8 | 6.021 9 | 6.691 0 |
| 43 | 137 | 223 | 317 | 4.092 0 | 4.774 0 | 5.456 0 | 6.138 0 | 6.820 0 |
| 44 | 136 | 224 | 316 | 4.168 2 | 4.862 9 | 5.557 6 | 6.252 3 | 6.947 0 |
| 45 | 135 | 225 | 315 | 4.242 6 | 4.949 7 | 5.656 8 | 6.363 9 | 7.071 0 |
| 46 | 134 | 226 | 314 | 4.315 8 | 5.035 1 | 5.754 4 | 6.473 7 | 7.193 0 |
| 47 | 133 | 227 | 313 | 4.388 4 | 5.119 8 | 5.851 2 | 6.582 6 | 7.314 0 |
| 48 | 132 | 228 | 312 | 4.458 6 | 5.201 7 | 5.944 8 | 6.687 9 | 7.431 0 |
| 49 | 131 | 229 | 311 | 4.528 2 | 5.282 9 | 6.037 6 | 6.792 3 | 7.547 0 |
| 50 | 130 | 230 | 310 | 4.596 0 | 5.362 0 | 6.128 0 | 6.894 0 | 7.660 0 |
| 51 | 129 | 231 | 309 | 4.662 6 | 5.439 7 | 6.216 8 | 6.993 9 | 7.771 0 |
| 52 | 128 | 232 | 308 | 4.728 0 | 5.516 0 | 6.304 0 | 7.092 0 | 7.880 0 |
| 53 | 127 | 233 | 307 | 4.791 6 | 5.590 2 | 6.388 8 | 7.187 4 | 7.986 0 |
| 54 | 126 | 234 | 306 | 4.854 0 | 5.663 0 | 6.472 0 | 7.281 0 | 8.090 0 |
| 55 | 125 | 235 | 305 | 4.915 2 | 5.734 4 | 6.553 6 | 7.372 8 | 8.192 0 |
| 56 | 124 | 236 | 304 | 4.974 0 | 5.803 0 | 6.632 0 | 7.461 0 | 8.290 0 |
| 57 | 123 | 237 | 303 | 5.032 2 | 5.870 9 | 6.709 6 | 7.508 3 | 8.387 0 |
| 58 | 122 | 238 | 302 | 5.088 0 | 5.936 0 | 6.784 0 | 7.632 0 | 8.480 0 |
| 59 | 121 | 239 | 301 | 5.143 2 | 6.000 4 | 6.857 6 | 7.714 8 | 8.572 0 |
| 60 | 120 | 240 | 300 | 5.196 0 | 6.062 0 | 6.928 0 | 7.794 0 | 8.660 0 |
| 61 | 119 | 241 | 299 | 5.247 6 | 6.122 2 | 6.996 8 | 7.871 4 | 8.746 0 |
| 62 | 118 | 242 | 298 | 5.297 4 | 6.180 3 | 7.063 2 | 7.946 1 | 8.829 0 |
| 63 | 117 | 243 | 297 | 5.346 0 | 6.237 0 | 7.128 0 | 8.019 0 | 8.910 0 |
| 64 | 116 | 244 | 296 | 5.392 8 | 6.291 6 | 7.190 4 | 8.089 2 | 8.988 0 |
| 65 | 115 | 245 | 295 | 5.437 8 | 6.344 1 | 7.250 4 | 8.156 7 | 9.063 0 |
| 66 | 114 | 246 | 294 | 5.481 0 | 6.394 5 | 7.308 0 | 8.221 5 | 9.135 0 |
| 67 | 113 | 247 | 293 | 5.523 0 | 6.443 5 | 7.364 0 | 8.284 5 | 9.205 0 |
| 68 | 112 | 248 | 292 | 5.563 2 | 6.490 4 | 7.417 6 | 8.344 8 | 9.272 0 |
| 69 | 111 | 249 | 291 | 5.601 6 | 6.535 2 | 7.468 8 | 8.402 4 | 9.336 0 |
| 70 | 110 | 250 | 290 | 5.638 2 | 6.577 9 | 7.517 6 | 8.457 3 | 9.397 0 |
| 71 | 109 | 251 | 289 | 5.673 0 | 6.618 5 | 7.564 0 | 8.509 5 | 9.455 0 |
| 72 | 108 | 252 | 288 | 5.706 6 | 6.657 7 | 7.608 8 | 8.559 9 | 9.511 0 |
| 73 | 107 | 253 | 287 | 5.737 8 | 6.694 1 | 7.650 4 | 8.606 7 | 9.563 0 |
| 74 | 106 | 254 | 286 | 5.767 8 | 6.729 1 | 7.690 4 | 8.651 7 | 9.612 0 |

*Table 3-3. Continued.*

| | Phase angle (degrees) | | | Peak voltage or current | | | | |
|---|---|---|---|---|---|---|---|---|
| | | | | 6 | 7 | 8 | 9 | 10 |
| 75 | 105 | 255 | 285 | 5.795 4 | 6.761 3 | 7.727 2 | 8.693 1 | 9.659 0 |
| 76 | 104 | 256 | 284 | 5.821 8 | 6.792 1 | 7.762 4 | 8.732 7 | 9.703 0 |
| 77 | 103 | 257 | 283 | 5.846 4 | 6.820 8 | 7.795 2 | 8.769 6 | 9.744 0 |
| 78 | 102 | 258 | 282 | 5.868 6 | 6.846 7 | 7.824 8 | 8.802 9 | 9.781 0 |
| 79 | 101 | 259 | 281 | 5.889 6 | 6.871 2 | 7.852 8 | 8.834 4 | 9.816 0 |
| 80 | 100 | 260 | 280 | 5.908 8 | 6.893 6 | 7.878 4 | 8.863 2 | 9.848 0 |
| 81 | 99 | 261 | 279 | 5.926 2 | 6.913 9 | 7.901 6 | 8.889 3 | 9.877 0 |
| 82 | 98 | 262 | 278 | 5.941 8 | 6.932 1 | 7.922 4 | 8.912 7 | 9.903 0 |
| 83 | 97 | 263 | 277 | 5.955 0 | 6.947 5 | 7.940 0 | 8.932 5 | 9.925 0 |
| 84 | 96 | 264 | 276 | 5.967 0 | 6.961 5 | 7.956 0 | 8.950 5 | 9.945 0 |
| 85 | 95 | 265 | 275 | 5.977 2 | 6.973 4 | 7.969 6 | 8.965 8 | 9.962 0 |
| 86 | 94 | 266 | 274 | 5.985 6 | 6.983 2 | 7.980 8 | 8.978 4 | 9.976 0 |
| 87 | 93 | 267 | 273 | 5.991 6 | 6.990 2 | 7.988 8 | 8.987 4 | 9.986 0 |
| 88 | 92 | 268 | 272 | 5.996 4 | 6.995 8 | 7.995 2 | 8.994 6 | 9.994 0 |
| 89 | 91 | 269 | 271 | 5.998 8 | 6.998 6 | 7.998 4 | 8.998 2 | 9.998 0 |
| 90 | 90 | 270 | 270 | 6.000 0 | 7.000 0 | 8.000 0 | 9.000 0 | 10.000 0 |

0.520 8 V, as shown in the respective columns headed 2 and 3 in Table 3-3. Using this technique, the instantaneous value of any sine wave or voltage can be found.

*Example*   What is the instantaneous value of a sine wave at 27° if the peak value of the wave is 138.2 V?

*Solution*   Locate 27° in the table. Move to the right and in column 1 find 0.454 0 V. This is the instantaneous value at 27° when the peak value is 1 V. For 138.2 V, multiply 0.454 0 by 138.2. That is, 0.454 0×138.2=62.742 8 V. A five-place table of natural trigonometric functions shows the value of 27° as 0.453 99. 0.453 99×138.2=62.741 418 V. Whether this greater accuracy is desirable depends on the work being done. The actual difference is 62.742 8−62.741 418=0.001 382 V. Although the example mentions peak in terms of volts, peak can be volts, millivolts, or microvolts, amperes, milliamperes, or microamperes.

*Example*   What is the instantaneous value of voltage at a phase angle of 37° when the peak value is 3 V?

*Solution*   Locate 37° in the left-hand column of the table. Move horizontally until the 3-V column is reached. The required voltage is 1.805 4 V.

*Example*   At what phase angles will the instantaneous voltage of a sine wave be 68% of its peak value?

*Solution*   Consider peak as 1 or 100%. Locate the nearest value to 68 in the column headed by the number 1. This value is 0.682 0. Move to the left of this number and you will see that the phase angle is 43°. Multiples of this value are also given so that you have 137° (180°−43°); 223° (43°+180°) and 317° (360°−43°).

*Example* What is the instantaneous value of a sine wave of current at a phase angle of 77° when its peak value is 30 mA?

*Solution* Locate 77° in the left-hand column of the table and move horizontally to the right to intercept 2.923 2 in the column headed by the number 3. Multiply 3 by 10 to obtain the peak value specified in the question. However, because 3 was changed to 30 by multiplying it by 10 (or by moving the decimal point one place to the right) the answer must be similarly treated. The value is 29.232 mA.

# Period and frequency

The time required for the completion of one complete cycle by a periodic function, such as a sine wave, is known as its period. The relationship between the period and the frequency of a wave is a reciprocal one and is shown in the formula:

$$T = \frac{1}{f}$$

*T*, the period of the wave, is the time required for the completion of one full cycle; *f* is the frequency in hertz (cycles per second).

Table 3-4 permits the rapid conversion between the period of a wave and its frequency. Values not given in the table can be obtained by moving the decimal point. However, since the relationship is an inverse one, the decimal point for frequency and for time will move in opposite directions. Thus, for a frequency of 10 Hz, the time is 0.1 s. For a frequency of 100 Hz, move the decimal point one place to the right, changing 10 to 100. For the corresponding value of time, however, move the decimal point one place to the left. This would change 0.1 s to 0.01 s.

*Example* The sine wave input to a power supply is 60 Hz. What is the period of this wave?

*Solution* Locate 60 in the frequency column. Immediately adjacent, you will see it requires 0.016 7 s to complete one single cycle of this waveform.

*Example* What is the period of a sine waveform having a frequency of 550 kHz?

*Solution* In Table 3-4, the frequency is given in hertz. In that table, 55 can be made to represent 550 kHz by multiplying it by 10 000 or by moving its decimal point four places to the right (550 000). However, as shown in the formula given earlier, *T* and *f* are inverse. Thus, if you move the decimal point to the right for the frequency column, you must move it to the left an equal number of places for the time column. For 55 Hz, the time is 0.018 2 s. For 550 000 Hz, the time is 0.000 001 82 s or 1.82 μs.

*Table 3-4. Period versus frequency.*

| Frequency (Hz) | Time (s) | Frequency (Hz) | Time (s) | Frequency (Hz) | Time (s) |
|---|---|---|---|---|---|
| 1 | 1.000 0 | 34 | 0.029 4 | 67 | 0.014 9 |
| 2 | 0.500 0 | 35 | 0.028 6 | 68 | 0.014 7 |
| 3 | 0.333 3 | 36 | 0.027 8 | 69 | 0.014 5 |
| 4 | 0.250 0 | 37 | 0.027 0 | 70 | 0.014 3 |
| 5 | 0.200 0 | 38 | 0.026 3 | 71 | 0.014 1 |
| 6 | 0.166 7 | 39 | 0.025 6 | 72 | 0.013 9 |
| 7 | 0.142 9 | 40 | 0.025 0 | 73 | 0.013 7 |
| 8 | 0.125 0 | 41 | 0.024 4 | 74 | 0.013 5 |
| 9 | 0.111 1 | 42 | 0.023 8 | 75 | 0.013 3 |
| 10 | 0.100 0 | 43 | 0.023 3 | 76 | 0.013 2 |
| 11 | 0.090 9 | 44 | 0.022 7 | 77 | 0.013 0 |
| 12 | 0.083 3 | 45 | 0.022 2 | 78 | 0.012 8 |
| 13 | 0.076 9 | 46 | 0.021 7 | 79 | 0.012 7 |
| 14 | 0.071 4 | 47 | 0.021 3 | 80 | 0.012 5 |
| 15 | 0.066 7 | 48 | 0.020 8 | 81 | 0.012 3 |
| 16 | 0.062 5 | 49 | 0.020 4 | 82 | 0.012 2 |
| 17 | 0.058 8 | 50 | 0.020 0 | 83 | 0.012 0 |
| 18 | 0.055 6 | 51 | 0.019 6 | 84 | 0.011 9 |
| 19 | 0.052 6 | 52 | 0.019 2 | 85 | 0.011 8 |
| 20 | 0.050 0 | 53 | 0.018 9 | 86 | 0.011 6 |
| 21 | 0.047 6 | 54 | 0.018 5 | 87 | 0.011 5 |
| 22 | 0.045 5 | 55 | 0.018 2 | 88 | 0.011 4 |
| 23 | 0.043 5 | 56 | 0.017 9 | 89 | 0.011 2 |
| 24 | 0.041 7 | 57 | 0.017 5 | 90 | 0.011 1 |
| 25 | 0.040 0 | 58 | 0.017 2 | 91 | 0.011 0 |
| 26 | 0.038 5 | 59 | 0.016 9 | 92 | 0.010 9 |
| 27 | 0.037 0 | 60 | 0.016 7 | 93 | 0.010 8 |
| 28 | 0.035 7 | 61 | 0.016 4 | 94 | 0.010 6 |
| 29 | 0.034 5 | 62 | 0.016 1 | 95 | 0.010 5 |
| 30 | 0.033 3 | 63 | 0.015 9 | 96 | 0.010 4 |
| 31 | 0.032 3 | 64 | 0.015 6 | 97 | 0.010 3 |
| 32 | 0.031 2 | 65 | 0.015 4 | 98 | 0.010 2 |
| 33 | 0.030 3 | 66 | 0.015 2 | 99 | 0.010 1 |
|  |  |  |  | 100 | 0.010 0 |

*Example* The time of a half wave is 61 $\mu$s. What is its frequency?

*Solution* Assuming the problem involves a sine wave, first multiply 61 by 2 to get the time of a full wave. $2 \times 61 = 122$. 122 $\mu$s corresponds to 0.000 122 s. The nearest value shown in the table is 0.012 2 s, and the frequency for this time value is 82 Hz. You can get 0.000 122 by moving the decimal point of 0.012 2 two places to the left. The decimal point for the frequency, then, should be moved an equivalent number of places to the right. This would give an answer of 8 200 Hz.

# Angular measurements of a sine wave

Because a sine wave can be generated by the rotating armature of an ac generator, the output of that generator can be plotted along an X axis, actually the straight line equivalent of the circumference produced by the armature (Fig. 3-4). This

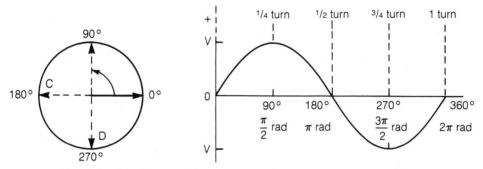

Fig. 3-4. *Development of a sine wave and its measurements in radians.*

line is 360° with sine wave peaks at 90° and 270°, and it represents one complete rotation of the armature of the generator. The next rotation would produce a repetition of the first sine wave and its X axis. The value of the sine wave at any instant is obtained by erecting a vertical line at any point on the axis.

# Radian measure of a sine wave

The radius of a circle is equal to the diameter divided by 2. Thus:

$$r=D/2$$

If the radius is superimposed on the circumference, it will subtend an angle of 57.3° (1 rad), as in Fig. 3-5. The total number of radians that can be positioned

Fig. 3-5. *The radius of a circle, measured on its circumference, is a radian.*

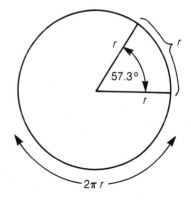

on the circumference is equal to $2\pi$. The relationships appear as in Table 3-5. The value of $\pi$ is obtained by dividing the circumference of a circle by its diameter. Table 3-6 shows sine wave relationships in terms of radians.

*Table 3-5.*
*Radian measure.*

| Degrees | Radians |
|---------|---------|
| 360 | $2\ \pi r$ |
| 270 | $\dfrac{3}{2}\ \pi r$ |
| 180 | $\dfrac{2}{2}\ \pi r = \pi r$ |
| 90 | $\dfrac{1}{2}\ \pi r$ |

*Table 3-6. Sine waves in terms of radians.*

| Angle $\theta$ Degrees | Radians | Sin $\theta$ | Voltage |
|---------|---------|-------|---------|
| 0 | 0 | 0 | Zero |
| 30 | $\dfrac{\pi}{6}$ | 0.500 | 50% of peak |
| 45 | $\dfrac{\pi}{4}$ | 0.707 | 70.7% of peak |
| 60 | $\dfrac{\pi}{3}$ | 0.866 | 86.6% of peak |
| 90 | $\dfrac{\pi}{2}$ | 1.000 | Positive peak |
| 180 | $\pi$ | 0 | Zero |
| 270 | $\dfrac{3\pi}{2}$ | $-1.000$ | Negative peak |
| 360 | $2\pi$ | 0 | Zero |

# Time units

The basic unit of time is the second, abbreviated as *s*. Submultiples of the basic unit are the millisecond, microsecond, and nanosecond, as indicated in Table 3-7.

# Wavelength

The wavelength of a single cycle of a sine wave is the distance from its start to its completion. Wavelength is represented by the Greek letter lambda, $\lambda$.

*Table 3-7. Time units.*

| Unit | Abbreviation | Value |
|------|-------------|-------|
| 1 millisecond | 1 ms | $1 \times 10^{-3}$ s |
| 1 microsecond | 1 $\mu$s | $1 \times 10^{-6}$ s |
| 1 nanosecond | 1 ns | $1 \times 10^{-9}$ s |

The wavelength of a sine wave can be represented by:

$$\lambda = \frac{\text{velocity}}{\text{frequency}}$$

The formula indicates that the wavelength and frequency of a sine wave have an inverse relationship, that is, the higher the frequency the shorter the wavelength. Conversely, the lower the frequency, the longer the wavelength.

# Frequency — wavelength conversions

The relationship between the frequency of a radio wave (f) in hertz or cycles per second and its wavelength, in meters, is supplied by the formula $\lambda = 300\,000\,000/f$. The number 300 000 000 in the numerator of the formula is the velocity of light (and of radio waves) in space, and is a constant. $\lambda$ is the wavelength in meters.

The velocity of light is in meters per second. Although it is frequently rounded off to 300 000 000, its more probable value is 299 820 000 m/s. When the frequency is in hertz, wavelength in meters equals $299\,820\,000/f$. When the frequency is in kilohertz, wavelength in meters equals $299\,820/f$. When the frequency is in megahertz, wavelength in meters equals $299.82/f$.

*Table 3-8. Frequency bands.*

| Frequency | Designation | Abbreviation |
|-----------|-------------|--------------|
| 30 to 300 Hz | extremely low frequencies (audio) | ELF |
| 20 Hz to 20 kHz | sound/frequencies | AF |
| 20 to 30 kHz | very-low frequency | VLF |
| 30 to 300 kHz | low frequency | LF |
| 300 to 3 000 kHz | medium frequency | MF |
| 3 000 to 30 000 kHz | high frequency | HF |
| 30 to 300 MHz | very-high frequency | VHF |
| 300 to 3 000 MHz | ultra-high frequency | UHF |
| 3 000 to 30 000 MHz | super-high frequency | SHF |
| 30 000 to 300 000 MHz (30 to 300 GHz) | extremely-high frequency | EHF |
| 300 to 3 000 GHz | no designation | |

Table 3-8 supplies the abbreviations and descriptions of waves whose frequency extends from 30 Hz to 3 000 GHz.

*Example*  A radio wave has a frequency of 500 kHz. What is its wavelength in meters?
*Solution*  Find the number 500 in the kHz column in Table 3-9. Move horizontally and you will see the corresponding wavelength of 599.6 m.

*Example*  One of the bands of a short-wave receiver covers the range from 3 000 kHz to 5 000 kHz. What wavelength range does this include?
*Solution*  Table 3-9 shows that 3 000 kHz corresponds to 99.94 m and that 5 000 kHz corresponds to 59.96 m. Thus, this particular band is from approximately 60 to 100 m.

*Table 3-9. Kilohertz (kHz) to meters (m) or meters to kilohertz.*

| kHz | m | kHz | m | kHz | m | kHz | m |
|---|---|---|---|---|---|---|---|
| 1 | 299 820 | 190 | 1 578 | 460 | 652.2 | 730 | 410.7 |
| 2 | 149 910 | 200 | 1 499 | 470 | 638.3 | 740 | 405.4 |
| 3 | 99 940 | 210 | 1 428 | 480 | 624.6 | 750 | 399.8 |
| 4 | 74 955 | 220 | 1 363 | 490 | 612.2 | 760 | 394.7 |
| 5 | 59 964 | 230 | 1 304 | 500 | 599.6 | 770 | 389.6 |
| 6 | 49 970 | 240 | 1 249 | 510 | 588.2 | 780 | 384.6 |
| 7 | 42 831 | 250 | 1 199 | 520 | 576.9 | 790 | 379.8 |
| 8 | 37 478 | 260 | 1 153 | 530 | 565.7 | 800 | 374.8 |
| 9 | 33 313 | 270 | 1 110 | 540 | 555.6 | 810 | 370.4 |
| 10 | 29 982 | 280 | 1 071 | 550 | 545.4 | 820 | 365.9 |
| 20 | 14 991 | 290 | 1 034 | 560 | 535.7 | 830 | 361.4 |
| 30 | 9 994 | 300 | 999.4 | 570 | 526.3 | 840 | 357.1 |
| 40 | 7 495 | 310 | 967.7 | 580 | 517.2 | 850 | 352.9 |
| 50 | 5 996 | 320 | 937.5 | 590 | 508.5 | 860 | 348.8 |
| 60 | 4 997 | 330 | 908.1 | 600 | 499.7 | 870 | 344.8 |
| 70 | 4 283 | 340 | 882.4 | 610 | 491.8 | 880 | 340.9 |
| 80 | 3 748 | 350 | 859.1 | 620 | 483.7 | 890 | 337.1 |
| 90 | 3 331 | 360 | 833.3 | 630 | 476.2 | 900 | 333.3 |
| 100 | 2 998 | 370 | 810.8 | 640 | 468.7 | 910 | 329.5 |
| 110 | 2 726 | 380 | 789.5 | 650 | 461.5 | 920 | 325.9 |
| 120 | 2 499 | 390 | 769.2 | 660 | 454.5 | 930 | 322.4 |
| 130 | 2 306 | 400 | 749.6 | 670 | 447.8 | 940 | 319.0 |
| 140 | 2 142 | 410 | 731.7 | 680 | 441.2 | 950 | 315.6 |
| 150 | 1 999 | 420 | 714.3 | 690 | 434.8 | 960 | 312.3 |
| 160 | 1 874 | 430 | 697.7 | 700 | 428.6 | 970 | 309.1 |
| 170 | 1 764 | 440 | 681.4 | 710 | 422.5 | 980 | 305.9 |
| 180 | 1 666 | 450 | 666.7 | 720 | 416.7 | 990 | 302.8 |

# Table 3-9. Continued.

| kHz | m | kHz | m | kHz | m | kHz | m |
|---|---|---|---|---|---|---|---|
| 1 000 | 299.8 | 1 420 | 211.1 | 1 840 | 162.9 | 2 260 | 132.7 |
| 1 010 | 296.9 | 1 430 | 209.7 | 1 850 | 162.1 | 2 270 | 132.1 |
| 1 020 | 293.9 | 1 440 | 208.2 | 1 860 | 161.2 | 2 280 | 131.5 |
| 1 030 | 291.1 | 1 450 | 206.8 | 1 870 | 160.3 | 2 290 | 130.9 |
| 1 040 | 288.3 | 1 460 | 205.4 | 1 880 | 159.5 | 2 300 | 130.4 |
| 1 050 | 285.5 | 1 470 | 204.0 | 1 890 | 158.6 | 2 310 | 129.8 |
| 1 060 | 282.8 | 1 480 | 202.6 | 1 900 | 157.8 | 2 320 | 129.2 |
| 1 070 | 280.2 | 1 490 | 201.2 | 1 910 | 157.0 | 2 330 | 128.7 |
| 1 080 | 277.6 | 1 500 | 199.9 | 1 920 | 156.2 | 2 340 | 128.1 |
| 1 090 | 275.1 | 1 510 | 198.6 | 1 930 | 155.3 | 2 350 | 127.6 |
| 1 100 | 272.6 | 1 520 | 197.2 | 1 940 | 154.5 | 2 360 | 127.0 |
| 1 110 | 270.1 | 1 530 | 196.0 | 1 950 | 153.8 | 2 370 | 126.5 |
| 1 120 | 267.7 | 1 540 | 194.7 | 1 960 | 153.0 | 2 380 | 126.0 |
| 1 130 | 265.3 | 1 550 | 193.4 | 1 970 | 152.2 | 2 390 | 125.4 |
| 1 140 | 263.0 | 1 560 | 192.2 | 1 980 | 151.4 | 2 400 | 124.9 |
| 1 150 | 260.7 | 1 570 | 191.0 | 1 990 | 150.7 | 2 410 | 124.4 |
| 1 160 | 258.5 | 1 580 | 189.8 | 2 000 | 149.9 | 2 420 | 123.9 |
| 1 170 | 256.3 | 1 590 | 188.6 | 2 010 | 149.2 | 2 430 | 123.4 |
| 1 180 | 254.1 | 1 600 | 187.4 | 2 020 | 148.4 | 2 440 | 122.9 |
| 1 190 | 251.9 | 1 610 | 186.2 | 2 030 | 147.7 | 2 450 | 122.4 |
| 1 200 | 249.9 | 1 620 | 185.1 | 2 040 | 147.0 | 2 460 | 121.9 |
| 1 210 | 247.8 | 1 630 | 183.9 | 2 050 | 146.3 | 2 470 | 121.4 |
| 1 220 | 245.8 | 1 640 | 182.8 | 2 060 | 145.5 | 2 480 | 120.9 |
| 1 230 | 243.8 | 1 650 | 181.7 | 2 070 | 144.8 | 2 490 | 120.4 |
| 1 240 | 241.8 | 1 660 | 180.6 | 2 080 | 144.1 | 2 500 | 119.9 |
| 1 250 | 239.9 | 1 670 | 179.5 | 2 090 | 143.5 | 2 510 | 119.5 |
| 1 260 | 238.0 | 1 680 | 178.5 | 2 100 | 142.8 | 2 520 | 119.0 |
| 1 270 | 236.1 | 1 690 | 177.4 | 2 110 | 142.1 | 2 530 | 118.5 |
| 1 280 | 234.2 | 1 700 | 176.4 | 2 120 | 141.4 | 2 540 | 118.0 |
| 1 290 | 232.4 | 1 710 | 175.3 | 2 130 | 140.8 | 2 550 | 117.6 |
| 1 300 | 230.6 | 1 720 | 174.3 | 2 140 | 140.1 | 2 560 | 117.1 |
| 1 310 | 228.9 | 1 730 | 173.3 | 2 150 | 139.5 | 2 570 | 116.7 |
| 1 320 | 227.1 | 1 740 | 172.3 | 2 160 | 138.8 | 2 580 | 116.2 |
| 1 330 | 225.4 | 1 750 | 171.3 | 2 170 | 138.2 | 2 590 | 115.8 |
| 1 340 | 223.7 | 1 760 | 170.4 | 2 180 | 137.5 | 2 600 | 115.3 |
| 1 350 | 222.1 | 1 770 | 169.4 | 2 190 | 136.9 | 2 610 | 114.9 |
| 1 360 | 220.5 | 1 780 | 168.4 | 2 200 | 136.3 | 2 620 | 114.4 |
| 1 370 | 218.8 | 1 790 | 167.5 | 2 210 | 135.7 | 2 630 | 114.0 |
| 1 380 | 217.3 | 1 800 | 166.6 | 2 220 | 135.1 | 2 640 | 113.6 |
| 1 390 | 215.7 | 1 810 | 165.6 | 2 230 | 134.4 | 2 650 | 113.1 |
| 1 400 | 214.2 | 1 820 | 164.7 | 2 240 | 133.8 | 2 660 | 112.7 |
| 1 410 | 212.6 | 1 830 | 163.8 | 2 250 | 133.3 | 2 670 | 112.3 |

## Table 3-9. Continued.

| kHz | m | kHz | m | kHz | m | kHz | m |
|---|---|---|---|---|---|---|---|
| 2 680 | 111.9 | 3 100 | 96.72 | 3 520 | 85.18 | 3 940 | 76.10 |
| 2 690 | 111.5 | 3 110 | 96.41 | 3 530 | 84.93 | 3 950 | 75.90 |
| 2 700 | 111.0 | 3 120 | 96.10 | | | 3 960 | 75.51 |
| 2 710 | 110.6 | 3 130 | 95.79 | 3 540 | 84.69 | 3 970 | 75.52 |
| 2 720 | 110.2 | | | 3 550 | 84.46 | 3 980 | 75.33 |
| 2 730 | 109.8 | 3 140 | 95.48 | 3 560 | 84.22 | | |
| | | 3 150 | 95.18 | 3 570 | 83.98 | 3 990 | 75.14 |
| 2 740 | 109.4 | 3 160 | 94.88 | 3 580 | 83.75 | 4 000 | 74.96 |
| 2 750 | 109.0 | 3 170 | 94.58 | | | 4 010 | 74.77 |
| 2 760 | 108.6 | 3 180 | 94.28 | 3 590 | 83.52 | 4 020 | 74.58 |
| 2 770 | 108.2 | | | 3 600 | 83.28 | 4 030 | 74.40 |
| 2 780 | 107.8 | 3 190 | 93.99 | 3 610 | 83.05 | | |
| | | 3 200 | 93.69 | 3 620 | 82.82 | 4 040 | 74.21 |
| 2 790 | 107.5 | 3 210 | 93.40 | 3 630 | 82.60 | 4 050 | 74.03 |
| 2 800 | 107.1 | 3 220 | 93.11 | | | 4 060 | 73.85 |
| 2 810 | 106.7 | 3 230 | 92.82 | 3 640 | 82.37 | 4 070 | 73.67 |
| 2 820 | 106.3 | | | 3 650 | 82.14 | 4 080 | 73.49 |
| 2 830 | 105.9 | 3 240 | 92.54 | 3 660 | 81.92 | | |
| | | 3 250 | 92.25 | 3 670 | 81.69 | 4 090 | 73.31 |
| 2 840 | 105.6 | 3 260 | 91.97 | 3 680 | 81.47 | 4 100 | 73.13 |
| 2 850 | 105.2 | 3 270 | 91.69 | | | 4 110 | 72.95 |
| 2 860 | 104.8 | 3 280 | 91.41 | 3 690 | 81.25 | 4 120 | 72.77 |
| 2 870 | 104.5 | | | 3 700 | 81.03 | 4 130 | 72.60 |
| 2 880 | 104.1 | 3 290 | 91.13 | 3 710 | 80.81 | | |
| | | 3 300 | 90.86 | 3 720 | 80.60 | 4 140 | 72.42 |
| 2 890 | 103.7 | 3 310 | 90.58 | 3 730 | 80.38 | 4 150 | 72.25 |
| 2 900 | 103.4 | 3 320 | 90.31 | | | 4 160 | 72.07 |
| 2 910 | 103.0 | 3 330 | 90.04 | 3 740 | 80.17 | 4 170 | 71.90 |
| 2 920 | 102.7 | | | 3 750 | 79.95 | 4 180 | 71.73 |
| 2 930 | 102.3 | 3 340 | 89.77 | 3 760 | 79.74 | | |
| | | 3 350 | 89.50 | 3 770 | 79.53 | 4 190 | 71.56 |
| 2 940 | 102.0 | 3 360 | 89.23 | 3 780 | 79.32 | 4 200 | 71.39 |
| 2 950 | 101.6 | 3 370 | 88.97 | | | 4 210 | 71.22 |
| 2 960 | 101.3 | 3 380 | 88.70 | 3 790 | 79.11 | 4 220 | 71.05 |
| 2 970 | 100.9 | | | 3 800 | 78.90 | 4 230 | 70.88 |
| 2 980 | 100.6 | 3 390 | 88.44 | 3 810 | 78.69 | | |
| | | 3 400 | 88.18 | 3 820 | 78.49 | 4 240 | 70.71 |
| 2 990 | 100.3 | 3 410 | 87.92 | 3 830 | 78.28 | 4 250 | 70.55 |
| 3 000 | 99.94 | 3 420 | 87.67 | | | 4 260 | 70.38 |
| 3 010 | 99.61 | 3 430 | 87.41 | 3 840 | 78.08 | 4 270 | 70.22 |
| 3 020 | 99.28 | | | 3 850 | 77.88 | 4 280 | 70.05 |
| 3 030 | 98.95 | 3 440 | 87.16 | 3 860 | 77.67 | | |
| | | 3 450 | 86.90 | 3 870 | 77.47 | 4 290 | 69.89 |
| 3 040 | 98.63 | 3 460 | 86.65 | 3 880 | 77.27 | 4 300 | 69.73 |
| 3 050 | 98.30 | 3 470 | 86.40 | | | 4 310 | 69.56 |
| 3 060 | 97.98 | 3 480 | 86.16 | 3 890 | 77.07 | 4 320 | 69.40 |
| 3 070 | 97.66 | | | 3 900 | 76.88 | 4 330 | 69.24 |
| 3 080 | 97.34 | 3 490 | 85.91 | 3 910 | 76.68 | | |
| 3 090 | 97.03 | 3 500 | 85.66 | 3 920 | 76.48 | 4 340 | 69.08 |
| | | 3 510 | 85.42 | 3 930 | 76.29 | 4 350 | 68.92 |

*Table 3-9. Continued.*

| kHz | m | kHz | m | kHz | m | kHz | m |
|---|---|---|---|---|---|---|---|
| 4 360 | 68.77 | 4 780 | 62.72 | 5 200 | 57.66 | 5 620 | 53.35 |
| 4 370 | 68.61 | 4 790 | 62.59 | 5 210 | 57.55 | 5 630 | 53.25 |
| 4 380 | 68.45 | 4 800 | 62.46 | 5 220 | 57.44 | 5 640 | 53.16 |
| 4 390 | 68.30 | 4 810 | 62.33 | 5 230 | 57.33 | 5 650 | 53.07 |
| 4 400 | 68.14 | 4 820 | 62.20 | 5 240 | 57.22 | 5 660 | 52.97 |
| 4 410 | 67.99 | 4 830 | 62.07 | 5 250 | 57.11 | 5 670 | 52.88 |
| 4 420 | 67.83 | 4 840 | 61.95 | 5 260 | 57.00 | 5 680 | 52.79 |
| 4 430 | 67.68 | 4 850 | 61.82 | 5 270 | 56.89 | 5 690 | 52.69 |
| 4 440 | 67.53 | 4 860 | 61.69 | 5 280 | 56.78 | 5 700 | 52.60 |
| 4 450 | 67.38 | 4 870 | 61.56 | 5 290 | 56.68 | 5 710 | 52.51 |
| 4 460 | 67.22 | 4 880 | 61.44 | 5 300 | 56.57 | 5 720 | 52.42 |
| 4 470 | 67.07 | 4 890 | 61.31 | 5 310 | 56.46 | 5 730 | 52.32 |
| 4 480 | 66.92 | 4 900 | 61.19 | 5 320 | 56.36 | 5 740 | 52.23 |
| 4 490 | 66.78 | 4 910 | 61.06 | 5 330 | 56.25 | 5 750 | 52.14 |
| 4 500 | 66.63 | 4 920 | 60.94 | 5 340 | 56.15 | 5 760 | 52.05 |
| 4 510 | 66.48 | 4 930 | 60.82 | 5 350 | 56.04 | 5 770 | 51.96 |
| 4 520 | 66.33 | 4 940 | 60.69 | 5 360 | 55.94 | 5 780 | 51.87 |
| 4 530 | 66.19 | 4 950 | 60.57 | 5 370 | 55.83 | 5 790 | 51.78 |
| 4 540 | 66.04 | 4 960 | 60.45 | 5 380 | 55.73 | 5 800 | 51.69 |
| 4 550 | 65.89 | 4 970 | 60.33 | 5 390 | 55.63 | 5 810 | 51.60 |
| 4 560 | 65.75 | 4 980 | 60.20 | 5 400 | 55.52 | 5 820 | 51.52 |
| 4 570 | 65.61 | 4 990 | 60.08 | 5 410 | 55.42 | 5 830 | 51.43 |
| 4 580 | 65.46 | 5 000 | 59.96 | 5 420 | 55.32 | 5 840 | 51.34 |
| 4 590 | 65.32 | 5 010 | 59.84 | 5 430 | 55.22 | 5 850 | 51.25 |
| 4 600 | 65.18 | 5 020 | 59.73 | 5 440 | 55.11 | 5 860 | 51.16 |
| 4 610 | 65.04 | 5 030 | 59.61 | 5 450 | 55.01 | 5 870 | 51.08 |
| 4 620 | 64.90 | 5 040 | 59.49 | 5 460 | 54.91 | 5 880 | 50.99 |
| 4 630 | 64.76 | 5 050 | 59.37 | 5 470 | 54.81 | 5 890 | 50.90 |
| 4 640 | 64.62 | 5 060 | 59.25 | 5 480 | 54.71 | 5 900 | 50.82 |
| 4 650 | 64.48 | 5 070 | 59.14 | 5 490 | 54.61 | 5 910 | 50.73 |
| 4 660 | 64.34 | 5 080 | 59.02 | 5 500 | 54.51 | 5 920 | 50.65 |
| 4 670 | 64.20 | 5 090 | 58.90 | 5 510 | 54.41 | 5 930 | 50.56 |
| 4 680 | 64.06 | 5 100 | 58.79 | 5 520 | 54.32 | 5 940 | 50.47 |
| 4 690 | 63.93 | 5 110 | 58.67 | 5 530 | 54.22 | 5 950 | 50.39 |
| 4 700 | 63.79 | 5 120 | 58.56 | 5 540 | 54.12 | 5 960 | 50.31 |
| 4 710 | 63.66 | 5 130 | 58.44 | 5 550 | 54.02 | 5 970 | 50.22 |
| 4 720 | 63.52 | 5 140 | 58.33 | 5 560 | 53.92 | 5 980 | 50.14 |
| 4 730 | 63.39 | 5 150 | 58.22 | 5 570 | 53.83 | 5 990 | 50.05 |
| 4 740 | 63.25 | 5 160 | 58.10 | 5 580 | 53.73 | 6 000 | 49.97 |
| 4 750 | 63.12 | 5 170 | 57.99 | 5 590 | 53.64 | 6 010 | 49.89 |
| 4 760 | 62.99 | 5 180 | 57.88 | 5 600 | 53.54 | 6 020 | 49.80 |
| 4 770 | 62.86 | 5 190 | 57.77 | 5 610 | 53.44 | 6 030 | 49.72 |

*Table 3-9. Continued.*

| kHz | m | kHz | m | kHz | m | kHz | m |
|-----|-----|-----|-----|-----|-----|-----|-----|
| 6 040 | 49.64 | 6 460 | 46.41 | 6 880 | 43.58 | 7 300 | 41.07 |
| 6 050 | 49.56 | 6 470 | 46.34 | 6 890 | 43.52 | 7 310 | 41.02 |
| 6 060 | 49.48 | 6 480 | 46.27 | 6 900 | 43.45 | 7 320 | 40.96 |
| 6 070 | 49.39 | | | 6 910 | 43.39 | 7 330 | 40.90 |
| 6 080 | 49.31 | 6 490 | 46.20 | 6 920 | 43.33 | | |
| | | 6 500 | 46.13 | 6 930 | 43.26 | 7 340 | 40.85 |
| 6 090 | 49.23 | 6 510 | 46.06 | | | 7 350 | 40.79 |
| 6 100 | 49.15 | 6 520 | 45.98 | 6 940 | 43.20 | 7 360 | 40.74 |
| 6 110 | 49.07 | 6 530 | 45.91 | 6 950 | 43.14 | 7 370 | 40.68 |
| 6 120 | 48.99 | | | 6 960 | 43.08 | 7 380 | 40.63 |
| 6 130 | 48.91 | 6 540 | 45.84 | 6 970 | 43.02 | | |
| | | 6 550 | 45.77 | 6 980 | 42.95 | 7 390 | 40.57 |
| 6 140 | 48.83 | 6 560 | 45.70 | | | 7 400 | 40.52 |
| 6 150 | 48.75 | 6 570 | 45.63 | 6 990 | 42.89 | 7 410 | 40.46 |
| 6 160 | 48.67 | 6 580 | 45.57 | 7 000 | 42.83 | 7 420 | 40.41 |
| 6 170 | 48.59 | | | 7 010 | 42.77 | 7 430 | 40.35 |
| 6 180 | 48.51 | 6 590 | 45.50 | 7 020 | 42.71 | | |
| | | 6 600 | 45.43 | 7 030 | 42.65 | 7 440 | 40.30 |
| 6 190 | 48.44 | 6 610 | 45.36 | | | 7 450 | 40.24 |
| 6 200 | 48.36 | 6 620 | 45.29 | 7 040 | 42.59 | 7 460 | 40.10 |
| 6 210 | 48.28 | 6 630 | 45.22 | 7 050 | 42.53 | 7 470 | 40.14 |
| 6 220 | 48.20 | | | 7 060 | 42.47 | 7 480 | 40.08 |
| 6 230 | 48.13 | 6 640 | 45.15 | 7 070 | 42.41 | | |
| | | 6 650 | 45.09 | 7 080 | 42.35 | 7 490 | 40.03 |
| 6 240 | 48.05 | 6 660 | 45.02 | | | 7 500 | 39.98 |
| 6 250 | 47.97 | 6 670 | 44.95 | 7 090 | 42.29 | 7 510 | 39.92 |
| 6 260 | 47.89 | 6 680 | 44.88 | 7 100 | 42.23 | 7 520 | 39.87 |
| 6 270 | 47.82 | | | 7 110 | 42.17 | 7 530 | 39.82 |
| 6 280 | 47.74 | 6 690 | 44.82 | 7 120 | 42.11 | | |
| | | 6 700 | 44.75 | 7 130 | 42.05 | 7 540 | 39.76 |
| 6 290 | 47.67 | 6 710 | 44.68 | | | 7 550 | 39.71 |
| 6 300 | 47.59 | 6 720 | 44.62 | 7 140 | 41.99 | 7 560 | 39.66 |
| 6 310 | 47.52 | 6 730 | 44.55 | 7 150 | 41.93 | 7 570 | 39.61 |
| 6 320 | 47.44 | | | 7 160 | 41.87 | 7 580 | 39.55 |
| 6 330 | 47.36 | 6 740 | 44.48 | 7 170 | 41.82 | | |
| | | 6 750 | 44.42 | 7 180 | 41.76 | 7 590 | 39.50 |
| 6 340 | 47.29 | 6 760 | 44.35 | | | 7 600 | 39.45 |
| 6 350 | 47.22 | 6 770 | 44.29 | 7 190 | 41.70 | 7 610 | 39.40 |
| 6 360 | 47.14 | 6 780 | 44.22 | 7 200 | 41.64 | 7 620 | 39.35 |
| 6 370 | 47.07 | | | 7 210 | 41.58 | 7 630 | 39.29 |
| 6 380 | 46.99 | 6 790 | 44.16 | 7 220 | 41.53 | | |
| | | 6 800 | 44.09 | 7 230 | 41.47 | 7 640 | 39.24 |
| 6 390 | 46.92 | 6 810 | 44.03 | | | 7 650 | 39.19 |
| 6 400 | 46.85 | 6 820 | 43.96 | 7 240 | 41.41 | 7 660 | 39.14 |
| 6 410 | 46.77 | 6 830 | 43.90 | 7 250 | 41.35 | 7 670 | 39.09 |
| 6 420 | 46.70 | | | 7 260 | 41.30 | 7 680 | 39.04 |
| 6 430 | 46.63 | 6 840 | 43.83 | 7 270 | 41.24 | | |
| | | 6 850 | 43.77 | 7 280 | 41.18 | 7 690 | 38.99 |
| 6 440 | 46.56 | 6 860 | 43.71 | | | 7 700 | 38.94 |
| 6 450 | 46.48 | 6 870 | 43.64 | 7 290 | 41.13 | 7 710 | 38.89 |

*Table 3-9. Continued.*

| kHz | m | kHz | m | kHz | m | kHz | m |
|---|---|---|---|---|---|---|---|
| 7 720 | 38.84 | 8 140 | 36.83 | 8 560 | 35.03 | 8 980 | 33.39 |
| 7 730 | 38.79 | 8 150 | 36.79 | 8 570 | 34.98 | 8 990 | 33.35 |
| | | 8 160 | 36.74 | 8 580 | 34.94 | 9 000 | 33.31 |
| 7 740 | 38.74 | 8 170 | 36.70 | | | 9 010 | 33.28 |
| 7 750 | 38.69 | 8 180 | 36.65 | 8 590 | 34.90 | 9 020 | 33.24 |
| 7 760 | 38.64 | | | 8 600 | 34.86 | 9 030 | 33.20 |
| 7 770 | 38.59 | 8 190 | 36.61 | 8 610 | 34.82 | | |
| 7 780 | 38.54 | 8 200 | 36.56 | 8 620 | 34.78 | 9 040 | 33.17 |
| 7 790 | 38.49 | 8 210 | 36.52 | 8 630 | 34.74 | 9 050 | 33.13 |
| 7 800 | 38.44 | 8 220 | 36.47 | | | 9 060 | 33.09 |
| 7 810 | 38.39 | 8 230 | 36.43 | 8 640 | 34.70 | 9 070 | 33.06 |
| 7 820 | 38.34 | | | 8 650 | 34.66 | 9 080 | 33.02 |
| 7 830 | 38.29 | 8 240 | 36.39 | 8 660 | 34.62 | | |
| | | 8 250 | 36.34 | 8 670 | 34.58 | 9 090 | 32.98 |
| 7 840 | 38.24 | 8 260 | 36.30 | 8 680 | 34.54 | 9 100 | 32.95 |
| 7 850 | 38.19 | 8 270 | 36.25 | | | 9 110 | 32.91 |
| 7 860 | 38.15 | 8 280 | 36.21 | 8 690 | 34.50 | 9 120 | 32.88 |
| 7 870 | 38.10 | | | 8 700 | 34.46 | 9 130 | 32.84 |
| 7 880 | 38.05 | 8 290 | 36.17 | 8 710 | 34.42 | | |
| | | 8 300 | 36.12 | 8 720 | 34.38 | 9 140 | 32.80 |
| 7 890 | 38.00 | 8 310 | 36.08 | 8 730 | 34.34 | 9 150 | 32.77 |
| 7 900 | 37.95 | 8 320 | 36.04 | | | 9 160 | 32.73 |
| 7 910 | 37.90 | 8 330 | 35.99 | 8 740 | 34.30 | 9 170 | 32.70 |
| 7 920 | 37.86 | | | 8 750 | 34.27 | 9 180 | 32.66 |
| 7 930 | 37.81 | 8 340 | 35.95 | 8 760 | 34.23 | | |
| | | 8 350 | 35.91 | 8 770 | 34.19 | 9 190 | 32.62 |
| 7 940 | 37.76 | 8 360 | 35.86 | 8 780 | 34.15 | 9 200 | 32.59 |
| 7 950 | 37.71 | 8 370 | 35.82 | | | 9 210 | 32.55 |
| 7 960 | 37.67 | 8 380 | 35.78 | 8 790 | 34.11 | 9 220 | 32.52 |
| 7 970 | 37.62 | | | 8 800 | 34.07 | 9 230 | 32.48 |
| 7 980 | 37.57 | 8 390 | 35.74 | 8 810 | 34.03 | | |
| | | 8 400 | 35.69 | 8 820 | 33.99 | 9 240 | 32.45 |
| 7 990 | 37.52 | 8 410 | 35.65 | 8 830 | 33.95 | 9 250 | 32.41 |
| 8 000 | 37.48 | 8 420 | 35.61 | | | 9 260 | 32.38 |
| 8 010 | 37.43 | 8 430 | 35.57 | 8 840 | 33.92 | 9 270 | 32.34 |
| 8 020 | 37.38 | | | 8 850 | 33.88 | 9 280 | 32.31 |
| 8 030 | 37.34 | 8 440 | 35.52 | 8 860 | 33.84 | | |
| | | 8 450 | 35.48 | 8 870 | 33.80 | 9 290 | 32.27 |
| 8 040 | 37.29 | 8 460 | 35.44 | 8 880 | 33.76 | 9 300 | 32.24 |
| 8 050 | 37.24 | 8 470 | 35.40 | | | 9 310 | 32.20 |
| 8 060 | 37.20 | 8 480 | 35.36 | 8 890 | 33.73 | 9 320 | 32.17 |
| 8 070 | 37.15 | | | 8 900 | 33.69 | 9 330 | 32.14 |
| 8 080 | 37.11 | 8 490 | 35.31 | 8 910 | 33.65 | | |
| | | 8 500 | 35.27 | 8 920 | 33.61 | 9 340 | 32.10 |
| 8 090 | 37.06 | 8 510 | 35.23 | 8 930 | 33.57 | 9 350 | 32.07 |
| 8 100 | 37.01 | 8 520 | 35.19 | | | 9 360 | 32.03 |
| 8 110 | 36.97 | 8 530 | 35.15 | 8 940 | 33.54 | 9 370 | 32.00 |
| 8 120 | 36.92 | | | 8 950 | 33.50 | 9 380 | 31.96 |
| 8 130 | 36.88 | 8 540 | 35.11 | 8 960 | 33.46 | 9 390 | 31.93 |
| | | 8 550 | 35.07 | 8 970 | 33.42 | | |

*Table 3-9. Continued.*

| kHz | m | kHz | m | kHz | m | kHz | m |
|---|---|---|---|---|---|---|---|
| 9 400 | 31.90 | 9 560 | 31.36 | 9 720 | 30.85 | 9 880 | 30.35 |
| 9 410 | 31.86 | 9 570 | 31.33 | 9 730 | 30.81 | | |
| 9 420 | 31.83 | 9 580 | 31.30 | | | 9 890 | 30.32 |
| 9 430 | 31.79 | | | 9 740 | 30.78 | 9 900 | 30.28 |
| | | 9 590 | 31.26 | 9 750 | 30.75 | 9 910 | 30.25 |
| 9 440 | 31.76 | 9 600 | 31.23 | 9 760 | 30.72 | 9 920 | 30.22 |
| 9 450 | 31.73 | 9 610 | 31.20 | 9 770 | 30.69 | 9 930 | 30.19 |
| 9 460 | 31.69 | 9 620 | 31.17 | 9 780 | 30.66 | | |
| 9 470 | 31.66 | 9 630 | 31.13 | | | 9 940 | 30.16 |
| 9 480 | 31.63 | | | 9 790 | 30.63 | 9 950 | 30.13 |
| | | 9 640 | 31.10 | 9 800 | 30.59 | 9 960 | 30.10 |
| 9 490 | 31.59 | 9 650 | 31.07 | 9 810 | 30.56 | 9 970 | 30.07 |
| 9 500 | 31.56 | 9 660 | 31.04 | 9 820 | 30.53 | 9 980 | 30.04 |
| 9 510 | 31.53 | 9 670 | 31.01 | 9 830 | 30.50 | | |
| 9 520 | 31.49 | 9 680 | 30.97 | | | 9 990 | 30.01 |
| 9 530 | 31.46 | | | 9 840 | 30.47 | 10 000 | 29.98 |
| | | 9 690 | 30.94 | 9 850 | 30.44 | | |
| 9 540 | 31.43 | 9 700 | 30.91 | 9 860 | 30.41 | | |
| 9 550 | 31.39 | 9 710 | 30.88 | 9 870 | 30.38 | | |

# Waveform variation

Just as there are basic units such as the volt, ampere, and ohm, the sine wave is considered the basic wave. The sine wave is just one of a tremendous variety, including the sawtooth, the square wave, and others. A square wave consists of a basic sine wave whose frequency is equal to the frequency of the square wave, plus a number of odd harmonics. If the fundamental frequency of the sine wave is 60 Hz, then:

$$3rd \text{ harmonic} = 3 \times fundamental = 3 \times 60 = 180 \text{ Hz}$$
$$5th \text{ harmonic} = 5 \times fundamental = 5 \times 60 = 300 \text{ Hz}$$
$$7th \text{ harmonic} = 7 \times fundamental = 7 \times 60 = 420 \text{ Hz}$$

The greater the number of harmonics used in the production of the square wave, the more closely its formation will approach that of a true square wave.

A symmetrical square wave has a negative pulse width equal to its positive pulse width as in Fig. 3-6. A rectangular pulse belongs to the square wave family, and consists of a square wave that might not have equal positive and negative pulses. The time of a representative rectangular wave is shown in Fig. 3-7.

A sawtooth wave rises linearly to a peak and then drops to its zero value extremely rapidly. The frequency is the number of times this voltage rises and falls per second. The wave is constructed from a fundamental sine wave plus a

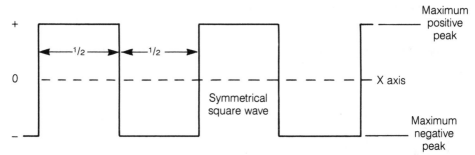

Fig. 3-6.   Symmetrical square wave.

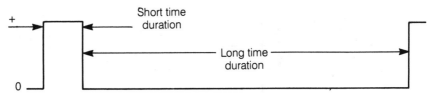

Fig. 3-7.   Time duration of a rectangular wave.

number of odd and even harmonics. If the fundamental frequency of the sine wave is 60 Hz, then:

$$2\text{nd harmonic} = 2 \times \text{fundamental} = 2 \times 60 = 120 \text{ Hz}$$
$$3\text{rd harmonic} = 3 \times \text{fundamental} = 3 \times 60 = 180 \text{ Hz}$$
$$4\text{th harmonic} = 4 \times \text{fundamental} = 4 \times 60 = 240 \text{ Hz}$$

These values are for a 60 Hz fundamental. The resulting sawtooth waveform, assuming the use of a large number of harmonics, is shown in Fig. 3-8.

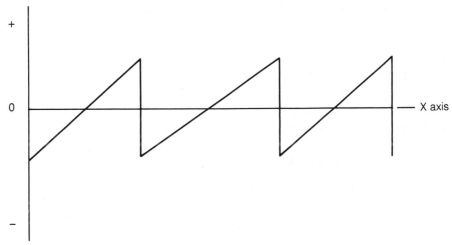

Fig. 3-8.   Sawtooth waveform.

# Angular velocity

In an ac generator, the rotating armature produces a voltage sine wave (Fig. 3-9). We can represent the armature by a rotating vector $r$. The angle through which this

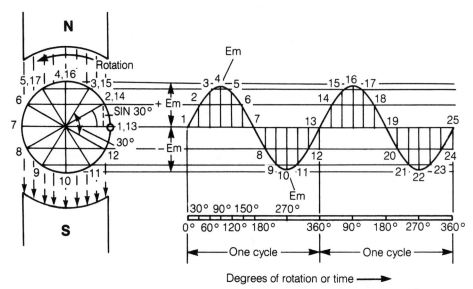

*Fig. 3-9. Production of a sine wave by a coil rotating in a fixed magnetic field.*

vector sweeps is usually indicated by the Greek letter $\theta$. The instantaneous values of the voltage produced are maximum when the phase angle, $\theta$, is 90° or 270° and zero when that phase angle is 0° or 180°.

The radius vector $r$ rotates about the origin, 0, and is taken to rotate in a counterclockwise direction. The angular velocity in radians per second of this rotating vector is the rate at which the angle, $\theta$, is produced by its rotation. Angular velocity is conveniently represented by $\omega$ or $2\pi f$ (radians per second) in which $2\pi$ is a constant and is equal to 6.28. The frequency $f$ is in hertz.

Angular velocity appears in formulas involving instantaneous values of sine waves of voltage or current ($e = E_{MAX} \sin \omega t$) (t is for the time) and in formulas involving inductive reactance ($X_L = 2\pi f L$) and capacitive reactance ($X_C = 1$ divided by $2\pi f C$).

The solution of problems involving angular velocity is simplified by Table 3-10. Here we have angular velocity corresponding to particular values of frequency.

Table 3-10 can be used for values of frequency other than those indicated by moving the decimal point. To find $\omega$ at a frequency of 4 kHz, find 40 in the $f$ column. To change 40 Hz to 4 kHz multiply 40 by 100. This means we must also multiply the value of 251.20 (which is the value of $\omega$ for 40) by 100. The answer is 25 120. To get megahertz values, multiply both $f$ and $\omega$ by 1 000 000.

*Table 3-10. Angular velocity (radians per second).*

| f (Hz) | ω | f (Hz) | ω | f (Hz) | ω | f (Hz) | ω |
|---|---|---|---|---|---|---|---|
| | | 26 | 163.28 | 51 | 320.28 | 76 | 477.28 |
| | | 27 | 169.56 | 52 | 326.56 | 77 | 483.56 |
| | | 28 | 175.84 | 53 | 332.84 | 78 | 489.84 |
| | | 29 | 182.12 | 54 | 339.12 | 79 | 496.12 |
| | | 30 | 188.40 | 55 | 345.40 | 80 | 502.40 |
| | | 31 | 194.68 | 56 | 351.68 | 81 | 508.68 |
| | | 32 | 200.96 | 57 | 357.96 | 82 | 514.96 |
| | | 33 | 207.24 | 58 | 364.24 | 83 | 521.24 |
| | | 34 | 213.52 | 59 | 370.52 | 84 | 527.52 |
| 10 | 62.8 | 35 | 219.80 | 60 | 376.80 | 85 | 533.80 |
| 11 | 69.08 | 36 | 226.08 | 61 | 383.08 | 86 | 540.08 |
| 12 | 75.36 | 37 | 232.36 | 62 | 389.36 | 87 | 546.36 |
| 13 | 81.64 | 38 | 238.64 | 63 | 395.64 | 88 | 552.65 |
| 14 | 87.92 | 39 | 244.92 | 64 | 401.92 | 89 | 558.92 |
| 15 | 94.20 | 40 | 251.20 | 65 | 408.20 | 90 | 565.20 |
| 16 | 100.48 | 41 | 257.48 | 66 | 414.48 | 91 | 571.48 |
| 17 | 106.76 | 42 | 263.76 | 67 | 420.76 | 92 | 577.76 |
| 18 | 113.04 | 43 | 270.04 | 68 | 427.04 | 93 | 584.04 |
| 19 | 119.32 | 44 | 276.32 | 69 | 433.32 | 94 | 590.32 |
| 20 | 125.60 | 45 | 282.60 | 70 | 439.60 | 95 | 596.60 |
| 21 | 131.88 | 46 | 288.88 | 71 | 445.88 | 96 | 602.88 |
| 22 | 138.16 | 47 | 295.16 | 72 | 452.16 | 97 | 609.16 |
| 23 | 144.44 | 48 | 301.44 | 73 | 458.44 | 98 | 615.44 |
| 24 | 150.72 | 49 | 307.72 | 74 | 464.72 | 99 | 621.72 |
| 25 | 157.00 | 50 | 314.00 | 75 | 471.00 | 100 | 628.00 |

*Example* What is the angular velocity of a generator, whose armature rotates at 180 rev/min?

*Solution* Divide 1 800 revolutions by 60 (60 s equals 1 min) to obtain the number of revolutions per second. 1 800/60 equals 30 rev/s. In 30 revolutions, the armature generates 30 complete sine wave cycles. The frequency, then, is 30. Locate 30 in the frequency (*f*) column of Table 3-10. The corresponding angular velocity is found to the right in the ω column. The answer is 188.4 rad/s.

# Ohm's law for ac

Ohm's law is as applicable for alternating-current circuits as it is for dc. Other than the substitution of impedance (*Z*) for resistance (*R*), the formulas are similar.

$$E = I \times Z$$

$$I = \frac{E}{Z}$$

$$Z=\frac{E}{I}$$

$Z$ is in ohms, $E$ is in volts and $I$ in amperes. $Z$, however, is more complex than the simple $R$ used in Ohm's law for dc.

# Power in ac circuits

Because power is the product of voltage and current, the phase angle between voltage and current is a factor that must enter such calculations. In a dc circuit, power is simply $E \times I$. Similarly, in a purely resistive ac circuit power, is also $E \times I$ because the phase angle, $\theta$, is zero.

In an ac circuit containing a reactive element (a coil or a capacitor),

$$P=E \times I \cos \theta$$

$P$ is the power in watts; $E$ and $I$ are the voltage and current, respectively. The cosine (or cos) is the ratio of the resistance to the impedance, or:

$$\cos \theta = \frac{R}{Z}$$

There are two conditions in an ac circuit when the phase angle is zero. These are a purely resistive circuit and a resonant circuit (either series or parallel type). Under these conditions:

$$\theta = 0$$
$$\cos \theta = 1$$

and

$$P=E \times I$$

No power is expended in a purely reactive circuit. In such a circuit, the power is returned to the source. Under these conditions:

$$\theta = 90°$$

$$\cos \theta = 0$$

and

$$P=0$$

In any ac circuit the power ranges somewhere between these two extremes and so $\cos \theta$ is generally some value less than one and greater than zero.

## Apparent power

If the power factor in a reactive ac circuit is disregarded, that is, if we say that $P=E \times I$, then the amount of power calculated in this way is larger than its true value. Known as apparent power or reactive power, it can be stated as:

$$\text{Apparent power} = E \times I$$

Multiplying apparent power $(E \times I)$ by the cosine of the phase angle will produce a number representing the true power in the circuit, or; $P = E \times I \times \cos \theta$. This can be written as:

$$\text{True power } (P) = E \times I \times \frac{R}{Z}$$

## Power factor

The ratio between the true power (also known as real power) and the apparent power is a value less than one since the true power is always less than the apparent power. This ratio is referred to as the power factor and is often abbreviated as $F_p$ or *pf*. Stated as a formula we have:

$$pf = \frac{\text{true power}}{\text{apparent power}}$$

True power $(P) = E \times I \times R/Z$

Apparent power $= E \times I$

Hence:

$$pf = \frac{E \times I \times R/Z}{E \times I}$$

$$pf = R/Z; \ R/Z = \cos \theta \text{ where } \theta \text{ is the phase angle}$$

$$pf = \cos \theta$$

Power factor is applicable to series and parallel RC, RL, and LC circuits. It is also a factor to be considered in the use of electrical devices, such as motors. Also, coils and capacitors, used alone, can have a power factor.

## Power factor of coils

The power factor of a coil is:

$$pf = \cos \theta_L \times \frac{R_L}{Z_L}$$

$Z_L$ is the impedance of the coil and can be calculated by:

$$Z_L = \sqrt{R_L^2 + X_L^2}$$

$R_L$ is the winding resistance of the coil and $X_L$ is the reactance of the coil at a particular frequency.

The formula for the impedance of a coil can be substituted for $Z_L$ in the power factor formula.

$$pf = \frac{R_L}{\sqrt{R_L{}^2 + X_L{}^2}}$$

Note that the resistance of the coil will remain constant regardless of frequency. However, the reactance of the coil, $X_L$, which is part of the impedance of the coil, will vary directly with frequency. The lowest frequency limit is 0 Hz. When this happens:

$$pf = \frac{R_L}{\sqrt{R_L{}^2}} = \frac{R_L}{R_L} = 1$$

In a circuit having resistance only, power factor is unity.

## Power factor of capacitors

Power factor for capacitors follows the same approach used for power factor for coils.

$$pf = \cos\,\theta_C = \frac{R_C}{Z_C} = \frac{R_C}{\sqrt{R_C{}^2 + X_C{}^2}}$$

Here the power factor is also dependent on frequency.

In ac circuits where power factor is not a consideration (such as in a purely resistive circuit) the formula for power is identical with that used in dc.

$$P = I^2 \times R$$

$$I = \sqrt{\frac{P}{R}}$$

Where the relationship between voltage and current (the phase angle) is a factor, Ohm's law for ac circuits can take a variety of forms. These can be expressed in terms of power, current, voltage or impedance.

Power

$$P = I^2\,Z\,\cos\,\theta$$

$$P = \frac{E^2\,\cos\,\theta}{Z}$$

Current

$$I = \frac{P}{E \cos \theta}$$

$$I = \sqrt{\frac{P}{Z \cos \theta}}$$

# Efficiency

No electrical device, or any other kind of device for that matter, is 100% efficient. Some engines have efficiencies of 5% to 15%. Transformers can have efficiencies of 90%, or possibly more. With a power transformer, its efficiency is a measure of how much power supplied to the primary winding reaches the secondary winding for delivery to a load. If a transformer has an efficiency of 90%, for every 10 W delivered to the primary, 9 W will be available through the secondary. The 1 W difference will be lost in the form of heat, or in the form of leakage flux, that is, in the form of energy lost in the core or windings. Whatever the reasons, less power is available from the secondary than is supplied to the primary. In terms of a formula:

Output = Input × Efficiency

*Example* A transformer connected to the 110 Vac line has a primary current of 1 A, an efficiency of 85%, and a 3:1 turns ratio. What is the voltage across the secondary winding? How much current flows through the secondary? How much power is delivered to the secondary?

*Solution* The power supplied to the primary winding by the power line can be calculated from the formula $P = E \times I$. Because $E = 110$ and $I = 1$, then $P = E \times I = 110 \times 1 = 110$-W primary power.

Since the transformer has an efficiency of 85%, the output can be calculated from:

output = input × efficiency
output = 110 × 0.85 = 93.5 W

available from the secondary winding.

The transformer has a turns ratio of 3:1. That is, $T_r = 3/1$ or 3:1. But $T_r = N_s/N_p$. And $N_s = 3$ and $N_p = 1$. However, $E_s/E_p = N_s/N_p$. $T_r$ is transformer ratio, subscript p is for primary, and subscript s is for secondary. Transposing:

$$E_s = \frac{E_p \times N_s}{N_p} = 110 \times 3/1 = 330 \text{ V}$$

The secondary voltage is 330; the secondary power is 93.5 W. The secondary current can be calculated from the formula $P=E\times I$ or, transposing, $I=P/E=93.5/330=0.283\ 33=0.28$ A.

# Summary of Ohm's law and power formulas for ac

Table 3-11 supplies a summary of Ohm's law and power formulas for ac.

*Table 3-11. Summary of power and Ohm's law formulas.*

| Watts | Amperes | Volts | Impedance |
|---|---|---|---|
| $P=$ | $I=$ | $E=$ | $Z=$ |
| $I^2R$ | $E/Z$ | $IZ$ | $E/I$ |
| $EI\cos\theta$ | $\dfrac{E}{E\cos\theta}$ | $\dfrac{P}{I\cos\theta}$ | $\dfrac{E^2\cos\theta}{P}$ |
| $\dfrac{E^2\cos\theta}{Z}$ | $\sqrt{\dfrac{P}{Z\cos\theta}}$ | $\sqrt{\dfrac{PZ}{\cos\theta}}$ | $\dfrac{P}{I^2\cos\theta}$ |
| $I^2Z\cos\theta$ | $\sqrt{\dfrac{P}{R}}$ | $\dfrac{\sqrt{PR}}{\cos\theta}$ | $\dfrac{R}{\cos\theta}$ |

# 4

CHAPTER

# Capacitance

## Capacitor complexity

From the viewpoint of variety, capacitors appear to be far more complex than resistors. For their dielectrics, they use air, paper, mica (including silvered mica), Teflon, Mylar, Amplifilm, polystyrene, polyethylene, tantalum oxide, aluminum oxide, ceramic, glass, and vitreous enamel. Their shapes include flat, rectangular, tubular, feedthrough, button, disc, standoff, bathtub, orange drop, cup type, can type, standard, tiny, midget, miniature, and subminiature. They can be polar or nonpolar. In function they work as blocking, buffer, bypass, coupling, filter, tuning, motor-starting, or temperature compensation. They have an astonishing range of capacitance, extending from a few picofarads or a million millionths of a farad to as much as a tenth of a farad. The resistance of a short length of wire is usually insignificant, but if the frequency is high enough, the capacitance between a pair of parallel wires can affect circuit performance seriously.

## Farad conversions

The basic unit of capacitance is the farad. Because it is such a large unit submultiples are used more often. Table 4-1 shows the conversion of farads to its submultiples using exponents. In decimal numbers:

$$1 \text{ F} = 1\ 000\ 000 \ \mu\text{F} = 1\ 000\ 000\ 000\ 000 \text{ pF}$$

$\mu$F is an abbreviation for microfarads; pF is an abbreviation for picofarads.

## Capacitors in series

The total capacitance of series capacitors (Fig. 4-1) is always less than that of the smallest capacitor in the series network. As a general rule of thumb, if two capac-

### Table 4-1. Capacitance conversions.

| Given this value | Multiply by this value to get | | |
|---|---|---|---|
| | **F** | **$\mu$F** | **pF** |
| F | — | $10^6$ | $10^{12}$ |
| $\mu$F | $10^{-6}$ | — | $10^6$ |
| pF | $10^{-12}$ | $10^{-6}$ | — |

Note: At one time, the picofarad was indicated as $\mu\mu$F or micromicrofarad. The picofarad (pF) is the preferred form.

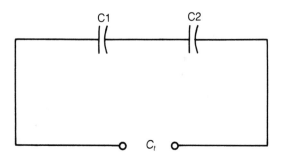

Fig. 4-1. Capacitors in series.

itors are in series, and one has ten or more times the value of the other, the resultant total capacitance can be considered as slightly less than or equal to the value of the smaller capacitor.

Where two series capacitors have equal values, the resultant capacitance is one-half that of either unit. When three capacitors in series have equal values, the resultant capacitance is one-third that of any of the units.

Table 4-2 gives the resultant capacitance of two capacitors in series. The table can be extended by moving the decimal point in the C1 and C2 columns an equal number of places to the right. The table can also be used for finding the total capacitance of three series capacitors by doing a two-step operation, i.e., determining the value of two capacitors and then combining the result with the remaining capacitor. $C_1$ and $C_2$ must be in similar units of microfarads or picofarads. The answers will then be microfarads or picofarads.

*Example* What is the capacitance of two series capacitors, having values of 47 pF and 15 pF?

*Solution* Locate 47 in the C2 column and move horizontally until you reach 11.37 in the C1 column headed by the number 15. The answer is 11.37 pF. You could also have solved this problem by locating 1.5 in the C2 column, and then

Table 4-2. Capacitors in series.

| $C_2$ | $C_1 \rightarrow$ 1 | 1.5 | 2 | 2.2 | 3 |
|---|---|---|---|---|---|
| 1 | 0.50 | 0.60 | 0.666 | 0.69 | 0.75 |
| 1.5 | 0.60 | 0.75 | 0.857 | 0.89 | 1.00 |
| 2 | 0.666 | 0.888 | 1.00 | 1.05 | 1.20 |
| 2.2 | 0.69 | 0.89 | 1.05 | 1.10 | 1.27 |
| 3 | 0.75 | 1.00 | 1.20 | 1.27 | 1.50 |
| 3.3 | 0.77 | 1.03 | 1.24 | 1.31 | 1.57 |
| 4 | 0.80 | 1.09 | 1.33 | 1.42 | 1.71 |
| 4.7 | 0.82 | 1.13 | 1.40 | 1.49 | 1.83 |
| 5 | 0.833 | 1.15 | 1.43 | 1.53 | 1.87 |
| 5.6 | 0.848 | 1.18 | 1.47 | 1.58 | 1.95 |
| 6.8 | 0.87 | 1.22 | 1.54 | 1.66 | 2.08 |
| 7.5 | 0.88 | 1.25 | 1.58 | 1.69 | 2.14 |
| 8.2 | 0.89 | 1.26 | 1.60 | 1.73 | 2.19 |
| 10 | 0.91 | 1.30 | 1.66 | 1.80 | 2.31 |
| 12 | 0.923 | 1.33 | 1.71 | 1.85 | 2.40 |
| 15 | 0.937 | 1.36 | 1.76 | 1.92 | 2.50 |
| 18 | 0.947 | 1.38 | 1.80 | 1.96 | 2.57 |
| 20 | 0.952 | 1.40 | 1.82 | 1.98 | 2.61 |
| 22 | 0.956 | 1.40 | 1.83 | 2.00 | 2.64 |
| 24 | 0.96 | 1.41 | 1.84 | 2.01 | 2.66 |
| 27 | 0.964 | 1.42 | 1.86 | 2.03 | 2.70 |
| 30 | 0.968 | 1.428 | 1.87 | 2.05 | 2.73 |
| 33 | 0.970 | 1.43 | 1.88 | 2.06 | 2.75 |
| 36 | 0.973 | 1.44 | 1.89 | 2.07 | 2.77 |
| 39 | 0.975 | 1.444 | 1.90 | 2.08 | 2.79 |
| 43 | 0.977 | 1.449 | 1.91 | 2.09 | 2.80 |
| 47 | 0.979 | 1.45 | 1.92 | 2.10 | 2.82 |
| 51 | 0.98 | 1.457 | 1.924 | 2.11 | 2.83 |
| 56 | 0.982 | 1.461 | 1.93 | 2.12 | 2.84 |
| 62 | 0.984 | 1.464 | 1.937 | 2.124 | 2.86 |
| 68 | 0.985 | 1.467 | 1.94 | 2.13 | 2.87 |
| 75 | 0.987 | 1.47 | 1.95 | 2.14 | 2.88 |
| 82 | 0.988 | 1.473 | 1.952 | 2.142 | 2.89 |
| 91 | 0.989 | 1.475 | 1.957 | 2.148 | 2.90 |
| 100 | 0.991 | 1.478 | 1.96 | 2.153 | 2.91 |

*Table 4-2. Continued.*

| $C_2$ | $C_1 \rightarrow$ 3.3 | 4 | 4.7 | 5 | 5.6 |
|---|---|---|---|---|---|
| 1 | 0.77 | 0.80 | 0.82 | 0.83 | 0.84 |
| 1.5 | 1.03 | 1.09 | 1.13 | 1.15 | 1.18 |
| 2 | 1.24 | 1.33 | 1.40 | 1.43 | 1.47 |
| 2.2 | 1.31 | 1.42 | 1.49 | 1.53 | 1.58 |
| 3 | 1.57 | 1.71 | 1.83 | 1.87 | 1.95 |
| 3.3 | 1.65 | 1.81 | 1.94 | 1.98 | 2.07 |
| 4 | 1.81 | 2.00 | 2.16 | 2.22 | 2.33 |
| 4.7 | 1.94 | 2.16 | 2.35 | 2.42 | 2.55 |
| 5 | 1.99 | 2.22 | 2.42 | 2.50 | 2.64 |
| 5.6 | 2.07 | 2.33 | 2.55 | 2.64 | 2.80 |
| 6.8 | 2.22 | 2.52 | 2.78 | 2.88 | 3.07 |
| 7.5 | 2.29 | 2.66 | 2.88 | 3.00 | 3.21 |
| 8.2 | 2.35 | 2.69 | 2.99 | 3.10 | 3.32 |
| 10 | 2.48 | 2.85 | 3.19 | 3.33 | 3.59 |
| 12 | 2.59 | 3.00 | 3.37 | 3.53 | 3.82 |
| 15 | 2.72 | 3.15 | 3.57 | 3.75 | 4.08 |
| 18 | 2.79 | 3.27 | 3.72 | 3.91 | 4.27 |
| 20 | 2.83 | 3.33 | 3.80 | 4.00 | 4.37 |
| 22 | 2.87 | 3.38 | 3.87 | 4.07 | 4.46 |
| 24 | 2.90 | 3.42 | 3.92 | 4.14 | 4.54 |
| 27 | 2.94 | 3.48 | 4.00 | 4.22 | 4.63 |
| 30 | 2.97 | 3.53 | 4.06 | 4.28 | 4.72 |
| 33 | 3.00 | 3.56 | 4.11 | 4.34 | 4.78 |
| 36 | 3.02 | 3.60 | 4.15 | 4.39 | 4.84 |
| 39 | 3.04 | 3.63 | 4.19 | 4.43 | 4.89 |
| 43 | 3.06 | 3.66 | 4.23 | 4.48 | 4.95 |
| 47 | 3.08 | 3.68 | 4.27 | 4.52 | 5.00 |
| 51 | 3.09 | 3.71 | 4.30 | 4.55 | 5.04 |
| 56 | 3.12 | 3.73 | 4.33 | 4.59 | 5.09 |
| 62 | 3.13 | 3.75 | 4.36 | 4.62 | 5.13 |
| 68 | 3.14 | 3.77 | 4.39 | 4.65 | 5.17 |
| 75 | 3.16 | 3.79 | 4.42 | 4.69 | 5.21 |
| 82 | 3.17 | 3.81 | 4.45 | 4.71 | 5.24 |
| 91 | 3.18 | 3.83 | 4.47 | 4.74 | 5.27 |
| 100 | 3.195 | 3.84 | 4.49 | 4.76 | 5.30 |

Table 4-2. Continued.

| $C_2$ | $C_1 \rightarrow$ | | | | |
|---|---|---|---|---|---|
| $\downarrow$ | 6.8 | 7.5 | 8.2 | 10 | 12 |
| 1 | 0.87 | 0.88 | 0.89 | 0.91 | 0.92 |
| 1.5 | 1.22 | 1.25 | 1.26 | 1.30 | 1.33 |
| 2 | 1.54 | 1.58 | 1.60 | 1.66 | 1.71 |
| 2.2 | 1.66 | 1.69 | 1.73 | 1.80 | 1.86 |
| 3 | 2.08 | 2.14 | 2.19 | 2.31 | 2.40 |
| 3.3 | 2.22 | 2.29 | 2.35 | 2.48 | 2.59 |
| 4 | 2.52 | 2.66 | 2.69 | 2.85 | 3.00 |
| 4.7 | 2.78 | 2.88 | 2.99 | 3.19 | 3.37 |
| 5 | 2.88 | 3.00 | 3.10 | 3.33 | 3.53 |
| 5.6 | 3.07 | 3.21 | 3.32 | 3.59 | 3.82 |
| 6.8 | 3.40 | 3.56 | 3.66 | 4.05 | 4.34 |
| 7.5 | 3.56 | 3.75 | 3.92 | 4.29 | 4.61 |
| 8.2 | 3.66 | 3.92 | 4.10 | 4.51 | 4.87 |
| 10 | 4.05 | 4.29 | 4.51 | 5.00 | 5.45 |
| 12 | 4.34 | 4.61 | 4.87 | 5.45 | 6.00 |
| 15 | 4.68 | 5.00 | 5.30 | 6.00 | 6.66 |
| 18 | 4.93 | 5.29 | 5.63 | 6.43 | 7.20 |
| 20 | 5.07 | 5.45 | 5.81 | 6.66 | 7.50 |
| 22 | 5.19 | 5.59 | 5.97 | 6.87 | 7.77 |
| 24 | 5.29 | 5.71 | 6.11 | 7.06 | 8.00 |
| 27 | 5.43 | 5.87 | 6.29 | 7.29 | 8.31 |
| 30 | 5.54 | 6.00 | 6.44 | 7.50 | 8.57 |
| 33 | 5.64 | 6.11 | 6.57 | 7.67 | 8.80 |
| 36 | 5.72 | 6.20 | 6.67 | 7.82 | 9.00 |
| 39 | 5.79 | 6.29 | 6.77 | 7.95 | 9.18 |
| 43 | 5.87 | 6.38 | 6.88 | 8.11 | 9.40 |
| 47 | 5.94 | 6.47 | 6.98 | 8.24 | 9.56 |
| 51 | 6.00 | 6.53 | 7.06 | 8.36 | 9.71 |
| 56 | 6.06 | 6.61 | 7.15 | 8.48 | 9.88 |
| 62 | 6.14 | 6.69 | 7.24 | 8.61 | 10.05 |
| 68 | 6.18 | 6.75 | 7.32 | 8.72 | 10.20 |
| 75 | 6.23 | 6.82 | 7.39 | 8.82 | 10.34 |
| 82 | 6.28 | 6.87 | 7.45 | 8.91 | 10.46 |
| 91 | 6.32 | 6.93 | 7.52 | 9.01 | 10.60 |
| 100 | 6.37 | 6.96 | 7.58 | 9.09 | 10.71 |

*Table 4-2. Continued.*

| | | | $C_1 \rightarrow$ | | |
|---|---|---|---|---|---|
| $C_2$ | 15 | 18 | 20 | 22 | 24 |
| $\downarrow$ | | | | | |
| 1 | 0.937 | 0.947 | 0.952 | 0.956 | 0.96 |
| 1.5 | 1.36 | 1.38 | 1.395 | 1.40 | 1.41 |
| 2 | 1.76 | 1.80 | 1.82 | 1.83 | 1.84 |
| 2.2 | 1.92 | 1.96 | 1.98 | 2.00 | 2.01 |
| 3 | 2.50 | 2.57 | 2.61 | 2.64 | 2.66 |
| 3.3 | 2.72 | 2.79 | 2.83 | 2.87 | 2.90 |
| 4 | 3.15 | 3.27 | 3.33 | 3.38 | 3.42 |
| 4.7 | 3.57 | 3.72 | 3.80 | 3.87 | 3.92 |
| 5 | 3.75 | 3.91 | 4.00 | 4.07 | 4.14 |
| 5.6 | 4.08 | 4.27 | 4.37 | 4.46 | 4.54 |
| 6.8 | 4.68 | 4.93 | 5.07 | 5.19 | 5.29 |
| 7.5 | 5.00 | 5.29 | 5.45 | 5.59 | 5.71 |
| 8.2 | 5.30 | 5.63 | 5.81 | 5.97 | 6.11 |
| 10 | 6.00 | 6.43 | 6.66 | 6.87 | 7.06 |
| 12 | 6.66 | 7.20 | 7.50 | 7.77 | 8.00 |
| 15 | 7.50 | 8.18 | 8.57 | 8.92 | 9.23 |
| 18 | 8.18 | 9.00 | 9.47 | 9.90 | 10.29 |
| 20 | 8.57 | 9.47 | 10.00 | 10.43 | 10.91 |
| 22 | 8.92 | 9.90 | 10.48 | 11.00 | 11.48 |
| 24 | 9.23 | 10.29 | 10.91 | 11.48 | 12.00 |
| 27 | 9.64 | 10.80 | 11.49 | 12.12 | 12.71 |
| 30 | 10.00 | 11.25 | 12.00 | 12.69 | 13.33 |
| 33 | 10.31 | 11.65 | 12.45 | 13.20 | 13.89 |
| 36 | 10.59 | 12.00 | 12.86 | 13.66 | 14.40 |
| 39 | 10.83 | 12.32 | 13.22 | 14.07 | 14.86 |
| 43 | 11.21 | 12.69 | 13.65 | 14.55 | 15.40 |
| 47 | 11.37 | 13.02 | 14.03 | 14.93 | 15.89 |
| 51 | 11.59 | 13.30 | 14.37 | 15.37 | 16.32 |
| 56 | 11.83 | 13.62 | 14.74 | 15.79 | 16.80 |
| 62 | 12.08 | 13.95 | 15.12 | 16.24 | 17.30 |
| 68 | 12.29 | 14.23 | 15.45 | 16.62 | 17.74 |
| 75 | 12.50 | 14.52 | 15.79 | 17.01 | 18.18 |
| 82 | 12.68 | 14.76 | 16.08 | 17.35 | 18.57 |
| 91 | 12.88 | 15.03 | 16.40 | 17.71 | 18.99 |
| 100 | 13.04 | 15.25 | 16.67 | 18.03 | 19.35 |
| 150 | 13.64 | 16.07 | 17.65 | 19.19 | 20.69 |
| 180 | 13.85 | 16.36 | 18.00 | 19.60 | 21.18 |
| 200 | 13.95 | 16.51 | 18.18 | 19.82 | 21.43 |
| 220 | 14.04 | 16.63 | 18.33 | 20.00 | 21.64 |
| 250 | 14.15 | 16.79 | 18.52 | 20.22 | 21.90 |

Table 4-2. Continued.

| | $C_1 \rightarrow$ | | | | | |
|---|---|---|---|---|---|---|
| $C_2$ | 47 | 51 | 56 | 62 | 68 | 75 |
| $\downarrow$ | | | | | | |
| 1 | 0.979 | 0.981 | 0.982 | 0.984 | 0.986 | 0.987 |
| 1.5 | 1.454 | 1.457 | 1.461 | 1.465 | 1.468 | 1.471 |
| 2 | 1.918 | 1.925 | 1.931 | 1.938 | 1.943 | 1.948 |
| 2.2 | 2.100 | 2.11 | 2.12 | 2.124 | 2.13 | 2.142 |
| 3 | 2.82 | 2.83 | 2.85 | 2.86 | 2.87 | 2.88 |
| 3.3 | 3.08 | 3.10 | 3.12 | 3.13 | 3.15 | 3.16 |
| 4 | 3.68 | 3.71 | 3.73 | 3.75 | 3.77 | 3.79 |
| 4.7 | 4.27 | 4.30 | 4.34 | 4.37 | 4.40 | 4.42 |
| 5 | 4.52 | 4.55 | 4.59 | 4.62 | 4.65 | 4.69 |
| 5.6 | 5.00 | 5.05 | 5.09 | 5.14 | 5.17 | 5.21 |
| 6.8 | 5.94 | 6.00 | 6.06 | 6.14 | 6.18 | 6.23 |
| 7.5 | 6.47 | 6.53 | 6.61 | 6.69 | 6.75 | 6.82 |
| 8.2 | 6.98 | 7.06 | 7.15 | 7.24 | 7.32 | 7.39 |
| 10 | 8.24 | 8.36 | 8.48 | 8.61 | 8.72 | 8.82 |
| 12 | 9.56 | 9.71 | 9.88 | 10.05 | 10.20 | 10.34 |
| 15 | 11.37 | 11.59 | 11.83 | 12.08 | 12.29 | 12.50 |
| 18 | 13.02 | 13.30 | 13.62 | 13.95 | 14.23 | 14.52 |
| 20 | 14.03 | 14.37 | 14.74 | 15.12 | 15.45 | 15.79 |
| 22 | 14.93 | 15.37 | 15.79 | 16.24 | 16.62 | 17.01 |
| 24 | 15.89 | 16.32 | 16.80 | 17.30 | 17.74 | 18.18 |
| 27 | 17.15 | 17.65 | 18.22 | 18.81 | 19.33 | 19.85 |
| 30 | 18.31 | 18.89 | 19.54 | 20.22 | 20.82 | 21.42 |
| 33 | 19.39 | 20.04 | 20.76 | 21.54 | 22.22 | 22.92 |
| 36 | 20.39 | 21.10 | 21.91 | 22.78 | 23.54 | 24.32 |
| 39 | 21.31 | 22.10 | 22.99 | 23.94 | 24.79 | 25.66 |
| 43 | 22.46 | 23.33 | 24.32 | 25.39 | 26.34 | 27.33 |
| 47 | 23.50 | 24.46 | 25.55 | 26.73 | 27.79 | 28.89 |
| 51 | 24.46 | 25.50 | 26.69 | 27.98 | 29.14 | 30.36 |
| 56 | 25.55 | 26.69 | 28.00 | 29.42 | 30.71 | 32.06 |
| 62 | 26.73 | 27.98 | 29.42 | 31.00 | 32.43 | 33.94 |
| 68 | 27.79 | 29.14 | 30.71 | 32.43 | 34.00 | 35.66 |
| 75 | 28.89 | 30.36 | 32.06 | 33.94 | 35.66 | 37.50 |
| 82 | 29.88 | 31.44 | 33.28 | 35.31 | 37.17 | 39.17 |
| 91 | 30.99 | 32.68 | 34.67 | 36.88 | 38.92 | 41.11 |
| 100 | 31.97 | 33.78 | 35.90 | 38.27 | 40.48 | 42.86 |

Table 4-2. Continued.

| | $C_1 \longrightarrow$ | | |
|---|---|---|---|
| $C_2$ | 82 | 91 | 100 |
| $\downarrow$ | | | |
| 1 | 0.988 | 0.989 | 0.990 |
| 1.5 | 1.473 | 1.476 | 1.478 |
| 2 | 1.952 | 1.957 | 1.961 |
| 2.2 | 2.142 | 2.148 | 2.153 |
| 3 | 2.89 | 2.90 | 2.91 |
| 3.3 | 3.17 | 3.19 | 3.20 |
| 4 | 3.81 | 3.83 | 3.84 |
| 4.7 | 4.45 | 4.47 | 4.49 |
| 5 | 4.71 | 4.74 | 4.76 |
| 5.6 | 5.24 | 5.28 | 5.30 |
| 6.8 | 6.28 | 6.32 | 6.37 |
| 7.5 | 6.87 | | |
| 8.2 | 7.45 | 7.52 | 7.58 |
| 10 | 8.91 | 9.01 | 9.09 |
| 12 | 10.46 | 10.60 | 10.71 |
| 15 | 12.68 | 12.88 | 13.04 |
| 18 | 14.76 | 15.03 | 15.25 |
| 20 | 16.08 | 16.40 | 16.67 |
| 22 | 17.35 | 17.71 | 18.03 |
| 24 | 18.57 | 18.99 | 19.35 |
| 27 | 20.31 | 20.82 | 21.26 |
| 30 | 21.96 | 22.56 | 23.08 |
| 33 | 23.53 | 24.22 | 24.81 |
| 36 | 25.02 | 25.80 | 26.47 |
| 39 | 26.43 | 27.30 | 28.06 |
| 43 | 28.21 | 29.20 | 30.07 |
| 47 | 29.88 | 30.99 | 31.97 |
| 51 | 31.44 | 32.68 | 33.78 |
| 56 | 33.28 | 34.67 | 35.90 |
| 62 | 35.31 | 36.88 | 38.27 |
| 68 | 37.17 | 38.92 | 40.48 |
| 75 | 39.17 | 41.11 | 42.86 |
| 82 | 41.00 | 43.13 | 45.06 |
| 91 | 43.13 | 45.50 | 47.64 |
| 100 | 45.06 | 47.64 | 50.00 |

moving across to reach 4.7 (column C1). The answer would, of course, be 1.13 pF. Moving the decimal point one place to the right supplies an answer of 11.3.

*Example* You have a number of capacitors available, but you do not have one with a capacitance of 6 pF—the value you require. What capacitor combination can you use in series to give you 6 pF?

*Solution* A value of exactly 6 pF is shown in the table. It can be made by connecting a 10 pF and a 15 pF in series or by connecting a 30 pF and 7.5 pF in series. Another combination which would result in a capacitance fairly close to 6 pF would be 33 pF and 7.5 pF, giving a total capacitance of 6.11 pF. Or you could use a pair of capacitors each having a value of 12 pF.

# Capacitors in parallel or series

For capacitors connected in parallel:

$$C_t = C_1 + C_2 + C_3. \ . \ .$$

$C_t$ is the total capacitance, $C_1$, $C_2$, $C_3$, etc., are the parallel (shunt) connected capacitors. To use the formula, the capacitors must be in the same units; that is, microfarads or picofarads. If not, their values must be changed accordingly, that is, microfarads to picofarads, or vice versa. The total capacitance, $C_t$, of capacitors in parallel is always larger than that of any individual capacitor in the network (Fig. 4-2).

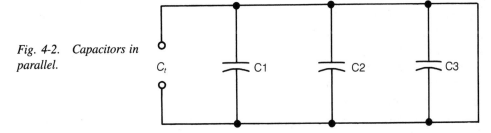

*Fig. 4-2. Capacitors in parallel.*

For capacitors connected in series:

$$C_t = \frac{C_1 \times C_2}{C_1 + C_2}$$

As in the case of the formula for parallel capacitors, all the capacitors indicated in the series formula must be in the same capacitance units.

The formula given here is intended only for two capacitors. For any number of series connected capacitors:

$$\frac{1}{C_t} = \frac{1}{C_1} + \frac{1}{C_2} + \frac{1}{C_3} \ . \ . \ .$$

Alternatively, the following formula can be used:

$$C_t = \cfrac{1}{\cfrac{1}{C_1} + \cfrac{1}{C_2} + \cfrac{1}{C_3}}$$

For capacitors in series, the total capacitance, $C_t$, is always less than the unit having the lowest value of capacitance in the series connected group.

# Capacitor codes and values

The letter $C$ is often used in connection with capacitor symbols in circuit diagrams. When a number of capacitors are used, they can be identified as C1, C2, C3 and so on. It is usual practice to start with capacitor C1 at the upper left-hand corner, but this isn't a practice that is always followed.

Sometimes the symbol without an accompanying value is used when the intent is simply to indicate the presence of capacitance. The value of capacitance might or might not follow the C number. In some diagrams all capacitances are in microfarads unless otherwise stated. Actual capacitance values can be indicated in an accompanying parts list.

# Dielectric materials

Dielectric materials, represented by the letter $k$, are used between the plates of a capacitor as a practical method for increasing the capacitance. A vacuum is the standard reference and is assigned a value of 1. The value of $k$ for all materials is compared to that of a vacuum. The dielectric constant of air has been found to be identical to that of a vacuum and so has a $k$ value of unity. The dielectric is inserted between the plates of a capacitor with its surface areas making good contact with the areas of the capacitor plates. The drawing in Fig. 4-3A shows the physical representation of a basic capacitor and its corresponding electronic symbol. The symbol in drawing (Fig. 4-3B) is that of a fixed capacitor. An older symbol, now discarded, consisted of a pair of straight lines as in Fig. 4-3C. The symbol for a variable capacitor is shown in Fig. 4-3D. Table 4-3 is a listing of some materials used as dielectrics.

In some instances, as in the case of electrolytic capacitors, the surface of the plates isn't smooth but is deliberately corrugated so as to increase the surface area, hence the amount of capacitance. Note also that the capacitance is directly proportional to the dielectric constant, $k$. A dielectric having a value of 100 will result in much more capacitance than a dielectric with a value of 1.

# Capacitor types

There are numerous capacitors with the most commonly used listed in Table 4-4. The values supplied in the table can vary, depending on the manufacturer.

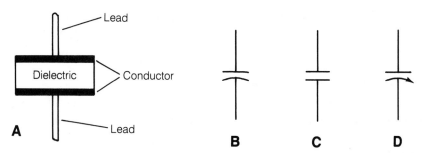

*Fig. 4-3. Dielectric—insulating material between parallel metallic plates (A); symbol for fixed capacitor (B); symbol formerly used for fixed capacitors (C); symbol for variable capacitor (D).*

*Table 4-3. Dielectrics and constants.*

| Material | Constant (k) |
|---|---|
| Vacuum | 1.000 0 |
| Air | 1.000 6 |
| Paraffin paper | 3.5 |
| Glass | 5 − 10 |
| Mica | 3 − 6 |
| Rubber | 2.5 − 35 |
| Wood | 2.5 − 8 |
| Glycerine (15 °C) | 56 |
| Petroleum | 2 |
| Pure water | 81 |

*Table 4-4. Capacitor types.*

| Type | Capacitance range | Voltage range |
|---|---|---|
| Ceramic | 1 pF to 2.2 $\mu$F | 3.3 V to 6 kVdc |
| Ceramic chip | 10 pF to 0.18 $\mu$F | |
| Mica | 1 pF to 0.1 $\mu$F | 100 to 2 500 Vdc |
| Mica chip | 1 to 10 000 pF | As high as 500 Vdc |
| Glass | 0.5 to 10 000 pF | 100 to 500 Vdc |
| Polycarbonate | As high as 50 $\mu$F | 1 kVdc (working voltage) |
| Polyester | 0.001 to 100 $\mu$F | As high as 1 500 Vdc |
| Polypropylene | To 10 $\mu$F | 400 Vdc |
| Tantalum | 0.001 to 1 000 $\mu$F | 6 to 120 Vdc |
| Electrolytic | 1 to 2 200 $\mu$F | 16 to 450 Vdc (working voltage) |

Less common types include the polysulfone, polyvinylidene fluoride, polyethylene terphthalate, metallized paper, polyester/paper, polyester foil, paper polypropylene, Teflon/Kapton, and Parylene. The latter three are trade names but these do not supply clues as to their construction. The listing does not include variable capacitors, which include glass, quartz, sapphire, plastic, ceramic, air, mica, and vacuum or gas. The materials named are used as dielectrics.

# Capacitor characteristics

There are variations in the electrical and physical characteristics of capacitors made by different manufacturers. Identifying particular capacitors can sometimes be difficult, especially in the case of capacitors that are very small. The connecting leads can be axial or radial.

## Ceramic capacitors

Ceramic capacitors have a round shape, and the dielectric is a ceramic material. Metal plates are attached to opposite faces of the dielectric. The ceramic is a clay material mixed with titanium dioxide or a clay mixture of magnesium and aluminum oxides. It is subjected to heat until it is very hard. Capacitance depends on area of the plates (made of silver) and the thickness of the dielectric. The complete capacitor is enclosed in a molded insulated case or dipped in a phenolic compound. The temperature coefficient is determined by the type of ceramic used and extends to 200 000 parts per million per Celsius degree. Common ceramic disc capacitors sometimes are called *pressed-part* capacitors. Others are the monolithic or chip types. Capacitance range is 1 pF to 2.2 $\mu$F, with a tolerance of 10% to 20%, and voltage ratings of 3.3 V to 6 kV.

## Mica capacitors

These are relatively large and available as dipped case and molded case types. The mica dielectric is very stable under variable voltage and temperature changes. Mica capacitors are available in a wide range of capacitance values from 1 to 100 000 pF. They have voltage ratings from 100 to 2.5 kVdc and a working temperature range of −55 °C to 150 °C.

## Glass capacitors

Glass can be used as the dielectric instead of mica. Consequently both types— mica and glass—are available in similar sizes and values. The glass is not the same as that used for windows but has physical properties that enhance its strength. Its dielectric constant can be improved or varied only slightly. Some glass-plate capacitors are made specifically for high-voltage use, with the complete assembly immersed in oil. The $Q$ of these capacitors is higher than that of mica types. The capacitance range extends from 0.5 to 10 000 pF. The dc working voltage rating is from 100 to 500 V.

### Metallized polycarbonate capacitors

These capacitors are especially designed for limited space applications and are manufactured in both oval and round packages. The capacitance ranges from 0.33 $\mu$F to 30 $\mu$F, all at 50 Vdc. The units are available in a variety of sizes, from 0.25 in outside diameter by 0.52 in long, to 0.75 in outside diameter by 1.68 in long.

### Tantalum-foil electrolytic capacitors

Tantalum foil capacitors are made of a pair of tantalum foil electrodes. These are separated by absorbent paper soaked with an electrolyte. The capacitor is polarized, and for its size has a large amount of capacitance. The high capacitance is mainly due to the high dielectric constant of tantalum oxide. Tantalum electrolytics are available in three types—solid, chip, and nonsolid. The foil can be etched, thus increasing the capacitance of the unit.

## Charge on a capacitor

The electric charge stored by a capacitor is determined by:

$$Q = C \times E$$

$Q$ is the electric charge in coulombs, $C$ is the capacitance in farads, and $E$ is the voltage placed across the terminals of the capacitor. When 1 coulomb (C) of electrons passes a fixed point in 1 s, the amount of current flow is 1 A.

The amount of energy stored in a charged capacitor is determined by:

$$W = \frac{1}{2}\, CE^2$$

$C$ is the capacitance in farads, $E$ is the applied dc voltage in volts, and $W$ is the energy in joules. The joule (J) is the standard unit of energy. A kilowatt-hour is equivalent to $3.6 \times 10^6$ J. Therefore, 1 W of electrical power is equal to 1 J/s.

## Capacitors as voltage dividers

When an ac voltage is impressed across capacitors connected in series, the voltage appearing across each individual capacitor is inversely proportional to its capacitance. If the capacitors, as in Fig. 4-4, have the same value of capacitance, there will be an equal division of voltage. In this example the source is 300 V and each capacitor is 1 $\mu$F. The voltage is divided equally and there is 150 V across each capacitor.

The circuit arrangement in Fig. 4-5 is the same but now one of the capacitors (C1) is 1 $\mu$F and the other (C2) is 3 $\mu$F. Without considering the way in which they are connected, the sum of the two capacitances is 4 $\mu$F. One of the capacitors, C2 in this case, will have one fourth of the total voltage, and the other, C1, will have

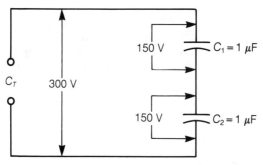

*Fig. 4-4. When capacitors of equal value are connected across a source voltage, the same amount of voltage appears across each capacitor.*

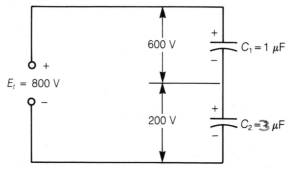

*Fig. 4-5. The voltage across series capacitors varies inversely with the capacitance.*

three fourths of the voltage. The total voltage is 800 V. The voltage across C1 is $3/4 \times 800 = 600$ V and that across C2 is $1/4 \times 800 = 200$ V.

## Capacitance of a capacitor

The capacitance of a capacitor is directly due to the physical elements of its construction, to the area of its plates, the amount of plate separation, the type of dielectric used, its thickness, and so on. Using the English system of measurement, the capacitance can be calculated by:

$$C = 0.223\,5\left(\frac{kA}{d}\right)(N-1)$$

$C$ is the capacitance in picofarads, $k$ is the dielectric constant, $A$ is the area of one plate in square inches, $d$ is the thickness of the dielectric, and N is the number of plates. The constant 0.223 5 can have a slightly different value, such as 0.224 9. This constant results from the conversion from metric to English units. For a two-plate capacitor the factor $(N-1)$ can be dropped from the formula.

*Example*  What is the capacitance of a two-plate capacitor using paper as the dielectric having a value of 3.5; a plate separation of 0.05 in, and the area of one of the plates is 12 in²?
*Solution*

$$C = 0.224\ 9 \left(\frac{kA}{d}\right)$$

$$= 0.224\ 9 \left(\frac{3.5 \times 12}{0.05}\right)$$

$$= 189 \text{ pF}$$

Using the metric system of measurements, the capacitance of a parallel-plate capacitor can be expressed as

$$C = 0.088\ 5\ k\ \frac{(N-1)A}{d} \text{ pF}$$

*A* is the area of one side of one plate in square centimeters, N is the number of plates, *d* is the thickness of the dielectric, and *k* is the dielectric constant.

# Capacitive reactance

The reactance of a capacitor, or its opposition to the flow of a varying or an alternating current, varies inversely with frequency and with capacitance. Capacitive reactance, expressed in ohms, is based on the ability of a capacitor to store a charge or counter-electromotive force. This emf (electromotive force), acting in opposition to the applied voltage, reduces the amount of current flowing in a circuit, hence produces an effect analogous to that of a resistor. With a resistor, though, the current through it and the voltage across it are in phase. However, the counter emf of a capacitor causes the voltage to lag behind the current. Ideally, the phase angle is 90°, but in practice the phase angle is less than this.

Table 4-5 gives the reactance of capacitors ranging from 0.001 $\mu$F to 0.000 5 $\mu$F for frequencies ranging from 10 to 5 000 kHz. Table 4-6 is for capacitors having values from 0.25 to 3 $\mu$F and for frequencies from 25 to 20 000 Hz. Both tables can be extended because doubling the frequency will halve the reactance. The same effect can be obtained by doubling the capacitance. Similarly, halving the frequency or capacitance will double the reactance. Naturally, other multiplication or division factors can be used.

*Example*  What is the reactance of a 0.01-$\mu$F capacitor at a frequency of 1 000 Hz? What will happen to this reactance if the frequency is increased to 10 000 Hz?

Table 4-5. Capacitive reactance (ohms).

| Frequency (kHz) | Capacitance (μF) | | | | | |
|---|---|---|---|---|---|---|
| | 0.001 | 0.000 15 | 0.000 2 | 0.000 25 | 0.000 3 | 0.000 5 |
| 10 | 159 236 | 106 157 | 79 618 | 63 694 | 53 078 | 31 847 |
| 20 | 79 618 | 53 079 | 39 809 | 31 848 | 26 539 | 15 924 |
| 30 | 53 079 | 35 836 | 26 539 | 21 232 | 17 693 | 10 616 |
| 40 | 39 809 | 26 540 | 19 905 | 15 924 | 13 270 | 7 962 |
| 50 | 31 847 | 21 230 | 15 924 | 12 740 | 10 616 | 6 370 |
| 60 | 26 539 | 17 693 | 13 270 | 10 616 | 8 847 | 5 308 |
| 70 | 22 748 | 15 165 | 11 374 | 9 098 | 7 852 | 4 549 |
| 80 | 19 905 | 13 270 | 9 953 | 7 962 | 6 635 | 3 981 |
| 90 | 17 693 | 11 795 | 8 847 | 7 078 | 5 897 | 3 539 |
| 100 | 15 924 | 10 615 | 7 962 | 6 370 | 5 308 | 3 185 |
| 150 | 10 616 | 7 077 | 5 308 | 4 246 | 3 539 | 2 123 |
| 200 | 7 962 | 5 308 | 3 981 | 3 186 | 2 654 | 1 593 |
| 250 | 6 369 | 4 246 | 3 185 | 2 548 | 2 123 | 1 274 |
| 300 | 5 308 | 3 538 | 2 654 | 2 124 | 1 770 | 1 062 |
| 350 | 4 550 | 3 033 | 2 275 | 1 820 | 1 516 | 910 |
| 400 | 3 981 | 2 654 | 1 991 | 1 594 | 1 326 | 797 |
| 450 | 3 539 | 2 539 | 1 769 | 1 414 | 1 179 | 707 |
| 500 | 3 185 | 2 123 | 1 592 | 1 274 | 1 062 | 637 |
| 550 | 2 895 | 1 930 | 1 448 | 1 158 | 965 | 579 |
| 600 | 2 654 | 1 769 | 1 327 | 1 062 | 885 | 531 |
| 650 | 2 450 | 1 633 | 1 225 | 980 | 816 | 490 |
| 700 | 2 275 | 1 516 | 1 138 | 910 | 758 | 455 |
| 750 | 2 123 | 1 416 | 1 062 | 850 | 708 | 425 |
| 800 | 1 991 | 1 327 | 896 | 798 | 663 | 399 |
| 850 | 1 873 | 1 249 | 937 | 750 | 624 | 375 |
| 900 | 1 769 | 1 179 | 885 | 708 | 589 | 354 |
| 950 | 1 676 | 1 117 | 838 | 670 | 559 | 335 |
| 1 000 | 1 592 | 1 062 | 796 | 637 | 530 | 319 |
| 2 000 | 796 | 531 | 398 | 319 | 265 | 159 |
| 2 500 | 637 | 425 | 319 | 255 | 212 | 127 |
| 3 000 | 531 | 354 | 266 | 212 | 177 | 106 |
| 3 500 | 455 | 303 | 228 | 182 | 152 | 91 |
| 4 000 | 398 | 265 | 199 | 159 | 133 | 80 |
| 4 500 | 354 | 254 | 177 | 141 | 118 | 71 |
| 5 000 | 319 | 212 | 159 | 127 | 106 | 64 |

**Solution** Table 4-6 shows a frequency of 1 000 Hz, but does not have a capacitance value marked 0.01. Locate 1 000 in the left-hand column and move horizontally to the right until you reach the number 159 under the heading of 1.0 μF. Change the heading of 1.0 μF to 0.01 by dividing it by 100. When you do this,

*Table 4-6. Capacitive reactance (ohms).*

| Frequency (Hz) | Capacitance (μF) | | | | |
|---|---|---|---|---|---|
| | 0.25 | 0.5 | 1.0 | 2.0 | 3.0 |
| 25 | 25 478 | 12 739 | 6 369 | 3 185 | 2 123 |
| 30 | 21 231 | 10 616 | 5 308 | 2 654 | 1 769 |
| 50 | 12 739 | 6 369 | 3 185 | 1 593 | 1 062 |
| 60 | 10 616 | 5 308 | 2 654 | 1 327 | 885 |
| 75 | 8 492 | 4 246 | 2 123 | 1 062 | 708 |
| 100 | 6 369 | 3 185 | 1 592 | 796 | 531 |
| 120 | 5 308 | 2 654 | 1 327 | 664 | 442 |
| 150 | 4 246 | 2 123 | 1 062 | 531 | 354 |
| 180 | 3 538 | 1 769 | 885 | 443 | 295 |
| 200 | 3 185 | 1 592 | 796 | 398 | 265 |
| 250 | 2 548 | 1 274 | 637 | 319 | 212 |
| 300 | 2 123 | 1 062 | 531 | 265 | 177 |
| 350 | 1 820 | 910 | 455 | 228 | 152 |
| 400 | 1 592 | 796 | 398 | 199 | 133 |
| 450 | 1 415 | 708 | 354 | 177 | 118 |
| 500 | 1 274 | 637 | 319 | 159 | 106 |
| 600 | 1 107 | 531 | 265 | 133 | 88 |
| 700 | 948 | 455 | 228 | 114 | 76 |
| 800 | 796 | 398 | 199 | 99 | 66 |
| 900 | 708 | 354 | 177 | 89 | 59 |
| 1 000 | 637 | 318 | 159 | 79 | 53 |
| 2 000 | 319 | 159 | 79 | 39 | 27 |
| 3 000 | 213 | 107 | 53 | 27 | 18 |
| 4 000 | 159 | 79 | 39 | 20 | 14 |
| 5 000 | 127 | 64 | 32 | 16 | 11 |
| 6 000 | 106 | 53 | 27 | 14 | 9 |
| 7 000 | 91 | 46 | 23 | 12 | 8 |
| 8 000 | 80 | 40 | 20 | 10 | 7 |
| 9 000 | 71 | 36 | 18 | 9 | 6 |
| 10 000 | 64 | 32 | 16 | 8 | 5 |
| 12 000 | 53 | 27 | 14 | 7 | 4.6 |
| 14 000 | 46 | 23 | 12 | 6 | 4 |
| 16 000 | 40 | 20 | 10 | 5 | 3.3 |
| 18 000 | 36 | 18 | 9 | 4.5 | 3 |
| 20 000 | 32 | 16 | 8 | 4 | 2.6 |

you must multiply 159 by 100. The answer is then 15 900 Ω. If you increase the frequency to 10 000 Hz, you will be multiplying the original frequency by a factor of 10. This means the reactance should be divided by a similar factor. The answer will be 15 900 divided by 10 or 1 590 Ω.

*Example*    What is the reactance of a 0.000 15 μF capacitor at a frequency of 3 000 kHz?

*Solution*    Table 4-5 shows that at this frequency the reactance is 354 Ω.

Table 4-7 supplies the capacitive reactance at various radio frequencies. Table 4-8 gives the capacitive reactance of various audio frequencies.

*Table 4-7. Capacitive reactance at radio frequencies.*

| Capacitance (μF) | Radio frequencies Reactance (Ω) | | | | | |
|---|---|---|---|---|---|---|
| | 175 Hz | 465 Hz | 550 Hz | 1 000 Hz | 1 500 Hz | 2 000 Hz |
| 0.000 05 | 18 200 | 6 850 | 5 800 | 3 180 | 2 120 | 1 590 |
| 0.000 1 | 9 100 | 3 420 | 2 900 | 1 590 | 1 060 | 795 |
| 0.000 25 | 3 640 | 1 370 | 1 160 | 637 | 424 | 319 |
| 0.000 5 | 1 820 | 685 | 579 | 318 | 212 | 159 |
| 0.001 | 910 | 342 | 290 | 159 | 106 | 80 |
| 0.005 | 182 | 68.5 | 57.9 | 31.8 | 21.2 | 15.9 |
| 0.01 | 91.0 | 34.2 | 28.9 | 15.9 | 10.6 | 8.0 |
| 0.02 | 45.5 | 17.1 | 14.5 | 7.96 | 5.31 | 3.95 |
| 0.05 | 18.2 | 6.85 | 4.79 | 3.18 | 2.12 | 1.59 |
| 0.1 | 9.10 | 3.42 | 2.89 | 1.59 | 1.06 | 0.80 |
| 0.25 | 3.64 | 1.37 | 1.16 | 0.637 | 0.424 | 0.319 |
| 0.5 | 1.82 | 0.685 | 0.579 | 0.318 | 0.212 | 0.159 |
| 1 | 0.910 | 0.342 | 0.289 | 0.159 | 0.106 | 0.080 |
| 2 | 0.455 | 0.171 | 0.145 | 0.079 6 | 0.053 1 | 0.039 8 |
| 4 | 0.227 | 0.085 6 | 0.072 3 | 0.039 8 | 0.026 5 | 0.019 9 |

## Calculation of capacitive reactance

Although the various tables that supply values of capacitive reactance save time and work, they do have limitations. They are suitable for specific capacitances and frequencies, but the tables will be unable to supply data for frequencies and capacitances that aren't listed. Capacitive reactance can be calculated from:

$$X_C = \frac{1}{2\pi f C}$$

$f$ is the frequency in hertz, $C$ is the capacitance in farads, and $X_C$ is the reactance in ohms.

Capacitances will ordinarily be supplied in microfarads or picofarads. These units will need to be converted to farads prior to use in the formula. To convert microfarads to farads, multiply microfarads by $10^{-6}$; to convert picofarads to farads multiply picofarads by $10^{-12}$.

*Table 4-8. Capacitive reactance at audio frequencies.*

| Capacitance (µF) | Audio frequencies Reactance (Ω) | | | | | |
|---|---|---|---|---|---|---|
| | 30 Hz | 60 Hz | 100 Hz | 400 Hz | 1 000 Hz | 5 000 Hz |
| 0.000 05 | — | — | — | — | — | 637 000 |
| 0.000 1 | — | — | — | — | 1 590 000 | 318 000 |
| 0.000 25 | — | — | — | 1 590 000 | 637 000 | 127 000 |
| 0.000 5 | — | — | 3 180 000 | 796 000 | 318 000 | 63 700 |
| 0.001 | — | 2 650 000 | 1 590 000 | 398 000 | 159 000 | 31 800 |
| 0.005 | 1 060 000 | 530 834 | 318 000 | 79 600 | 31 800 | 6 370 |
| 0.01 | 531 000 | 265 000 | 159 000 | 39 800 | 15 900 | 3 180 |
| 0.02 | 263 000 | 132 500 | 79 600 | 19 900 | 7 960 | 1 590 |
| 0.05 | 106 000 | 53 083 | 31 800 | 7 960 | 3 180 | 637 |
| 0.1 | 53 100 | 26 500 | 15 900 | 3 980 | 1 590 | 318 |
| 0.25 | 21 200 | 10 584 | 6 370 | 1 590 | 637 | 127 |
| 0.5 | 10 600 | 5 308 | 3 180 | 796 | 318 | 63.7 |
| 1 | 5 310 | 2 650 | 1 590 | 389 | 159 | 31.8 |
| 2 | 2 650 | 1 325 | 796 | 199 | 79.6 | 15.9 |
| 4 | 1 310 | 663 | 398 | 99.5 | 39.8 | 7.96 |
| 8 | 663 | 332 | 199 | 49.7 | 19.9 | 3.98 |
| 16 | 332 | 166 | 99.5 | 24.9 | 9.95 | 1.99 |
| 20 | 262 | 133 | 80 | 20 | 8 | 1.6 |
| 25 | 212 | 106 | 63.7 | 15.9 | 6.37 | 1.27 |
| 35 | 152 | 86 | 45.5 | 11.4 | 4.55 | 0.910 |
| 40 | 131 | 66 | 39.8 | 9.9 | 3.9 | 0.8 |

**Example** What is the capacitive reactance of a capacitor working at a frequency of 60 Hz with a capacitance of 133 µF?

**Solution** As a first step convert 133 µF to farads by multiplying it by $10^{-6}$. This results in $133 \times 10^{-6}$ equivalent to $1.33 \times 10^{-4}$.

$$X_C = \frac{1}{2\pi f C} = \frac{1}{6.28 \times 60 \times 1.33 \times 10^{-4}} = 20 \ \Omega$$

# 5
## CHAPTER

# Inductance

## Units of inductance

The basic unit of inductance is the henry. Submultiples are the millihenry and the microhenry. It is sometimes necessary to move from the basic unit of inductance to submultiples. Table 5-1 supplies the multiplication factors for making these conversions.

## Inductors

Like capacitors and resistors, inductors (coils) are one of the basic components of electronic circuitry. The unusual feature of inductors is that they are surrounded by a magnetic field when an electric current flows through them. Inductors can be connected in series, in parallel, or in series-parallel combinations. When connecting inductors, you must consider the magnetic fields, because they affect the inductance of the inductors.

Like capacitors, inductors are capable of storing electrical energy, a characteristic not shared by resistors. An inductor consists of wire usually wound in circular fashion, often around a form, which might or might not contain a core made of iron, steel or an alloy, and are referred to as iron-core coils. In the absence of a core the coil is designated as an air-core type.

Because coils are made of wire, and wire has resistance, that resistance is regarded as part of the electrical characteristics of the coil, even though that resistance does not exist as a separate component. (Resistance is also inherent in capacitors.) The built-in resistance of a coil is considered as being in series with it. In a circuit, this can be shown as a coil followed by a series resistor, although that resistance isn't physically tangible.

*Table 5-1. Inductance conversions.*

| Given this value | Multiply by this value to get | | |
|---|---|---|---|
| | Henries | Millihenries | Microhenries |
| Henries | — | $10^3$ | $10^6$ |
| Millihenries | $10^{-3}$ | — | $10^{-3}$ |
| Microhenries | $10^{-6}$ | $10^{-3}$ | — |

# Coefficient of magnetic coupling

If two coils are magnetically coupled, the mutual inductance depends on:

The physical dimensions of the coils
The number of turns used by each coil
The distance between the two coils
The angular displacement of the two coils
The permeability of the cores of the coils
Whether the coils are magnetically shielded and the effectiveness of the shielding
The polarity of the magnetic field around each coil

# Unity coefficient of coupling

The unity condition exists when all the lines of magnetic force of one of the coils surrounds and cuts across the turns of a closely adjacent coil. This condition would be extremely unusual, but 90% coupling is achievable. If the lines of magnetic force of one coil cut across half the turns of wire of the second coil, the coefficient of coupling is 50%.

The mutual inductance existing between a pair of magnetically coupled coils can be expressed as:

$$M = k \sqrt{L_1 L_2}$$

where the percentage of $k$ is expressed as a decimal, $L_1$ and $L_2$ can be supplied in henries, millihenries, or microhenries, with both coils using the same inductance units. The submultiple selected also applies to the mutual inductance, $M$.

The coefficient of coupling ($k$) is based on how completely the flux lines of one coil link or interact with the flux lines of the other. The coefficient of coupling is a decimal and has a value of less than 1. When two coils are wound so that the wire turns are immediately adjacent the coefficient of coupling can sometimes be greater than 95%. The coefficient of coupling can be calculated from:

$$k = \frac{M}{\sqrt{L_1 \times L_2}}$$

## Inductance of a coil

The inductance of an iron-core coil can be calculated by:

$$L = \frac{0.4\pi \; N^2 \mu A \times 10^{-8}}{\ell}$$

where $L$ is the inductance of the coil in henries, N is the number of turns of the coil, $\mu$ is the permeability of the iron core, $A$ is the cross-sectional area of the core in square centimeters, cm, and $\ell$ is the mean length of the core in centimeters.

## Mutual inductance

The magnetic fields of a pair of coils connected in series may aid each other or may oppose depending on the direction of current flow through them. Figure 5-1A indicates that the current flow is in the same direction; Fig. 5-1B shows the

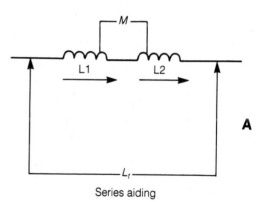

Series aiding

*Fig. 5-1. Coils in series aiding (A); in series opposing (B).*

A

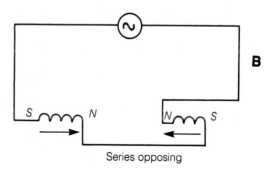

Series opposing

B

same coils connected in series opposing. The effect of the magnetic fields surrounding the two coils is referred to as mutual inductance. When the inductances are wired in series aiding, the total inductance is increased by the amount of mutual inductance, as in:

$$L_t = L_1 + L_2 + 2M$$

$L_t$ is the total inductance of the two coils, $L_1$ and $L_2$ are the inductances of the two series-connected coils, and $M$ is the mutual inductance. All inductance units must be in the same values, that is, in henries, millihenries, or microhenries.

When the magnetic fields of the two coils oppose each other:

$$L_t = L_1 + L_2 - 2M$$

# Inductors in series

Inductors (coils) can be connected in series, as shown in Fig. 5-2. There is no magnetic coupling between the inductors, something that can be done by separat-

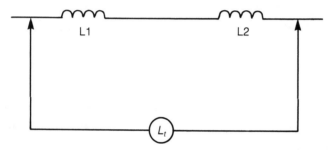

Fig. 5-2.   Separating coils reduces mutual inductance.

ing them widely, mounting them at right angles to each other, or by electromagnetic shielding. The total inductance is then equal to

$$L_t = L_1 + L_2 + L_3 + \ldots$$

To be able to add the individual values they must all be in similar terms—henries, millihenries, or microhenries. The formula is similar to that of resistors in series.

# Inductors in parallel

Inductors can be connected in parallel as shown in Fig. 5-3. The formula for finding the total inductance is given by:

$$L_t = \frac{L_1 \times L_2}{L_1 + L_2}$$

and can be used for two inductors. There is no electromagnetic coupling between the coils, and (as in the case of series inductors) this can be done by physically sepa-

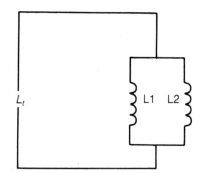

Fig. 5-3.   *Inductors in parallel.*   $L_t$

rating the coils, mounting them at right angles, or using magnetic shields. The formula is similar to that for three or more resistors in parallel. The individual inductances must all be in the same units: henries, millihenries, or microhenries.

For more than two inductors in parallel, an alternative formula can be used. The formula assumes no electromagnetic coupling between the coils.

$$L_t = \frac{1}{\dfrac{1}{L_1} + \dfrac{1}{L_2} + \dfrac{1}{L_3} + \ \cdots}$$

## Inductive reactance

The reactance of a coil (inductor) varies directly with frequency and with inductance. Like capacitive reactance, coil reactance or inductive reactance is measured in ohms. Inductive reactance is an effect produced by the counter emf (electromotive force) induced across the coil. This voltage, acting in opposition to the applied voltage, reduces the amount of circuit current.

As in the case of capacitors, the coil produces a phase shift. The current lags the voltage and in a hypothetical coil (one containing no resistance) the phase angle would be 90°. In practice, the phase angle is less than 90°.

## Calculating inductive reactance

It is possible to calculate the inductive reactance of a coil from:

$$X_L = 2\pi f \times L$$

$X_L$ is the reactance in ohms, $f$ is the frequency in hertz, and $L$ is the inductance in henries. If multiples or submultiples of the frequency or inductance are used, these must first be converted to basic units. In some instances the Greek letter omega, $\omega$, is used as the equivalent of $2\pi f$. In that case the formula for inductive reactance becomes:

$$X_L = \omega L$$

The reactance of a coil is directly proportional to the frequency of the ac source voltage. If the frequency is reduced to zero, the source voltage is equivalent to pure dc, and the reactance also becomes zero. In that case, the only opposition to the flow of current through the coil is its dc resistance.

*Example* What is the reactance of a 40 mH coil at a frequency of 2.4 kHz?
*Solution*

$$X_L = 2\pi f L$$

$$2.4 \text{ kHz} = 2\,400 \text{ Hz} = 24 \times 10^2 \text{ Hz}$$

$$40 \text{ mH} = 40 \times 10^{-3} \text{ H}$$

$$2\pi = 6.28$$

$$X_L = 6.28 \times 24 \times 10^2 \times 40 \times 10^{-3}$$

$$X_L = 602.88 \ \Omega$$

Table 5-2 supplies the inductive reactance of coils ranging from 10 to 100 mH at frequencies ranging from 1 to 1 000 kHz. Table 5-3 supplies the reactance of coils from 0.001 H to 10 H at frequencies of 25 to 1 000 Hz.

The behavior of an inductor is opposite that of a capacitor. Increasing the frequency or the inductance increases the reactance proportionately, and vice versa. Thus, doubling either the frequency or the inductance will double the reactance. See also Table 5-4.

*Example* What is the inductive reactance of a 20 mH coil at a frequency of 400 kHz?
*Solution* Locate 400 in the frequency column in Table 5-2. Move horizontally to the row headed by 20 mH. The reactance is shown as 50 240 $\Omega$.

*Example* What is the reactance of a coil having an inductance of 0.005 H at a frequency of 1 kHz?
*Solution* Locate 1 000 Hz in the frequency column in Table 5-3. Move to the right to the column headed by 0.005. The inductive reactance is 31.40 $\Omega$.

*Example* What is the reactance of a 5-H choke coil at a frequency of 60 Hz?
*Solution* Locate 60 Hz in the frequency column in Table 5-3 and move to the right, finding an inductive reactance of 1 884 $\Omega$ in the column headed by the number 5. Note that the reactance of this coil at twice the frequency (120 Hz) is 3 768 $\Omega$, double its original value.

*Example* What is the reactance of a 1-H coil at a frequency of 150 Hz?
*Solution* A frequency of 150 Hz is not listed in Table 5-3. Locate 50 Hz, find the corresponding reactance of 314 $\Omega$ and multiply this value by 3 to get your

## Table 5-2. Inductive reactance (ohms).

| Frequency (kHz) | Inductance (mH) | | | | |
|---|---|---|---|---|---|
| | 10 | 20 | 30 | 40 | 50 |
| 1 | 62.8 | 125.6 | 188.4 | 251.2 | 314 |
| 2 | 125.6 | 251.2 | 376.8 | 502.4 | 628 |
| 3 | 188.4 | 376.8 | 565.2 | 753.6 | 942 |
| 4 | 251.2 | 502.4 | 753.6 | 1 004.8 | 1 256 |
| 5 | 314 | 628 | 942 | 1 256 | 1 570 |
| 6 | 376.8 | 753.6 | 1 130.4 | 1 507.2 | 1 884 |
| 7 | 439.6 | 879.2 | 1 318.8 | 1 758.4 | 2 198 |
| 8 | 502.4 | 1 004.8 | 1 507.2 | 2 009.6 | 2 512 |
| 9 | 565.2 | 1 130.4 | 1 695.6 | 2 260.8 | 2 826 |
| 10 | 628 | 1 256 | 1 884 | 2 512 | 3 140 |
| 20 | 1 256 | 2 512 | 3 768 | 5 024 | 6 280 |
| 25 | 1 570 | 3 140 | 4 710 | 6 280 | 7 850 |
| 30 | 1 884 | 3 768 | 5 652 | 7 536 | 9 420 |
| 40 | 2 512 | 5 024 | 7 536 | 10 048 | 12 560 |
| 50 | 3 140 | 6 280 | 9 420 | 12 560 | 15 700 |
| 60 | 3 768 | 7 536 | 11 304 | 15 072 | 18 840 |
| 70 | 4 396 | 8 792 | 13 188 | 17 584 | 21 980 |
| 80 | 5 024 | 10 048 | 15 072 | 20 096 | 25 120 |
| 90 | 5 652 | 11 304 | 16 956 | 22 608 | 28 260 |
| 100 | 6 280 | 12 560 | 18 840 | 25 120 | 31 400 |
| 150 | 9 420 | 18 840 | 28 260 | 37 680 | 47 100 |
| 200 | 12 560 | 25 120 | 37 680 | 50 240 | 62 800 |
| 250 | 15 700 | 31 400 | 47 100 | 62 800 | 78 500 |
| 300 | 18 840 | 37 680 | 56 520 | 75 360 | 94 200 |
| 350 | 21 980 | 43 960 | 65 940 | 87 920 | 109 900 |
| 400 | 25 120 | 50 240 | 75 360 | 100 480 | 125 600 |
| 450 | 28 260 | 56 520 | 84 780 | 113 040 | 141 300 |
| 500 | 31 400 | 62 800 | 94 200 | 125 600 | 157 000 |
| 550 | 34 540 | 69 080 | 103 620 | 138 160 | 172 700 |
| 600 | 37 680 | 75 360 | 113 040 | 150 720 | 188 400 |
| 650 | 40 820 | 81 640 | 122 460 | 163 280 | 204 100 |
| 700 | 43 960 | 87 920 | 131 880 | 175 840 | 219 800 |
| 800 | 50 240 | 100 480 | 150 720 | 200 960 | 251 200 |
| 900 | 56 520 | 113 040 | 168 560 | 226 080 | 282 600 |
| 1 000 | 62 800 | 125 600 | 188 400 | 251 200 | 314 000 |

Table 5-2. Continued.

| Frequency (kHz) | Inductance (mH) | | | | |
|---|---|---|---|---|---|
| | 60 | 70 | 80 | 90 | 100 |
| 1 | 376.8 | 439.6 | 502.4 | 565.2 | 628 |
| 2 | 753.6 | 879.2 | 1 004.8 | 1 130.4 | 1 256 |
| 3 | 1 130.4 | 1 318.8 | 1 507.2 | 1 695.6 | 1 884 |
| 4 | 1 507.2 | 1 758.4 | 2 009.6 | 2 260.8 | 2 512 |
| 5 | 1 884 | 2 198 | 2 512 | 2 826 | 3 140 |
| 6 | 2 260.8 | 2 637.6 | 3 014.4 | 3 391.2 | 3 768 |
| 7 | 2 637.6 | 3 077.2 | 3 516.8 | 3 956.4 | 4 396 |
| 8 | 3 014.4 | 3 516.8 | 4 019.2 | 4 521.6 | 5 024 |
| 9 | 3 391.2 | 3 856.4 | 4 521.6 | 5 086.8 | 5 652 |
| 10 | 3 768 | 4 396 | 5 024 | 5 652 | 6 280 |
| 20 | 7 536 | 8 792 | 10 048 | 11 304 | 12 560 |
| 25 | 9 420 | 10 990 | 12 560 | 14 130 | 15 700 |
| 30 | 11 304 | 13 188 | 15 072 | 16 956 | 18 840 |
| 40 | 15 072 | 17 584 | 20 096 | 22 608 | 25 120 |
| 50 | 18 840 | 21 980 | 25 120 | 28 260 | 31 400 |
| 60 | 22 608 | 26 376 | 30 144 | 33 912 | 37 680 |
| 70 | 26 376 | 30 772 | 35 168 | 39 564 | 43 960 |
| 80 | 30 144 | 35 168 | 40 192 | 45 216 | 50 240 |
| 90 | 33 912 | 38 564 | 45 216 | 50 868 | 56 520 |
| 100 | 37 680 | 43 960 | 50 240 | 56 520 | 62 800 |
| 150 | 56 520 | 65 940 | 75 360 | 84 780 | 94 200 |
| 200 | 75 360 | 87 920 | 100 480 | 113 040 | 125 600 |
| 250 | 94 200 | 109 900 | 125 600 | 141 300 | 157 000 |
| 300 | 113 040 | 131 880 | 150 720 | 169 560 | 188 400 |
| 350 | 131 880 | 153 860 | 175 840 | 197 820 | 219 800 |
| 400 | 150 720 | 175 840 | 200 960 | 226 080 | 251 200 |
| 450 | 169 560 | 192 820 | 226 080 | 254 340 | 282 600 |
| 500 | 188 400 | 219 800 | 251 200 | 282 600 | 314 000 |
| 550 | 207 240 | 241 780 | 276 320 | 310 860 | 345 400 |
| 600 | 226 080 | 263 760 | 301 440 | 339 120 | 376 800 |
| 650 | 244 920 | 285 740 | 326 560 | 367 380 | 408 200 |
| 700 | 263 760 | 307 720 | 351 680 | 395 640 | 439 600 |
| 800 | 301 440 | 351 680 | 401 920 | 452 160 | 502 400 |
| 900 | 339 120 | 385 640 | 452 160 | 508 680 | 565 200 |
| 1 000 | 376 800 | 439 600 | 502 400 | 565 200 | 628 000 |

### Table 5-3. Inductive reactance (ohms).

| Frequency (Hz) | Inductance (H) | | | | |
|---|---|---|---|---|---|
| | 0.001 | 0.002 | 0.003 | 0.005 | 0.01 |
| 25 | 0.157 0 | 0.314 0 | 0.471 0 | 0.785 | 1.570 |
| 30 | 0.188 4 | 0.376 8 | 0.565 2 | 0.942 | 1.884 |
| 40 | 0.251 2 | 0.502 4 | 0.753 6 | 1.256 | 2.512 |
| 45 | 0.282 6 | 0.565 2 | 0.847 8 | 1.413 | 2.826 |
| 50 | 0.314 0 | 0.628 0 | 0.942 0 | 1.570 | 3.140 |
| 55 | 0.345 4 | 0.690 8 | 1.036 2 | 1.727 | 3.454 |
| 60 | 0.376 8 | 0.753 6 | 1.130 4 | 1.884 | 3.768 |
| 65 | 0.408 2 | 0.816 4 | 1.224 6 | 2.041 | 4.082 |
| 70 | 0.439 6 | 0.879 2 | 1.318 8 | 2.198 | 4.396 |
| 75 | 0.471 0 | 0.942 0 | 1.413 0 | 2.355 | 4.710 |
| 80 | 0.502 4 | 1.004 8 | 1.507 2 | 2.512 | 5.024 |
| 85 | 0.533 8 | 1.067 6 | 1.601 4 | 2.669 | 5.338 |
| 90 | 0.565 2 | 1.130 4 | 1.695 6 | 2.826 | 5.652 |
| 95 | 0.596 6 | 1.193 2 | 1.789 8 | 2.983 | 5.966 |
| 100 | 0.628 0 | 1.256 0 | 1.884 0 | 3.140 | 6.280 |
| 110 | 0.690 8 | 1.381 6 | 2.072 4 | 3.454 | 6.908 |
| 120 | 0.753 6 | 1.507 2 | 2.260 8 | 3.768 | 7.536 |
| 150 | 0.942 | 1.884 0 | 2.826 | 4.710 | 9.420 |
| 175 | 1.099 | 2.198 0 | 3.297 | 5.495 | 10.990 |
| 200 | 1.256 | 2.512 0 | 3.768 | 6.280 | 12.560 |
| 250 | 1.570 | 3.140 | 4.710 | 7.85 | 15.70 |
| 300 | 1.884 | 3.768 | 5.652 | 9.42 | 18.84 |
| 350 | 2.198 | 4.396 | 6.594 | 10.99 | 21.98 |
| 400 | 2.512 | 5.024 | 7.536 | 12.56 | 25.12 |
| 500 | 3.140 | 6.280 | 9.420 | 15.70 | 31.40 |
| 550 | 3.454 | 6.908 | 10.362 | 17.27 | 34.54 |
| 600 | 3.768 | 7.536 | 11.304 | 18.84 | 37.68 |
| 650 | 4.082 | 8.164 | 12.246 | 20.41 | 40.82 |
| 700 | 4.396 | 8.792 | 13.188 | 21.98 | 43.96 |
| 750 | 4.710 | 9.420 | 14.130 | 23.55 | 47.10 |
| 800 | 5.024 | 10.048 | 15.072 | 25.12 | 50.24 |
| 850 | 5.338 | 10.676 | 16.014 | 26.69 | 53.38 |
| 900 | 5.652 | 11.304 | 16.956 | 28.26 | 56.52 |
| 950 | 5.966 | 11.932 | 17.898 | 29.83 | 59.66 |
| 1 000 | 6.280 | 12.560 | 18.840 | 31.40 | 62.80 |

*Table 5-3. Continued.*

| Frequency (Hz) | Inductance (H) | | | | |
|---|---|---|---|---|---|
| | **1** | **2** | **3** | **4** | **5** |
| 25 | 157.0 | 314.0 | 471.0 | 628.0 | 785 |
| 30 | 188.4 | 376.8 | 565.2 | 753.6 | 942 |
| 35 | 219.8 | 439.6 | 659.4 | 879.2 | 1 099 |
| 40 | 251.2 | 502.4 | 753.6 | 1 004.8 | 1 256 |
| 45 | 282.6 | 565.2 | 847.8 | 1 130.4 | 1 413 |
| 50 | 314.0 | 628.0 | 942.0 | 1 256.0 | 1 570 |
| 55 | 345.4 | 690.8 | 1 036.2 | 1 318.6 | 1 727 |
| 60 | 376.8 | 753.6 | 1 130.4 | 1 507.2 | 1 884 |
| 65 | 408.2 | 816.4 | 1 224.6 | 1 632.8 | 2 041 |
| 70 | 439.6 | 879.2 | 1 318.8 | 1 758.4 | 2 198 |
| 75 | 471.0 | 942.0 | 1 413.0 | 1 884.0 | 2 355 |
| 80 | 502.4 | 1 004.8 | 1 507.2 | 2 009.6 | 2 512 |
| 85 | 533.8 | 1 067.6 | 1 601.4 | 2 135.2 | 2 669 |
| 90 | 565.2 | 1 130.4 | 1 695.6 | 2 260.8 | 2 826 |
| 95 | 596.6 | 1 193.2 | 1 789.8 | 2 386.4 | 2 983 |
| 100 | 628.0 | 1 256.0 | 1 884.0 | 2 512.0 | 3 140 |
| 120 | 753.6 | 1 507.2 | 2 260.8 | 3 014.4 | 3 768 |
| 200 | 1 256.0 | 2 512.0 | 3 768.0 | 5 024.0 | 6 280 |
| 240 | 1 507.2 | 3 014.4 | 4 521.6 | 6 028.8 | 7 536 |
| 250 | 1 570.0 | 3 140.0 | 4 710.0 | 6 280.0 | 7 850 |
| 300 | 1 884 | 3 768 | 5 652 | 7 536 | 9 420 |
| 350 | 2 198 | 4 396 | 6 594 | 8 792 | 10 990 |
| 360 | 2 260 | 4 522 | 6 783 | 9 044 | 11 305 |
| 400 | 2 512 | 5 024 | 7 536 | 10 048 | 12 560 |
| 450 | 2 826 | 5 652 | 8 478 | 11 304 | 14 130 |
| 500 | 3 140 | 6 280 | 9 420 | 12 560 | 15 700 |
| 550 | 3 454 | 6 908 | 10 362 | 13 816 | 17 270 |
| 600 | 3 768 | 7 536 | 11 304 | 15 072 | 18 840 |
| 650 | 4 082 | 8 164 | 12 246 | 16 328 | 20 410 |
| 700 | 4 396 | 8 792 | 13 188 | 17 584 | 21 980 |
| 750 | 4 710 | 9 420 | 14 130 | 18 840 | 23 550 |
| 800 | 5 024 | 10 048 | 15 072 | 20 096 | 25 120 |
| 850 | 5 338 | 10 676 | 16 014 | 21 352 | 26 690 |
| 900 | 5 652 | 11 304 | 16 956 | 22 608 | 28 260 |
| 1 000 | 6 280 | 12 560 | 18 840 | 25 120 | 31 400 |

*Table 5-3. Continued.*

| Frequency (kHz) | Inductance (H) | | | | |
|---|---|---|---|---|---|
| | **6** | **7** | **8** | **9** | **10** |
| 25 | 942.0 | 1 099.0 | 1 256.0 | 1413.0 | 1 570 |
| 30 | 1 130.4 | 1 318.8 | 1 507.2 | 1 695.6 | 1 884 |
| 35 | 1 318.8 | 1 538.6 | 1 758.4 | 1 978.2 | 2 198 |
| 40 | 1 507.2 | 1 758.4 | 2 009.6 | 2 260.8 | 2 512 |
| 45 | 1 695.6 | 1 978.2 | 2 260.8 | 2 543.4 | 2 826 |
| 50 | 1 884.0 | 2 198.0 | 2 512.0 | 2 826.0 | 3 140 |
| 55 | 2 072.4 | 2 417.8 | 2 763.2 | 3 108.6 | 3 454 |
| 60 | 2 260.8 | 2 637.6 | 3 014.4 | 3 391.2 | 3 768 |
| 65 | 2 449.2 | 2 857.4 | 3 265.6 | 3 673.8 | 4 082 |
| 70 | 2 637.6 | 3 077.2 | 3 516.8 | 3 956.4 | 4 396 |
| 75 | 2 826.0 | 3 297.0 | 3 768.0 | 4 239.0 | 4 710 |
| 80 | 3 014.4 | 3 516.8 | 4 019.2 | 4 521.6 | 5 024 |
| 85 | 3 202.8 | 3 736.6 | 4 270.4 | 4 804.2 | 5 338 |
| 90 | 3 391.2 | 3 956.4 | 4 521.6 | 5 086.8 | 5 652 |
| 95 | 3 579.6 | 4 176.2 | 4 772.8 | 5 396.4 | 5 966 |
| 100 | 3 768.0 | 4 396.0 | 5 024.0 | 5 652.0 | 6 280 |
| 120 | 4 521.6 | 5 275.2 | 6 028.8 | 6 782.4 | 7 536 |
| 200 | 7 536.0 | 8 792.0 | 10 048.0 | 11 304 | 12 560 |
| 240 | 9 043.2 | 10 550.4 | 12 057.6 | 13 565 | 15 072 |
| 250 | 9 420.0 | 10 990.0 | 12 460.0 | 14 130 | 15 700 |
| 300 | 11 304 | 13 188 | 15 072 | 16 956 | 18 840 |
| 350 | 13 188 | 15 386 | 17 584 | 19 782 | 21 980 |
| 360 | 13 414 | 15 826 | 18 086 | 20 347 | 22 608 |
| 400 | 15 072 | 17 584 | 20 096 | 22 608 | 25 120 |
| 450 | 16 956 | 19 782 | 22 608 | 25 434 | 28 260 |
| 500 | 18 840 | 21 980 | 25 120 | 28 260 | 31 400 |
| 550 | 20 724 | 24 178 | 27 632 | 31 086 | 34 540 |
| 600 | 22 608 | 26 376 | 30 124 | 33 912 | 37 680 |
| 650 | 24 492 | 28 574 | 32 656 | 36 738 | 40 820 |
| 700 | 26 376 | 30 772 | 35 168 | 39 564 | 43 960 |
| 750 | 28 260 | 32 970 | 37 680 | 42 390 | 47 100 |
| 800 | 30 144 | 35 168 | 40 192 | 45 216 | 50 240 |
| 850 | 32 028 | 37 366 | 42 704 | 48 042 | 53 380 |
| 900 | 33 912 | 39 564 | 45 216 | 50 868 | 56 520 |
| 1 000 | 37 680 | 43 960 | 50 240 | 56 520 | 62 800 |

Table 5-4. Inductive reactance at spot frequencies.

| Induction | 50 Hz | 100 Hz | 1 kHz | 10 kHz | 100 kHz | 1 MHz | 10 MHz | 100 MHz |
|---|---|---|---|---|---|---|---|---|
| | | | | | **Frequency** | | | |
| 1 µH | — | — | — | — | 0.63 | 6.3 | 63 | 630 |
| 5 µH | — | — | — | 0.31 | 3.1 | 31 | 310 | 3.1 k |
| 10 µH | — | — | — | 0.63 | 6.3 | 63 | 630 | 6.3 k |
| 50 µH | — | — | 0.31 | 3.1 | 31 | 310 | 3.1 k | 31 k |
| 100 µH | — | — | 0.63 | 6.3 | 63 | 630 | 6.3 k | 63 k |
| 250 µH | — | 0.16 | 1.6 | 16 | 160 | 1.6 k | 16 k | 160 k |
| 1 mH | 0.31 | 0.63 | 6.3 | 63 | 630 | 6.3 k | 63 k | 630 k |
| 2.5 mH | 0.8 | 1.6 | 16 | 160 | 1.6 k | 16 k | 160 k | 1.6 M |
| 10 mH | 3.1 | 6.3 | 63 | 630 | 6.3 k | 63 k | 630 k | 6.3 M |
| 25 mH | 8 | 16 | 160 | 1.6 k | 16 k | 160 k | 1.6 M | — |
| 100 mH | 31 | 63 | 630 | 6.3 k | 63 k | 630 k | 6.3 M | — |
| 1 H | 310 | 630 | 6.3 k | 63 k | 630 k | 6.3 M | — | — |
| 5 H | 1.5 k | 3.1 k | 31 k | 310 k | 3.1 M | — | — | — |
| 10 H | 3.1 k | 6.3 k | 63 k | 630 k | 6.3 M | — | — | — |
| 100 H | 31 k | 63 k | 630 k | 6.3 M | — | — | — | — |

Values in ohms.

answer. You could also get the same result by locating 75 Hz in the table, finding the reactance of 471 Ω and multiplying this result by 2. In either case the answer is 942 Ω.

# Single-layer air-core coils

There are various ways of determining the inductance of a coil. Possibly the easiest is to use an inductance bridge. If one isn't available, it is possible to get a fair approximation from the physical dimensions of the coil. Figure 5-4 shows a single-layer air-core coil. Its inductance can be calculated from:

$$L = \frac{(N \times r)^2}{9r + 10\ell}$$

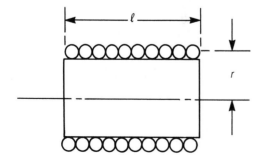

Fig. 5-4.   Single-layer air-core coil.

$L$ is the inductance of the coil in microhenries, N is the number of turns of the wire, $r$ is the radius of the coil form including the wire radius (measured in inches), and $\ell$ is the length of the coil in inches from the outside of the first turn to the outside of the last turn.

In some instances the inductance is known, but the dimensions of the coil need to be calculated. This can be done from:

$$N = \frac{\sqrt{L(9r+10\ell)}}{r}$$

## Multilayer air-core coils

If a coil is to have a number of layers (Fig. 5-5), its approximate value of inductance can be determined from:

$$L = \frac{0.8(N \times r)^2}{6r + 9\ell + 10b}$$

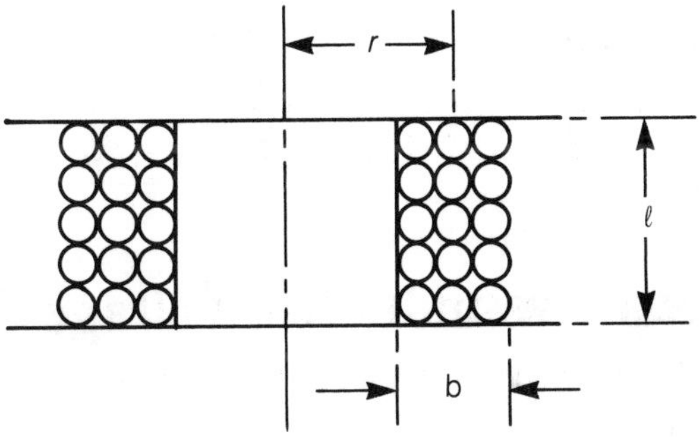

*Fig. 5-5. Multilayer air-core coil.*

$L$ is the inductance in microhenries, N is the number of turns, $r$ is the radius of the coil (including the radius of all layers), $\ell$ is the length of the coil, and $b$ is the depth of the coil winding. All measurements are in inches.

## Ohm's law for an inductive circuit

Inductive reactance, measured in ohms, can be substituted directly into the Ohm's law formula:

$$E_t = I \times X_L$$

The voltage across a coil is equal to the alternating current through it multiplied by the inductive reactance. To find the current through the coil, rearrange the formula to read:

$$I = E_L / X_L = E_L / 6.28 \times f \times L$$

and

$$X_L = E_t / I$$

# Inductive voltage divider

When two coils are connected in series and there is no interaction between their magnetic fields, the voltage division of the source voltage is:

$$E_2 = E_1 \times \frac{L_1}{L_1 + L_2}$$

The voltage is in volts and the inductance is in henries. The arrangement of the voltage divider is shown in Fig. 5-6.

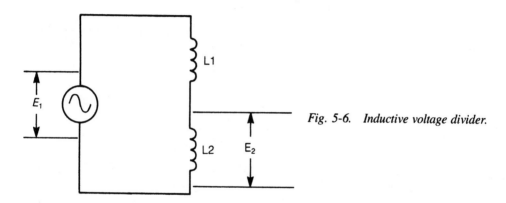

Fig. 5-6.   *Inductive voltage divider.*

# Resistive-inductive voltage divider

If $L_1$ in Fig. 5-6 is replaced by a resistor, the voltage across $L_2$ becomes:

$$E_2 = E_1 \times \frac{X_1}{\sqrt{R^2 + X_L^2}}$$

The circuit is shown in Fig. 5-7.

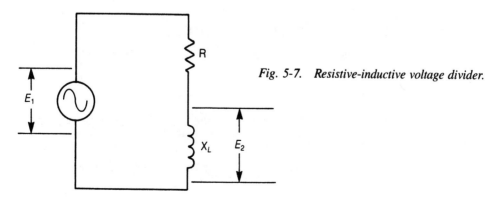

Fig. 5-7.   *Resistive-inductive voltage divider.*

# Capacitive-inductive voltage divider

Figure 5-8 shows a capacitive-inductive voltage divider. The voltage across the coil is:

$$E_2 = E_1 \times \frac{X_L}{X_L - X_C}$$

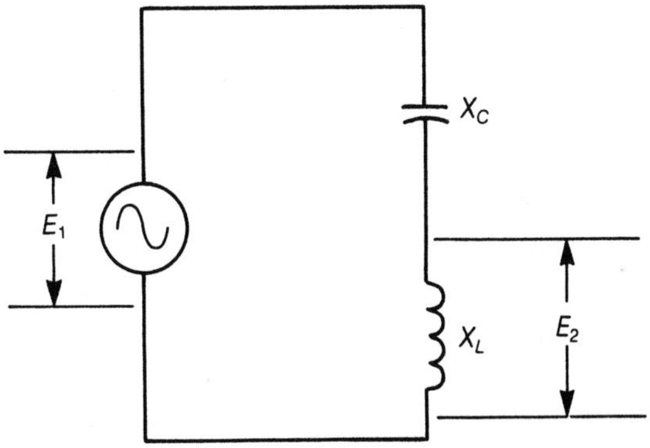

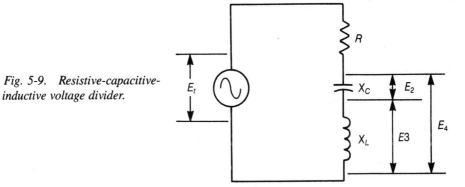

*Fig. 5-8.   Inductive-capacitive voltage divider.*

# Resistive-capacitive-inductive voltage divider

In a series circuit (Fig. 5-9), the voltages across the individual reactances and also the total reactive voltage is expressed in these formulas:

$$E_2 = E_1 \times \frac{X_C}{\sqrt{R^2 + (X_L - X_C)^2}}$$

*Fig. 5-9.   Resistive-capacitive-inductive voltage divider.*

Note that the denominator of this expression is the impedance of the series circuit and so this formula can be simplified to:

$$E_2 = E_1 \times \frac{X_C}{Z}$$

Thus, the voltage across the capacitor is equal to the source voltage, $E_1$, multiplied by the ratio of the capacitive reactance to the impedance. The voltage across the coil is $E_3$.

$$E_3 = E_1 \times \frac{X_L}{\sqrt{R^2 + (X_L - X_C)^2}}$$

As in the case of the voltage across the capacitor, the voltage across the coil is the source voltage multiplied by the ratio of inductive reactance to the series impedance. This formula can also be simplified to:

$$E_3 = E_1 \times \frac{X_L}{Z}$$

The voltage, $E_4$, across the two reactances, the coil and the capacitor is:

$$E_4 = E_1 \times \frac{X_L - X_C}{\sqrt{R^2 + (X_L - X_C)^2}}$$

$X_L - X_C$ is the net reactance. If the capacitive reactance is greater than the inductive reactance the numerator becomes $X_C - X_L$. In either case the net reactance is represented by $X$. The formula simplifies to:

$$E_4 = E_1 \times \frac{X}{Z}$$

# Inductive-capacitive voltage divider

In an LC voltage divider (Fig. 5-10) a proportion can be set up between voltages and reactances.

$$E_C : E_S = X_C : X_t$$

$E_S$ is the voltage from the cathode (K) of the rectifier to the minus ($-$) bus. The voltage is dc with an ac component. $X_t$ is the total reactance. $X_t = X_L - X_C$. The rip-

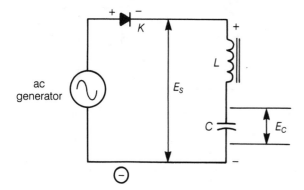

Fig. 5-10. Inductive-capacitive voltage divider following rectifier circuit.

ple frequency in hertz from K to the minus bus is the same as the ac generator frequency.

# Voltages in a series resistive-capacitive circuit

Whether ac or dc, the current flowing in a series RC circuit is the same in all parts of the circuit. If each of the terms in the formula is multiplied by $I$ (representing the current in the circuit):

$$IZ=\sqrt{I^2R^2+I^2X_C^2}$$

or

$$E_{source}=\sqrt{E_R^2+E_C^2}$$

$IZ$ is the generator or source voltage. That is:

$$E=I\times Z$$

By transposing, two other forms of this formula are obtained:

$$I=E/Z$$

and

$$Z=E/I$$

# Resistive-capacitive voltage divider

Figure 5-11 shows an RC voltage divider. In this arrangement, the formula for the voltage across the capacitor is:

$$E_2=\frac{X_C}{\sqrt{R^2+X_C^2}}\times E_1$$

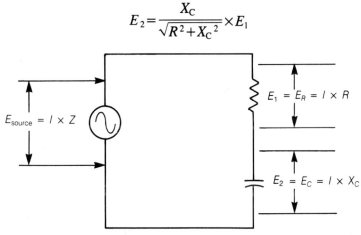

Fig. 5-11. *Voltage distribution in a series RC circuit. The source voltage is equal to the vector sum of the voltages across the resistor and capacitor.*

The denominator is the impedance of this circuit and so this formula can be simplified to:

$$E_2 = \frac{X_C}{Z} \times E_1$$

# Inductive-capacitive product for resonance

Table 5-5 shows the relationship between the wavelength in meters, the frequency in kilohertz and the inductive-capacitance product ($L \times C$) required to produce resonance. The inductance is in microhenries and the capacitance is in microfarads.

Table 5-5. Inductive-capacitive product for resonance.

| Wave-length (m) | Frequency (kHz) | $L \times C$ | Wave-length (m) | Frequency (kHz) | $L \times C$ |
|---|---|---|---|---|---|
| 1 | 300 000 | 0.000 000 3 | 200 | 1 500 | 0.011 26 |
| 2 | 150 000 | 0.000 011 1 | 210 | 1 429 | 0.012 41 |
| 3 | 100 000 | 0.000 001 8 | | | |
| 4 | 75 000 | 0.000 004 5 | 220 | 1 364 | 0.013 62 |
| 5 | 60 000 | 0.000 005 7 | 230 | 1 304 | 0.014 89 |
| | | | 240 | 1 250 | 0.016 21 |
| 6 | 50 000 | 0.000 010 1 | 250 | 1 200 | 0.017 59 |
| 7 | 42 900 | 0.000 013 8 | 260 | 1 154 | 0.019 03 |
| 8 | 37 500 | 0.000 018 0 | | | |
| 9 | 33 333 | 0.000 022 8 | 270 | 1 112 | 0.020 5 |
| 10 | 30 000 | 0.000 028 2 | 280 | 1 071 | 0.022 1 |
| | | | 290 | 1 034 | 0.023 7 |
| 20 | 15 000 | 0.000 112 9 | 300 | 1 000 | 0.025 3 |
| 30 | 10 000 | 0.000 253 0 | 310 | 968 | 0.027 0 |
| 40 | 7 500 | 0.000 450 0 | | | |
| 50 | 6 000 | 0.000 704 0 | 320 | 938 | 0.028 8 |
| 60 | 5 000 | 0.001 014 0 | 330 | 909 | 0.030 6 |
| | | | 340 | 883 | 0.032 5 |
| 70 | 4 290 | 0.001 378 0 | 350 | 857 | 0.034 5 |
| 80 | 3 750 | 0.001 801 0 | 360 | 834 | 0.036 5 |
| 90 | 3 333 | 0.002 280 0 | | | |
| 100 | 3 000 | 0.002 82 | 370 | 811 | 0.038 5 |
| 110 | 2 727 | 0.003 41 | 380 | 790 | 0.040 6 |
| | | | 390 | 769 | 0.042 8 |
| 120 | 2 500 | 0.004 05 | 400 | 750 | 0.045 0 |
| 130 | 2 308 | 0.004 76 | 410 | 732 | 0.047 3 |
| 140 | 2 143 | 0.005 52 | | | |
| 150 | 2 000 | 0.006 33 | 420 | 715 | 0.049 6 |
| 160 | 1 875 | 0.007 21 | 430 | 698 | 0.052 0 |
| | | | 440 | 682 | 0.054 5 |
| 170 | 1 764 | 0.008 13 | 450 | 667 | 0.057 0 |
| 180 | 1 667 | 0.009 12 | 460 | 652 | 0.059 6 |
| 190 | 1 579 | 0.010 15 | 470 | 639 | 0.062 2 |

*Table 5-5. Continued.*

| Wave-length (m) | Frequency (kHz) | $L \times C$ | Wave-length (m) | Frequency (kHz) | $L \times C$ |
|---|---|---|---|---|---|
| 480 | 625 | 0.064 9 | 695 | 432 | 0.136 0 |
| 490 | 612 | 0.067 6 | 700 | 429 | 0.137 9 |
| 500 | 600 | 0.070 4 | 705 | 426 | 0.139 9 |
| 505 | 594 | 0.071 8 | 710 | 423 | 0.141 9 |
| 510 | 588 | 0.073 2 | 715 | 420 | 0.143 9 |
| 515 | 583 | 0.074 7 | 720 | 417 | 0.145 9 |
| 520 | 577 | 0.076 1 | 725 | 414 | 0.147 9 |
| 525 | 572 | 0.077 6 | 730 | 411 | 0.150 0 |
| 530 | 566 | 0.079 1 | 735 | 408 | 0.152 1 |
| 535 | 561 | 0.080 6 | 740 | 405 | 0.154 1 |
| 540 | 556 | 0.082 1 | 745 | 403 | 0.156 2 |
| 545 | 551 | 0.083 6 | 750 | 400 | 0.158 3 |
| 550 | 546 | 0.085 2 | 755 | 397 | 0.160 4 |
| 555 | 541 | 0.086 7 | 760 | 395 | 0.162 6 |
| 560 | 536 | 0.088 3 | 765 | 392 | 0.164 7 |
| 565 | 531 | 0.089 9 | 770 | 390 | 0.166 9 |
| 570 | 527 | 0.091 5 | 775 | 387 | 0.169 0 |
| 575 | 522 | 0.093 1 | 780 | 385 | 0.171 2 |
| 580 | 517 | 0.094 7 | 785 | 382 | 0.173 4 |
| 585 | 513 | 0.096 3 | 790 | 380 | 0.175 6 |
| 590 | 509 | 0.098 0 | 795 | 377 | 0.177 9 |
| 595 | 504 | 0.099 6 | 800 | 375 | 0.180 1 |
| 600 | 500 | 0.101 3 | 805 | 373 | 0.182 4 |
| 605 | 496 | 0.103 0 | 810 | 370 | 0.184 7 |
| 610 | 492 | 0.104 7 | 815 | 368 | 0.187 0 |
| 615 | 488 | 0.106 5 | 820 | 366 | 0.189 3 |
| 620 | 484 | 0.108 2 | 825 | 364 | 0.191 6 |
| 625 | 480 | 0.110 0 | 830 | 361 | 0.193 9 |
| 630 | 476 | 0.111 7 | 835 | 359 | 0.196 2 |
| 635 | 472 | 0.113 5 | 840 | 357 | 0.198 6 |
| 640 | 469 | 0.115 3 | 845 | 355 | 0.201 |
| 645 | 465 | 0.117 1 | 850 | 353 | 0.203 |
| 650 | 462 | 0.118 9 | 855 | 351 | 0.206 |
| 655 | 458 | 0.120 8 | 860 | 349 | 0.208 |
| 660 | 455 | 0.122 6 | 865 | 347 | 0.211 |
| 665 | 451 | 0.124 5 | 870 | 345 | 0.213 |
| 670 | 448 | 0.126 4 | 875 | 343 | 0.216 |
| 675 | 444 | 0.128 3 | 880 | 341 | 0.218 |
| 680 | 441 | 0.130 2 | 885 | 339 | 0.220 |
| 685 | 438 | 0.132 1 | 890 | 337 | 0.223 |
| 690 | 435 | 0.134 0 | 895 | 335 | 0.225 |

Table 5-5. Continued.

| Wave-length (m) | Frequency (kHz) | $L \times C$ | Wave-length (m) | Frequency (kHz) | $L \times C$ |
|---|---|---|---|---|---|
| 900 | 333 | 0.228 | 955 | 314 | 0.257 |
| 905 | 331 | 0.231 | | | |
| | | | 960 | 313 | 0.260 |
| 910 | 330 | 0.233 | 965 | 311 | 0.262 |
| 915 | 328 | 0.236 | 970 | 309 | 0.265 |
| 920 | 326 | 0.238 | 975 | 308 | 0.268 |
| 925 | 324 | 0.241 | 980 | 306 | 0.270 |
| 930 | 323 | 0.243 | | | |
| | | | 985 | 305 | 0.273 |
| 935 | 321 | 0.246 | 990 | 303 | 0.276 |
| 940 | 319 | 0.249 | 995 | 302 | 0.279 |
| 945 | 317 | 0.251 | 1 000 | 300 | 0.282 |
| 950 | 316 | 0.254 | | | |

In an LC circuit a condition of resonance is reached when the reactive elements are equal—that is, when the inductive and capacitive reactances are identical. When the inductance and the capacitance are both in microunits (microhenries and micro-farads), the resonant frequency will be in kilohertz.

To find the resonant frequency without resorting to formulas, multiply the values of $L$ and $C$, first converting these to microhenries and microfarads. Knowing the $LC$ product, find the resonant frequency by using Table 5-5. At the same time the table also supplies the wavelength in meters.

*Example* What is the wavelength in meters of a circuit when the inductance is 221 $\mu$H and the capacitance is 100 pF?

*Solution* Table 5-5 requires that the capacitance be in microfarads. 100 pF is equivalent to 0.000 1 $\mu$F. Multiply 221 by 0.000 1. The answer will be 0.022 1. Locate this value in the $L \times C$ column in the table. You will find that the frequency in kilohertz is 1 071 and the corresponding wavelength is 280 m.

*Example* What is the resonant frequency of a circuit whose capacitance has a value of 250 pF and an inductance of 136 $\mu$H?

*Solution* The inductance value can be used as it is, but the capacitance must be changed to microfarads. 250 pF is equivalent to 0.000 25 $\mu$F. The product of 0.000 25 and 136 is 0.034 0. The table does not list such a value, but it does have a value that is close. The $L \times C$ column shows this to be 0.034 5 with a corresponding frequency in kilohertz of 857. The wavelength is 350 m.

*Example* You want a circuit that will be resonant at 5 MHz. What must be the value of the $LC$ product?

*Solution* 5 MHz is equal to 5 000 kHz. The table shows that the *LC* product is 0.001 014 0. Any combination of inductance and capacitance producing this product will be resonant at 5 MHz.

# Resistive-capacitive time constants

RC networks are used in a variety of applications, including differentiating and integrating networks; emphasis and de-emphasis circuits; time delay elements in radio control receivers; oscillators; electronic switches and light flashers.

The time required to charge or discharge a capacitor depends on the amount of capacitance and the value of the associated resistor. The formula for the charge or discharge of a capacitor through a resistor is

$$t = R \times C$$

The resistance is in ohms and the capacitance is in farads. The time, t, is the time in seconds for the capacitor to charge to 63% of its maximum, or for the charge on the capacitor to drop to 37% of its maximum. Multiples are shown in Table 5-6.

Table 5-7 supplies the time constants for a capacitance range of 0.1 to 1 $\mu$F and from 0.1 to 10 s. Note that increasing the value of $R$ or $C$ will increase the time constant proportionately.

*Table 5-6. Multiples of*
*units for time constants.*

| Where<br>*R* is in | *C* is in | *L* is in | *T* is in |
|---|---|---|---|
| $\Omega$ | F | H | s |
| $\Omega$ | $\mu$F | $\mu$H | $\mu$s |
| k$\Omega$ | $\mu$F | H | ms |
| k$\Omega$ | — | mH | $\mu$s |
| M$\Omega$ | $\mu$F | — | s |
| M$\Omega$ | pF | H | $\mu$s |

Although the formula for RC time constants requires basic units, multiples can be used if conversions are made in accordance with Table 5-6.

*Example* How long will it take to charge a capacitor having a value of 0.2 $\mu$F through a 500 k$\Omega$ resistor?

*Solution* Locate 0.2 in the capacitance column in Table 5-7. Immediately beneath this value locate 500 k$\Omega$. Move to the left and you will see that it will take 0.1 s.

## Table 5-7. Resistive-capacitive time constants.

| Time (s) | Capacitance (μF) | | | | |
|---|---|---|---|---|---|
| | 0.1 | 0.2 | 0.3 | 0.4 | 0.5 |
| 0.1 | 1.0 MΩ | 500 kΩ | 333 kΩ | 250 kΩ | 200 kΩ |
| 0.15 | 1.5 MΩ | 750 kΩ | 500 kΩ | 375 kΩ | 300 kΩ |
| 0.2 | 2.0 MΩ | 1.00 MΩ | 666 kΩ | 500 kΩ | 400 kΩ |
| 0.25 | 2.5 MΩ | 1.25 MΩ | 833 kΩ | 625 kΩ | 500 kΩ |
| 0.3 | 3.0 MΩ | 1.50 MΩ | 1.00 MΩ | 750 kΩ | 600 kΩ |
| 0.35 | 3.5 MΩ | 1.75 MΩ | 1.17 MΩ | 875 kΩ | 700 kΩ |
| 0.4 | 4.0 MΩ | 2.00 MΩ | 1.33 MΩ | 1.00 MΩ | 800 kΩ |
| 0.45 | 4.5 MΩ | 2.25 MΩ | 1.50 MΩ | 1.13 MΩ | 900 kΩ |
| 0.5 | 5.0 MΩ | 2.50 MΩ | 1.67 MΩ | 1.25 MΩ | 1.0 MΩ |
| 0.55 | 5.5 MΩ | 2.75 MΩ | 1.83 MΩ | 1.38 MΩ | 1.1 MΩ |
| 0.6 | 6.0 MΩ | 3.00 MΩ | 2.00 MΩ | 1.50 MΩ | 1.2 MΩ |
| 0.65 | 6.5 MΩ | 3.25 MΩ | 2.17 MΩ | 1.63 MΩ | 1.3 MΩ |
| 0.7 | 7.0 MΩ | 3.50 MΩ | 2.33 MΩ | 1.75 MΩ | 1.4 MΩ |
| 0.75 | 7.5 MΩ | 3.75 MΩ | 2.50 MΩ | 1.88 MΩ | 1.5 MΩ |
| 0.8 | 8.0 MΩ | 4.00 MΩ | 2.67 MΩ | 2.00 MΩ | 1.6 MΩ |
| 0.85 | 8.5 MΩ | 4.25 MΩ | 2.83 MΩ | 2.13 MΩ | 1.7 MΩ |
| 0.9 | 9.0 MΩ | 4.50 MΩ | 3.00 MΩ | 2.25 MΩ | 1.8 MΩ |
| 0.95 | 9.5 MΩ | 4.75 MΩ | 3.17 MΩ | 2.38 MΩ | 1.9 MΩ |
| 1.0 | 10.0 MΩ | 5.00 MΩ | 3.33 MΩ | 2.50 MΩ | 2.0 MΩ |
| 1.5 | 15.0 MΩ | 7.50 MΩ | 5.00 MΩ | 3.75 MΩ | 3.0 MΩ |
| 2.0 | 20.0 MΩ | 10.00 MΩ | 6.66 MΩ | 5.00 MΩ | 4.0 MΩ |
| 2.5 | 25.0 MΩ | 12.50 MΩ | 8.33 MΩ | 6.25 MΩ | 5.0 MΩ |
| 3.0 | 30.0 MΩ | 15.00 MΩ | 10.00 MΩ | 7.50 MΩ | 6.0 MΩ |
| 3.5 | 35.0 MΩ | 17.50 MΩ | 11.66 MΩ | 8.75 MΩ | 7.0 MΩ |
| 4.0 | 40.0 MΩ | 20.00 MΩ | 13.33 MΩ | 10.00 MΩ | 8.0 MΩ |
| 4.5 | 45.0 MΩ | 22.50 MΩ | 15.00 MΩ | 11.25 MΩ | 9.0 MΩ |
| 5.0 | 50.0 MΩ | 25.00 MΩ | 16.67 MΩ | 12.50 MΩ | 10.0 MΩ |
| 5.5 | 55.0 MΩ | 27.50 MΩ | 18.33 MΩ | 13.75 MΩ | 11.0 MΩ |
| 6.0 | 60.0 MΩ | 30.00 MΩ | 20.00 MΩ | 15.00 MΩ | 12.0 MΩ |
| 6.5 | 65.0 MΩ | 32.50 MΩ | 21.67 MΩ | 16.25 MΩ | 13.0 MΩ |
| 7.0 | 70.0 MΩ | 35.00 MΩ | 23.33 MΩ | 17.50 MΩ | 14.0 MΩ |
| 7.5 | 75.0 MΩ | 37.50 MΩ | 25.00 MΩ | 18.75 MΩ | 15.0 MΩ |
| 8.0 | 80.0 MΩ | 40.00 MΩ | 26.67 MΩ | 20.00 MΩ | 16.0 MΩ |
| 9.0 | 90.0 MΩ | 45.00 MΩ | 30.00 MΩ | 22.50 MΩ | 18.0 MΩ |
| 10.0 | 100.0 MΩ | 50.00 MΩ | 33.33 MΩ | 25.00 MΩ | 20.0 MΩ |

kΩ = kilohms         MΩ = megohms

*Table 5-7. Continued.*

| Time (s) | Capacitance ($\mu$F) | | | | |
|---|---|---|---|---|---|
| | **0.6** | **0.7** | **0.8** | **0.9** | **1.0** |
| 0.1 | 166 kΩ | 143 kΩ | 125 kΩ | 111 kΩ | 100 kΩ |
| 0.15 | 250 kΩ | 214 kΩ | 188 kΩ | 167 kΩ | 150 kΩ |
| 0.2 | 333 kΩ | 286 kΩ | 250 kΩ | 222 kΩ | 200 kΩ |
| 0.25 | 417 kΩ | 357 kΩ | 313 kΩ | 278 kΩ | 250 kΩ |
| 0.3 | 500 kΩ | 429 kΩ | 375 kΩ | 333 kΩ | 300 kΩ |
| 0.35 | 583 kΩ | 500 kΩ | 438 kΩ | 389 kΩ | 350 kΩ |
| 0.4 | 666 kΩ | 571 kΩ | 500 kΩ | 444 kΩ | 400 kΩ |
| 0.45 | 750 kΩ | 643 kΩ | 563 kΩ | 500 kΩ | 450 kΩ |
| 0.5 | 833 kΩ | 714 kΩ | 625 kΩ | 555 kΩ | 500 kΩ |
| 0.55 | 917 kΩ | 786 kΩ | 688 kΩ | 611 kΩ | 550 kΩ |
| 0.6 | 1.00 MΩ | 857 kΩ | 750 kΩ | 666 kΩ | 600 kΩ |
| 0.65 | 1.08 MΩ | 929 kΩ | 813 kΩ | 722 kΩ | 650 kΩ |
| 0.7 | 1.17 MΩ | 1.00 MΩ | 875 kΩ | 778 kΩ | 700 kΩ |
| 0.75 | 1.25 MΩ | 1.07 MΩ | 938 kΩ | 833 kΩ | 750 kΩ |
| 0.8 | 1.33 MΩ | 1.14 MΩ | 1.00 MΩ | 889 kΩ | 800 kΩ |
| 0.85 | 1.42 MΩ | 1.21 MΩ | 1.06 MΩ | 944 kΩ | 850 kΩ |
| 0.9 | 1.50 MΩ | 1.29 MΩ | 1.13 MΩ | 1.00 MΩ | 900 kΩ |
| 0.95 | 1.58 MΩ | 1.36 MΩ | 1.19 MΩ | 1.06 MΩ | 950 kΩ |
| 1.0 | 1.67 MΩ | 1.43 MΩ | 1.25 MΩ | 1.11 MΩ | 1.0 MΩ |
| 1.5 | 2.50 MΩ | 2.14 MΩ | 1.88 MΩ | 1.67 MΩ | 1.5 MΩ |
| 2.0 | 3.33 MΩ | 2.86 MΩ | 2.50 MΩ | 2.22 MΩ | 2.0 MΩ |
| 2.5 | 4.17 MΩ | 3.57 MΩ | 3.13 MΩ | 2.78 MΩ | 2.5 MΩ |
| 3.0 | 5.00 MΩ | 4.29 MΩ | 3.75 MΩ | 3.33 MΩ | 3.0 MΩ |
| 3.5 | 5.83 MΩ | 5.00 MΩ | 4.38 MΩ | 3.89 MΩ | 3.5 MΩ |
| 4.0 | 6.66 MΩ | 5.71 MΩ | 5.00 MΩ | 4.44 MΩ | 4.0 MΩ |
| 4.5 | 7.50 MΩ | 6.43 MΩ | 5.63 MΩ | 5.00 MΩ | 4.5 MΩ |
| 5.0 | 8.33 MΩ | 7.14 MΩ | 6.25 MΩ | 5.55 MΩ | 5.0 MΩ |
| 5.5 | 9.17 MΩ | 7.86 MΩ | 6.88 MΩ | 6.11 MΩ | 5.5 MΩ |
| 6.0 | 10.00 MΩ | 8.57 MΩ | 7.50 MΩ | 6.66 MΩ | 6.0 MΩ |
| 6.5 | 10.83 MΩ | 9.29 MΩ | 8.13 MΩ | 7.22 MΩ | 6.5 MΩ |
| 7.0 | 11.67 MΩ | 10.00 MΩ | 8.75 MΩ | 7.78 MΩ | 7.0 MΩ |
| 7.5 | 12.50 MΩ | 10.71 MΩ | 9.38 MΩ | 8.33 MΩ | 7.5 MΩ |
| 8.0 | 13.33 MΩ | 11.43 MΩ | 10.00 MΩ | 8.89 MΩ | 8.0 MΩ |
| 9.0 | 15.00 MΩ | 12.86 MΩ | 11.25 MΩ | 10.00 MΩ | 9.0 MΩ |
| 10.0 | 16.66 MΩ | 14.28 MΩ | 12.50 MΩ | 11.11 MΩ | 10.0 MΩ |

kΩ = kilohms          MΩ = megohms

*Example*   What combination of resistance and capacitance can be used to obtain a 1-s time constant?

*Solution*   Locate 1 s in the *time in seconds* column in Table 5-7. If you will now move to the right you will see that a number of possible combinations are available—10 MΩ and 0.1 μF; 5 MΩ and 0.2 μF; 3.3 MΩ and 0.3 μF, etc.

Time constants can be had for values other than those shown in the left-hand column of Table 5-6. Thus if the value of capacitance is reduced by half or if the value of resistance is similarly reduced (but not both at the same time) the value of the time constant is also lowered by half.

*Example*   What is the time constant for a value of capacitance of 0.05 μF and a resistance of 1 MΩ?

*Solution*   There is no column headed by 0.05 μF in Table 5-7 but there is a column marked 0.1. Divide this by 2 and you will have 0.05 μF. Move down from 0.1 (now considered as 0.05) until you reach 1.0 MΩ, the given value of resistance. Move directly to the left and you will see a time constant of 0.1 s. Divide this by 2 and the time constant for the values of $R$ and $C$ that were given will be 0.05 s. To get a time constant of 0.05 s, either reduce the resistance 1.0 MΩ to 500 kΩ or the capacitance to 0.05 μF. If you reduce both $R$ and $C$ by 50%, the value of the time constant becomes one-fourth of the value given in Table 5-7. Thus, for a value of 500 kΩ and a capacitance of 0.05 μF, the time constant is 0.025 s.

# Resistive-inductive time constants

RL circuits—circuits consisting of a resistor in series with an inductor—are used in timing circuits or in relays where the relay must make or break at predetermined times. The time constant of an RL circuit is based on the formula:

$$t = L/R$$

$t$ is the time in seconds, $L$ is the inductance in henries, and $R$ is the resistance in ohms. The time in seconds is that required for the current to reach 63% of its maximum value, or to fall to 37% of its maximum. Time constants of RL circuits are listed in Table 5-8.

The resistance of the coil wire must be considered when calculating the time constant. This resistance is regarded as acting in series with the coil. Thus, if the resistance of a coil is 10 Ω and a 100-Ω series resistor is required for an RL circuit, a value close to 90 Ω should be used. As a rule of thumb, if the resistance of the coil is 10% or more of the required resistance, it should be taken into consideration. In the case just mentioned, subtracting the 10 Ω of the coil from 100 Ω shows that a 90-Ω resistor would be required. If the value of the external resistor is in the order of kilohms, a coil whose resistance is just a few ohms would not seriously affect the time constant.

*Table 5-8. Resistive-inductive constants.*

| Time (s) | Inductance (H) | | | | |
|---|---|---|---|---|---|
| | **10** | **20** | **30** | **40** | **50** |
| 0.1 | 100.0 | 200.0 | 300.0 | 400.0 | 500.0 |
| 0.15 | 66.7 | 133.3 | 200.0 | 266.7 | 333.3 |
| 0.2 | 50.0 | 100.0 | 150.0 | 200.0 | 250.0 |
| 0.25 | 40.0 | 80.0 | 125.0 | 160.0 | 200.0 |
| 0.3 | 33.3 | 66.7 | 100.0 | 133.3 | 166.7 |
| 0.35 | 28.6 | 57.1 | 86.6 | 114.3 | 142.9 |
| 0.4 | 25.0 | 50.0 | 75.0 | 100.0 | 125.0 |
| 0.45 | 22.2 | 44.4 | 66.7 | 88.9 | 111.1 |
| 0.5 | 20.0 | 40.0 | 60.0 | 80.0 | 100.0 |
| 0.55 | 18.2 | 36.4 | 54.5 | 72.7 | 90.9 |
| 0.6 | 16.7 | 33.3 | 50.0 | 66.7 | 83.3 |
| 0.65 | 15.4 | 30.8 | 46.2 | 61.5 | 76.9 |
| 0.7 | 14.3 | 28.6 | 42.9 | 57.1 | 71.4 |
| 0.75 | 13.3 | 26.7 | 40.0 | 53.3 | 66.7 |
| 0.8 | 12.5 | 25.0 | 37.5 | 50.0 | 62.5 |
| 0.85 | 11.8 | 23.5 | 35.3 | 47.1 | 58.8 |
| 0.9 | 11.1 | 22.2 | 33.3 | 44.4 | 55.5 |
| 0.95 | 10.5 | 21.1 | 31.6 | 42.1 | 52.6 |
| 1.0 | 10.0 | 20.0 | 30.0 | 40.0 | 50.0 |
| 1.5 | 6.7 | 13.3 | 20.0 | 26.7 | 33.3 |
| 2.0 | 5.0 | 10.0 | 15.0 | 20.0 | 25.0 |
| 2.5 | 4.0 | 8.0 | 12.0 | 16.0 | 20.0 |
| 3.0 | 3.3 | 6.7 | 10.0 | 13.3 | 16.7 |
| 3.5 | 2.9 | 5.7 | 8.7 | 11.4 | 14.3 |
| 4.0 | 2.5 | 5.0 | 7.5 | 10.0 | 12.5 |
| 4.5 | 2.2 | 4.4 | 6.7 | 8.9 | 11.1 |
| 5.0 | 2.0 | 4.0 | 6.0 | 8.0 | 10.0 |
| 5.5 | 1.8 | 3.6 | 5.5 | 7.3 | 9.1 |
| 6.0 | 1.7 | 3.3 | 5.0 | 6.7 | 8.3 |
| 6.5 | 1.5 | 3.1 | 4.6 | 6.2 | 7.7 |
| 7.0 | 1.4 | 2.9 | 4.3 | 5.7 | 7.1 |
| 7.5 | 1.3 | 2.7 | 4.0 | 5.3 | 6.7 |
| 8.0 | 1.2 | 2.5 | 3.8 | 5.0 | 6.3 |
| 9.0 | 1.1 | 2.2 | 3.3 | 4.4 | 5.5 |
| 10.0 | 1.0 | 2.0 | 3.0 | 4.0 | 5.0 |

All resistance values in ohms

*Table 5-8. Continued.*

| Time | Inductance (H) | | | | |
|------|------|------|------|------|------|
| (s) | 60 | 70 | 80 | 90 | 100 |
| 0.1 | 600.0 | 700.0 | 800.0 | 900.0 | 1 000.0 |
| 0.15 | 400.0 | 466.7 | 533.3 | 600.0 | 666.7 |
| 0.2 | 300.0 | 350.0 | 400.0 | 450.0 | 500.0 |
| 0.25 | 240.0 | 280.0 | 320.0 | 360.0 | 400.0 |
| 0.3 | 200.0 | 233.3 | 266.6 | 300.0 | 333.3 |
| 0.35 | 171.4 | 200.0 | 228.6 | 257.1 | 285.7 |
| 0.4 | 150.0 | 175.0 | 200.0 | 225.0 | 250.0 |
| 0.45 | 133.3 | 155.6 | 177.8 | 200.0 | 222.2 |
| 0.5 | 120.0 | 140.0 | 160.0 | 180.0 | 200.0 |
| 0.55 | 109.1 | 127.3 | 145.5 | 163.6 | 181.8 |
| 0.6 | 100.0 | 116.7 | 133.3 | 150.0 | 166.7 |
| 0.65 | 92.3 | 107.7 | 123.1 | 138.5 | 153.8 |
| 0.7 | 85.7 | 100.0 | 114.3 | 128.7 | 142.9 |
| 0.75 | 80.0 | 93.3 | 106.7 | 120.0 | 133.3 |
| 0.8 | 75.0 | 87.5 | 100.0 | 112.5 | 125.0 |
| 0.85 | 70.6 | 82.3 | 94.1 | 105.9 | 117.6 |
| 0.9 | 66.6 | 77.8 | 88.9 | 100.0 | 111.1 |
| 0.95 | 63.2 | 73.7 | 84.2 | 94.7 | 105.3 |
| 1.0 | 60.0 | 70.0 | 80.0 | 90.0 | 100.0 |
| 1.5 | 40.0 | 46.7 | 53.3 | 60.0 | 66.7 |
| 2.0 | 30.0 | 35.0 | 40.0 | 45.0 | 50.0 |
| 2.5 | 24.0 | 28.0 | 32.0 | 36.0 | 40.0 |
| 3.0 | 20.0 | 23.3 | 26.7 | 30.0 | 33.3 |
| 3.5 | 17.1 | 20.0 | 22.9 | 25.7 | 28.6 |
| 4.0 | 15.0 | 17.5 | 20.0 | 22.5 | 25.0 |
| 4.5 | 13.3 | 15.6 | 17.8 | 20.0 | 22.2 |
| 5.0 | 12.0 | 14.0 | 16.0 | 18.0 | 20.0 |
| 5.5 | 10.9 | 12.7 | 14.6 | 16.4 | 18.2 |
| 6.0 | 10.0 | 11.7 | 13.3 | 15.0 | 16.7 |
| 6.5 | 9.2 | 10.8 | 12.3 | 13.9 | 15.4 |
| 7.0 | 8.6 | 10.0 | 11.4 | 12.9 | 14.3 |
| 7.5 | 8.0 | 9.3 | 10.7 | 12.0 | 13.3 |
| 8.0 | 7.5 | 8.8 | 10.0 | 11.3 | 12.5 |
| 9.0 | 6.7 | 7.8 | 8.9 | 10.0 | 11.1 |
| 10.0 | 6.0 | 7.0 | 8.0 | 9.0 | 10.0 |

All resistance values in ohms

*Example* A relay coil whose internal resistance is negligible, and whose inductance is 0.05 H, will make (contacts will close) when the current through the coil reaches 63% of its peak. What value of series resistor is needed to have the relay close 0.008 s after the circuit is on?

*Solution* Note that Table 5-8 does not include a time of 0.008 s nor an inductance of 0.05 H. However, it is easy to extend the table by moving the decimal point. Change 0.008 to 8 by moving its decimal point three places to the right. The inductance, 0.05 H, will then become 50. Locate the number 8 in the time column. Move across to the right to reach the 50 column. The answer is 6.3 Ω. The decimal point does not need to be moved in the answer. The reason for this is based on the time-constant formula, $t=L/R$. Equal increases or decreases in $L$ and $R$ will have no effect on the time constant.

*Example* An RL circuit has an inductance of 30 H and a resistance of 42.9 Ω. What is its time constant?

*Solution* In Table 5-8 locate the column marked with the number 30 at the top. Move down in this column until you reach the number 42.9. Then move to the left to the time (s) column, and you will find the time constant of this circuit is 0.7 s.

If you were to do this problem by using the formula, $t=L/R$, you would have t=30/42.9=0.699 3 s. The difference between the two answers is 0.7−0.699 3 =0.000 7 s, an extremely small percentage of error.

# Q of a coil

The $Q$ of a coil is not expressed in any particular unit but simply appears as a whole decimal number. There are no multiples or submultiples, nor does it appear in conversion tables. The letter $Q$ is an abbreviation for quality, or it is sometimes referred to as the figure of merit of a coil. The formula for $Q$ is:

$$Q=\frac{X_L}{R}$$

where $X_L$ is the reactance of the coil at a particular frequency, and $R$ is the internal resistance of the coil.

# ac versus dc resistance

The resistance of a coil, as measured with an ohmmeter, is its dc resistance. This is the amount of resistance that opposes the flow of current through the coil, whether the source voltage is ac or dc. The ac resistance of a coil is due to factors such as hysteresis loss, eddy current and skin effect. It exists only when the source voltage is ac and is affected by frequency. It cannot be measured directly.

For a parallel LC circuit, with negligible resistance in the wiring, $R$ can be the resistance of the coil. For low frequencies, the resistance of the coil can be the dc resistance, that is, the resistance measured with a bridge or ohmmeter. However, at higher frequencies, skin effect takes place in which current tends to move through the outer portions of the wire of the coil, and so the ac resistance becomes a factor.

Circuit $Q$ is dependent on center frequency and bandwidth, measured at the half-power points. These points are 3 dB below peak value, equivalent to the 70.7% point on the curve, as shown in Fig. 5-12. If the center frequency of the circuit and

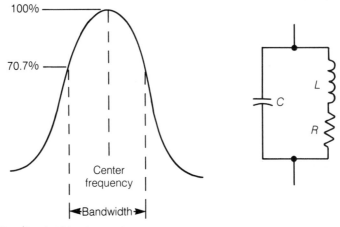

Fig. 5-12. *Bandwidth of circuit is measured between the 70.7% points on the curve.*

the bandwidth are known, the circuit $Q$ can be determined by using the 3-scale nomogram shown in N 5-1. The slanted line in the drawing illustrates a typical example. If digit 1 on the left scale represents 10 kHz, then 45, the point at which the line crosses the left scale, is equivalent to 455 kHz. On the center scale, 1 corresponds to 1 kHz and so, by interpolation, we can see that the straightedge is cutting across at about 12 kHz, the bandwidth of the circuit. The circuit $Q$, then, is approximately 38.

Once the circuit $Q$ is known, the ac resistance can be found by using the 4-scale nomogram in N 5-2. The capacitor in this example has a value of 0.001 $\mu$F and the coil has an inductance of 500 $\mu$H. A straightedge crossing these two points will intersect a vertical center line known as a *turning scale*. This scale is not numbered and serves solely as a pivot on which to turn the straightedge.

Assume that the circuit $Q$ was previously determined by using the nomogram N 5-1 and that the value of $Q$ was about 11. The straightedge is now placed between the pivot point on the turning scale and the circuit $Q$ scale on N 5-2. It intersects at about 65 on the series resistance scale.

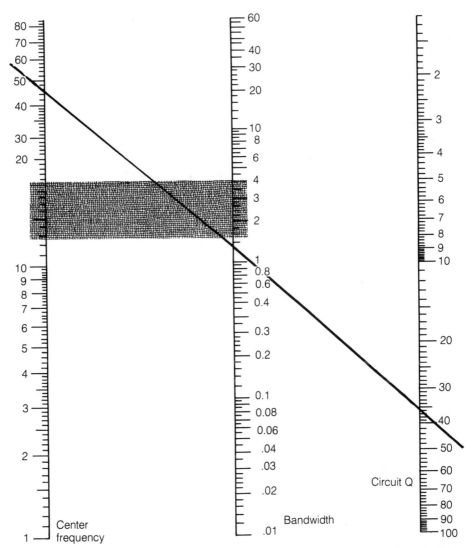

*N 5-1. Nomograph for determining circuit Q when the center frequency and bandwidth are known.*

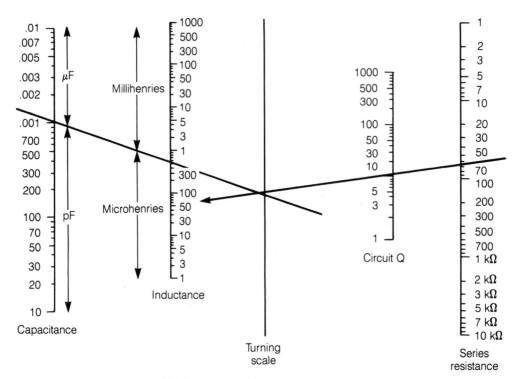

*N 5-2. Nomograph using a turning scale.*

# Impedance

## Phase

The phase relationships of resistive and reactive components (such as inductors and capacitors) can be indicated by lines (vectors) terminating in arrows. For series resistors, as shown in Fig. 6-1A, the phase difference is zero. Voltages appearing across these resistors will be in phase.

For a coil and resistor (Fig. 6-1B) the phase angle is a maximum of 90°. This is indicated by drawing the vectors for the two components at a right angle to each other. The length of the vector line should be comparable to the value, in ohms, of each component. Thus, if the resistor is 10 $\Omega$ and the inductive reactance ($X_L$) of the coil is 20 $\Omega$, the vector line for the coil should have twice the length of the line for the resistor.

Vectors can also be drawn for a series circuit consisting of $R$ and $C$, as indicated in Fig. 6-1C. For a series circuit consisting of $R$, $C$, and $L$, as in Fig. 6-1D, the final vector diagram indicates the difference between $X_L$ and $X_C$. If $X_L$ is larger than $X_C$, then the amount of capacitive reactance is subtracted from inductive reactance, reducing it accordingly. The capacitive reactance can be larger than the inductive reactance, though, in which case $X_L$ is subtracted from $X_C$.

If you have both types of reactance, then, subtract one from the other to get the value of $X$. It makes no difference which reactive component (inductance or capacitance) has the larger reactance. Subtract the value of the smaller reactance from that of the larger to obtain the value of $X$. When $X_L$ is equal to $X_C$, the two reactances effectively cancel and the impedance is equal to $R$.

$$X = X_L - X_C$$

or

$$X = X_C - X_L$$

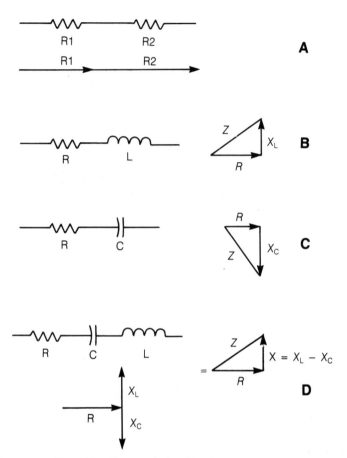

Fig. 6-1. *Vector relationships in series circuits.*

$$Z=\sqrt{R^2+X^2}$$

The impedance of a series RC, RL, or RLC circuit is the vector sum of the individual reactance and resistance. Table 6-1 supplies the impedance in ohms, when the values of $R$ and $X$ are known. $X$ can represent either inductive or capacitive reactance, or $X$ may be the vector sum of these reactive components when both are present in the circuit.

Table 6-1 covers a range of $X$ and of $R$ from 1 to 35 $\Omega$. The table can be extended by moving the decimal point an equal number of places for $R$, $X$, and the answer. Thus, if you had a resistor with a value of 30 $\Omega$ and a reactance with a value of 90 $\Omega$, you could consider 3 in the $R$ row as 30, and 9 in the $X$ row as 90. The value of impedance would be shown in the table as 9.49 $\Omega$, but moving the decimal point one place to the right gives an impedance of 94.9 $\Omega$.

## Table 6-1. Impedance (ohms) for series R and X.

| | | | R | | | |
|---|---|---|---|---|---|---|
| X | 1 | 2 | 3 | 4 | 5 | 6 |
| 1 | 1.41 | 2.24 | 3.16 | 4.12 | 5.10 | 6.08 |
| 2 | 2.24 | 2.83 | 3.61 | 4.47 | 5.39 | 6.32 |
| 3 | 3.16 | 3.61 | 4.24 | 5.00 | 5.83 | 6.71 |
| 4 | 4.12 | 4.47 | 5.00 | 5.57 | 6.40 | 7.21 |
| 5 | 5.10 | 5.39 | 5.83 | 6.40 | 7.07 | 7.81 |
| 6 | 6.08 | 6.32 | 6.71 | 7.21 | 7.81 | 8.48 |
| 7 | 7.07 | 7.28 | 7.62 | 8.06 | 8.60 | 9.22 |
| 8 | 8.06 | 8.25 | 8.54 | 8.94 | 9.43 | 10.00 |
| 9 | 9.06 | 9.22 | 9.49 | 9.85 | 10.29 | 10.81 |
| 10 | 10.05 | 10.19 | 10.44 | 10.77 | 11.18 | 11.66 |
| 11 | 11.04 | 11.18 | 11.40 | 11.70 | 12.08 | 12.52 |
| 12 | 12.04 | 12.16 | 12.36 | 12.64 | 13.00 | 13.41 |
| 13 | 13.03 | 13.15 | 13.34 | 13.60 | 13.92 | 14.31 |
| 14 | 14.03 | 14.14 | 14.31 | 14.56 | 14.86 | 15.23 |
| 15 | 15.03 | 15.13 | 15.29 | 15.52 | 15.81 | 16.15 |
| 16 | 16.03 | 16.12 | 16.28 | 16.49 | 16.76 | 17.09 |
| 17 | 17.03 | 17.12 | 17.26 | 17.46 | 17.72 | 18.03 |
| 18 | 18.03 | 18.11 | 18.25 | 18.44 | 18.68 | 18.97 |
| 19 | 19.03 | 19.10 | 19.24 | 19.42 | 19.65 | 19.92 |
| 20 | 20.02 | 20.05 | 20.22 | 20.40 | 20.62 | 20.88 |
| 21 | 21.02 | 21.10 | 21.21 | 21.38 | 21.59 | 21.84 |
| 22 | 22.02 | 22.09 | 22.20 | 22.36 | 22.56 | 22.80 |
| 23 | 23.02 | 23.09 | 23.19 | 23.35 | 23.54 | 23.75 |
| 24 | 24.02 | 24.08 | 24.19 | 24.33 | 24.52 | 24.74 |
| 25 | 25.02 | 25.08 | 25.18 | 25.32 | 25.50 | 25.71 |
| 26 | 26.02 | 26.08 | 26.17 | 26.31 | 26.48 | 26.68 |
| 27 | 27.02 | 27.07 | 27.16 | 27.29 | 27.46 | 27.66 |
| 28 | 28.02 | 28.07 | 28.16 | 28.28 | 28.44 | 28.64 |
| 29 | 29.02 | 29.07 | 29.15 | 29.27 | 29.43 | 29.61 |
| 30 | 30.02 | 30.07 | 30.15 | 30.27 | 30.41 | 30.59 |
| 31 | 31.02 | 31.06 | 31.14 | 31.26 | 31.40 | 31.58 |
| 32 | 32.01 | 32.06 | 32.14 | 32.24 | 32.39 | 32.55 |
| 33 | 33.01 | 33.06 | 33.14 | 33.24 | 33.38 | 33.54 |
| 34 | 34.01 | 34.06 | 34.14 | 34.24 | 34.37 | 34.53 |
| 35 | 35.01 | 35.06 | 35.13 | 35.23 | 35.36 | 35.51 |

*Table 6-1. Continued.*

|   | $R$ | | | | | |
|---|---|---|---|---|---|---|
| $X$ | 7 | 8 | 9 | 10 | 11 | 12 |
| 1 | 7.07 | 8.06 | 9.06 | 10.05 | 11.04 | 12.04 |
| 2 | 7.28 | 8.25 | 9.22 | 10.19 | 11.18 | 12.16 |
| 3 | 7.62 | 8.54 | 9.49 | 10.44 | 11.40 | 12.36 |
| 4 | 8.06 | 8.94 | 9.85 | 10.77 | 11.70 | 12.64 |
| 5 | 8.60 | 9.43 | 10.29 | 11.18 | 12.08 | 13.00 |
| 6 | 9.22 | 10.00 | 10.81 | 11.66 | 12.52 | 13.41 |
| 7 | 9.89 | 10.63 | 11.40 | 12.20 | 13.03 | 13.89 |
| 8 | 10.63 | 11.31 | 12.04 | 12.80 | 13.60 | 14.42 |
| 9 | 11.40 | 12.04 | 12.72 | 13.45 | 14.21 | 15.00 |
| 10 | 12.20 | 12.80 | 13.45 | 14.14 | 14.86 | 15.62 |
| 11 | 13.03 | 13.60 | 14.21 | 14.86 | 15.55 | 16.27 |
| 12 | 13.89 | 14.42 | 15.00 | 15.62 | 16.27 | 16.97 |
| 13 | 14.76 | 15.26 | 15.81 | 16.40 | 17.02 | 17.69 |
| 14 | 15.65 | 16.12 | 16.64 | 17.20 | 17.80 | 18.43 |
| 15 | 16.55 | 17.00 | 17.49 | 18.02 | 18.60 | 19.20 |
| 16 | 17.46 | 17.89 | 18.36 | 18.87 | 19.42 | 20.00 |
| 17 | 18.38 | 18.79 | 19.24 | 19.72 | 20.25 | 20.81 |
| 18 | 19.31 | 19.70 | 20.12 | 20.59 | 21.10 | 21.63 |
| 19 | 20.25 | 20.62 | 21.02 | 21.47 | 21.95 | 22.47 |
| 20 | 21.19 | 21.54 | 21.93 | 22.36 | 22.83 | 23.32 |
| 21 | 22.14 | 22.47 | 22.85 | 23.26 | 23.71 | 24.19 |
| 22 | 23.09 | 23.41 | 23.77 | 24.17 | 24.60 | 25.06 |
| 23 | 24.04 | 24.35 | 24.70 | 25.08 | 25.50 | 25.94 |
| 24 | 25.00 | 25.30 | 25.63 | 26.00 | 26.40 | 26.83 |
| 25 | 25.96 | 26.25 | 26.57 | 26.93 | 27.31 | 27.73 |
| 26 | 26.93 | 27.20 | 27.51 | 27.86 | 28.23 | 28.64 |
| 27 | 27.89 | 28.16 | 28.46 | 28.79 | 29.15 | 29.55 |
| 28 | 28.86 | 29.12 | 29.41 | 29.73 | 30.08 | 30.46 |
| 29 | 29.83 | 30.08 | 30.36 | 30.68 | 31.02 | 31.38 |
| 30 | 30.81 | 31.05 | 31.32 | 31.62 | 31.96 | 32.31 |
| 31 | 31.78 | 31.93 | 32.25 | 32.56 | 32.90 | 33.24 |
| 32 | 32.75 | 32.98 | 33.24 | 33.52 | 33.85 | 34.18 |
| 33 | 33.74 | 33.96 | 34.21 | 34.49 | 34.79 | 35.12 |
| 34 | 34.72 | 34.93 | 35.17 | 35.44 | 35.74 | 36.06 |
| 35 | 35.70 | 35.90 | 36.14 | 36.40 | 36.69 | 37.00 |

Table 6-1. Continued.

| X | R 13 | 14 | 15 | 16 | 17 | 18 |
|---|---|---|---|---|---|---|
| 1 | 13.03 | 14.03 | 15.03 | 16.03 | 17.03 | 18.03 |
| 2 | 13.15 | 14.14 | 15.13 | 16.12 | 17.12 | 18.11 |
| 3 | 13.34 | 14.31 | 15.29 | 16.28 | 17.26 | 18.25 |
| 4 | 13.60 | 14.56 | 15.52 | 16.49 | 17.46 | 18.44 |
| 5 | 13.92 | 14.86 | 15.81 | 16.76 | 17.72 | 18.68 |
| 6 | 14.31 | 15.23 | 16.15 | 17.09 | 18.03 | 18.97 |
| 7 | 14.76 | 15.65 | 16.55 | 17.47 | 18.38 | 19.31 |
| 8 | 15.26 | 16.12 | 17.00 | 17.89 | 18.79 | 19.70 |
| 9 | 15.81 | 16.64 | 17.49 | 18.36 | 19.24 | 20.12 |
| 10 | 16.40 | 17.20 | 18.02 | 18.87 | 19.72 | 20.59 |
| 11 | 17.02 | 17.80 | 18.60 | 19.42 | 20.25 | 21.10 |
| 12 | 17.69 | 18.43 | 19.20 | 20.00 | 20.81 | 21.63 |
| 13 | 18.38 | 19.10 | 19.84 | 20.62 | 21.40 | 22.20 |
| 14 | 19.10 | 19.79 | 20.51 | 21.26 | 22.02 | 22.80 |
| 15 | 19.84 | 20.51 | 21.21 | 21.93 | 22.67 | 23.43 |
| 16 | 20.62 | 21.26 | 21.93 | 22.63 | 23.35 | 24.08 |
| 17 | 21.40 | 22.02 | 22.67 | 23.35 | 24.04 | 24.76 |
| 18 | 22.20 | 22.80 | 23.43 | 24.08 | 24.76 | 25.46 |
| 19 | 23.02 | 23.60 | 24.20 | 24.84 | 25.50 | 26.17 |
| 20 | 23.85 | 24.41 | 25.00 | 25.61 | 26.25 | 26.91 |
| 21 | 24.70 | 25.24 | 25.81 | 26.40 | 27.02 | 27.66 |
| 22 | 25.55 | 26.08 | 26.63 | 27.20 | 27.80 | 28.43 |
| 23 | 26.42 | 26.93 | 27.46 | 28.02 | 28.60 | 29.21 |
| 24 | 27.29 | 27.78 | 28.30 | 28.84 | 29.41 | 30.00 |
| 25 | 28.18 | 28.65 | 29.15 | 29.68 | 30.23 | 30.81 |
| 26 | 29.07 | 29.53 | 30.02 | 30.53 | 31.06 | 31.62 |
| 27 | 29.97 | 30.41 | 30.89 | 31.38 | 31.91 | 32.45 |
| 28 | 30.87 | 31.30 | 31.77 | 32.35 | 32.76 | 33.29 |
| 29 | 31.78 | 32.21 | 32.65 | 33.12 | 33.62 | 34.14 |
| 30 | 32.70 | 33.11 | 33.54 | 34.00 | 34.48 | 34.99 |
| 31 | 33.62 | 34.01 | 34.44 | 34.89 | 35.36 | 35.85 |
| 32 | 34.54 | 35.50 | 35.90 | 36.33 | 36.78 | 37.26 |
| 33 | 35.47 | 35.85 | 36.25 | 36.67 | 37.12 | 37.59 |
| 34 | 36.40 | 36.77 | 37.16 | 37.58 | 38.01 | 38.47 |
| 35 | 37.34 | 37.60 | 38.08 | 38.48 | 38.91 | 39.36 |

Table 6-1. Continued.

| | | | R | | | |
|---|---|---|---|---|---|---|
| X | 19 | 20 | 21 | 22 | 23 | 24 |
| 1 | 19.03 | 20.02 | 21.02 | 22.02 | 23.02 | 24.02 |
| 2 | 19.10 | 20.10 | 21.10 | 22.09 | 23.09 | 24.08 |
| 3 | 19.24 | 20.22 | 21.21 | 22.20 | 23.19 | 24.19 |
| 4 | 19.42 | 20.40 | 21.38 | 22.36 | 23.35 | 24.33 |
| 5 | 19.65 | 20.62 | 21.59 | 22.56 | 23.54 | 24.52 |
| 6 | 19.92 | 20.88 | 21.84 | 22.80 | 23.77 | 24.74 |
| 7 | 20.25 | 21.18 | 22.14 | 23.09 | 24.04 | 25.00 |
| 8 | 20.62 | 21.54 | 22.47 | 23.41 | 24.35 | 25.30 |
| 9 | 21.02 | 21.93 | 22.85 | 23.77 | 24.70 | 25.63 |
| 10 | 21.47 | 22.36 | 23.26 | 24.17 | 25.08 | 26.00 |
| 11 | 21.95 | 22.82 | 23.71 | 24.60 | 25.50 | 26.40 |
| 12 | 22.47 | 23.32 | 24.19 | 25.06 | 25.94 | 26.83 |
| 13 | 23.02 | 23.85 | 24.70 | 25.55 | 26.42 | 27.29 |
| 14 | 23.60 | 24.41 | 25.24 | 26.08 | 26.93 | 27.78 |
| 15 | 24.20 | 25.00 | 25.81 | 26.63 | 27.46 | 28.30 |
| 16 | 24.84 | 25.61 | 26.40 | 27.20 | 28.02 | 28.84 |
| 17 | 25.50 | 26.25 | 27.02 | 27.80 | 28.60 | 29.41 |
| 18 | 26.17 | 26.91 | 27.66 | 28.43 | 29.21 | 30.00 |
| 19 | 26.86 | 27.59 | 28.32 | 29.07 | 29.83 | 30.61 |
| 20 | 27.59 | 28.28 | 29.00 | 29.73 | 30.48 | 31.24 |
| 21 | 28.32 | 29.00 | 29.70 | 30.41 | 31.14 | 31.89 |
| 22 | 29.07 | 29.73 | 30.41 | 31.11 | 31.83 | 32.56 |
| 23 | 29.83 | 30.48 | 31.14 | 31.83 | 32.52 | 33.24 |
| 24 | 30.61 | 31.24 | 31.89 | 32.56 | 33.24 | 33.94 |
| 25 | 31.40 | 32.02 | 32.65 | 33.30 | 33.97 | 34.66 |
| 26 | 32.20 | 32.80 | 33.42 | 34.06 | 34.71 | 35.38 |
| 27 | 33.02 | 33.60 | 34.21 | 34.83 | 35.47 | 36.13 |
| 28 | 33.85 | 34.41 | 35.01 | 35.61 | 36.24 | 36.88 |
| 29 | 34.67 | 35.23 | 35.81 | 36.40 | 37.01 | 37.64 |
| 30 | 35.51 | 36.06 | 36.62 | 37.20 | 37.80 | 38.42 |
| 31 | 36.36 | 36.89 | 37.44 | 38.01 | 38.60 | 39.21 |
| 32 | 37.22 | 37.73 | 38.28 | 38.83 | 39.41 | 40.00 |
| 33 | 38.08 | 38.59 | 39.12 | 39.66 | 40.22 | 40.81 |
| 34 | 38.95 | 39.45 | 39.96 | 40.40 | 41.05 | 41.62 |
| 35 | 39.82 | 40.31 | 40.82 | 41.34 | 41.88 | 42.44 |

*Table 6-1. Continued.*

|   | R | | | | | |
|---|---|---|---|---|---|---|
| X | 25 | 26 | 27 | 28 | 29 | 30 |
| 1 | 25.02 | 26.02 | 27.02 | 28.02 | 29.02 | 30.02 |
| 2 | 25.08 | 26.08 | 27.07 | 28.07 | 29.07 | 30.07 |
| 3 | 25.18 | 26.17 | 27.17 | 28.16 | 29.15 | 30.15 |
| 4 | 25.32 | 26.31 | 27.29 | 28.28 | 29.27 | 30.27 |
| 5 | 25.50 | 26.46 | 27.46 | 28.44 | 29.43 | 30.41 |
| 6 | 25.71 | 26.68 | 27.66 | 28.64 | 29.61 | 30.59 |
| 7 | 25.98 | 26.93 | 27.89 | 28.86 | 29.83 | 30.81 |
| 8 | 26.25 | 27.20 | 28.16 | 29.12 | 30.08 | 31.05 |
| 9 | 26.57 | 27.51 | 28.46 | 29.41 | 30.36 | 31.32 |
| 10 | 26.93 | 27.86 | 28.79 | 29.73 | 30.68 | 31.62 |
| 11 | 27.31 | 28.23 | 29.15 | 30.08 | 31.02 | 31.95 |
| 12 | 27.73 | 28.64 | 29.55 | 30.46 | 31.38 | 32.31 |
| 13 | 28.18 | 29.07 | 29.97 | 30.87 | 31.78 | 32.60 |
| 14 | 28.65 | 29.53 | 30.41 | 31.30 | 32.20 | 33.11 |
| 15 | 29.15 | 30.02 | 30.89 | 31.77 | 32.64 | 33.54 |
| 16 | 29.68 | 30.53 | 31.38 | 32.25 | 33.12 | 33.80 |
| 17 | 30.23 | 31.06 | 31.91 | 32.76 | 33.62 | 34.48 |
| 18 | 30.81 | 31.62 | 32.45 | 33.29 | 34.13 | 34.99 |
| 19 | 31.40 | 32.20 | 33.02 | 33.85 | 34.67 | 35.51 |
| 20 | 32.02 | 32.80 | 33.60 | 34.41 | 35.23 | 36.06 |
| 21 | 32.65 | 33.42 | 34.21 | 35.01 | 35.81 | 36.62 |
| 22 | 33.30 | 34.06 | 34.83 | 35.61 | 36.40 | 37.20 |
| 23 | 33.97 | 34.71 | 35.47 | 36.24 | 37.01 | 37.80 |
| 24 | 34.66 | 35.38 | 36.13 | 36.88 | 37.64 | 38.42 |
| 25 | 35.36 | 36.07 | 36.70 | 37.54 | 38.29 | 39.05 |
| 26 | 36.07 | 36.77 | 37.48 | 38.21 | 38.95 | 39.60 |
| 27 | 36.70 | 37.48 | 38.18 | 38.80 | 39.62 | 40.36 |
| 28 | 37.54 | 38.21 | 38.71 | 39.50 | 40.31 | 41.04 |
| 29 | 38.29 | 38.95 | 39.62 | 40.31 | 41.01 | 41.73 |
| 30 | 39.05 | 39.60 | 40.36 | 41.04 | 41.73 | 42.43 |
| 31 | 39.81 | 40.46 | 41.11 | 41.77 | 42.44 | 43.14 |
| 32 | 40.61 | 41.23 | 41.87 | 42.52 | 43.19 | 43.86 |
| 33 | 41.40 | 42.01 | 42.64 | 43.28 | 43.93 | 44.50 |
| 34 | 42.20 | 42.80 | 43.42 | 44.05 | 44.69 | 45.34 |
| 35 | 43.01 | 43.60 | 44.20 | 44.82 | 45.45 | 46.00 |

Table 6-1. Continued.

| | | | R | | |
|---|---|---|---|---|---|
| X | 31 | 32 | 33 | 34 | 35 |
| 1 | 31.01 | 32.02 | 33.02 | 34.02 | 35.01 |
| 2 | 31.06 | 32.06 | 33.06 | 34.06 | 35.06 |
| 3 | 31.14 | 32.14 | 33.14 | 34.13 | 35.13 |
| 4 | 31.26 | 32.25 | 33.24 | 34.23 | 35.23 |
| 5 | 31.40 | 32.39 | 33.38 | 34.37 | 35.36 |
| 6 | 31.57 | 32.56 | 33.54 | 34.53 | 35.51 |
| 7 | 31.78 | 32.76 | 33.74 | 34.71 | 35.69 |
| 8 | 32.02 | 32.99 | 33.96 | 34.93 | 35.90 |
| 9 | 32.28 | 33.24 | 34.21 | 35.17 | 36.14 |
| 10 | 32.57 | 33.53 | 34.48 | 35.36 | 36.40 |
| 11 | 32.89 | 33.85 | 34.79 | 35.74 | 36.69 |
| 12 | 33.24 | 34.18 | 35.11 | 36.06 | 36.80 |
| 13 | 33.62 | 34.53 | 35.48 | 36.40 | 37.34 |
| 14 | 34.02 | 34.93 | 35.85 | 36.77 | 37.60 |
| 15 | 34.44 | 35.34 | 36.25 | 37.16 | 38.08 |
| 16 | 34.89 | 35.78 | 36.67 | 37.58 | 38.48 |
| 17 | 35.37 | 36.24 | 37.12 | 38.01 | 38.90 |
| 18 | 35.85 | 36.72 | 37.59 | 38.47 | 39.36 |
| 19 | 36.36 | 37.22 | 38.08 | 38.95 | 39.82 |
| 20 | 36.89 | 37.74 | 38.59 | 39.46 | 40.31 |
| 21 | 37.44 | 38.28 | 39.12 | 39.96 | 40.82 |
| 22 | 38.01 | 38.83 | 39.66 | 40.40 | 41.34 |
| 23 | 38.60 | 39.41 | 40.22 | 41.05 | 41.88 |
| 24 | 39.21 | 40.00 | 40.80 | 41.63 | 42.44 |
| 25 | 39.81 | 40.61 | 41.40 | 42.20 | 43.01 |
| 26 | 40.46 | 41.23 | 42.01 | 42.80 | 43.60 |
| 27 | 41.11 | 41.87 | 42.64 | 43.42 | 44.20 |
| 28 | 41.77 | 42.52 | 43.28 | 44.05 | 44.82 |
| 29 | 42.44 | 43.19 | 43.93 | 44.69 | 45.45 |
| 30 | 43.14 | 43.86 | 44.50 | 45.34 | 46.00 |
| 31 | 43.84 | 44.55 | 45.28 | 46.01 | 46.76 |
| 32 | 44.55 | 45.26 | 45.97 | 46.69 | 47.42 |
| 33 | 45.28 | 45.98 | 46.68 | 47.38 | 48.10 |
| 34 | 46.01 | 46.69 | 47.38 | 48.08 | 48.70 |
| 35 | 46.75 | 47.42 | 48.10 | 48.70 | 49.40 |

*Example*   A circuit consists of a 5-Ω resistor, a coil having an inductive reactance of 35 Ω and a capacitor having a capacitive reactance of 21 Ω. What is the impedance of this circuit?

*Solution*   Subtract the value of the smaller reactance from that of the larger. 35−21=14 Ω. This is the net reactance and constitutes the value of *X*. Using this value, locate 14 in the *X* column in Table 6-1. Move horizontally to reach the 5 row (*R*). The impedance is shown as 14.86 Ω.

*Example*   You require an impedance of 30 Ω for an RLC circuit. What values of *R* and *X* can you use to obtain this impedance?

*Solution*   Table 6-1 shows you can get this impedance by using an 18-Ω resistor and a reactance having a value of 24 Ω. This reactance may be a coil having an inductive reactance of this value, or a capacitor, or both components whose net reactance is 24 Ω. Of course you could also use a 24-Ω resistor and a reactance of 18 Ω. There are many other combinations in the table that are very close to this desired 30-Ω impedance.

# Vector conversion

Table 6-2 can be used to change the form of vector quantities from j-notation to polar notation, or from polar to j-notation. The first column is the ratio of reactance, *X*, to resistance, *R*. The second column is the phase angle of the polar vector and is the angle whose tangent is *X/R*. The third column, *Z*, is the absolute magnitude of the vector in terms of *X*.

*Example*   The impedance of a circuit, expressed in polar form, is *Z*= 3 000−j4 000. What is the absolute magnitude of the impedance and the phase angle?

*Solution*   3 000 represents the resistance *R*; 4 000 is the reactance *X*. The ratio *X/R* is 4 000/3 000=1.333 3. The closest value of *X/R* in Table 6-2 is 1.331 9. Immediately to the right is the phase angle of 53.1°. However, because the reactance is negative and is given as −j4 000, the phase angle is also negative and is −53.1°. The value for *Z* in the column directly to the right of the phase angle is 1.250 5*X*. Because *X* is 4 000, *Z*=1.250 5*X*, or 5 002 Ω. The answer, then, is that *Z* is 5 002 Ω, and the phase angle is −53.1°.

*Example*   Take the example just given and work it backward. Suppose you are told that the circuit impedance is 5 000 Ω (5 002 rounded off to 5 000) and the phase angle is −53.1°. How could you express this in j-notation?

*Solution*   Start with the phase angle. In the third column 1.25*X*=*Z*. But *Z*=5 000 Ω. Because 1.25*X*=5 000, then *X*=5 000/1.25, or 4 000 Ω. Now move back to the first column. Here *X/R*=1.331 9. Thus, 4 000/*R*=1.331 9.

## Table 6-2. Vector conversion.

| $\frac{X}{R}$ | $\theta$ (degrees) | Z | $\frac{X}{R}$ | $\theta$ (degrees) | Z |
|---|---|---|---|---|---|
| 0.000 00 | 0.0 | Z=R | 0.071 68 | 4.1 | 13.986X |
| 0.001 75 | 0.1 | 572.96X | 0.073 44 | 4.2 | 13.654X |
| 0.003 49 | 0.2 | 286.48X | 0.075 19 | 4.3 | 13.337X |
| 0.005 24 | 0.3 | 190.98X | 0.076 94 | 4.4 | 13.034X |
| 0.006 98 | 0.4 | 143.24X | | | |
| | | | 0.078 70 | 4.5 | 12.745X |
| 0.008 73 | 0.5 | 114.59X | 0.080 46 | 4.6 | 12.469X |
| 0.010 47 | 0.6 | 95.495X | 0.082 21 | 4.7 | 12.204X |
| 0.012 22 | 0.7 | 81.853X | 0.083 97 | 4.8 | 11.950X |
| 0.013 96 | 0.8 | 71.622X | 0.085 73 | 4.9 | 11.707X |
| 0.015 71 | 0.9 | 63.664X | | | |
| | | | 0.087 70 | 5.0 | 11.474X |
| 0.017 46 | 1.0 | 57.299X | 0.089 25 | 5.1 | 11.249X |
| 0.019 20 | 1.1 | 52.090X | 0.091 01 | 5.2 | 11.033X |
| 0.020 95 | 1.2 | 47.750X | 0.092 77 | 5.3 | 10.826X |
| 0.022 69 | 1.3 | 44.077X | 0.094 53 | 5.4 | 10.626X |
| 0.024 44 | 1.4 | 40.930X | | | |
| | | | 0.096 29 | 5.5 | 10.433X |
| 0.026 18 | 1.5 | 38.201X | 0.098 05 | 5.6 | 10.248X |
| 0.027 93 | 1.6 | 35.814X | 0.099 81 | 5.7 | 10.068X |
| 0.029 68 | 1.7 | 33.708X | 0.101 58 | 5.8 | 9.895 5X |
| 0.031 43 | 1.8 | 31.836X | 0.103 44 | 5.9 | 9.728 3X |
| 0.033 17 | 1.9 | 30.161X | | | |
| | | | 0.105 10 | 6.0 | 9.566 8X |
| 0.034 92 | 2.0 | 28.654X | 0.106 87 | 6.1 | 9.410 5X |
| 0.036 67 | 2.1 | 27.290X | 0.108 63 | 6.2 | 9.259 3X |
| 0.038 42 | 2.2 | 26.050X | 0.110 40 | 6.3 | 9.112 9X |
| 0.040 16 | 2.3 | 24.918X | 0.112 17 | 6.4 | 8.971 1X |
| 0.041 91 | 2.4 | 23.880X | | | |
| | | | 0.113 94 | 6.5 | 8.833 7X |
| 0.043 66 | 2.5 | 22.925X | 0.115 70 | 6.6 | 8.700 4X |
| 0.045 41 | 2.6 | 22.044X | 0.117 47 | 6.7 | 8.571 1X |
| 0.047 16 | 2.7 | 21.228X | 0.119 24 | 6.8 | 8.445 7X |
| 0.048 91 | 2.8 | 20.471X | 0.121 07 | 6.9 | 8.323 8X |
| 0.050 66 | 2.9 | 19.766X | | | |
| | | | 0.122 78 | 7.0 | 8.205 5X |
| 0.052 41 | 3.0 | 19.107X | 0.124 56 | 7.1 | 8.090 5X |
| 0.054 16 | 3.1 | 18.491X | 0.126 63 | 7.2 | 7.978 7X |
| 0.055 91 | 3.2 | 17.914X | 0.128 10 | 7.3 | 7.780 0X |
| 0.057 66 | 3.3 | 17.372X | 0.129 88 | 7.4 | 7.764 2X |
| 0.059 41 | 3.4 | 16.861X | | | |
| | | | 0.131 65 | 7.5 | 7.661 3X |
| 0.061 16 | 3.5 | 16.380X | 0.133 43 | 7.6 | 7.561 1X |
| 0.062 91 | 3.6 | 15.926X | 0.135 21 | 7.7 | 7.463 4X |
| 0.064 67 | 3.7 | 15.496X | 0.136 98 | 7.8 | 7.368 3X |
| 0.066 42 | 3.8 | 15.089X | 0.138 76 | 7.9 | 7.275 6X |
| 0.068 17 | 3.9 | 14.702X | | | |
| | | | 0.140 54 | 8.0 | 7.011 2X |
| 0.069 93 | 4.0 | 14.335X | 0.145 88 | 8.1 | 7.087 2X |

Table 6-2. Continued.

| $\dfrac{X}{R}$ | $\theta$ (degrees) | Z | $\dfrac{X}{R}$ | $\theta$ (degrees) | Z |
|---|---|---|---|---|---|
| 0.144 10 | 8.2 | 7.011 2X | 0.218 03 | 12.3 | 4.694 2X |
| 0.145 88 | 8.3 | 6.927 3X | 0.219 86 | 12.4 | 4.656 9X |
| 0.147 67 | 8.4 | 6.845 4X | 0.221 69 | 12.5 | 4.620 1X |
| 0.149 45 | 8.5 | 6.775 5X | 0.223 53 | 12.6 | 4.584 1X |
| 0.151 24 | 8.6 | 6.687 4X | 0.225 36 | 12.7 | 4.548 6X |
| 0.153 02 | 8.7 | 6.611 1X | 0.227 19 | 12.8 | 4.513 7X |
| 0.154 81 | 8.8 | 6.536 5X | 0.229 03 | 12.9 | 4.479 3X |
| 0.156 60 | 8.9 | 6.463 7X | 0.230 87 | 13.0 | 4.445 4X |
| 0.158 38 | 9.0 | 6.392 4X | 0.232 70 | 13.1 | 4.412 1X |
| 0.160 17 | 9.1 | 6.322 8X | 0.234 55 | 13.2 | 4.379 2X |
| 0.161 96 | 9.2 | 6.254 6X | 0.236 93 | 13.3 | 4.346 9X |
| 0.163 76 | 9.3 | 6.188 0X | 0.238 23 | 13.4 | 4.315 0X |
| 0.165 55 | 9.4 | 6.122 7X | 0.240 08 | 13.5 | 4.283 6X |
| 0.167 34 | 9.5 | 6.058 8X | 0.241 92 | 13.6 | 4.252 7X |
| 0.169 14 | 9.6 | 5.996 3X | 0.243 77 | 13.7 | 4.222 3X |
| 0.170 93 | 9.7 | 5.935 1X | 0.245 62 | 13.8 | 4.192 3X |
| 0.172 73 | 9.8 | 5.875 1X | 0.247 47 | 13.9 | 4.162 7X |
| 0.174 53 | 9.9 | 5.816 3X | 0.249 33 | 14.0 | 4.133 6X |
| 0.176 33 | 10.0 | 5.758 8X | 0.251 18 | 14.1 | 4.104 8X |
| 0.178 13 | 10.1 | 5.702 3X | 0.253 04 | 14.2 | 4.076 5X |
| 0.179 93 | 10.2 | 5.647 0X | 0.254 90 | 14.3 | 4.048 6X |
| 0.181 73 | 10.3 | 5.592 8X | 0.256 76 | 14.4 | 4.021 1X |
| 0.183 53 | 10.4 | 5.539 6X | 0.258 62 | 14.5 | 3.993 9X |
| 0.185 34 | 10.5 | 5.487 4X | 0.260 48 | 14.6 | 3.967 2X |
| 0.187 14 | 10.6 | 5.436 2X | 0.262 34 | 14.7 | 3.940 8X |
| 0.188 95 | 10.7 | 5.386 0X | 0.264 21 | 14.8 | 3.914 7X |
| 0.190 76 | 10.8 | 5.336 7X | 0.266 08 | 14.9 | 3.889 0X |
| 0.192 57 | 10.9 | 5.288 3X | 0.267 95 | 15.0 | 3.863 7X |
| 0.194 38 | 11.0 | 5.240 8X | 0.269 82 | 15.1 | 3.838 7X |
| 0.196 19 | 11.1 | 5.194 2X | 0.271 69 | 15.2 | 3.814 0X |
| 0.198 00 | 11.2 | 5.148 4X | 0.273 57 | 15.3 | 3.789 7X |
| 0.199 82 | 11.3 | 5.103 4X | 0.275 54 | 15.4 | 3.765 7X |
| 0.201 63 | 11.4 | 5.059 3X | 0.277 32 | 15.5 | 3.742 0X |
| 0.203 45 | 11.5 | 5.105 8X | 0.279 20 | 15.6 | 3.718 6X |
| 0.205 27 | 11.6 | 5.973 2X | 0.281 09 | 15.7 | 3.695 5X |
| 0.207 09 | 11.7 | 5.931 3X | 0.282 97 | 15.8 | 3.672 7X |
| 0.208 91 | 11.8 | 5.890 1X | 0.284 86 | 15.9 | 3.650 2X |
| 0.210 73 | 11.9 | 5.849 6X | 0.286 74 | 16.0 | 3.627 9X |
| 0.212 56 | 12.0 | 4.809 7X | 0.288 63 | 16.1 | 3.606 0X |
| 0.214 38 | 12.1 | 4.770 6X | 0.290 53 | 16.2 | 3.914 7X |
| 0.216 21 | 12.2 | 4.732 0X | 0.292 42 | 16.3 | 3.562 9X |

*Table 6-2. Continued.*

| $\dfrac{X}{R}$ | $\theta$ (degrees) | Z | $\dfrac{X}{R}$ | $\theta$ (degrees) | Z |
|---|---|---|---|---|---|
| 0.294 32 | 16.4 | 3.541 8X | 0.373 88 | 20.5 | 2.855 4X |
| 0.296 21 | 16.5 | 3.520 9X | 0.375 87 | 20.6 | 2.842 2X |
| 0.298 11 | 16.6 | 3.500 3X | 0.377 87 | 20.7 | 0.829 0X |
| 0.300 01 | 16.7 | 3.479 9X | 0.379 86 | 20.8 | 2.816 0X |
| 0.301 92 | 16.8 | 3.459 8X | 0.381 86 | 20.9 | 2.803 2X |
| 0.303 82 | 16.9 | 3.439 9X | | | |
| | | | 0.383 86 | 21.0 | 2.790 4X |
| 0.305 73 | 17.0 | 3.420 3X | 0.385 87 | 21.1 | 2.777 8X |
| 0.307 64 | 17.1 | 3.400 9X | 0.387 87 | 21.2 | 2.765 3X |
| 0.309 95 | 17.2 | 3.381 7X | 0.389 88 | 21.3 | 2.752 9X |
| 0.311 46 | 17.3 | 3.362 7X | 0.391 89 | 21.4 | 2.740 6X |
| 0.313 38 | 17.4 | 3.344 0X | | | |
| | | | 0.393 91 | 21.5 | 2.728 5X |
| 0.315 30 | 17.5 | 3.325 5X | 0.395 93 | 21.6 | 2.716 5X |
| 0.317 22 | 17.6 | 3.307 2X | 0.397 95 | 21.7 | 2.704 5X |
| 0.319 14 | 17.7 | 3.289 1X | 0.399 97 | 21.8 | 2.692 7X |
| 0.321 06 | 17.8 | 3.271 2X | 0.402 00 | 21.9 | 2.681 0X |
| 0.322 99 | 17.9 | 3.253 5X | | | |
| | | | 0.404 03 | 22.0 | 2.669 5X |
| 0.324 92 | 18.0 | 3.236 1X | 0.406 06 | 22.1 | 2.658 0X |
| 0.326 85 | 18.1 | 3.218 8X | 0.408 09 | 22.2 | 2.646 6X |
| 0.328 78 | 18.2 | 3.201 7X | 0.410 13 | 22.3 | 2.635 3X |
| 0.330 72 | 18.3 | 3.184 8X | 0.412 17 | 22.4 | 2.624 2X |
| 0.332 65 | 18.4 | 3.168 1X | | | |
| | | | 0.414 21 | 22.5 | 2.613 1X |
| 0.334 59 | 18.5 | 3.151 5X | 0.416 26 | 22.6 | 2.602 2X |
| 0.336 54 | 18.6 | 3.135 2X | 0.418 31 | 22.7 | 2.591 3X |
| 0.338 48 | 18.7 | 3.119 0X | 0.420 36 | 22.8 | 2.580 5X |
| 0.340 43 | 18.8 | 3.103 0X | 0.422 42 | 22.9 | 2.569 9X |
| 0.342 38 | 18.9 | 3.087 2X | | | |
| | | | 0.424 47 | 23.0 | 2.559 3X |
| 0.344 33 | 19.0 | 3.071 5X | 0.426 54 | 23.1 | 2.548 8X |
| 0.346 28 | 19.1 | 3.056 1X | 0.428 60 | 23.2 | 2.538 4X |
| 0.348 24 | 19.2 | 3.040 7X | 0.430 67 | 23.3 | 2.528 1X |
| 0.350 19 | 19.3 | 3.025 6X | 0.432 74 | 23.4 | 2.517 9X |
| 0.352 15 | 19.4 | 3.010 6X | | | |
| | | | 0.434 81 | 23.5 | 2.507 8X |
| 0.354 12 | 19.5 | 2.995 7X | 0.436 89 | 23.6 | 2.497 8X |
| 0.356 08 | 19.6 | 2.981 0X | 0.438 97 | 23.7 | 2.487 9X |
| 0.358 05 | 19.7 | 2.966 5X | 0.441 05 | 23.8 | 2.478 0X |
| 0.360 02 | 19.8 | 2.952 1X | 0.443 14 | 23.9 | 2.468 3X |
| 0.361 99 | 19.9 | 2.937 9X | | | |
| | | | 0.445 23 | 24.0 | 2.458 6X |
| 0.363 97 | 20.0 | 2.923 8X | 0.447 32 | 24.1 | 2.449 0X |
| 0.365 95 | 20.1 | 2.909 8X | 0.449 42 | 24.2 | 2.439 5X |
| 0.367 93 | 20.2 | 2.896 0X | 0.451 52 | 24.3 | 2.430 0X |
| 0.369 91 | 20.3 | 2.882 4X | 0.453 62 | 24.4 | 2.420 7X |
| 0.371 90 | 20.4 | 2.868 8X | 0.455 73 | 24.5 | 2.411 4X |

*Table 6-2. Continued.*

| $\dfrac{X}{R}$ | $\theta$ (degrees) | Z | $\dfrac{X}{R}$ | $\theta$ (degrees) | Z |
|---|---|---|---|---|---|
| 0.457 83 | 24.6 | 2.402 2X | 0.547 48 | 28.7 | 2.082 4X |
| 0.459 95 | 24.7 | 2.393 1X | 0.549 75 | 28.8 | 2.075 7X |
| 0.462 06 | 24.8 | 2.384 1X | 0.552 03 | 28.9 | 2.069 2X |
| 0.464 18 | 24.9 | 2.375 1X | | | |
| | | | 0.554 31 | 29.0 | 2.062 7X |
| 0.466 31 | 25.0 | 2.366 2X | 0.554 31 | 29.1 | 2.056 2X |
| 0.468 43 | 25.1 | 2.357 4X | 0.558 88 | 29.2 | 2.049 8X |
| 0.470 56 | 25.2 | 2.348 6X | 0.561 17 | 29.3 | 2.043 4X |
| 0.472 70 | 25.3 | 2.339 9X | 0.563 47 | 29.4 | 2.037 0X |
| 0.474 83 | 25.4 | 2.331 3X | | | |
| | | | 0.565 77 | 29.5 | 2.030 8X |
| 0.476 97 | 25.5 | 2.322 8X | 0.568 08 | 29.6 | 2.024 5X |
| 0.479 12 | 25.6 | 2.314 3X | 0.570 39 | 29.7 | 2.018 3X |
| 0.481 27 | 25.7 | 2.305 9X | 0.572 70 | 29.8 | 2.012 2X |
| 0.483 42 | 25.8 | 2.297 6X | 0.575 02 | 29.9 | 2.006 1X |
| 0.485 57 | 25.9 | 2.289 4X | | | |
| | | | 0.577 35 | 30.0 | 2.000 0X |
| 0.487 73 | 26.0 | 2.281 2X | 0.579 68 | 30.1 | 1.994 0X |
| 0.489 89 | 26.1 | 2.273 0X | 0.582 01 | 30.2 | 1.988 0X |
| 0.492 06 | 26.2 | 2.265 0X | 0.584 35 | 30.3 | 1.982 0X |
| 0.494 23 | 26.3 | 2.257 0X | 0.586 70 | 30.4 | 1.976 1X |
| 0.496 40 | 26.4 | 2.249 0X | | | |
| | | | 0.589 04 | 30.5 | 1.970 3X |
| 0.498 58 | 26.5 | 2.241 1X | 0.591 40 | 30.6 | 1.964 5X |
| 0.500 76 | 26.6 | 2.233 3X | 0.593 76 | 30.7 | 1.958 7X |
| 0.502 95 | 26.7 | 2.225 6X | 0.596 12 | 30.8 | 1.953 0X |
| 0.505 14 | 26.8 | 2.217 9X | 0.598 49 | 30.9 | 1.947 3X |
| 0.507 33 | 26.9 | 2.210 3X | | | |
| | | | 0.600 86 | 31.0 | 1.941 6X |
| 0.509 52 | 27.0 | 2.202 7X | 0.603 24 | 31.1 | 1.936 0X |
| 0.511 72 | 27.1 | 2.195 2X | 0.605 62 | 31.2 | 1.930 4X |
| 0.513 93 | 27.2 | 2.187 7X | 0.608 01 | 31.3 | 1.924 8X |
| 0.516 14 | 27.3 | 2.180 3X | 0.610 40 | 31.4 | 1.919 3X |
| 0.518 35 | 27.4 | 2.173 0X | | | |
| | | | 0.612 80 | 31.5 | 1.913 9X |
| 0.520 57 | 27.5 | 2.165 7X | 0.615 20 | 31.6 | 1.908 4X |
| 0.522 79 | 27.6 | 2.158 4X | 0.617 61 | 31.7 | 1.903 0X |
| 0.525 01 | 27.7 | 2.151 3X | 0.620 03 | 31.8 | 1.897 7X |
| 0.527 24 | 27.8 | 2.144 1X | 0.622 44 | 31.9 | 1.892 4X |
| 0.529 47 | 27.9 | 2.137 1X | | | |
| | | | 0.624 87 | 32.0 | 1.887 1X |
| 0.531 71 | 28.0 | 2.130 0X | 0.627 30 | 32.1 | 1.881 8X |
| 0.533 95 | 28.1 | 2.123 1X | 0.629 73 | 32.2 | 1.876 6X |
| 0.536 19 | 28.2 | 2.116 2X | 0.632 17 | 32.3 | 1.871 4X |
| 0.538 44 | 28.3 | 2.109 3X | 0.634 62 | 32.4 | 1.866 3X |
| 0.540 70 | 28.4 | 2.102 5X | | | |
| | | | 0.637 07 | 32.5 | 1.861 1X |
| 0.542 95 | 28.5 | 2.095 7X | 0.639 53 | 32.6 | 1.856 1X |
| 0.545 22 | 28.6 | 2.089 0X | 0.641 99 | 32.7 | 1.851 0X |

*Table 6-2. Continued.*

| $\dfrac{X}{R}$ | $\theta$ (degrees) | Z | $\dfrac{X}{R}$ | $\theta$ (degrees) | Z |
|---|---|---|---|---|---|
| 0.644 66 | 32.8 | 1.846 0X | 0.750 82 | 36.9 | 1.665 5X |
| 0.646 93 | 32.9 | 1.841 0X | | | |
| | | | 0.753 55 | 37.0 | 1.661 6X |
| 0.649 41 | 33.0 | 1.836 1X | 0.756 29 | 37.1 | 1.657 8X |
| 0.651 89 | 33.1 | 1.831 1X | 0.759 04 | 37.2 | 1.654 0X |
| 0.654 38 | 33.2 | 1.826 3X | 0.761 79 | 37.3 | 1.650 2X |
| 0.656 88 | 33.3 | 1.821 4X | 0.765 46 | 37.4 | 1.646 4X |
| 0.659 38 | 33.4 | 1.816 6X | | | |
| | | | 0.767 33 | 37.5 | 1.642 7X |
| 0.661 88 | 33.5 | 1.811 8X | 0.770 10 | 37.6 | 1.638 9X |
| 0.664 40 | 33.6 | 1.807 0X | 0.772 89 | 37.7 | 1.635 2X |
| 0.666 92 | 33.7 | 1.802 3X | 0.775 68 | 37.8 | 1.631 6X |
| 0.669 44 | 33.8 | 1.797 6X | 0.778 48 | 37.9 | 1.627 9X |
| 0.671 97 | 33.9 | 1.792 9X | | | |
| | | | 0.781 28 | 38.0 | 1.624 3X |
| 0.674 51 | 34.0 | 1.788 3X | 0.784 10 | 38.1 | 1.620 6X |
| 0.677 05 | 34.1 | 1.783 7X | 0.786 92 | 38.2 | 1.617 0X |
| 0.671 60 | 34.2 | 1.779 1X | 0.789 75 | 38.3 | 1.613 5X |
| 0.682 15 | 34.3 | 1.774 5X | 0.792 59 | 38.4 | 1.609 9X |
| 0.684 71 | 34.4 | 1.770 0X | | | |
| | | | 0.795 43 | 38.5 | 1.606 4X |
| 0.687 28 | 34.5 | 1.765 5X | 0.798 29 | 38.6 | 1.602 9X |
| 0.689 85 | 34.6 | 1.761 0X | 0.801 15 | 38.7 | 1.599 4X |
| 0.692 43 | 34.7 | 1.756 6X | 0.804 02 | 38.8 | 1.595 9X |
| 0.695 02 | 34.8 | 1.752 2X | 0.806 90 | 38.9 | 1.592 4X |
| 0.697 61 | 34.9 | 1.747 8X | | | |
| | | | 0.809 78 | 39.0 | 1.589 0X |
| 0.700 21 | 35.0 | 1.743 4X | 0.812 68 | 39.1 | 1.585 6X |
| 0.702 81 | 35.1 | 1.739 1X | 0.815 58 | 39.2 | 1.582 2X |
| 0.705 42 | 35.2 | 1.734 8X | 0.818 49 | 39.3 | 1.578 8X |
| 0.708 04 | 35.3 | 1.730 5X | 0.821 41 | 39.4 | 1.575 5X |
| 0.710 66 | 35.4 | 1.726 3X | | | |
| | | | 0.824 34 | 39.5 | 1.572 1X |
| 0.713 29 | 35.5 | 1.722 0X | 0.827 27 | 39.6 | 1.568 8X |
| 0.715 93 | 35.6 | 1.717 8X | 0.830 22 | 39.7 | 1.565 5X |
| 0.718 57 | 35.7 | 1.713 7X | 0.833 17 | 39.8 | 1.562 2X |
| 0.721 22 | 35.8 | 1.709 5X | 0.836 13 | 39.9 | 1.559 0X |
| 0.723 88 | 35.9 | 1.705 4X | | | |
| | | | 0.839 10 | 40.0 | 1.555 7X |
| 0.726 54 | 36.0 | 1.701 3X | 0.842 08 | 40.1 | 1.552 5X |
| 0.729 21 | 36.1 | 1.697 2X | 0.845 06 | 40.2 | 1.549 3X |
| 0.731 89 | 36.2 | 1.693 2X | 0.848 06 | 40.3 | 1.546 1X |
| 0.734 57 | 36.3 | 1.689 1X | 0.851 07 | 40.4 | 1.542 9X |
| 0.737 26 | 36.4 | 1.685 1X | | | |
| | | | 0.854 08 | 40.5 | 1.539 8X |
| 0.739 96 | 36.5 | 1.681 2X | 0.857 10 | 40.6 | 1.536 6X |
| 0.742 66 | 36.6 | 1.677 2X | 0.860 13 | 40.7 | 1.533 5X |
| 0.745 38 | 36.7 | 1.673 3X | 0.863 18 | 40.8 | 1.530 4X |
| 0.748 09 | 36.8 | 1.669 4X | 0.866 23 | 40.9 | 1.527 3X |

*Table 6-2. Continued.*

| $\dfrac{X}{R}$ | $\theta$ (degrees) | Z | $\dfrac{X}{R}$ | $\theta$ (degrees) | Z |
|---|---|---|---|---|---|
| 0.869 29 | 41.0 | 1.524 2X | 1.003 5 | 45.1 | 1.411 7X |
| 0.872 35 | 41.1 | 1.521 2X | 1.007 0 | 45.2 | 1.409 3X |
| 0.875 43 | 41.2 | 1.518 2X | 1.010 5 | 45.3 | 1.406 9X |
| 0.878 52 | 41.3 | 1.515 1X | 1.014 1 | 45.4 | 1.404 0X |
| 0.881 62 | 41.4 | 1.512 1X | 1.017 6 | 45.5 | 1.402 0X |
| 0.884 72 | 41.5 | 1.509 2X | 1.021 2 | 45.6 | 1.399 6X |
| 0.887 84 | 41.6 | 1.506 2X | 1.024 7 | 45.7 | 1.397 2X |
| 0.890 97 | 41.7 | 1.503 2X | 1.028 3 | 45.8 | 1.394 9X |
| 0.894 10 | 41.8 | 1.500 3X | 1.031 9 | 45.9 | 1.392 5X |
| 0.897 25 | 41.9 | 1.497 4X | 1.035 5 | 46.0 | 1.390 2X |
| 0.900 40 | 42.0 | 1.494 5X | 1.039 1 | 46.1 | 1.387 8X |
| 0.903 57 | 42.1 | 1.491 6X | 1.042 8 | 46.2 | 1.385 5X |
| 0.906 74 | 42.2 | 1.488 7X | 1.046 4 | 46.3 | 1.383 2X |
| 0.909 93 | 42.3 | 1.485 8X | 1.050 1 | 46.4 | 1.380 9X |
| 0.913 12 | 42.4 | 1.483 0X | 1.053 8 | 46.5 | 1.378 6X |
| 0.916 33 | 42.5 | 1.480 2X | 1.057 5 | 46.6 | 1.376 3X |
| 0.919 55 | 42.6 | 1.477 4X | 1.061 2 | 46.7 | 1.374 0X |
| 0.922 77 | 42.7 | 1.474 6X | 1.064 9 | 46.8 | 1.371 8X |
| 0.926 01 | 42.8 | 1.471 8X | 1.068 6 | 46.9 | 1.369 5X |
| 0.929 26 | 42.9 | 1.469 0X | 1.072 4 | 47.0 | 1.367 3X |
| 0.932 51 | 43.0 | 1.466 3X | 1.076 1 | 47.1 | 1.365 1X |
| 0.935 78 | 43.1 | 1.463 5X | 1.079 9 | 47.2 | 1.362 9X |
| 0.939 06 | 43.2 | 1.460 8X | 1.083 7 | 47.3 | 1.360 7X |
| 0.942 35 | 43.3 | 1.458 1X | 1.087 5 | 47.4 | 1.358 5X |
| 0.945 35 | 43.4 | 1.455 4X | 1.091 3 | 47.5 | 1.356 3X |
| 0.948 96 | 43.5 | 1.452 7X | 1.095 1 | 47.6 | 1.354 2X |
| 0.952 29 | 43.6 | 1.450 1X | 1.099 0 | 47.7 | 1.352 0X |
| 0.955 62 | 43.7 | 1.447 4X | 1.102 8 | 47.8 | 1.349 9X |
| 0.958 96 | 43.8 | 1.444 8X | 1.106 7 | 47.9 | 1.347 7X |
| 0.962 32 | 43.9 | 1.442 2X | 1.110 6 | 48.0 | 1.345 6X |
| 0.965 69 | 44.0 | 1.439 5X | 1.114 5 | 48.1 | 1.343 5X |
| 0.969 07 | 44.1 | 1.437 0X | 1.118 4 | 48.2 | 1.339 3X |
| 0.972 46 | 44.2 | 1.434 4X | 1.122 4 | 48.3 | 1.339 2X |
| 0.975 86 | 44.3 | 1.431 8X | 1.126 3 | 48.4 | 1.337 2X |
| 0.979 27 | 44.4 | 1.429 2X | 1.138 3 | 48.5 | 1.331 1X |
| 0.982 70 | 44.5 | 1.426 7X | 1.134 3 | 48.6 | 1.333 1X |
| 0.986 13 | 44.6 | 1.424 2X | 1.138 3 | 48.7 | 1.331 1X |
| 0.989 58 | 44.7 | 1.421 7X | 1.142 3 | 48.8 | 1.329 0X |
| 0.993 04 | 44.8 | 1.419 2X | 1.146 3 | 48.9 | 1.327 0X |
| 0.996 51 | 44.9 | 1.416 7X | 1.150 4 | 49.0 | 1.325 0X |
| 1.000 0 | 45.0 | 1.414 2X | 1.154 4 | 49.1 | 1.323 0X |

*Table 6-2. Continued.*

| $\dfrac{X}{R}$ | $\theta$ (degrees) | $Z$ | $\dfrac{X}{R}$ | $\theta$ (degrees) | $Z$ |
|---|---|---|---|---|---|
| 1.158 5 | 49.2 | 1.321 0X | 1.346 5 | 53.4 | 1.245 6X |
| 1.162 6 | 49.3 | 1.319 0X | | | |
| 1.166 7 | 49.4 | 1.317 0X | 1.351 4 | 53.5 | 1.244 0X |
| | | | 1.356 4 | 53.6 | 1.242 4X |
| 1.170 8 | 49.5 | 1.315 1X | 1.361 3 | 53.7 | 1.240 8X |
| 1.175 0 | 49.6 | 1.313 1X | 1.366 3 | 53.8 | 1.239 2X |
| 1.179 1 | 49.7 | 1.311 2X | 1.371 3 | 53.9 | 1.237 6X |
| 1.183 3 | 49.8 | 1.309 2X | | | |
| 1.187 5 | 49.9 | 1.307 3X | 1.376 4 | 54.0 | 1.236 1X |
| | | | 1.381 4 | 54.1 | 1.234 5X |
| 1.191 7 | 50.0 | 1.305 4X | 1.386 5 | 54.2 | 1.232 9X |
| 1.196 0 | 50.1 | 1.303 5X | 1.391 6 | 54.3 | 1.231 4X |
| 1.200 2 | 50.2 | 1.301 6X | 1.396 8 | 54.4 | 1.229 8X |
| 1.204 5 | 50.3 | 1.299 7X | | | |
| 1.208 8 | 50.4 | 1.297 8X | 1.401 9 | 54.5 | 1.228 3X |
| | | | 1.407 1 | 54.6 | 1.226 8X |
| 1.213 1 | 50.5 | 1.296 0X | 1.412 3 | 54.7 | 1.225 3X |
| 1.217 4 | 50.6 | 1.294 1X | 1.417 6 | 54.8 | 1.223 8X |
| 1.221 8 | 50.7 | 1.292 2X | 1.422 8 | 54.9 | 1.222 3X |
| 1.226 1 | 50.8 | 1.290 4X | | | |
| 1.230 5 | 50.9 | 1.288 6X | 1.428 1 | 55.0 | 1.220 8X |
| | | | 1.433 5 | 55.1 | 1.219 3X |
| 1.234 9 | 51.0 | 1.286 7X | 1.438 8 | 55.2 | 1.217 8X |
| 1.239 3 | 51.1 | 1.284 9X | 1.444 2 | 55.3 | 1.216 3X |
| 1.243 7 | 51.2 | 1.283 1X | 1.449 6 | 55.4 | 1.214 9X |
| 1.248 2 | 51.3 | 1.281 3X | | | |
| 1.252 7 | 51.4 | 1.279 5X | 1.455 0 | 55.5 | 1.213 4X |
| | | | 1.460 5 | 55.6 | 1.211 9X |
| 1.257 2 | 51.5 | 1.277 8X | 1.465 9 | 55.7 | 1.210 5X |
| 1.261 7 | 51.6 | 1.276 0X | 1.471 4 | 55.8 | 1.209 1X |
| 1.266 2 | 51.7 | 1.274 2X | 1.477 0 | 55.9 | 1.207 6X |
| 1.270 8 | 51.8 | 1.272 5X | | | |
| 1.275 3 | 51.9 | 1.270 7X | 1.482 6 | 56.0 | 1.206 2X |
| | | | 1.488 1 | 56.1 | 1.204 8X |
| 1.279 9 | 52.0 | 1.269 0X | 1.493 8 | 56.2 | 1.203 4X |
| 1.284 5 | 52.1 | 1.267 3X | 1.499 4 | 56.3 | 1.202 0X |
| 1.289 2 | 52.2 | 1.265 6X | 1.505 1 | 56.4 | 1.200 6X |
| 1.293 8 | 52.3 | 1.263 9X | | | |
| 1.298 5 | 52.4 | 1.262 2X | 1.510 8 | 56.5 | 1.199 2X |
| | | | 1.516 6 | 56.6 | 1.197 8X |
| 1.303 2 | 52.5 | 1.260 5X | 1.522 3 | 56.7 | 1.196 4X |
| 1.307 9 | 52.6 | 1.258 8X | 1.522 3 | 56.8 | 1.195 1X |
| 1.312 7 | 52.7 | 1.257 1X | 1.534 0 | 56.9 | 1.193 7X |
| 1.317 4 | 52.8 | 1.255 4X | | | |
| 1.322 2 | 52.9 | 1.253 8X | 1.539 9 | 57.0 | 1.192 4X |
| | | | 1.545 8 | 57.1 | 1.191 0X |
| 1.327 0 | 53.0 | 1.252 1X | 1.551 7 | 57.2 | 1.189 7X |
| 1.331 9 | 53.1 | 1.250 5X | 1.557 7 | 57.3 | 1.188 3X |
| 1.336 7 | 53.2 | 1.248 8X | 1.563 6 | 57.4 | 1.187 0X |
| 1.341 6 | 53.3 | 1.247 2X | | | |

Table 6-2. Continued.

| $\dfrac{X}{R}$ | $\theta$ (degrees) | Z | $\dfrac{X}{R}$ | $\theta$ (degrees) | Z |
|---|---|---|---|---|---|
| 1.569 7 | 57.5 | 1.185 7X | 1.849 5 | 61.6 | 1.136 8X |
| 1.575 7 | 57.6 | 1.184 4X | 1.857 2 | 61.7 | 1.135 7X |
| 1.581 8 | 57.7 | 1.183 1X | 1.865 0 | 61.8 | 1.134 7X |
| 1.588 0 | 57.8 | 1.181 8X | 1.872 8 | 61.9 | 1.133 6X |
| 1.594 1 | 57.9 | 1.180 5X | | | |
| | | | 1.880 7 | 62.0 | 1.132 6X |
| 1.600 3 | 58.0 | 1.179 2X | 1.888 7 | 62.1 | 1.131 5X |
| 1.600 6 | 58.1 | 1.177 9X | 1.896 7 | 62.2 | 1.130 5X |
| 1.612 8 | 58.2 | 1.176 6X | 1.904 7 | 62.3 | 1.129 4X |
| 1.619 1 | 58.3 | 1.175 3X | 1.912 8 | 62.4 | 1.128 4X |
| 1.625 5 | 58.4 | 1.174 1X | | | |
| | | | 1.921 0 | 62.5 | 1.127 4X |
| 1.631 8 | 58.5 | 1.172 8X | 1.929 2 | 62.6 | 1.126 4X |
| 1.638 3 | 58.6 | 1.171 6X | 1.937 5 | 62.7 | 1.125 3X |
| 1.644 7 | 58.7 | 1.170 3X | 1.945 8 | 62.8 | 1.124 3X |
| 1.651 2 | 58.8 | 1.169 1X | 1.954 2 | 62.9 | 1.123 3X |
| 1.657 7 | 58.9 | 1.167 8X | | | |
| | | | 1.962 6 | 63.0 | 1.122 3X |
| 1.664 3 | 59.0 | 1.166 6X | 1.971 1 | 63.1 | 1.121 3X |
| 1.670 9 | 59.1 | 1.165 4X | 1.979 7 | 63.2 | 1.120 3X |
| 1.677 5 | 59.2 | 1.164 2X | 1.988 3 | 63.3 | 1.119 3X |
| 1.684 2 | 59.3 | 1.163 0X | 1.996 9 | 63.4 | 1.118 4X |
| 1.690 9 | 59.4 | 1.161 8X | | | |
| | | | 2.005 7 | 63.5 | 1.117 4X |
| 1.697 7 | 59.5 | 1.160 6X | 2.014 5 | 63.6 | 1.116 4X |
| 1.704 4 | 59.6 | 1.159 4X | 2.023 3 | 63.7 | 1.115 5X |
| 1.711 3 | 59.7 | 1.158 2X | 2.032 3 | 63.8 | 1.114 5X |
| 1.718 2 | 59.8 | 1.157 0X | 2.041 2 | 63.9 | 1.113 5X |
| 1.725 1 | 59.9 | 1.155 9X | | | |
| | | | 2.050 3 | 64.0 | 1.112 6X |
| 1.732 0 | 60.0 | 1.154 7X | 2.059 4 | 64.1 | 1.111 6X |
| 1.739 0 | 60.1 | 1.153 5X | 2.068 6 | 64.2 | 1.110 7X |
| 1.746 1 | 60.2 | 1.152 4X | 2.077 8 | 64.3 | 1.109 8X |
| 1.753 2 | 60.3 | 1.151 2X | 2.087 2 | 64.4 | 1.108 8X |
| 1.760 3 | 60.4 | 1.150 1X | | | |
| | | | 2.096 5 | 64.5 | 2.096 9X |
| 1.767 5 | 60.5 | 1.148 9X | 2.106 0 | 64.6 | 1.107 0X |
| 1.774 7 | 60.6 | 1.147 8X | 2.111 5 | 64.7 | 1.106 1X |
| 1.782 0 | 60.7 | 1.146 7X | 2.125 1 | 64.8 | 1.105 2X |
| 1.789 3 | 60.8 | 1.145 6X | 2.134 8 | 64.9 | 1.104 3X |
| 1.796 6 | 60.9 | 1.144 5X | | | |
| | | | 2.144 5 | 65.0 | 1.103 4X |
| 1.804 0 | 61.0 | 1.143 3X | 2.154 3 | 65.1 | 1.102 5X |
| 1.811 5 | 61.1 | 1.142 2X | 2.164 2 | 65.2 | 1.101 6X |
| 1.819 0 | 61.2 | 1.141 1X | 2.174 1 | 65.3 | 1.100 7X |
| 1.826 5 | 61.3 | 1.140 1X | 2.184 2 | 65.4 | 1.099 8X |
| 1.834 1 | 61.4 | 1.139 0X | | | |
| | | | 2.194 3 | 65.5 | 1.098 9X |
| 1.841 8 | 61.5 | 1.137 9X | 2.204 5 | 65.6 | 1.098 1X |
| | | | 2.214 7 | 65.7 | 1.097 2X |

Table 6-2. Continued.

| $\dfrac{X}{R}$ | $\theta$ (degrees) | Z | $\dfrac{X}{R}$ | $\theta$ (degrees) | Z |
|---|---|---|---|---|---|
| 2.225 1 | 65.8 | 1.096 3X | 2.732 6 | 69.9 | 1.064 8X |
| 2.235 5 | 65.9 | 1.095 5X | | | |
| | | | 2.747 5 | 70.0 | 1.064 2X |
| 2.246 0 | 66.0 | 1.094 6X | 2.762 5 | 70.1 | 1.063 5X |
| 2.256 6 | 66.1 | 1.093 8X | 2.777 6 | 70.2 | 1.062 8X |
| 2.267 3 | 66.2 | 1.092 9X | 2.792 9 | 70.3 | 1.062 2X |
| 2.278 1 | 66.3 | 1.092 1X | 2.808 3 | 70.4 | 1.061 5X |
| 2.288 9 | 66.4 | 1.091 3X | | | |
| | | | 2.823 9 | 70.5 | 1.060 8X |
| 2.299 8 | 66.5 | 1.090 4X | 2.839 6 | 70.6 | 1.060 2X |
| 2.310 9 | 66.6 | 1.089 6X | 2.855 5 | 70.7 | 1.059 5X |
| 2.322 0 | 66.7 | 1.088 8X | 2.871 6 | 70.8 | 1.058 9X |
| 2.333 2 | 66.8 | 1.088 0X | 2.887 8 | 70.9 | 1.058 2X |
| 2.344 5 | 66.9 | 1.087 2X | | | |
| | | | 2.904 2 | 71.0 | 1.057 6X |
| 2.355 8 | 67.0 | 1.086 4X | 2.920 8 | 71.1 | 1.057 0X |
| 2.367 3 | 67.1 | 1.085 5X | 2.937 5 | 71.2 | 1.056 3X |
| 2.378 9 | 67.2 | 1.084 7X | 2.954 4 | 71.3 | 1.055 7X |
| 2.390 6 | 67.3 | 1.084 0X | 2.971 4 | 71.4 | 1.055 1X |
| 2.402 3 | 67.4 | 1.083 2X | | | |
| | | | 2.988 7 | 71.5 | 1.054 5X |
| 2.414 2 | 67.5 | 1.082 4X | 3.006 1 | 71.6 | 1.053 9X |
| 2.426 2 | 67.6 | 1.081 6X | 3.023 7 | 71.7 | 1.053 3X |
| 2.438 2 | 67.7 | 1.080 8X | 3.041 5 | 71.8 | 1.052 7X |
| 2.450 4 | 67.8 | 1.080 1X | 3.059 5 | 71.9 | 1.052 1X |
| 2.462 7 | 67.9 | 1.079 3X | | | |
| | | | 3.077 7 | 72.0 | 1.051 5X |
| 2.475 1 | 68.0 | 1.078 5X | 3.096 0 | 72.1 | 1.050 9X |
| 2.487 6 | 68.1 | 1.077 8X | 3.114 6 | 72.2 | 1.050 3X |
| 2.500 2 | 68.2 | 1.077 0X | 3.133 4 | 72.3 | 1.049 7X |
| 2.512 9 | 68.3 | 1.076 3X | 3.152 4 | 72.4 | 1.049 1X |
| 2.525 7 | 68.4 | 1.075 5X | | | |
| | | | 3.171 6 | 72.5 | 1.048 5X |
| 2.538 6 | 68.5 | 1.074 8X | 3.191 0 | 72.6 | 1.047 9X |
| 2.551 7 | 68.6 | 1.074 0X | 3.210 6 | 72.7 | 1.047 4X |
| 2.564 9 | 68.7 | 1.073 3X | 3.230 5 | 72.8 | 1.046 8X |
| 2.578 1 | 68.8 | 1.072 6X | 3.250 5 | 72.9 | 1.046 2X |
| 2.591 6 | 68.9 | 1.071 9X | | | |
| | | | 3.270 8 | 73.0 | 1.045 7X |
| 2.605 1 | 69.0 | 1.071 1X | 3.291 4 | 73.1 | 1.045 1X |
| 2.618 7 | 69.1 | 1.070 4X | 3.312 1 | 73.2 | 1.046 6X |
| 2.632 5 | 69.2 | 1.069 7X | 3.333 2 | 73.3 | 1.044 0X |
| 2.646 4 | 69.3 | 1.069 0X | 3.354 4 | 73.4 | 1.043 5X |
| 2.660 4 | 69.4 | 1.068 3X | | | |
| | | | 3.375 9 | 73.5 | 1.042 9X |
| 2.674 6 | 69.5 | 1.067 6X | 3.397 7 | 73.6 | 1.042 4X |
| 2.688 9 | 69.6 | 1.066 9X | 3.419 7 | 73.7 | 1.041 9X |
| 2.703 3 | 69.7 | 1.066 2X | 3.442 0 | 73.8 | 1.041 3X |
| 2.717 9 | 69.8 | 1.065 5X | 3.464 6 | 73.9 | 1.040 8X |

Table 6-2. Continued.

| $\frac{X}{R}$ | $\theta$ (degrees) | Z | $\frac{X}{R}$ | $\theta$ (degrees) | Z |
|---|---|---|---|---|---|
| 3.487 4 | 74.0 | 1.040 3X | 4.745 3 | 78.1 | 1.022 0X |
| 3.510 5 | 74.1 | 1.039 8X | 4.786 7 | 78.2 | 1.021 6X |
| 3.533 9 | 74.2 | 1.039 3X | 4.828 8 | 78.3 | 1.021 2X |
| 3.557 6 | 74.3 | 1.038 7X | 4.871 6 | 78.4 | 1.020 8X |
| 3.581 6 | 74.4 | 1.038 2X | 4.915 1 | 78.5 | 1.020 5X |
| 3.605 9 | 74.5 | 1.037 7X | 4.959 4 | 78.6 | 1.020 1X |
| 3.630 5 | 74.6 | 1.037 2X | 5.004 5 | 78.7 | 1.019 8X |
| 3.655 4 | 74.7 | 1.036 7X | 5.050 4 | 78.8 | 1.019 4X |
| 3.680 6 | 74.8 | 1.036 2X | 5.097 0 | 78.9 | 1.019 1X |
| 3.706 2 | 74.9 | 1.035 8X | 5.144 5 | 79.0 | 1.018 7X |
| 3.732 0 | 75.0 | 1.035 3X | 5.192 9 | 79.1 | 1.018 4X |
| 3.758 3 | 75.1 | 1.034 8X | 5.242 2 | 79.2 | 1.018 0X |
| 3.784 8 | 75.2 | 1.034 3X | 5.292 3 | 79.3 | 1.017 7X |
| 3.811 8 | 75.3 | 1.033 8X | 5.343 4 | 79.4 | 1.017 4X |
| 3.839 0 | 75.4 | 1.033 4X | 5.395 5 | 79.5 | 1.017 0X |
| 3.866 7 | 75.5 | 1.032 9X | 5.448 6 | 79.6 | 1.016 7X |
| 3.894 7 | 75.6 | 1.032 4X | 5.502 6 | 79.7 | 1.016 4X |
| 3.923 1 | 75.7 | 1.032 0X | 5.557 8 | 79.8 | 1.016 0X |
| 3.952 0 | 75.8 | 1.031 5X | 5.614 0 | 79.9 | 1.015 7X |
| 3.981 2 | 75.9 | 1.031 1X | 5.671 3 | 80.0 | 1.015 4X |
| 4.010 8 | 76.0 | 1.030 6X | 5.729 7 | 80.1 | 1.015 1X |
| 4.040 8 | 76.1 | 1.030 2X | 5.789 4 | 80.2 | 1.014 8X |
| 4.071 3 | 76.2 | 1.029 7X | 5.850 2 | 80.3 | 1.014 5X |
| 4.102 2 | 76.3 | 1.029 3X | 5.912 3 | 80.4 | 1.014 2X |
| 4.133 5 | 76.4 | 1.028 8X | 5.975 8 | 80.5 | 1.013 9X |
| 4.165 3 | 76.5 | 1.028 4X | 6.040 5 | 80.6 | 1.013 6X |
| 4.197 6 | 76.6 | 1.028 0X | 6.106 6 | 80.7 | 1.013 3X |
| 4.230 3 | 76.7 | 1.027 6X | 6.174 2 | 80.8 | 1.013 0X |
| 4.263 5 | 76.8 | 1.027 1X | 6.243 2 | 80.9 | 1.012 7X |
| 4.297 2 | 76.9 | 1.026 7X | 6.313 7 | 81.0 | 1.012 5X |
| 4.331 5 | 77.0 | 1.026 3X | 6.385 9 | 81.1 | 1.012 2X |
| 4.366 2 | 77.1 | 1.025 9X | 6.459 6 | 81.2 | 1.011 9X |
| 4.401 5 | 77.2 | 1.022 5X | 6.535 0 | 81.3 | 1.011 6X |
| 4.437 3 | 77.3 | 1.025 1X | 6.612 2 | 81.4 | 1.011 4X |
| 4.473 7 | 77.4 | 1.024 7X | 6.691 1 | 81.5 | 1.011 1X |
| 4.510 7 | 77.5 | 1.024 3X | 6.772 0 | 81.6 | 1.010 8X |
| 4.548 3 | 77.6 | 1.023 9X | 6.854 7 | 81.7 | 1.010 6X |
| 4.586 4 | 77.7 | 1.023 5X | 6.939 5 | 81.8 | 1.010 3X |
| 4.625 2 | 77.8 | 1.023 1X | 7.026 4 | 81.9 | 1.010 1X |
| 4.664 6 | 77.9 | 1.022 7X | 7.115 4 | 82.0 | 1.009 8X |
| 4.704 6 | 78.0 | 1.022 3X | 7.206 6 | 82.1 | 1.009 6X |

*Table 6-2. Continued.*

| $\dfrac{X}{R}$ | $\theta$ (degrees) | Z | $\dfrac{X}{R}$ | $\theta$ (degrees) | Z |
|---|---|---|---|---|---|
| 7.300 2 | 82.2 | 1.009 3X | 15.056 | 86.2 | 1.002 2X |
| 7.396 1 | 82.3 | 1.009 1X | 15.464 | 86.3 | 1.002 1X |
| 7.494 6 | 82.4 | 1.008 9X | 15.894 | 86.4 | 1.002 0X |
| 7.595 7 | 82.5 | 1.008 6X | 16.350 | 86.5 | 1.001 9X |
| 7.699 6 | 82.6 | 1.008 4X | 16.832 | 86.6 | 1.001 8X |
| 7.806 2 | 82.7 | 1.008 2X | 17.343 | 86.7 | 1.001 7X |
| 7.915 8 | 82.8 | 1.007 9X | 17.886 | 86.8 | 1.001 6X |
| 8.028 5 | 82.9 | 1.007 7X | 18.464 | 86.9 | 1.001 5X |
| 8.144 3 | 83.0 | 1.007 5X | 19.081 | 87.0 | 1.001 4X |
| 8.263 5 | 83.1 | 1.007 3X | 19.740 | 87.1 | 1.001 3X |
| 8.386 2 | 83.2 | 1.007 1X | 20.446 | 87.2 | 1.001 2X |
| 8.512 6 | 83.3 | 1.006 9X | 21.205 | 87.3 | 1.001 1X |
| 8.642 7 | 83.4 | 1.006 7X | 22.022 | 87.4 | 1.001 0X |
| 8.776 9 | 83.5 | 1.006 5X | 23.904 | 87.5 | 1.000 9X |
| 8.915 2 | 83.6 | 1.006 3X | 24.859 | 87.6 | 1.000 9X |
| 9.057 9 | 83.7 | 1.006 1X | 24.898 | 87.7 | 1.000 8X |
| 9.205 1 | 83.8 | 1.005 9X | 26.031 | 87.8 | 1.000 7X |
| 9.357 2 | 83.9 | 1.005 7X | 27.271 | 87.9 | 1.000 7X |
| 9.514 4 | 84.0 | 1.005 5X | 28.636 | 88.0 | 1.000 6X |
| 9.676 8 | 84.1 | 1.005 3X | 30.145 | 88.1 | 1.000 5X |
| 9.844 8 | 84.2 | 1.005 1X | 31.820 | 88.2 | 1.000 5X |
| 10.019 | 84.3 | 1.005 0X | 33.693 | 88.3 | 1.000 4X |
| 10.199 | 84.4 | 1.004 8X | 35.800 | 88.4 | 1.000 4X |
| 10.385 | 84.5 | 1.004 6X | 38.188 | 88.5 | 1.000 3X |
| 10.579 | 84.6 | 1.004 4X | 40.917 | 88.6 | 1.000 3X |
| 10.780 | 84.7 | 1.004 3X | 44.066 | 88.7 | 1.000 2X |
| 10.988 | 84.8 | 1.004 1X | 47.739 | 88.8 | 1.000 2X |
| 11.205 | 84.9 | 1.004 0X | 52.081 | 88.9 | 1.000 2X |
| 11.430 | 85.0 | 1.003 8X | 57.290 | 89.0 | 1.000 1X |
| 11.664 | 85.1 | 1.003 7X | 63.657 | 89.1 | 1.000 1X |
| 11.909 | 85.2 | 1.003 5X | 71.615 | 89.2 | 1.000 1X |
| 12.163 | 85.3 | 1.003 4X | 81.847 | 89.3 | 1.000 0X |
| 12.429 | 85.4 | 1.003 2X | 95.489 | 89.4 | 1.000 0X |
| 12.706 | 85.5 | 1.003 1X | 114.59 | 89.5 | 1.000 0X |
| 12.996 | 85.6 | 1.002 9X | 143.24 | 89.6 | 1.000 0X |
| 13.229 | 85.7 | 1.002 8X | 190.98 | 89.7 | 1.000 0X |
| 13.617 | 85.8 | 1.002 7X | 286.48 | 89.8 | 1.000 0X |
| 13.951 | 85.9 | 1.002 6X | 572.96 | 89.9 | 1.000 0X |
| 14.301 | 86.0 | 1.002 4X | $R=0$ | 90.0 | 1.000 0X |
| 14.668 | 86.1 | 1.002 3X | | | |

Solving for $R$ you get $R=3\ 000\ \Omega$. And, since you know that the phase angle is negative, you also know that your j term will also be negative. Our answer, then, is $3\ 000-j4\ 000$.

***Example*** What is the impedance of a circuit whose resistance, $R$, is 1 000 $\Omega$ and whose reactance, $X$, is 775 $\Omega$?

***Solution*** The ratio, $X/R$ is 775/1 000 or 0.775. Locating the nearest equivalent number in the first column of Table 6-2 gives 0.775 68. To the right of this number, the phase angle is 37.8° and, continuing to the right, $Z=1.631\ 6X$. Thus, the impedance, $Z=1.631\ 6\times775$, or 1 264.49 $\Omega$.

## Impedance and turns ratio

Transformers are conveniently used as impedance transformation devices (Fig. 6-2). The impedance of a transformer varies as the square of the turns ratio. These ratios, in steps of 1 to 100, are given in Table 6-3.

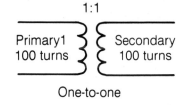

1:1

Primary1
100 turns

Secondary
100 turns

One-to-one

*Fig. 6-2. The turns ratio is the number of turns in the secondary winding compared to the number of turns in the primary.*

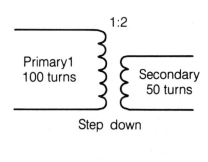

1:2

Primary1
100 turns

Secondary
50 turns

Step down

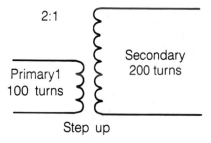

2:1

Secondary
200 turns

Primary1
100 turns

Step up

*Table 6-3. Impedance ratio and turns ratio of a transformer.*

| Turns ratio | Impedance ratio | Turns ratio | Impedance ratio |
|---|---|---|---|
| 100:1 | 10 000:1 | 58:1 | 3 364:1 |
| 99:1 | 9 801:1 | 57:1 | 3 249:1 |
| 98:1 | 9 604:1 | 56:1 | 3 136:1 |
| 97:1 | 9 409:1 | 55:1 | 3 025:1 |
| 96:1 | 9 216:1 | 54:1 | 2 916:1 |
| 95:1 | 9 025:1 | 53:1 | 2 809:1 |
| 94:1 | 8 836:1 | 52:1 | 2 704:1 |
| 93:1 | 8 649:1 | 51:1 | 2 601:1 |
| 92:1 | 8 464:1 | 50:1 | 2 500:1 |
| 91:1 | 8 281:1 | 49:1 | 2 401:1 |
| 90:1 | 8 100:1 | 48:1 | 2 304:1 |
| 89:1 | 7 921:1 | 47:1 | 2 209:1 |
| 88:1 | 7 744:1 | 46:1 | 2 116:1 |
| 87:1 | 7 569:1 | 45:1 | 2 025:1 |
| 86:1 | 7 396:1 | 44:1 | 1 936:1 |
| 85:1 | 7 225:1 | 43:1 | 1 849:1 |
| 84:1 | 7 056:1 | 42:1 | 1 764:1 |
| 83:1 | 6 889:1 | 41:1 | 1 681:1 |
| 82:1 | 6 724:1 | 40:1 | 1 600:1 |
| 81:1 | 6 571:1 | 39:1 | 1 521:1 |
| 80:1 | 6 400:1 | 38:1 | 1 444:1 |
| 79:1 | 6 241:1 | 37:1 | 1 369:1 |
| 78:1 | 6 084:1 | 36:1 | 1 296:1 |
| 77:1 | 5 929:1 | 35:1 | 1 225:1 |
| 76:1 | 5 776:1 | 34:1 | 1 156:1 |
| 75:1 | 5 625:1 | 33:1 | 1 089:1 |
| 74:1 | 5 476:1 | 32:1 | 1 024:1 |
| 73:1 | 5 329:1 | 31:1 | 961:1 |
| 72:1 | 5 184:1 | 30:1 | 900:1 |
| 71:1 | 5 041:1 | 29:1 | 841:1 |
| 70:1 | 4 900:1 | 28:1 | 784:1 |
| 69:1 | 4 761:1 | 27:1 | 729:1 |
| 68:1 | 4 624:1 | 26:1 | 676:1 |
| 67:1 | 4 489:1 | 25:1 | 625:1 |
| 66:1 | 4 356:1 | 24:1 | 576:1 |
| 65:1 | 4 225:1 | 23:1 | 529:1 |
| 64:1 | 4 096:1 | 22:1 | 484:1 |
| 63:1 | 3 969:1 | 21:1 | 441:1 |
| 62:1 | 3 844:1 | 20:1 | 400:1 |
| 61:1 | 3 721:1 | 19:1 | 361:1 |
| 60:1 | 3 600:1 | 18:1 | 324:1 |
| 59:1 | 3 481:1 | 17:1 | 289:1 |

*Table 6-3. Continued.*

| Turns ratio | Impedance ratio | Turns ratio | Impedance ratio |
|---|---|---|---|
| 16:1 | 256:1 | 1:75 | 1:5 625 |
| | | 1:74 | 1:5 476 |
| 15:1 | 225:1 | 1:73 | 1:5 329 |
| 14:1 | 196:1 | 1:72 | 1:5 184 |
| 13:1 | 169:1 | 1:71 | 1:5 041 |
| 12:1 | 144:1 | | |
| 11:1 | 121:1 | 1:70 | 1:4 900 |
| | | 1:69 | 1:4 761 |
| 10:1 | 100:1 | 1:68 | 1:4 624 |
| 9:1 | 81:1 | 1:67 | 1:4 489 |
| 8:1 | 64:1 | 1:66 | 1:4 356 |
| 7:1 | 49:1 | | |
| 6:1 | 36:1 | 1:65 | 1:4 225 |
| | | 1:64 | 1:4 096 |
| 5:1 | 25:1 | 1:63 | 1:3 969 |
| 4:1 | 16:1 | 1:62 | 1:3 844 |
| 3:1 | 9:1 | 1:61 | 1:3 721 |
| 2:1 | 4:1 | | |
| 1:1 | 1:1 | 1:60 | 1:3 600 |
| | | 1:59 | 1:3 481 |
| | | 1:58 | 1:3 364 |
| 1:100 | 1:10 000 | 1:57 | 1:3 249 |
| 1:99 | 1:9 801 | 1:56 | 1:3 136 |
| 1:98 | 1:9 604 | | |
| 1:97 | 1:9 409 | 1:55 | 1:3 025 |
| 1:96 | 1:9 216 | 1:54 | 1:2 916 |
| | | 1:53 | 1:2 809 |
| 1:95 | 1:9 025 | 1:52 | 1:2 704 |
| 1:94 | 1:8 836 | 1:51 | 1:2 601 |
| 1:93 | 1:8 649 | | |
| 1:92 | 1:8 464 | 1:50 | 1:2 500 |
| 1:91 | 1:8 281 | 1:49 | 1:2 401 |
| | | 1:48 | 1:2 304 |
| 1:90 | 1:8 100 | 1:47 | 1:2 209 |
| 1:89 | 1:7 921 | 1:46 | 1:2 116 |
| 1:88 | 1:7 744 | | |
| 1:87 | 1:7 569 | 1:45 | 1:2 025 |
| 1:86 | 1:7 396 | 1:44 | 1:1 936 |
| | | 1:43 | 1:1 849 |
| 1:85 | 1:7 225 | 1:42 | 1:1 764 |
| 1:84 | 1:7 056 | 1:41 | 1:1 681 |
| 1:83 | 1:6 889 | | |
| 1:82 | 1:6 724 | 1:40 | 1:1 600 |
| 1:81 | 1:6 571 | 1:39 | 1:1 521 |
| | | 1:38 | 1:1 444 |
| 1:80 | 1:6 400 | 1:37 | 1:1 369 |
| 1:79 | 1:6 241 | 1:36 | 1:1 296 |
| 1:78 | 1:6 084 | | |
| 1:77 | 1:5 929 | 1:35 | 1:1 225 |
| 1:76 | 1:5 776 | 1:34 | 1:1 156 |

*Table 6-3. Continued.*

| Turns ratio | Impedance ratio | Turns ratio | Impedance ratio |
|---|---|---|---|
| 1:33 | 1:1 089 | 1:16 | 1:256 |
| 1:32 | 1:1 024 | | |
| 1:31 | 1:961 | 1:15 | 1:225 |
| 1:30 | 1:900 | 1:14 | 1:196 |
| | | 1:13 | 1:169 |
| 1:29 | 1:841 | 1:12 | 1:144 |
| 1:28 | 1:784 | 1:11 | 1:121 |
| 1:27 | 1:729 | | |
| 1:26 | 1:676 | 1:10 | 1:100 |
| 1:25 | 1:625 | | |
| | | 1:9 | 1:81 |
| 1:24 | 1:576 | 1:8 | 1:64 |
| | | 1:7 | 1:49 |
| 1:23 | 1:529 | | |
| 1:22 | 1:484 | 1:6 | 1:36 |
| 1:21 | 1:441 | 1:5 | 1:25 |
| | | 1:4 | 1:16 |
| 1:20 | 1:400 | 1:3 | 1:9 |
| 1:19 | 1:361 | 1:2 | 1:4 |
| 1:18 | 1:324 | | |
| 1:17 | 1:289 | 1:1 | 1:1 |

*Example*  A step-down transformer has a turns ratio of 37 to 1. What is the impedance ratio?

*Solution*  Locate the turns ratio, 37:1 in Table 6-3. Immediately alongside is the impedance ratio of 1 369:1.

*Example*  What is the impedance transformation of a step-up transformer having a turns ratio of 1:53?

*Solution*  Table 6-3 shows that the impedance transformation for this turns ratio is 1:2 809.

The turns ratio of a transformer is specifically that—a ratio—and gives no indication of the actual number of turns used by the transformer. A transformer having 50 primary turns and 50 secondary turns has a 1:1 ratio. So does a transformer having 5 000 primary and 5 000 secondary turns.

To get the ratio, divide the larger number of turns by the smaller. Table 6-3 will then supply the impedance ratio.

*Example*  A step-down transformer has 2 465 primary turns and 85 secondary turns. What is the impedance ratio?

*Solution*  Divide 2 465 by 85. This supplies a ratio of 29:1. Table 6-3 shows the impedance ratio is 841:1. If the transformer had 85 primary turns and 2 465 secondary turns, the impedance ratio would be 1:841.

The turns ratio of a transformer and its primary and secondary impedances can be calculated from:

$$N = \frac{\sqrt{Z_s}}{\sqrt{Z_p}}$$

$$N^2 = \frac{Z_s}{Z_p}$$

$$Z_s = Z_p \, N^2$$

$$Z_p = \frac{Z_s}{N^2}$$

$$\frac{Z_s}{Z_p} = \frac{N_s^2}{N_p^2} = N^2$$

N is the turns ratio, $Z_s$ is the secondary impedance, $Z_p$ is the primary impedance, $N_s$ is the number of secondary turns, and $N_p$ is the number of primary turns.

## Summary of impedance formulas

A summary of impedance formulas is supplied in Table 6-4.

*Table 6-4. Summary of impedance formulas.*

Impedance of R circuit:

$$Z = R$$

Impedance of series RC circuit:

$$Z = \sqrt{R_2 + X_C^2}$$

Impedance of series RL circuit:

$$Z = \sqrt{R_2 + X_L^2}$$

Impedance of series RLC circuit:

$$Z = \sqrt{R_2 + (X_L - X_C)^2} \quad X_L \text{ is larger than } X_C$$
$$Z = \sqrt{R_2 + (X_C - X_L)^2} \quad X_C \text{ is larger than } X_L$$
$$Z = \sqrt{R_2 + X^2}$$

Vector addition of reactances:

$$X = X_L - X_C$$
$$X = X_C - X_L$$

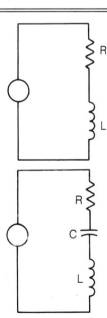

*Table 6-4. Continued.*

Impedance of parallel RL circuit:

$$Z = \frac{RX_L}{\sqrt{R_2 + X_L{}^2}}$$

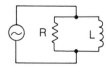

Impedance of parallel RC circuit:

$$Z = \frac{RX_C}{\sqrt{R_2 + X_C{}^2}}$$

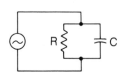

Impedance of parallel RLC circuit:

$$Z = \frac{RX_L \, X_C}{\sqrt{X_L{}^2 \, X_C{}^2 + R_2 \, (X_L - X_C)^2}}$$

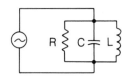

Impedance and phase angle of series RL shunted by C:

$$Z = X_C \sqrt{\frac{R_2 \, X_L{}^2}{R_2 + (X_L - X_C)^2}}$$

$$\phi = \text{arc tan} \frac{X_L \, (X_C - X_L) \, - R_2}{RX_C}$$

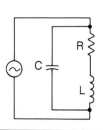

# 7
CHAPTER

# Magnetics

## The English, CGS, and MKS systems

The three systems of units used in magnetic circuit calculations are the so-called practical or English, the CGS (centimeter-gram-second), and the MKS (meter-kilogram-second). The last two are both metric, while the practical system uses English units. The basic difference between the CGS and the MKS is size of units. The CGS (Table 7-1) uses the centimeter for its measurements, and the MKS (Table 7-2) uses meters. The unit of length in the English system (Table 7-3) is the inch, for example, a magnetizing force ($H$) is one ampere-turn per inch.

It is sometimes difficult to avoid confusion when working with magnetic problems since the different systems can use identical letters. The letter $H$, representing magnetomotive force, is used in all three systems. In the English system,

*Table 7-1. CGS units.*

Centimeter
Oersted
Gram
Dyne
Gilbert
Erg
Ergs per second
Grams per cubic centimeter
Dynes per square centimeter
Centimeters per second
Dynes per centimeter
Gauss
Second
Maxwell

*Table 7-2. MKS units.*

Meter
Kilogram
Joule
Kilograms per cubic centimeter
Newtons per square meter
Webers per square meter
Tesla (webers per square meter)
Second
Ampere-turns per meter
Ampere-turns per weber
Webers per ampere-turn

*Table 7-3. English units.*

Inch
Inch$^2$
Foot
Lines per square inch
Amperes
Ampere-turns per inch

*Table 7-4. Metric-to-English*
*and English-to-metric conversions.*

| | **Metric units to English equivalents** | | |
|---|---|---|---|
| Lengths | 1 mm | 0.039 37 | in |
| | 1 cm | 0.393 7 | in |
| | 1 m | 39.37 | in or 1.093 6 yd |
| | 1 km | 1 093.61 | yd or 0.621 4 mi |
| Areas | 1 mm$^2$ | 0.001 55 | in$^2$ |
| | 1 cm$^2$ | 0.155 | in$^2$ |
| | 1 m$^2$ | 10.764 | ft$^2$ |
| | | 1.196 | yd$^2$ |
| | 1 km$^2$ | 0.386 1 | mi$^2$ |
| Volumes | | 0.000 061 | in$^3$ |
| | 1 mm$^3$ | 0.061 | in$^3$ |
| | 1 cm$^3$ | 61.025 | in$^3$ |
| | 1 L | 35.314 | ft$^3$ or |
| | 1 m$^3$ | 1.307 9 | yd$^3$ |
| | **English system units to metric equivalents** | | |
| Lengths | 1 in | 25.4 | mm or |
| | | 2.54 | cm |
| | 1 ft | 0.304 8 | m |
| | 1 yd | 0.914 4 | m |
| | 1 mi | 1.609 3 | km |
| Areas | 1 in$^2$ | 645.16 | mm$^2$ |
| | | 6.452 | cm$^2$ |
| | 1 ft$^2$ | 0.092 9 | m$^2$ |
| | 1 yd$^2$ | 0.836 1 | m$^2$ |
| | 1 mi$^2$ | 2.59 | km$^2$ |
| Volumes | 1 in$^3$ | 16 387.2 | mm$^3$ |
| | | 16.387 2 | cm$^3$ |
| | 1 ft$^3$ | 0.028 32 | m$^3$ |
| | 1 yd$^3$ | 0.764 6 | m$^3$ |

this letter is always written in Roman style (H), but it is italicized in the metric systems (CGS or MKS).

There is still another, lesser-known system, referred to as the *Georgi* system. It is similar to the MKS system. Both the MKS and the Georgi system come under the general heading of SI (Système Internationale or International System of Measurement).

It is sometimes necessary to work between metric and English units. Table 7-4 supplies the conversion factors.

## Conversion factors

It might seem odd to have three different systems of units for working problems in magnetics. Basically, there are only two: English and metric. In the metric system, two subsystems are required: MKS for working with large number problems and CGS for small number problems. Sometimes it is necessary to work back and forth. For this purpose it will be helpful to use the conversion factors in Table 7-5.

*Table 7-5. Conversion factors for MKS, CGS and English units.*

| Multiply | by | To obtain |
|---|---|---|
| $F$ in A-turns | $0.4 \pi = 1.257$ | $F$ in Gb |
| $F$ in Gb | $1/0.4 \pi = 0.796$ | $F$ in A-turns |
| $H$ in A-turns/in | $0.4 \pi/2.54 = 0.495$ | $H$ in Oe |
| $H$ in Oe | $2.54/0.4 \pi = 2.02$ | $H$ in A-turns/in |
| $\beta$ in Mx/in$^2$ <br> $\beta$ in G <br> $\beta$ in Mx/cm$^2$ } | $1/6.45 = 0.155$ <br> $6.45$ | $\beta$ in G <br> $\beta$ in Mx/in$^2$ |
| $\beta$ in Wb/m$^2$ | $10^4$ | $\beta$ in G |
| $\beta$ in G | $10^{-4}$ | $\beta$ in Wb/m$^2$ |
| $\beta$ in Mx/in$^2$ | $10^{-4}/6.45 = 0.155 \times 10^{-4}$ | $\beta$ in Wb/m$^2$ |
| $\beta$ in Wb/m$^2$ | $6.45 \times 10^4$ | $\beta$ in Mx/in$^2$ |
| $\phi$ in Mx <br> $\phi$ in lines of flux } | $10^{-8}$ | $\phi$ in Wb |
| $\phi$ in Wb | $10^8$ | $\phi$ in Mx |

$F$ = magnetomotive force; $H$ = field intensity; $\beta$ = flux density; $\phi$ = flux

## Linear measurements

The physical measurements of a magnet are sometimes involved in problems. It is sometimes necessary to convert from a larger metric unit to a smaller one or vice versa. Additionally, it might be helpful to work back and forth between English and metric units. These are listed in Table 7-6.

*Table 7-6.*
*Linear measurements
in English and metric.*

| | | |
|---|---|---|
| 1 μm | = | 0.001 mm |
| 1 μm | = | 0.000 001 m |
| 1 mm | = | 0.039 370 0 in |
| 1 mm | = | 0.003 28 ft |
| 1 cm | = | 10 mm |
| 1 cm | = | 0.393 700 in |
| 1 cm | = | 0.032 808 ft |
| 1 cm | = | 0.010 936 11 yd |
| 1 m | = | 39.370 0 in |
| 1 m | = | 3.280 833 333 ft |
| 1 m | = | 1.093 61 yd |
| 1 dm | = | 10 cm |
| 1 dm | = | 3.937 in |
| 1 m | = | 10 dm |
| 1 m | = | 100 cm |
| 1 m | = | 1 000 mm |
| 1 dam | = | 10 m |
| 1 dam | = | 393.7 in |
| 1 hm | = | 10 dam |
| 1 hm | = | 328 ft, 1 in |
| 1 km | = | 10 hm |
| 1 km | = | 0.621 37 mi |
| 1 in | = | 25.400 05 mm |
| 1 in | = | 2.540 005 cm |
| 1 in | = | 0.025 400 05 m |
| 1 ft | = | 30.480 06 cm |

# Terms and formulas for magnetics

Just as terms are defined and formulas are available for work in dc and ac circuits, these are also required for work involving the production and use of magnetic fields. Working with magnetic fields is not a discipline that is separate and distinct for dc and ac. The two are interrelated. Magnetism can be used to produce a current; the flow of current in a straight wire or coil, whether dc or ac, is always accompanied by a magnetic field.

## Flux

*Flux* consists of magnetic lines of force. These lines are continuous and a partial line does not exist. Lines of flux surround all magnets, whether temporary or permanent. The total number of lines of flux is always a whole number; there are no fractional lines. The number of lines of flux can increase or decrease, but the total number of lines external to a magnet is the same as the total number in the magnet. Lines of flux can expand or contract. They can move away from a pre-existing position but they do not flow.

## The maxwell

A single line of flux is identified as a maxwell. The maxwells or lines of flux surrounding a magnet form a magnetic field. The maxwell is a CGS unit.

## Flux density

Flux density is the number of maxwells or magnetic lines of force per unit area, with that area considered as perpendicular to the lines. In the CGS system, that area is always the square centimeter ($cm^2$). In this system, the area need not always be specified. A line of flux per square centimeter is known as a gauss and is represented by the Greek letter beta ($\beta$). Thus, if a magnet has a strength of 200 gauss (G), the statement is equivalent to 200 lines of flux or maxwells per square centimeter. As a comparison, a small magnet used for holding papers to a refrigerator has a strength of about 100 G.

The identifying letter for lines of flux is the Greek letter phi ($\phi$) and is used for flux just as the letter $I$ is used for current. The total number of lines of flux can be expressed by:

$$\phi = \beta \times A$$

where $\phi$ is the lines of flux, $\beta$ is the flux density, and $A$ is the area in square centimeters. The terms in this formula can be transposed to read:

$$\beta = \frac{\phi}{A}$$

For each line of magnetic flux inside a magnet, there is a continuing line of flux outside that magnet. The lines of flux, however, are more concentrated inside the magnet. Thus, the area they occupy is smaller. For magnetic lines of force there are two areas to consider—the area of the magnet and the area outside it. As a result, there are two ways of measuring flux density.

The measurement of flux density in the area surrounding a magnet is the oersted. A magnetic field intensity of 100 oersteds (Oe) is the same as 100 lines of magnetic flux per square centimeter. Thus, there are two units for describing field intensity, the gauss and the oersted.

## Ampere-turns

The number of ampere-turns of a coil consists of the current flowing through the coil, in amperes, multiplied by the number of turns comprising the coil. If the current flow data is in microamperes or milliamperes, these must be converted to amperes. Arranged as a formula:

$$\text{ampere-turns} = N \times I$$

N is the number of turns and $I$ is the current in amperes.

If a current of 500 mA flows through a coil having 22 turns, the number of ampere-turns is equal to $0.5 \times 22 = 11$ ampere-turns. Prior to doing the arithmetic, 500 mA was first converted to 0.5 A. If the current is increased to 1 A and the number of turns of the coil is decreased to 11, then the number of ampere-turns remains the same:

$$\text{ampere-turns} = 1 \times 11 = 11 \text{ ampere-turns}$$

By itself the ampere-turn does not include any length units. Further, the $NI$ value of a coil is applicable only to air-core types. For the CGS system, the ampere-turn per centimeter is 1.257 Oe; in the English system, the ampere-turn per inch is 0.495 Oe.

If, for example, the length of a coil is 10 in using 90 turns and the coil has a current of 1 000 mA flowing through it, the $NI$ value is $90 \times 1$ (1 000 mA = 1 A) = 90 ampere-turns or 9 ampere-turns per inch. The magnetic field intensity of this coil, indicated by the letter $H$, is $90 \times 0.495 = 44.55$ Oe. This is the magnetic field intensity of the entire coil. To determine the value of $H$ per unit length, divide the answer in oersteds by the length of the coil.

The length of the coil in the English system can be changed to the CGS system by converting inches to centimeters. 1 in = 2.54 cm; 10 in = 25.4 cm. The ampere-turn per centimeter = 1.257 Oe. The ampere-turn per inch = 0.495 Oe.

Actually, the magnetic field intensity is the same in both instances, even though the answers using the English and the CGS system are different. Be sure to decide on the system to use and to stay with that system throughout any calculations.

## Reluctance

Reluctance in a magnetic circuit is comparable to resistance in an electric circuit. Lines of magnetic flux exist more easily in certain metallic substances such as iron or steel but meet much more opposition in air, paper, plastic, and similar insulating, nonelectric current-carrying substances.

The opposition to the existence of magnetic lines of flux is known as reluctance and its symbol is the script letter ($\Re$). Just as the ohm is the unit of resistance in electric circuits, the rel is the unit of reluctance in magnetic circuits. However, the comparison has its limitations. In a resistor used in an electric circuit the amount of resistance remains fixed. In magnetic material, such as iron or cobalt, the amount of reluctance depends on the total number of flux lines and is a variable. For nonmagnetic substances, such as air, the reluctance remains constant and is not dependent on flux density. *Reluctivity* (Table 7-7) is a measure of reluctance per unit volume.

*Table 7-7. Reluctivity of magnetic substances.*

| | |
|---|---|
| Cobalt | 0.005 88 |
| Iron-cobalt alloy | 0.000 076 9 |
| Iron, commercial annealed | 0.000 166 6 to 0.000 124 |
| Nickel | 0.002 5 to 0.001 |
| Permalloy | 0.000 012 5 |
| Perminvar | 0.000 5 |
| Sendust | 0.000 033 33 to 0.000 008 333 |
| Silicon steel | 0.000 2 to 0.0001 |
| Steel cast | 0.000 666 6 |
| Steel, open-hearth | 0.000 333 to 0.000 142 857 |

## Magnetomotive force

Magnetomotive force (mmf) is comparable to voltage in an electric circuit, and is the force by which a magnetic field is produced. The mmf can be stated in two ways, depending on which system of measurements is used, the English or the metric. In the English system of measurements it can be stated in terms of ampere-turns per inch and in the CGS system in gilberts per centimeter or in oersteds. The gilbert is a unit of magnetomotive force and is represented by the letter $H$ or H. Because this letter is used in both the English and metric systems, it is italicized in the metric system, but not in the English.

The magnetomotive force, using the metric system, can be calculated from:

$$H = 1.257 N \times I$$

$H$ is the mmf in gilberts per centimeter, 1.257 is a constant and is equal to $0.4\pi$, N is the number of turns of wire of the coil, and $I$ is the electric current in amperes. This formula is applicable only to air-core coils. The constant 1.257 is sometimes represented by the letter $k$, and so the formula for mmf becomes:

$$H = k \times N \times I$$

The constant, $k$, is used only for air-core coils. There is a different constant for coils having a metallic core. The value of $k$ depends on the type of core metal.

Although the use of the letter $H$ is widely accepted as representing magneto-motive force, the italic letter $F$ is sometimes substituted in the MKS system:

$$F = 0.4\pi N I = 1.257 N I$$

## Magnetic relationships

Just as Ohm's law is used in electric circuits, a comparable arrangement finds application in magnetic circuits. In magnetic circuits the relationship between mmf, lines of flux, and reluctance can be expressed in three equations:

$$\phi = \frac{F}{\mathcal{R}}$$

$$F = \phi \mathcal{R}$$

and

$$\mathcal{R} = \frac{F}{\phi}$$

where $\phi$ represents the lines of flux in maxwells, $\mathcal{R}$ is the reluctance in rels and $F$ is the magnetomotive force in gilberts. $F$ is directly proportional to the amount of flux, and so the greater the magnetomotive force the larger the number of magnetic lines. The amount of flux, though, is inversely proportional to the reluctance. The smaller the reluctance, the greater the number of magnetic lines.

The strength of the mmf is affected by the structure of the coil and the amount of current flowing through it as indicated by this formula:

$$H = 2\pi NI/10r$$

$H$ is the field strength in oersteds, N is the number of turns of wire forming the coil, $I$ is the current flow through the coil in amperes, and $r$ is the radius of the coil in centimeters. Pi ($\pi$) is a constant and is usually written as 3.141 6. The product $NI$ is sometimes referred to as ampere-turns.

*Example*   A 10-turn coil has a diameter of 3 in and carries a direct current of 2 000 mA. What is the magnetic field strength of this coil in oersteds?

*Solution*   The diameter of the coil, supplied in inches, must first be converted to centimeters. 1 in=2.54 cm. 3 in=3×2.54=7.62 cm. The radius is $^1/_2$ this amount and equals 7.62/2=3.81 cm. 2 000 mA=2 000/1 000=2 A.

$$H = 2\pi NI/10r$$

$$= \frac{(6.28)(10)(2)}{(10)(3.81)}$$

$$= 3.296 \ \text{Oe}$$

## Temporary and permanent magnets

Whenever a current flows through a conductor, whether that conductor is in the shape of a coil or simply a straight wire, a magnetic field accompanies the current. If the straight wire is wound in the form of a coil, the magnetic field is intensified. It is further strengthened if the coil has an iron core instead of air. The metal has a much lower reluctance, that is, it permits the easier existence of magnetic lines. But in all these cases, the magnetic field disappears as soon as the current flow is discontinued, and the wire or coil is referred to as a temporary magnet.

If a metallic core is removed from a current carrying coil, that core will retain some of the magnetism. It will do so, to a greater or lesser degree, depending on the metal the core is made of, the strength of the original magnetic field, and the length of time the core was subjected to the magnetic field. Magnets produced in this way are referred to as permanent. Another method that has limited applications is to put an unmagnetized metal, such as iron or steel, in close contact with a strong magnet.

### The weber

The weber is a unit of magnetic flux and consists of 1 line of flux in the meter-kilogram-second (MKS) system of measurement. When large magnetic units are involved, use of the weber is preferred to the gauss.

### Permeability

The permeability of a substance is a measure of its ability to permit the existence of magnetic lines of force. Permeability is the ratio of flux density in gauss ($\beta$) to a magnetizing force, $H$, in oersteds. In terms of a formula permeability is expressed as:

$$\mu = \beta/H$$

$\mu$ is the permeability, $\beta$ is the flux density in gauss, and $H$ is the field intensity in oersteds. The permeability of air is considered as unity. Table 7-8 lists the permeability of various magnetic materials.

*Table 7-8. Permeability of magnetic materials.*

| | |
|---|---|
| Cobalt | 170 |
| Iron-cobalt alloy (Co 34%) | 13 000 |
| Iron, commercial annealed | 6 000 to 8 000 |
| Nickel | 400 to 1 000 |
| Permalloy (Ni 78.5%, Fe 21.5%) | over 80 000 |
| Perminvar (Ni 45%, Fe 30%, Co 25%) | 2 000 |
| Sendust | 30 000 to 120 000 |
| Silicon steel (Si 4%) | 5 000 to 10 000 |
| Steel, cast | 1 500 |
| Steel, open hearth | 3 000 to 7 000 |

## Magnetic materials

There are three classifications of magnetic materials.

Diamagnetic
Paramagnetic
Ferromagnetic

**Diamagnetic**   Diamagnetic substances have a permeability less than that of air. Air is used as a reference value of 1 or unity. This value is approximate, because air can be slightly magnetized. Diamagnetic materials include:

Antimony
Bismuth
Copper
Gold
Mercury
Phosphorus
Silver

Bismuth has the lowest permeability known (0.998).

**Paramagnetic**   Paramagnetic materials can be only slightly magnetized and have permeabilities of a little more than 1. Paramagnetic materials include:

Aluminum
Chromium
Platinum

**Ferromagnetic**   Ferromagnetic substances are characterized by having permeabilities greater than unity. The list includes:

Cobalt
Iron
Nickel
Steel

The list also includes many alloys using either trade names or names which supply some clue as to their construction. Alnico is one material and is made of aluminum, nickel, and cobalt. Another uses a trade name, Sendust.

## Electric and magnetic units

Lack of familiarity with magnetic units may make it difficult to work with them. Table 7-9 is a comparison between electric and magnetic units and might make working with magnetic units easier.

## Table 7-9. Comparison of electric and magnetic circuits.

| | Electric circuit | Magnetic circuit |
|---|---|---|
| Force | V, $E$, or emf | Gb, $\mathcal{F}$, or mmf |
| Flow | A, $I$ | Flux, $\phi$, in Mx |
| Resistance | Ohms, $R$ | Reluctance,$\mathcal{R}$, in rels |
| Law | Ohm's Law, $I=\dfrac{E}{R}$ | Rowland's Law $\phi=\dfrac{\mathcal{F}}{\mathcal{R}}$ |
| Intensity of force | V/cm of length | $H=\dfrac{1.257\ IN}{l}$ gilberts per centimeter of length |
| Density | Current density—for example, A/cm$^2$ | Flux density—for example, lines per square centimeter, or gauss. The tesla (T) is sometimes used as the unit of flux density. |

### Voltage induced across a conductor

This formula can be used for calculating the electromotive force induced across a conductor as it moves through a magnetic field. The assumption is made that the conductor is at right angles to all the lines of flux.

$$e=Blv\times10^{-8}$$

$e$ is the induced emf in volts; $B$ is the flux density in lines per square centimeter; $l$ is the length of the conductor in centimeters; $v$ is the relative velocity between the flux and the conductor in centimeters per second.

### Unit poles

Unlike magnetic poles attract each other, similar poles repel. If a pair of magnetic poles are separated by a distance of 1 cm and if the force of attraction or repulsion is equal to 1 dyne, then the strength of each of the magnets is said to be 1 unit pole. This concept is illustrated in Fig. 7-1. The separation is in air or a vacuum.

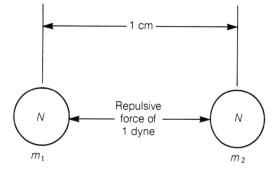

*Fig. 7-1. Force existing between unit poles.*

Table 7-10. Summary of magnetic units in CGS, MKS, and English systems.

| Term | Description | Symbol | CGS units | MKS units | English units | Notes |
|---|---|---|---|---|---|---|
| Flux | Total no. of lines | $\phi = \dfrac{\text{mmf}}{\mathfrak{R}}$ | 1 Mx=1 line | 1 Wb=$10^8$ lines | 1 kiloline=$10^3$ =1 000 lines | Comparable to electric current |
| Flux density | Lines per unit area | $\beta = \dfrac{\phi}{A}$ | 1 G=1 Mx/cm$^2$ | $T = \dfrac{\text{Wb}}{\text{m}^2}$ | kilolines per square inch | $1\ m^2 = 10^4\ \text{cm}^2$ $1\ \text{in}^2 = 6.45\ \text{cm}^2$ |
| Magnetomotive force | Total force producing flux | $\mathfrak{F} = \phi \times \mathfrak{R}$ | Gb=1.256 × ampere-turns | ampere-turn | ampere-turn | Corresponds to voltage, independent of length: $0.796 = 1/1.256$ |
| Field intensity, or magnetizing force | Force per unit length of flux path | $H$ | 1 Oe=1 Gb/cm | ampere-turn per meter | ampere-turn per inch | Corresponds to voltage per unit length: 1 in= 2.54 cm |
| Reluctance | Opposition to flux | $\mathfrak{R} = \dfrac{\text{mmf}}{\phi}$ | Gb or Mx | ampere-turn per weber | ampere-turn per kiloline | Corresponds to resistance |
| Permeability | Ability to concentrate flux | $\mu = \dfrac{B}{H}$ | G or Oe | $1.256 \times 10^{-6}\ \mu*$ | $3.2 \times 10^{-3}\ \mu*$ | $\mu$ of air or vacuum is 1 |

*Multiply $\mu$ by these factors to calculate $\beta$ from $H$ in MKS or English units.

Polarity can be assigned to unit poles. If the solution to a problem involving unit poles is positive, the force acting between the poles is repulsive; if negative, the force is attractive. The force, represented by $F$, is:

$$F = \frac{m_1 m_2}{\mu d^2}$$

$F$ is the force between the unit poles and is expressed in dynes, $m_1$ and $m_2$ are the magnetic poles of the two magnets express in terms of unit poles, $\mu$ is the permeability of the material between the unit poles (for air it has a value of 1), and $d$ is the distance between the two poles in cm.

The force of attraction and repulsion between magnetic poles finds application in the design of motors.

# Summary of magnetic units in CGS, MKS, and English systems

Table 7-10 is a summary of magnetic units in the three magnetic systems.

# 8
CHAPTER

# Microphones

## Microphone types

All microphones are transducers and convert one form of energy—sound—to another form—electrical. Not all microphones have the same features, but microphone types are basically:

Carbon
Piezoelectric
Ceramic
Dynamic
Condenser
Electret
Pressure zone

Microphones are sometimes known by the way they are used. A lavalier microphone (mike or mic) is a very small type used when a mic is needed but must be inconspicuous. It can be mounted on a lapel or a tie or fastened to a dress. Still another is a shotgun mic, a highly directional mic mounted at one end of a long tube. Still another type is the differential or noise-cancelling mic, useful for communications in noisy situations. The type of mic helps define a microphone more closely, thus, a mic could be referred to as a dynamic, supercardioid, shotgun. The first word, dynamic, indicates it uses a moving coil, the next word, supercardioid, specifies its response pattern, and the last word, shotgun, reveals its physical shape.

### Carbon microphone

The carbon microphone is the oldest and still the most widely used type. It consists of a small cylinder, known as a button, packed with carbon granules (Fig.

8-1). Pressing against the carbon granules in the supporting button is a metallic diaphragm held in position by a ring around its circumference. The diaphragm is free to vibrate by all points along its surface except at its edge. Sound can cause the diaphragm to vibrate; when it does so it exerts more or less pressure against

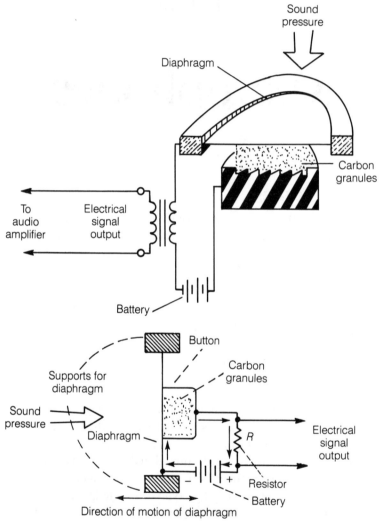

Fig. 8-1.  *Structure of the carbon microphone.*

the carbon granules, changing its resistance accordingly. As a result, more or less current, supplied by a battery, flows through the button.

This current path, as indicated in Fig. 8-1, is from the battery, through the carbon button, through a resistor $R$, and then back to the battery. The varying current through the resistor develops a voltage across it. The current varies in step with the sound input to the diaphragm and is an audio frequency voltage.

## Piezoelectric microphone

The basic structure is illustrated in Fig. 8-2. The face of the microphone consists of a diaphragm capable of vibrating when subjected to sound pressure. The inside center is equipped with a small drive pin making contact with a piezoelectric element. The resulting pressure against the element causes it to deform slightly,

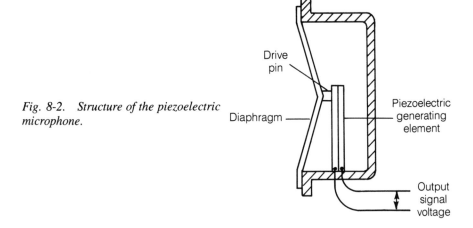

*Fig. 8-2. Structure of the piezoelectric microphone.*

Drive pin

Diaphragm

Piezoelectric generating element

Output signal voltage

resuming its shape when the pressure is removed. The shape change of the crystal produces an alternating voltage across it, at the rate of sound pressure change.

This mic supplies a moderately high output signal voltage for a given sound input. The crystal, though, is temperature and humidity sensitive, and can be easily damaged by excessive levels. It has poor frequency response and is primarily used in voice communications.

## Ceramic microphone

The transducing element in this mic is barium titanate, a substance that has piezoelectric properties. As in the case of piezoelectric microphones, the element develops a voltage when subjected to a varying pressure.

The microphone supplies a high signal output, but its frequency response is poor. Technical improvements in the response can be made, but at the expense of the output. For a given frequency response its output is comparable to that of the crystal mic.

## Dynamic microphone

This is one of the most widely used microphone types. As shown in Fig. 8-3, it consists of a coil of wire whose turns are insulated from each other. The coil is free to move on a supporting form. The moving element consists of a diaphragm supported and kept in position around its edges. The remainder of the diaphragm

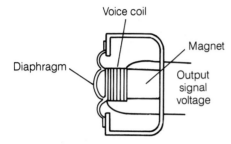

Fig. 8-3. Structure of the dynamic microphone.

is free to move. Part of the diaphragm is attached to the coil, and its movement causes the coil to slide back and forth.

The coil is surrounded by a magnet, and the movement of the coil in a magnetic field induces a voltage across the coil. Because the motion of the coil is caused by sound impinging on the diaphragm, the coil induces an audio-frequency voltage. The principle used in the dynamic mic is similar to that found in dynamic loudspeakers. In intercom systems, the dynamic speaker is also used as a microphone.

Dynamic microphones, also called moving-coil microphones, are the most rugged and have a smooth and extended frequency response. They have good transient response, acceptable output signal level, and are widely used in home and studio recording.

### Condenser microphone

The basic structure of this mic is essentially that of a capacitor (once referred to as a condenser). When a capacitor is connected across a voltage source it will take an electrical charge. The amount of charge depends on the area of the two plates forming the capacitor, the distance between them, and the type of dielectric between the plates. The closer the plates, the larger the possible charge. Conversely, as the plates are separated, the smaller the possible charge.

The basic condenser microphone (Fig. 8-4) consists of two plates; one plate, called the back plate, is fixed in position. Facing this plate is one that is capable of moving back and forth, resulting in a variable-voltage device.

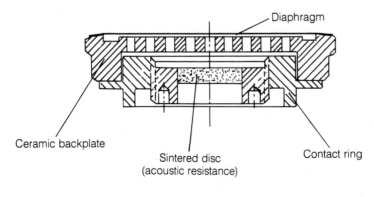

Fig. 8-4. Structure of the condenser microphone.

The moving diaphragm or element in a condenser mic is a metal plate, tightly stretched, but capable of movement. It is often made of a plastic material coated with an extremely thin covering of gold. The diaphragm then forms one of the plates with its plastic backing working as a dielectric. The dielectric faces the second plate, a fixed-position, gold-plated ceramic backplate.

A charging source is required. Known as a polarizing voltage, it is a dc potential ranging between 50 and 200 V. When sound strikes the moving element it moves, varying the electric charge. As it moves toward the back plate, the condenser spacing decreases and the electric charge increases. Conversely, as the spacing increases, the electric charge decreases. The result is a voltage change that varies in accordance with the sound pressure on the movable plate.

The resulting varying voltage is extremely small and cannot be used directly. A solid-state amplifier is built directly into the microphone housing. Power for the amplifier can be supplied by small batteries integral with the mic, or it can be taken from external electronic equipment (phantom powering).

Although the direct output of this mic is very small the unit is capable of excellent frequency response. Its disadvantage is that it requires two operating voltages—one for the amplifier and another as a polarizing voltage.

### Electret microphone

The electret is a variation of the condenser mic. The unit uses a self-polarizing element, electrically charged during the time it is manufactured. It still requires a built-in audio amplifier (also known as an impedance converter) and a battery. However, no further polarizing voltage is needed for the microphone.

# Polar patterns

A microphone can respond to sound coming at it from all directions—front, back, above, and below. Not all microphones are equally sensitive, but the response can be plotted by a graph known as a polar pattern or diagram, illustrated in Figs. 8-5 and 8-6.

The preparation of this graph requires a sound generator using a selected audio frequency of 125, 250, 500, 1 000, 2 000 Hz, and so on. This sound source is kept in a fixed position while the mic is moved in a complete circle around it. The distance between the two, the revolving mic and the sound source, is usually 1 m. The circular movement of the mic is a trip along the surface area of an imaginary sphere. This trip can be horizontal to the sound source, vertical to it, or on any other path (Fig. 8-7).

Figure 8-5 is used for plotting the response sensitivity of the microphone. It consists of a series of equally spaced concentric circles, marked 0, −5, −10 and so on with these representing negative values in decibels. The vertical line extending from 0 to 180° is the axis of the diagram. Sounds arriving directly at the front

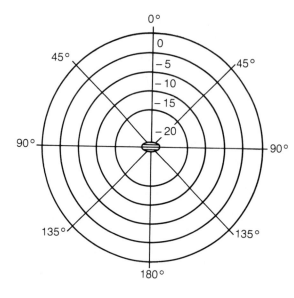

Fig. 8-5. Polar graph. Small ellipse at center is the mic.

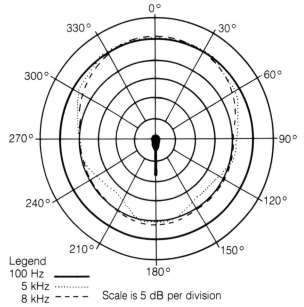

Fig. 8-6. Polar pattern of an omnidirectional (omni) mic.

Legend
100 Hz ———
5 kHz ··········
8 kHz ― ― ―    Scale is 5 dB per division

of the mic are said to be on axis; those arriving from the rear are directly off axis. The graph is further divided into angles of 45, 90, and 135°.

## The omni and cardioid patterns

By modifying the design of a microphone a manufacturer can produce a large variety of response patterns. The omni (Fig. 8-6) is one of the most common; the cardioid is another.

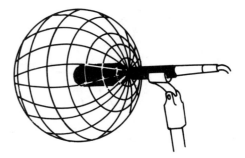

Fig. 8-7. *Three-dimensional response of an omni mic.*

The omnidirectional, more often referred to as an omni, will respond to sound coming at it from all directions: in front, in back, above, below, and all other angles. Although the polar pattern is plotted in two dimensions, the mic has a three-dimensional response, as indicated in Fig. 8-7.

The cardioid response, shown in Fig. 8-8, is called *cardioid* because of its fancied resemblance to the shape of a heart. The cardioid mic has its maximum response on axis. As the sound reaches the mic from other angles the sensitivity

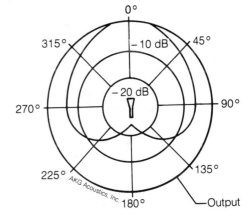

Fig. 8-8. *Polar response of a cardioid mic.*

of the mic becomes less and is minimum when the mic is completely off axis, at 180°. The amount of decrease in sensitivity is shown at the point where the response cuts through the circles marked in decibels. Thus, at 180°, the completely off-axis point, the sensitivity of the mic is down about 20 dB compared with on-axis at 0°. The cardioid is sometimes mistakenly called unidirectional but it is not since it is still capable of picking up sound off axis.

It is essential to regard the cardioid, and all other mics in a three-dimensional manner, as shown in Fig. 8-9. The basic cardioid pattern can be modified by design so that it is more responsive to sound in certain directions. In the case of the omni the mic will pick up audience sound even though the head of the mic is pointed toward the musicians. The cardioid, though, because of its response, is much less sensitive to audience noise.

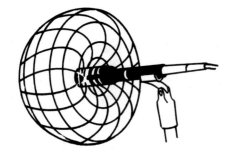

*Fig. 8-9. Three-dimensional response of a cardioid mic.*

## The bidirectional pattern

A microphone can be designed to have maximum response both on axis and off axis, with minimum response at the sides, as shown in the polar pattern in Fig. 8-10. The pattern is known as bidirectional or figure 8.

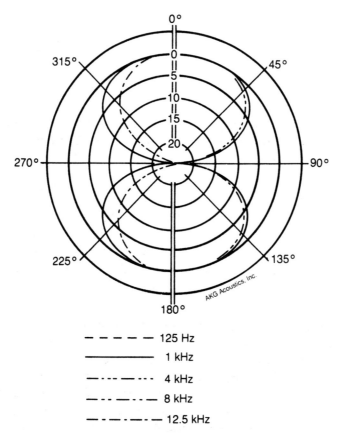

————— 125 Hz

————————— 1 kHz

—···—··· 4 kHz

—··—··— 8 kHz

—·—·—·— 12.5 kHz

*Fig. 8-10. Polar pattern of bidirectional mic.*

# Microphone specifications

The specifications (specs) of a mic are an indication of how the microphone will perform under various working conditions. These can be fairly extensive, but some of the more important include impedance, pickup pattern, sensitivity, frequency response, proximity effect, transient response, efficiency, and output level.

## Impedance

The output impedance of mic can range from 50 Ω to 10 kΩ or more. The impedance isn't usually indicated as a precise amount, but rather as either low or high. A low-impedance mic is one whose output impedance ranges from as little as 50 Ω to as much as 250 Ω. A high-impedance mic is one that has an impedance in excess of 10 kΩ.

## Pickup pattern

Pickup pattern is the response of the microphone and is an indication of the directional sensitivity.

## Sensitivity

This is a measure of the response of a microphone to a given SPL (sound pressure level). The output voltage of a mic is supplied in decibels referenced to 1 mW at an SPL of 10 dynes/cm at a specific frequency.

## Frequency response

Ideally, a microphone should have a flat, wide-range response to sound arriving from any direction, including sounds that it is suppressing and that are weaker than direct pickup. A problem is that instruments on and off axis do not sound the same, resulting in sound that is colored. The side sound will be dull.

## Proximity effect

Except in the case of omni microphones, the frequency response of a mic can become distorted when used too close to the sound. The result is rasping, spitting, popping, booming, or exaggeration of low frequencies.

Proximity effect is the increase in low-frequency response produced in most cardioid and bidirectional mics when the distance from the sound source is decreased. The effect is most noticeable at a distance of less than 2 ft. For vocals, proximity effect can result in up to a 16-dB bass boost when the mic is positioned about 1 in from the mouth. The voltage output can jump many times the catalog sensitivity rating. A mic with a catalog rating of −55 dB can put out a signal of −19 dB.

### Transient response

Transient response is the ability of a mic to respond to a sound having very short or no sustain time.

### Efficiency

Efficiency is the ratio of the output of the mic compared to the input.

### Output level

Output level is the amount of voltage generated by sound pressure driving a microphone diaphragm.

### The bar

The bar is the basic unit of atmospheric pressure and at sea level is 14.7 psi. The sound pressure put on a mic diaphragm is much less and is measured in millionths of an atmosphere (microbars, written as $\mu$bar). The threshold of hearing, the point at which sound first becomes perceptible to humans, is equivalent to 0.000 2 $\mu$bar.

### Sound pressure level

Sound pressure level (SPL) is measured in decibels and is used as an alternative form to the microbar. It requires a very small change in SPL to produce audible sound. Sound that is barely audible is caused by an increase or decrease in normal air pressure of about one millionth of 1%. A variation of normal air pressure of about one-tenth of 1% puts sound at the threshold of pain.

### Microphone reference levels

Microphone output voltage is usually referenced to 1 V and power output to 1 mW. This does not mean that these references must be smaller than the mic output, for in many instances they are larger. In both examples, though, the ratio of the mic output to the reference is expressed in decibels and, because the reference is larger, the ratio in decibels is preceded by a minus sign. The smaller the number of minus decibels the larger the output. A mic with an output of − 15 dB is larger than one with an output of −20 dB. This is the basis for marking the polar pattern in minus decibels. When the mic output is equal to the reference, the ratio of output to input is 1 to 1 and is then designated as 0 dB. The minus sign used in connection with output in decibels does not indicate subtraction. Instead, it is just an indication of how much weaker the output signal of the mic is compared to the reference voltage or power.

A specification listing the output of a mic in decibels is an indication of its sensitivity. The smaller the number in minus decibels the more sensitive the mic. In comparing mic sensitivities, it is essential that the same reference level be used.

### The dyne

There are three ways of describing the amount of pressure applied to a mic diaphragm. These are:

The microbar ($\mu$bar)
SPL (sound pressure level)
The dyne

The dyne is a unit of force and is the force that will produce a velocity of 1 cm/s when acting on a mass of 1 g. The dyne per square centimeter is equivalent to 1 $\mu$bar or 74 dB SPL. $-10$ dynes/cm$^2$ is equivalent to 10 $\mu$bar or 94 dB SPL. Thus, a mic manufacturer might, in a specification sheet, supply the sensitivity of a particular mic as:

$$-74 \text{ dB re } 1 \text{ V/}\mu\text{bar}$$

When translated, this means that this mic has an output of $-74$ dB SPL or 1 $\mu$bar. The abbreviation *re* represents the word reference. 1 V/$\mu$bar is the amount of the reference. Another mic might indicate its sensitivity in a spec sheet as:

$$-56 \text{ dB re } 1 \text{ mW/10 dynes/cm}^2$$

The output of this microphone is $-56$ dB, equivalent to 10 dynes/cm$^2$ with a reference of 1 mW. The output is 1 mW/10 $\mu$bar.

The disadvantage of having a number of different ways to indicate mic sensitivity is that it becomes difficult to make comparisons among microphones. As an example, consider this listing:

$-74$ dB re 1 V/$\mu$bar, or $-74$ dB referenced to 1 V/$\mu$bar
$-58$ dB re 1 mW/10 $\mu$bar, or $-58$ dB referenced to 1 mW/10 $\mu$bar and
$-58$ dB re 1 mW/10 dynes/cm$^2$, or $-58$ dB referenced to 1 mW/10 dynes/cm$^2$

All of these microphones have an identical output. The selection of a particular reference level can make one mic appear to have more output than another, that is, it might seem to be more sensitive. Microphone output level is meaningless unless the reference level is supplied along with the SPL, the impedance, and the frequency response.

## Microphone quality

The fact that a particular mic has a high output doesn't immediately make it desirable from the viewpoint of quality. A mic might develop a stronger output simply because its magnet structure has more lines of flux per unit area. But this does not answer the question of linearity and s/n (signal-to-noise) ratio.

## Microphone reach

There are undeniable advantages in relatively high output mics. The mic might be able to supply the voltage input requirements of a following amplifier, and if there is some danger of overloading, it is always possible to use a pad or attenuator. High output can mean a better s/n ratio. But if the mic has very low output, these options might not be available.

The greater the output of the mic, the higher its sensitivity and the greater its reach. This means the mic will be able to pick up wanted sounds at a greater distance. With a less sensitive mic, the output might be buried in the noise floor.

## Microphone overload

Distortion in a sound recording system is sometimes blamed on the mic. If the performer works close in with lips practically touching the mic grille, the result can be up to 100 dB SPL. Loudness impacts on the mic diaphragm from a screaming vocalist or a musical instrument with amplified output can reach an average SPL of 120 or 130 dB on peaks. A loudness of 130 dB is at the threshold of pain, the point at which sound is not only heard but produces pain as well. Some professional mics can be used in the mouth of a trumpet, which can produce a loudness of up to 146 dB.

Surprisingly, even under such working conditions, the output of the mic can remain clean, or undistorted. Any distortion that is produced is from overdriving a following mic amplifier.

An amplifier or recorder input sensitivity rating describes the absolute maximum amount of voltage or power it can handle before it overloads and distorts. Components with a −60-dB input signal rating, can handle a signal of this level before introducing distortion.

There is no standardization for microphones and there cannot be because mics are never used at the same distance. Also, sound-pressure sources are of an infinite variety of loudness. As an example, consider an SPL of 94 dB applied to the diaphragm of a mic. This will result in a catalog output rating of −53 dB. But if the SPL is raised by 36 dB, the mic output will now be the original output of −53 dB plus 36 dB more. The output is now −17 dB. Smaller negative numbers mean higher output, getting closer to the 1 V or 1 mW reference point, or 0 dB. The output of the mic is now −17 dB. Assume an amplifier with a −60-dB input rating. This means that the mic output is now 43 dB higher than the amplifier input rating. Unless a pad or attenuator is used, the inevitable result is distortion.

# Reading the mic specification sheet

The data on a spec sheet should include output, internal impedance, and sound pressure effects. The output of the mic in either a voltage or power ratio. The output is often supplied in the form of decibels and that is always a ratio or comparison with some reference level. The reference level should be supplied, otherwise the output in decibels is meaningless. The internal impedance of a mic is supplied in ohms. The actual amount isn't critical, and a description such as low or high impedance is adequate. The amount of sound pressure applied can be supplied in several different ways such as 1 mW/10 $\mu$bar. This corresponds to 94-dB SPL.

European manufacturers often specify mic output voltage in terms of microvolt per microbar when the mic isn't loaded, that is, not connected to a following amplifier. U.S. manufacturers specify output voltage in maximum power output, as expressed in decibels compared to 1 mW (dBm) when the mic is loaded with its characteristic impedance, that is, when the impedance of the mic and the following amplifier are fairly similar. A dBm is the decibel referred to 1 mW across 600 $\Omega$. The sound pressure applied often is expressed as referenced to 1 mW/10 dynes/cm$^2$. This is equal to 1 mW/10 $\mu$bar and this, in turn, is equal to 94 dB SPL.

Most low-impedance mics are in the 150-$\Omega$ range, but the actual impedance extends from 100 to 300 $\Omega$. Such mics are specified with 94-dB SPL applied. When the output voltage of the mic is specified as $-50$ dBm output, the mic is 10 dB more sensitive than one rated at $-60$ dBm. Because conversation at a distance of 1 ft from the mic is about 67 dB SPL, the 94 dB SPL used for a mic rating is typical of a pop vocalist or instrumentalist.

## Microphone source impedance

Microphones are ac generators. To emphasize this fact it is possible to connect a sensitive, very low-wattage electric light across its output. This light will flicker as the mic receives input from a sound source.

Mics have both internal resistance and reactance, forming electrical impedance. Source impedance is usually measured at 1 kHz. Impedance is frequency sensitive and to say a component has an impedance of 200 $\Omega$ simply means that this is its impedance at a particular frequency.

Because all dynamic mics are low-impedance types, a transformer is needed if the mic is to impedance match a high-impedance input. Some mics have built-in switchable transformers, and are low/high impedance types.

## Frequency and polar patterns

The polar pattern of a selected mic often will show the response of that mic at a single selected frequency. This does not mean that the pattern will be representative at other frequencies in the audio range. In some instances, as in Fig. 8-11, a

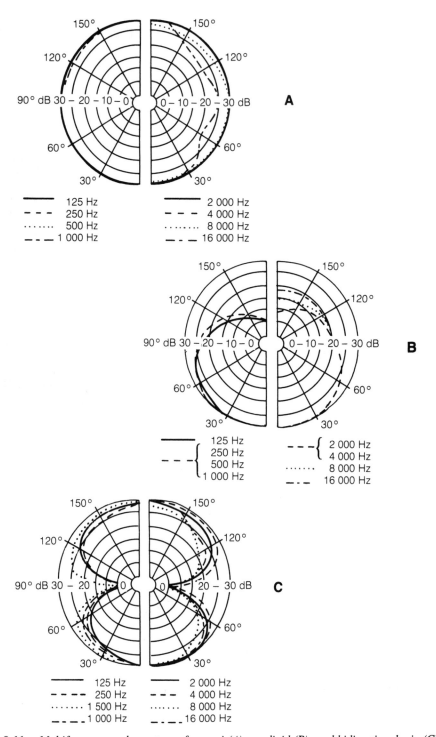

Fig. 8-11.   *Multifrequency polar patterns for omni (A); cardioid (B); and bidirectional mic (C).*

number of polar patterns will be shown at selected frequencies. This approach is better than the single-frequency pattern, but it does present a problem. If a large number of frequencies are chosen, the polar pattern can become very confusing.

An alternative method, but one requiring much more work, is to have a separate pattern for each frequency. Figure 8-12 shows eight patterns for an omni microphone at 125, 250, 500, 1 000, 2 000, 4 000, 8 000, and 16 000 Hz. This

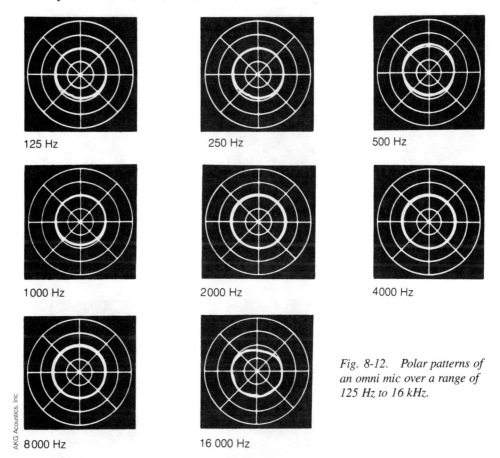

125 Hz          250 Hz          500 Hz

1 000 Hz          2 000 Hz          4 000 Hz

8 000 Hz          16 000 Hz

AKG Acoustics, Inc.

*Fig. 8-12.   Polar patterns of an omni mic over a range of 125 Hz to 16 kHz.*

represents a doubling of frequency starting at 125 Hz. This series of patterns shows that the mic has a uniform sensitivity with the on-axis response decreasing a bit at 16 000 Hz. Figure 8-13 shows the responses of a cardioid mic over the same frequency range. The off-axis response shows a decrease in sensitivity starting at 8 000 Hz. Figure 8-14 illustrates the responses of a bidirectional mic. Beginning at about 4 000 Hz the response begins to deteriorate. Figure 8-15 shows the polar patterns of a shotgun mic taken at four different frequencies: 250 Hz (A); 1 000 Hz (B); 4 000 Hz (C); and 8 000 Hz (D). These graphs have one other interesting characteristic. The on-axis point, the 0° point, is at the bottom of the graph.

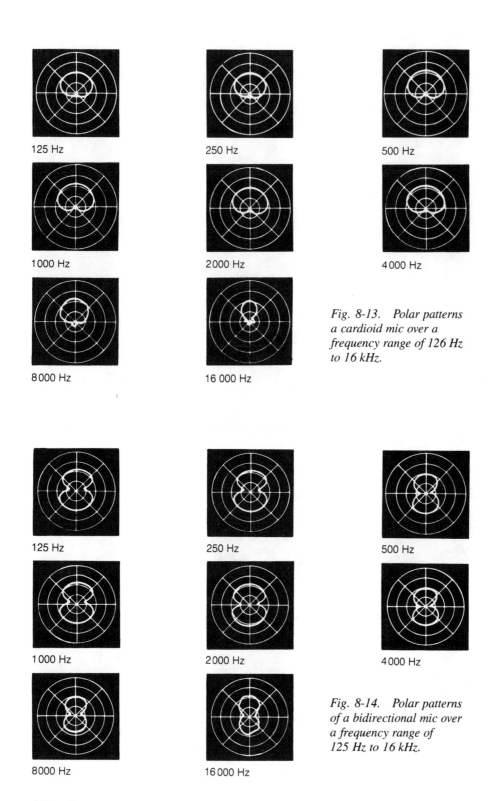

125 Hz

250 Hz

500 Hz

1000 Hz

2000 Hz

4000 Hz

8000 Hz

16 000 Hz

*Fig. 8-13. Polar patterns a cardioid mic over a frequency range of 126 Hz to 16 kHz.*

125 Hz

250 Hz

500 Hz

1000 Hz

2000 Hz

4000 Hz

8000 Hz

16 000 Hz

*Fig. 8-14. Polar patterns of a bidirectional mic over a frequency range of 125 Hz to 16 kHz.*

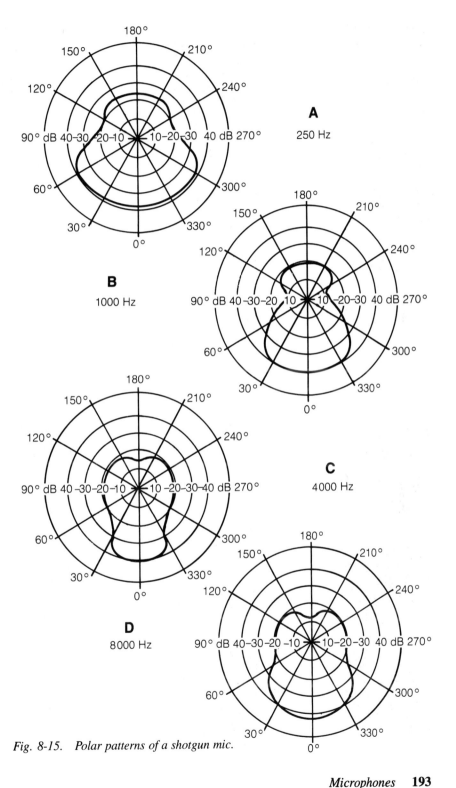

*Fig. 8-15. Polar patterns of a shotgun mic.*

# 9
## CHAPTER

# Decibels

## Voltage and current ratios versus power ratios and decibels

Table 9-1 supplies voltage or current ratios vs. power (watts) ratios for decibels (dB) ranging from 0.1 to 50. A decibel (one-tenth of a bel) is a ratio, a means of comparing the relative strengths of two currents, voltages, or powers. In itself, a decibel is not indicative of any particular amount of power, voltage, or current.

Because the decibel is a comparison unit, and not an absolute value, some reference level must be indicated. A common reference, also called zero level, is 1 mW. Other reference levels can be used, but in any event, the reference should be specified. If 1 mW is the reference, the letter *m* is added to dB, the unit being called the dBm (m for milliwatt). Thus, if an amplifier (assuming equal input and output resistances) has an output of 1 W, the relationship of the power output to the power input is:

$$dB = 10 \log P_2/P_1$$

$$dB = 10 \log 1/0.001 = 10 \log 1\ 000 = 30\ dB$$

Note: The logarithm (log) of 1 000 is 3.

Instead of using the formula, the result could have been obtained by consulting Table 9-1. The power ratio is 1 000—that is, an output of 1 W is 1 000 times greater than the reference level of 1 mW. Locate 1 000 under the heading of power ratio in the table. Move to the left and note a gain of 30 dB.

*Example* The input voltage to an amplifier is 1 V, and the output is 20 V. What is the voltage gain of the amplifier in decibels?

*Solution* The ratio of the two voltages is 20. The nearest comparable value in Table 9-1 is 19.95. The gain is 26 dB.

*Table 9-1. Voltage or current ratios versus power ratios and decibels.*

| Voltage or current ratio | Power ratio | −db+ | Voltage or current ratio | Power ratio |
|---|---|---|---|---|
| 1.000 | 1.000 | 0 | 1.000 0 | 1.000 0 |
| 0.989 | 0.977 | 0.1 | 1.011 6 | 1.023 3 |
| 0.977 | 0.955 | 0.2 | 1.023 3 | 1.047 1 |
| 0.966 | 0.933 | 0.3 | 1.035 1 | 1.071 5 |
| 0.955 | 0.912 | 0.4 | 1.047 1 | 1.096 5 |
| 0.944 | 0.891 | 0.5 | 1.059 3 | 1.122 0 |
| 0.933 | 0.871 | 0.6 | 1.071 5 | 1.148 2 |
| 0.923 | 0.851 | 0.7 | 1.083 9 | 1.174 9 |
| 0.912 | 0.832 | 0.8 | 1.096 5 | 1.202 3 |
| 0.902 | 0.813 | 0.9 | 1.109 2 | 1.230 3 |
| 0.891 | 0.794 | 1.0 | 1.122 0 | 1.258 9 |
| 0.881 | 0.776 | 1.1 | 1.135 | 1.288 |
| 0.871 | 0.759 | 1.2 | 1.148 2 | 1.318 3 |
| 0.861 | 0.741 | 1.3 | 1.161 | 1.349 |
| 0.851 | 0.724 | 1.4 | 1.175 | 1.380 |
| 0.841 | 0.708 | 1.5 | 1.189 | 1.413 |
| 0.832 | 0.692 | 1.6 | 1.202 | 1.445 |
| 0.822 | 0.676 | 1.7 | 1.216 | 1.479 |
| 0.813 | 0.661 | 1.8 | 1.230 | 1.514 |
| 0.803 | 0.646 | 1.9 | 1.245 | 1.549 |
| 0.749 | 0.631 | 2.0 | 1.258 9 | 1.584 9 |
| 0.776 | 0.603 | 2.2 | 1.288 | 1.660 |
| 0.759 | 0.575 | 2.4 | 1.318 | 1.738 |
| 0.750 | 0.562 | 2.5 | 1.334 | 1.778 |
| 0.724 | 0.525 | 2.8 | 1.380 | 1.905 |
| 0.708 | 0.501 | 3.0 | 1.412 5 | 1.995 3 |
| 0.692 | 0.479 | 3.2 | 1.445 | 2.089 |
| 0.676 | 0.457 | 3.4 | 1.479 | 2.188 |
| 0.668 | 0.447 | 3.5 | 1.496 2 | 2.238 7 |
| 0.661 | 0.436 | 3.6 | 1.514 | 2.291 |
| 0.646 | 0.417 | 3.8 | 1.549 | 2.399 |
| 0.631 | 0.398 | 4.0 | 1.584 9 | 2.511 9 |
| 0.596 | 0.355 | 4.5 | 1.678 8 | 2.818 4 |
| 0.562 | 0.316 | 5.0 | 1.778 3 | 3.162 3 |
| 0.531 | 0.282 | 5.5 | 1.883 6 | 3.548 1 |
| 0.501 | 0.251 | 6.0 | 1.995 3 | 3.981 1 |
| 0.473 | 0.224 | 6.5 | 2.113 | 4.467 |
| 0.447 | 0.200 | 7.0 | 2.239 | 5.012 |
| 0.422 | 0.178 | 7.5 | 2.371 | 5.623 |
| 0.398 | 0.159 | 8.0 | 2.512 | 6.310 |
| 0.376 | 0.141 | 8.5 | 2.661 | 7.079 |

*Table 9-1. Continued.*

| Voltage or current ratio | Power ratio | −db+ | Voltage or current ratio | Power ratio |
|---|---|---|---|---|
| 0.355 | 0.126 | 9.0 | 2.818 | 7.943 |
| 0.335 | 0.112 | 9.5 | 2.985 | 8.913 |
| 0.316 | 0.100 | 10 | 3.162 | 10.00 |
| 0.282 | 0.079 4 | 11 | 3.55 | 12.6 |
| 0.251 | 0.063 1 | 12 | 3.98 | 15.9 |
| 0.224 | 0.050 1 | 13 | 4.47 | 20.0 |
| 0.200 | 0.039 8 | 14 | 5.01 | 25.1 |
| 0.178 | 0.031 6 | 15 | 5.62 | 31.6 |
| 0.159 | 0.025 1 | 16 | 6.31 | 39.8 |
| 0.141 | 0.020 0 | 17 | 7.08 | 50.1 |
| 0.126 | 0.015 9 | 18 | 7.94 | 63.1 |
| 0.112 | 0.012 6 | 19 | 8.91 | 79.4 |
| 0.100 00 | 0.010 0 | 20 | 10.00 | 100.0 |
| 0.089 13 | 0.007 9 | 21 | 11.22 | 125.9 |
| 0.079 43 | 0.006 3 | 22 | 12.59 | 158.5 |
| 0.070 79 | 0.005 0 | 23 | 14.13 | 199.5 |
| 0.063 10 | 0.003 98 | 24 | 15.85 | 251.2 |
| 0.056 23 | 0.031 62 | 25 | 17.78 | 316.2 |
| 0.050 12 | 0.002 512 | 26 | 19.95 | 398.1 |
| 0.044 67 | 0.001 995 | 27 | 22.39 | 501.2 |
| 0.039 81 | 0.001 585 | 28 | 25.12 | 631.0 |
| 0.035 48 | 0.001 259 | 29 | 28.18 | 794.3 |
| 0.031 62 | 0.001 000 | 30 | 31.62 | 1 000 |
| 0.028 18 | 0.000 794 | 31 | 35.48 | 1 259 |
| 0.025 12 | 0.000 631 | 32 | 39.81 | 1 585 |
| 0.022 39 | 0.000 501 | 33 | 44.67 | 1 995 |
| 0.019 95 | 0.000 398 | 34 | 50.12 | 2 512 |
| 0.017 78 | 0.000 316 | 35 | 56.23 | 3 162 |
| 0.015 85 | 0.000 251 | 36 | 63.10 | 3 981 |
| 0.014 13 | 0.000 199 | 37 | 70.79 | 5 012 |
| 0.012 59 | 0.000 158 | 38 | 79.43 | 6 310 |
| 0.011 22 | 0.000 126 | 39 | 89.13 | 7 943 |
| 0.010 00 | 0.000 100 | 40 | 100.00 | 10 000 |
| 0.008 91 | 0.000 079 | 41 | 112.2 | 12 590 |
| 0.007 94 | 0.000 063 | 42 | 125.9 | 15 850 |
| 0.007 08 | 0.000 050 | 43 | 141.3 | 19 950 |
| 0.006 31 | 0.000 040 | 44 | 158.5 | 25 120 |
| 0.005 62 | 0.000 032 | 45 | 177.8 | 31 620 |
| 0.005 01 | 0.000 025 | 46 | 199.5 | 39 810 |
| 0.004 47 | 0.000 020 | 47 | 223.9 | 50 120 |
| 0.003 98 | 0.000 016 | 48 | 251.2 | 63 100 |

*Table 9-1. Continued.*

| Voltage or current ratio | Power ratio | −db+ | Voltage or current ratio | Power ratio |
|---|---|---|---|---|
| 0.003 55 | 0.000 013 | 49 | 281.8 | 79 430 |
| 0.003 16 | 0.000 010 | 50 | 316.2 | 100 000 |

***Example***   An amplifier has a gain of 20 dB. What is its output in watts?

***Solution***   Locate the number 20 in the decibel column in Table 9-1. Move to the right to see that this represents a power ratio of 100. If no power input is given and a reference level of 1 mW is indicated, then the output is 100 mW. If an input power is specified, multiply the input power by 100 to get the value of output power.

Note that the first two columns in Table 9-1 are the reciprocals of the last two columns. It makes no difference whether you divide the output power of an amplifier by the input or vice versa. If you put the larger number in the numerator of the decibel power formula, the answer will be a whole number and you will work with the two columns at the right in Table 9-1. If you use the smaller power value in the numerator, the answer will be a decimal, as indicated in the first two columns in the table. In either case, the answer in decibels will be the same. If you have an electronic device with a power ratio of 1 to 1 000 or, conversely, 1 000 to 1, the gain or loss in decibels will be 30 in either case. Some technicians and engineers prefer working with whole numbers and put a minus sign in front of their answer to indicate a loss.

It is essential to remember that the decibel has no absolute value. All it does is indicate how many times greater a power, voltage, or current is than some reference level (or how many times smaller).

Still another fact is that the relationship between decibel levels isn't linear. You cannot regard 6 dB as simply being twice as much as 3 dB. Arithmetically, 6 dB is twice as much as 3 dB, but the values these numbers represent do not have a 2-to-1 relationship. For example:

10 dB is 3.1×the reference level

20 dB is 10×the reference level

30 dB is 31.6×the reference level

40 dB is 100×the reference level

50 dB is 316×the reference level

60 dB is 1 000×the reference level

# Decibel references

No reference is needed when a pair of powers, voltages, or currents are being compared. However, when a single value is used, a reference is required. The value of the reference can be established in two ways. First, a specific reference can be stated as part of the data of the problem; second, by using an abbreviation such as one of those supplied in Table 9-2.

*Table 9-2. Decibel references.*

| dB reference | Explanation |
|---|---|
| dB | When used alone, it indicates that a pair of voltages, currents, or power are being compared. |
| dBf | 0 dBmV = 1 000 $\mu$V   This is used as a measurement of receiver signal input sensitivity in terms of power. |
| dBm | Ratio of watts with 1 mW as the reference. |
| dBmV | Decibel referenced to 1 mV or 1 000 $\mu$V. Thus, 0 dBmV = 1 000 $\mu$V. |
| dB, SPL | Ratio of sound pressures referenced to 20 $\mu$Pa. |
| dBU | Ratio of volts referred to 0.774 6 V. |
| dBv | Ratio of volts referenced to 1 V. |
| dBw | Ratio of watts referenced to 1 W. |

# Conversion table for microvolts to decibels (dBf)

At one time, the sensitivity of a radio receiver was measured only in terms of microvolts per meter. Subsequently, the decibel (dBf) was used instead. If the sensitivity is in decibels, the reference is to a voltage measurement. Decibels (dBf) is used if the measurement is made in terms of power. While the present trend in receiver specification sheets is to list sensitivity in decibels (dBf), there are a few that still use decibels. The letter *f* is an abbreviation for the femtowatt. The prefix femto is $10^{-15}$, hence a femtowatt is $10^{-15}$ W.

*Example*   A tuner has an input sensitivity of 40 $\mu$V when set for monaural reception. What is its sensitivity in decibels (dBf)?
*Solution*

$$dBf = 20 \log (\mu V/0.55)$$
$$= 20 \log 40/0.55$$
$$= 20 \log 72.727\ 2$$
$$\log 72.727\ 2 = 1.861\ 5$$
$$dBf = (20)\ (1.861\ 5) = 37.23\ dBf$$

Using Table 9-3, locate 40 $\mu$V in the left column. Move across to the right for an answer of 37.23 decibels (dBf).

Previously prepared tables where various references are concerned can be very helpful. Table 9-3 is a conversion table for microvolts to decibels (dBf). Still another table, Table 9-4 uses 1 W as the reference.

Table 9-3. Conversion of
microvolts to decibels (dBf).

| $\mu$V | dBf | $\mu$V | dBf |
|---|---|---|---|
| 1.5 | 8.71 | 3.5 | 16.07 |
| 1.6 | 9.28 | 3.6 | 16.3 |
| 1.7 | 9.8 | 3.7 | 16.548 |
| 1.8 | 10.3 | 3.8 | 16.776 |
| 1.9 | 10.77 | 3.9 | 17.012 |
| 2.0 | 11.198 | 4.0 | 17.2 |
| 2.1 | 11.6 | 4.5 | 18.256 |
| 2.2 | 12.04 | 5.0 | 19.17 |
| 2.3 | 12.424 | 6.0 | 20.748 |
| 2.4 | 12.8 | 7.0 | 22.075 |
| 2.5 | 13.15 | 8.0 | 22.984 |
| 2.6 | 13.478 | 9.0 | 24.296 |
| 2.7 | 13.804 | 10.0 | 25.19 |
| 2.8 | 14.134 | 30.0 | 34.74 |
| 2.9 | 14.436 | 32.0 | 35.3 |
| 3.0 | 14.74 | 40.0 | 37.23 |
| 3.1 | 15.01 | 50.0 | 39.17 |
| 3.2 | 15.3 | 55.0 | 40.0 |
| 3.3 | 15.7 | 100.0 | 45.19 |
| 3.4 | 15.8 | 1 000.0 | 65.154 |

Table 9-4. Decibels with 1 W as the reference (dBW).

| W | dBW | W | dBW | W | dBW | W | dBW |
|---|---|---|---|---|---|---|---|
| 1.00 | 0 | 6.3 | 8 | 40 | 16 | 250 | 24 |
| 1.25 | 1 | 8.0 | 9 | 50 | 17 | 320 | 25 |
| 1.6 | 2 | 10.0 | 10 | 63 | 18 | 400 | 26 |
| 2.0 | 3 | 12.5 | 11 | 80 | 19 | 500 | 27 |
| 2.5 | 4 | 16 | 12 | 100 | 20 | 630 | 28 |
| 3.2 | 5 | 20 | 13 | 125 | 21 | 800 | 29 |
| 4.0 | 6 | 25 | 14 | 160 | 22 | | |
| 5.0 | 7 | 32 | 15 | 200 | 23 | | |

# Receiver or tuner sensitivity

Sensitivity is the ability of a receiver or tuner to respond to signals, previously specified in microvolts ($\mu$V). These specs have been changed from voltage to power ratings. Sensitivity in microvolts can be converted to dBf by:

$$dBf = 20 \log (\mu V/0.55)$$

Log is log to the base 10 and $\mu$V is microvolts based on a 300-$\Omega$ antenna input. Sensitivity is not the same for monaural and stereo so tuners and receivers should include separate decibel (dBf) figures for each.

*Example*  A tuner has an input sensitivity of 2 $\mu$V when set in the stereo mode. What is its sensitivity in decibels (dBf)?
*Solution*

$$dBf = 20 \log (\mu V/0.55)$$
$$= 20 \log 2/0.55$$
$$= 20 \log 3.636\ 363$$
$$\log 3.636\ 363 = 0.559\ 9$$
$$dBf = (20)\ (0.559\ 9) = 11.198\ dBf$$

Consulting Table 9-3 would have eliminated the need for doing this work.

# Power and voltage gain

Table 9-5 shows power gain in watts and voltage gain (in volts) when expressed in terms of decibels. A gain of 3 dB means a doubling of power, or, expressed another way, a doubling of power means a 3 dB increase. As an example, locate 0.2 in the left-hand column. This number represents 0.2 W. Doubling the power results in 0.4 W. But if 0.2 W corresponds to 3 dB then 0.4 W corresponds to 6 dB.

In terms of voltage, find 1.6 V in the right-hand column. Doubling this voltage results in 3.2 V. In doubling the voltage you have a 6 dB increase because you go from the corresponding 24 dB to 30 dB. Hence, a doubling of voltage means a 6 dB increase.

The reverse, of course, is also valid. A decrease in voltage by 50% is a 6 dB decrease. A 50% decrease in power is a 3 dB decrease.

In Table 9-6, the reference level, 0 dB, is 1 000 $\mu$V. Because the decibel is referenced to the millivolt, it is written as dBmV. This is shown in the heading of the left-hand column. In the column to its right, the heading is microvolts ($\mu$V).

Any voltage below 1 000 $\mu$V is less than the reference and so dBmV are indicated by a minus sign placed in front of the number. For any values above the 1 000 $\mu$V level, decibels (dBmV) are expressed as positive numbers.

*Table 9-5. Power and voltage gain in decibels.*

| Power (W) | dB gain (reference) | V |
|---|---|---|
| 0.1 | 0 | 0.1 |
| 0.2 | 3 | |
| 0.4 | 6 | 0.2 |
| 0.8 | 9 | |
| 1.6 | 12 | 0.4 |
| 3.2 | 15 | |
| 6.4 | 18 | 0.8 |
| 12.8 | 21 | |
| 25.6 | 24 | 1.6 |
| 51.2 | 27 | |
| 102.4 | 30 | 3.2 |
| | 33 | |
| | 36 | 6.4 |
| | 39 | |
| | 42 | 12.8 |
| | 45 | |
| | 48 | 25.6 |
| | 51 | |
| | 54 | 51.2 |
| | 57 | |
| | 60 | 102.4 |

This table is useful when making comparisons between voltages that are less than 1 V. These are expressed in microvolts to avoid using decimal numbers. Thus, 60 dBmV in the chart is a comparison between 1.0 V (shown in the right-hand column under the heading of $\mu$V) and the reference of 1 000 $\mu$V or 0 dBmV.

Immediately above 60 dBmV is 59 dBmV, and the number to its right is 900 000. This is 900 000 $\mu$V. This could also have been written decimally as 0.9 V.

*Example* An amplifier has an output of 0.08 V. What is its output when referenced to 1 000 $\mu$V (0 dBmV)?

*Solution* 0.08 V is equal to 80 000 $\mu$V. In Table 9-6 locate 80 000 in the column headed $\mu$V. The closest amount to 80 000 in the table is 79 430. To the left of this number find 38. This is 38 dBmV.

# Decibels to nepers conversions

The neper, like the decibel, is a dimensionless unit. The decibel is derived from common logarithms (logarithms to the base 10), and the neper is used to express

*Table 9-6. Decibels (dBmV) versus microvolts.*

| dBmV | μV | dBmV | μV | dBmV | μV | dBmV | μV |
|------|------|------|--------|------|---------|------|------------|
| −40 | 10.00 | −9 | 354.8 | 21 | 11 220 | 52 | 398 100 |
| −39 | 11.22 | −8 | 398.1 | 22 | 12 590 | 53 | 446 700 |
| −38 | 12.59 | −7 | 446.7 | 23 | 14 130 | 54 | 501 200 |
| −37 | 14.13 | −6 | 501.2 | 24 | 15 850 | 55 | 562 300 |
| −36 | 15.85 | −5 | 562.3 | 25 | 17 780 | 56 | 631 000 |
| −35 | 17.78 | −4 | 631.0 | 26 | 19 950 | 57 | 707 900 |
| −34 | 19.95 | −3 | 707.9 | 27 | 22 390 | 58 | 794 300 |
| −33 | 22.39 | −2 | 794.3 | 28 | 25 120 | 59 | 891 300 |
| −32 | 25.12 | −1 | 891.3 | 29 | 28 180 | 60 | 1 000 000 |
| −31 | 28.18 | −0 | 1 000.0 | 30 | 31 620 | 61 | 1 122 000 |
| −30 | 31.62 | 0 | 1 000 | 31 | 35 480 | 62 | 1 259 000 |
| −29 | 35.48 | 1 | 1 122 | 32 | 39 810 | 63 | 1 413 000 |
| −28 | 39.81 | 2 | 1 259 | 33 | 44 670 | 64 | 1 585 000 |
| −27 | 44.67 | 3 | 1 413 | 34 | 50 120 | 65 | 1 778 000 |
| −26 | 50.12 | 4 | 1 585 | 35 | 56 230 | 66 | 1 995 000 |
| −25 | 56.23 | 5 | 1 778 | 36 | 63 100 | 67 | 2 239 000 |
| −24 | 63.10 | 6 | 1 995 | 37 | 70 790 | 68 | 2 512 000 |
| −23 | 70.79 | 7 | 2 239 | 38 | 79 430 | 69 | 2 818 000 |
| −22 | 79.43 | 8 | 2 512 | 39 | 89 130 | 70 | 3 162 000 |
| −21 | 89.13 | 9 | 2 818 | 40 | 100 000 | 71 | 3 548 000 |
| −20 | 100.0 | 10 | 3 162 | 41 | 110 200 | 72 | 3 981 000 |
| −19 | 112.2 | 11 | 3 548 | 42 | 125 900 | 73 | 4 467 000 |
| −18 | 125.9 | 12 | 3 981 | 43 | 141 300 | 74 | 5 012 000 |
| −17 | 141.3 | 13 | 4 467 | 44 | 158 500 | 75 | 5 623 000 |
| −16 | 158.5 | 14 | 5 012 | 45 | 177 800 | 76 | 6 310 000 |
| −15 | 177.8 | 15 | 5 623 | 46 | 199 500 | 77 | 7 079 000 |
| −14 | 199.5 | 16 | 6 310 | 47 | 223 900 | 78 | 7 943 000 |
| −13 | 223.9 | 17 | 7 079 | 48 | 251 200 | 79 | 8 913 000 |
| −12 | 251.2 | 18 | 7 943 | 49 | 281 800 | 80 | 10 000 000 |
| −11 | 281.8 | 19 | 8 913 | 50 | 316 200 | | |
| −10 | 316.2 | 20 | 10 000 | 51 | 354 800 | | |

the ratio of two power levels using the natural system of logarithms—logarithms to the base e (e equals 2.718 28). The formula for finding the number of nepers is:

$$\text{nepers} = {}^{1}\!/_{2} \log_e P_2/P_1$$

The relationships between decibels and nepers are as follows:

1 dB = 0.1 B
1 dB = 0.115 1 Np
1 B  = 1.151 Np
1 B  = 10 dB
1 Np = 0.868 6 B
1 Np = 8.686 dB

As in the case of decibels, nepers must be used with some reference level if just one value of power, either input or output, is specified. Table 9-7 supplies the conversion between decibels and nepers. Table 9-8 gives nepers to decibels.

*Table 9-7. Decibels versus nepers (Np, nepers; dB, decibels).*

| dB | Np | dB | Np | dB | Np | dB | Np |
|----|-------|----|--------|----|--------|-----|---------|
| 1 | 0.115 1 | 26 | 2.992 6 | 51 | 5.870 1 | 76 | 8.747 6 |
| 2 | 0.230 2 | 27 | 3.107 7 | 52 | 5.985 2 | 77 | 8.862 7 |
| 3 | 0.345 3 | 28 | 3.222 8 | 53 | 6.100 3 | 78 | 8.977 8 |
| 4 | 0.460 4 | 29 | 3.337 9 | 54 | 6.215 4 | 79 | 9.092 9 |
| 5 | 0.575 5 | 30 | 3.453 0 | 55 | 6.330 5 | 80 | 9.208 0 |
| 6 | 0.690 6 | 31 | 3.568 1 | 56 | 6.445 6 | 81 | 9.323 1 |
| 7 | 0.805 7 | 32 | 3.683 2 | 57 | 6.560 7 | 82 | 9.438 2 |
| 8 | 0.920 8 | 33 | 3.798 3 | 58 | 6.675 8 | 83 | 9.553 3 |
| 9 | 1.035 9 | 34 | 3.913 4 | 59 | 6.790 9 | 84 | 9.668 4 |
| 10 | 1.151 0 | 35 | 4.028 5 | 60 | 6.906 0 | 85 | 9.783 5 |
| 11 | 1.266 1 | 36 | 4.143 6 | 61 | 7.021 1 | 86 | 9.898 6 |
| 12 | 1.381 2 | 37 | 4.258 7 | 62 | 7.136 2 | 87 | 10.013 7 |
| 13 | 1.496 3 | 38 | 4.373 8 | 63 | 7.251 3 | 88 | 10.128 8 |
| 14 | 1.611 4 | 39 | 4.488 9 | 64 | 7.366 4 | 89 | 10.243 9 |
| 15 | 1.726 5 | 40 | 4.604 0 | 65 | 7.481 5 | 90 | 10.359 0 |
| 16 | 1.841 6 | 41 | 4.719 1 | 66 | 7.596 6 | 91 | 10.474 1 |
| 17 | 1.956 7 | 42 | 4.834 2 | 67 | 7.711 7 | 92 | 10.589 2 |
| 18 | 2.071 8 | 43 | 4.949 3 | 68 | 7.826 8 | 93 | 10.704 3 |
| 19 | 2.186 9 | 44 | 5.064 4 | 69 | 7.941 9 | 94 | 10.819 4 |
| 20 | 2.302 0 | 45 | 5.179 5 | 70 | 8.057 0 | 95 | 10.934 5 |
| 21 | 2.417 1 | 46 | 5.294 6 | 71 | 8.172 1 | 96 | 11.049 6 |
| 22 | 2.532 2 | 47 | 5.409 7 | 72 | 8.287 2 | 97 | 11.164 7 |
| 23 | 2.647 3 | 48 | 5.524 8 | 73 | 8.402 3 | 98 | 11.279 8 |
| 24 | 2.762 4 | 49 | 5.639 9 | 74 | 8.517 4 | 99 | 11.394 9 |
| 25 | 2.877 5 | 50 | 5.755 0 | 75 | 8.632 5 | 100 | 11.510 0 |

*Example* An amplifier has a gain of 10 dB. What is its gain in nepers?

*Solution* Locate the number 10 in the dB column in Table 9-7. To the right of this number you will find 1.151 0 Np.

*Example* Assuming a zero reference level of 1 mW, what is the gain in nepers of an amplifier whose output is 50 mW?

*Solution* The power ratio in the problem is 50 to 1. Locate the nearest number to this in Table 9-1. This is shown as 50:1. The gain in decibels is 17. Now consult Table 9-7. Locate 17 dB in the left-hand column. The number of nepers corresponding to 17 dB is 1.956 7.

*Table 9-8. Neper versus decibel conversion.*

| Np | dB | Np | dB | Np | dB | Np | dB |
|----|-----|----|-----|----|-----|----|-----|
| 1 | 8.686 | 26 | 225.836 | 51 | 442.986 | 76 | 660.136 |
| 2 | 17.372 | 27 | 234.522 | 52 | 451.672 | 77 | 668.822 |
| 3 | 26.058 | 28 | 243.208 | 53 | 460.358 | 78 | 677.508 |
| 4 | 34.744 | 29 | 251.894 | 54 | 469.044 | 79 | 686.194 |
| 5 | 43.430 | 30 | 260.580 | 55 | 477.730 | 80 | 694.880 |
| 6 | 52.116 | 31 | 269.266 | 56 | 486.416 | 81 | 703.556 |
| 7 | 60.802 | 32 | 277.952 | 57 | 495.102 | 82 | 712.252 |
| 8 | 69.488 | 33 | 286.638 | 58 | 503.788 | 83 | 720.938 |
| 9 | 78.174 | 34 | 295.324 | 59 | 512.474 | 84 | 729.624 |
| 10 | 86.860 | 35 | 304.010 | 60 | 521.160 | 85 | 738.310 |
| 11 | 95.546 | 36 | 312.696 | 61 | 529.846 | 86 | 746.996 |
| 12 | 104.232 | 37 | 321.382 | 62 | 538.532 | 87 | 755.682 |
| 13 | 112.918 | 38 | 330.068 | 63 | 547.218 | 88 | 764.368 |
| 14 | 121.604 | 39 | 338.754 | 64 | 555.904 | 89 | 773.054 |
| 15 | 130.290 | 40 | 347.440 | 65 | 564.590 | 90 | 781.740 |
| 16 | 138.976 | 41 | 356.126 | 66 | 573.276 | 91 | 790.426 |
| 17 | 147.662 | 42 | 364.812 | 67 | 581.962 | 92 | 799.112 |
| 18 | 156.348 | 43 | 373.498 | 68 | 590.648 | 93 | 807.798 |
| 19 | 165.034 | 44 | 382.184 | 69 | 599.334 | 94 | 816.484 |
| 20 | 173.720 | 45 | 390.870 | 70 | 608.020 | 95 | 825.170 |
| 21 | 182.406 | 46 | 399.556 | 71 | 616.706 | 96 | 833.856 |
| 22 | 191.092 | 47 | 408.242 | 72 | 625.392 | 97 | 842.542 |
| 23 | 199.778 | 48 | 416.928 | 73 | 634.078 | 98 | 851.228 |
| 24 | 208.464 | 49 | 425.614 | 74 | 642.764 | 99 | 859.914 |
| 25 | 217.150 | 50 | 434.300 | 75 | 651.450 | 100 | 868.600 |

# Nomogram

Although the formula for dB gain or loss does not involve much arithmetic, it does require determining the log of two ratios. A simpler way, but one that isn't more accurate because it can require some interpolation, involves the use of a nomogram, such as the one shown in N 9-1. A ruler or straightedge is put across the nomogram so that it touches the amount of input voltage on the left scale and the output voltage on the right scale. If, for example, the straightedge is put on 10 on the left and on 100 on the right, it will lie directly across the center scale, from which you can read the voltage ratio and the decibel ratio. In this case it would be a voltage ratio of 10 equivalent to 20 dB.

The slant line across the nomogram shows how a straightedge can be used. On the left the straightedge is at 30, representing the input voltage. On the right the straightedge is at 300, the output voltage. The straightedge crosses the center scale at 10 for the voltage ratio and is at 20 dB.

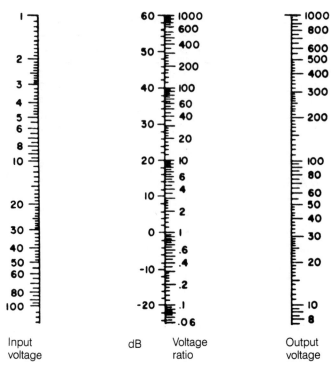

Fig. N 9-1. Nomogram for dB gain or loss.

There are two great advantages supplied by nomograms that do not apply to any equivalent formulas. A nomogram can be used to supply a large number of solutions, requiring nothing more than swinging the straightedge from one position to another. Also, the nomogram permits working back from a given solution to the values that will produce that solution. As an example, assume you want to construct an amplifier with a gain of 100, but do not know the amount of required input voltage, nor the amount of output voltage that will be available. To use the nomogram, put the straightedge on 100 on the voltage ratio scale in the center. Using this as a pivot point, you can immediately determine a large number of possible input voltages and their corresponding output voltages.

# 10
## CHAPTER

# Sound and acoustics

## Wavelengths of sound

On a sine wave, the distance between two successive positive peaks, two successive negative peaks, or between any two corresponding points is known as its wavelength, often represented by the Greek letter lambda $\lambda$. This description is not only applicable to radio-frequency waves, but also to sound waves. The reference here is not to a complex sound waveform, but to a pure sine wave only.

A sound wave of constant velocity (represented by the Greek letter $\mu$) will travel a distance of one wavelength in a one-period interval. This is more concisely stated in the formula $\mu = \lambda/T$. But the period of a wave has an inverse relationship to the frequency. Thus, $T = 1/f$. By substituting in the formula $\mu = \lambda/T$, you get $\mu = f\lambda$. You can rearrange this formula to read $\lambda = \mu/f$.

The velocity of sound in air at a temperature of 20 °C (68 °F) is 1 130 ft/s. Using this information, Table 10-1 gives the relationship between the frequency of sound in air in hertz and the wavelengths of sound in feet.

*Example* What is the wavelength, in feet, of a 60-Hz sine wave?
*Solution* Locate the number 60 in the left-hand column of Table 10-1. The corresponding value is shown as 18.83 ft.

*Example* A sound wave has a length of 7 ft. What is its frequency in hertz?
*Solution* The closest value given in Table 10-1 is 7.06 ft. The frequency of this sound wave, then, is approximately 160 Hz.

## Range of musical instruments

The fundamental range of musical instruments is limited. At the low-frequency end, few musical instruments can produce tones below 50 Hz. The human voice

## Table 10-1. Sound wavelengths.
### (1 130 ft/s in air, at 20°C; 68°F)

| Frequency (Hz) | Wavelength (ft) | Frequency (Hz) | Wavelength (ft) | Frequency (Hz) | Wavelength (ft) | Frequency (Hz) | Wavelength (ft) |
|---|---|---|---|---|---|---|---|
| 20 | 56.50 | 140 | 8.07 | 380 | 2.97 | 1 000 | 13.56 |
| 25 | 45.20 | 150 | 7.53 | 400 | 2.83 | 2 000 | 6.78 |
| 30 | 37.67 | 160 | 7.06 | 420 | 2.69 | 3 000 | 4.52 |
| 35 | 32.29 | 170 | 6.65 | 440 | 2.57 | 4 000 | 3.39 |
| 40 | 28.25 | 180 | 6.28 | 460 | 2.46 | 5 000 | 2.71 |
| 45 | 25.11 | 190 | 5.95 | 480 | 2.35 | 6 000 | 2.26 |
| 50 | 22.60 | 200 | 5.65 | 500 | 2.26 | 7 000 | 1.94 |
| 55 | 20.55 | 210 | 5.38 | 525 | 2.15 | 8 000 | 1.70 |
| 60 | 18.83 | 220 | 5.14 | 550 | 2.05 | 9 000 | 1.51 |
| 65 | 17.38 | 230 | 4.91 | 575 | 1.97 | 10 000 | 1.36 |
| 70 | 16.14 | 240 | 4.71 | 600 | 1.88 | 11 000 | 1.23 |
| 75 | 15.07 | 250 | 4.52 | 650 | 1.74 | 12 000 | 1.13 |
| 80 | 14.13 | 260 | 4.35 | 700 | 1.61 | 13 000 | 1.04 |
| 85 | 13.29 | 270 | 4.19 | 750 | 1.51 | 14 000 | 0.97 |
| 90 | 12.56 | 280 | 4.04 | 800 | 1.41 | 15 000 | 0.90 |
| 95 | 11.89 | 290 | 3.90 | 850 | 1.33 | 16 000 | 0.85 |
| 100 | 11.30 | 300 | 3.77 | 900 | 1.26 | 17 000 | 0.80 |
| 110 | 10.27 | 320 | 3.53 | 950 | 1.19 | 18 000 | 0.75 |
| 120 | 9.42 | 340 | 3.32 | 975 | 1.16 | 19 000 | 0.71 |
| 130 | 8.69 | 360 | 3.14 | 990 | 1.14 | 20 000 | 0.68 |

doesn't go much lower than 70 Hz. At the high-frequency end, all musical instruments and voices are under 5 kHz in fundamental frequency. It is the fundamental frequency that determines the pitch of a tone. See Fig. 10-1.

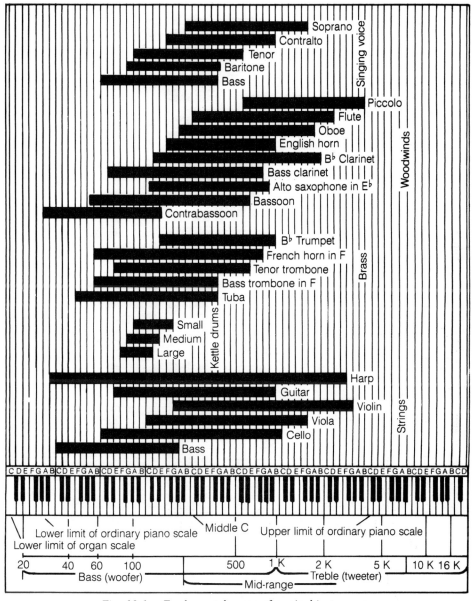

Fig. 10-1. Fundamental range of musical instruments.

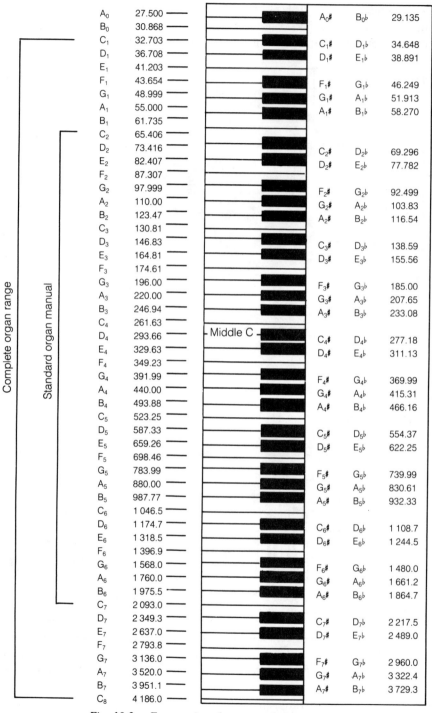

Fig. 10-2. Frequencies of organ and piano tones.

What you can hear depends on your age, sex, the physical condition of your ears and brain, and musical training. Although the audio spectrum is assumed to have a range of 20 Hz to 20 kHz, few people have a hearing capability that goes down to 20 Hz and equally few can hear as high as 20 kHz. The pipe organ, contrabassoon and the harp can reach below 40 Hz. Natural sounds include hardly any low frequencies, and what you may find at such frequencies is noise. See Fig. 10-2.

## The octave

An octave is a doubling of frequency. From 30 Hz to 60 Hz could be called an octave, 60 Hz to 120 Hz as another. You do not start with 0 Hz for that is actually dc. As a practical beginning 32 Hz often is selected, but we can start with 16 Hz as a bottom limit. If you select 16 Hz as your starting point, you can have 10 octaves up to approximately 16 kHz.

The ten octaves in Table 10-2 are of particular interest because they roughly represent a substantial part of the range of human hearing capability. Sounds below 64 Hz and above 16 kHz cannot be heard by most. Consider 16 Hz and 16 kHz as the outermost hearing limits.

*Table 10-2.*
*Frequency range in octaves.*

| Frequency range (Hz) | Octave |
|---|---|
| 16 to 32 | first |
| 32 to 64 | second |
| 64 to 128 | third |
| 128 to 256 | fourth |
| 256 to 512 | fifth |
| 512 to 1 024 | sixth |
| 1 024 to 2 048 | seventh |
| 2 048 to 4 096 | eighth |
| 4 096 to 8 192 | ninth |
| 8 192 to 16 384 | tenth |

## Frequency range of the piano

Middle C on the piano (Table 10-3) is indicated by the letter C (not followed by a number) and corresponds to a frequency of 261.63 Hz. From key A3 to A4 is an octave; from key A3 to A2 is also an octave. Figure 10-2 shows the frequencies of organ and piano tones.

The frequency range of other musical instruments appears in Table 10-4, and Table 10-5 supplies the frequency range of musical voices.

Table 10-3. Frequency range of the piano.

| Key | Frequency (Hz) | Key | Frequency (Hz) | Key | Frequency (Hz) | Key | Frequency (Hz) |
|-----|------|-----|------|-----|------|-----|------|
| A4 | 27.50 | G2 | 98.00 | F | 349.23 | E2 | 1 318.50 |
| B4 | 30.87 | A2 | 110.00 | G | 392.00 | F2 | 1 396.90 |
| C3 | 32.70 | B2 | 123.47 | A | 440.00 | G2 | 1 568.00 |
| D3 | 36.71 | C1 | 130.81 | B | 493.88 | A2 | 1 760.00 |
| E3 | 41.20 | D1 | 146.83 | C1 | 523.25 | B2 | 1 975.50 |
| F3 | 43.65 | E1 | 164.81 | D1 | 587.33 | C3 | 2 023.00 |
| G3 | 49.00 | F1 | 174.61 | E1 | 659.26 | D3 | 2 349.30 |
| A3 | 55.00 | G1 | 196.00 | F1 | 698.46 | E3 | 2 637.00 |
| F3 | 61.74 | A1 | 220.00 | G1 | 783.99 | F3 | 2 793.80 |
| C2 | 65.41 | B1 | 246.94 | A1 | 880.00 | G3 | 3 136.00 |
| D2 | 73.42 | C | 261.63 | B1 | 987.77 | A3 | 3 520.00 |
| E2 | 82.41 | D | 293.66 | C2 | 1 046.50 | B3 | 3 951.10 |
| F2 | 87.31 | E | 329.63 | D2 | 1 174.70 | C4 | 4 186.00 |

Table 10-4. Frequency range of musical instruments.

| Instrument | Low Hz | High Hz |
|-----|------|------|
| Bass clarinet | 82.41 | 493.88 |
| Bass tuba | 43.65 | 349.23 |
| Bass viola | 41.20 | 246.94 |
| Bassoon | 61.74 | 493.88 |
| Cello | 130.81 | 698.46 |
| Clarinet | 164.81 | 1 567.00 |
| Flute | 261.63 | 3 349.30 |
| French horn | 110.00 | 880.00 |
| Trombone | 82.41 | 493.88 |
| Trumpet | 164.81 | 987.77 |
| Oboe | 261.63 | 1 568.00 |
| Violin | 130.81 | 1 174.70 |
| Violin | 196.00 | 3 136.00 |

Table 10-5. Frequency range of musical voices.

| | Low Hz | High Hz |
|-----|------|------|
| Alto | 130.81 | 698.46 |
| Baritone | 98.00 | 392.00 |
| Bass | 87.31 | 349.23 |
| Soprano | 246.94 | 1 174.70 |
| Tenor | 130.81 | 493.88 |

Figure 10-3 is a comparison of the range of fundamental frequencies of various instruments and male and female voices.

# Velocity of sound in air

Sound velocity can be calculated from either of the following equations:

$$V = 49\sqrt{459.4 + °F} \text{ ft/s}$$

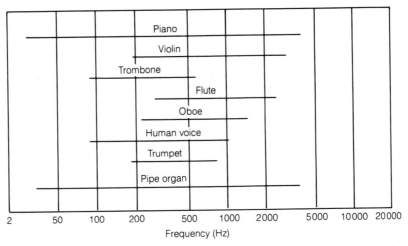

*Fig. 10-3. Comparison of the range of fundamental frequencies of instruments and voices.*

or

$$V = 20.06\sqrt{273 + °C}\ \text{m/s}$$

*V* is the velocity in feet/second or meters/second, °F is the temperature in degrees Fahrenheit, °C the temperature in degrees Celsius.

The velocity of sound isn't affected by frequency, somewhat slightly by humidity, much more so by temperature, and varies considerably depending on the material (medium) through which it moves. In solid substances, such as brick or steel, the velocity of sound is far greater than in air. In air, the velocity of sound increases by about 2 ft/s for each increase of 1 °C. Refer to Tables 10-6 and 10-7.

*Table 10-6. Velocity of sound
in air at various temperatures.*

| °F | Speed (ft/s) | °C | Speed (m/s) |
|----|-----|----|-----|
| 32 | 1 087 | 0 | 331.32 |
| 50 | 1 107 | 10 | 337.42 |
| 59 | 1 117 | 15 | 340.47 |
| 68 | 1 127 | 20 | 343.51 |
| 86 | 1 147 | 30 | 349.61 |

# Relative volume levels of ordinary sounds

The threshold of hearing is 0 dB and the threshold of hearing pain is 130 dB (Fig. 10-4). Table 10-8 indicates the intensity levels of various musical instruments measured at a distance of 10 ft. These are referenced to the threshold of hearing or 0 dB.

## Table 10-7. Velocity of sound in liquids and solids.

| Material | Sound velocity (ft/s) | (m/s) | Material | Sound velocity (ft/s) | (m/s) |
|---|---|---|---|---|---|
| Alcohol | 4 724 | 1 440 | Mercury | 4 790 | 1 460 |
| Aluminum | 20 407 | 6 220 | Nickel | 18 372 | 5 600 |
| Brass | 14 530 | 4 430 | Polystyrene | 8 760 | 2 670 |
| Copper | 15 157 | 4 620 | Quartz | 18 865 | 5 750 |
| Glass | 17 716 | 5 400 | Steel | 20 046 | 6 110 |
| Lead | 7 972 | 2 430 | Water | 4 757 | 1 450 |
| Magnesium | 17 487 | 5 330 | | | |

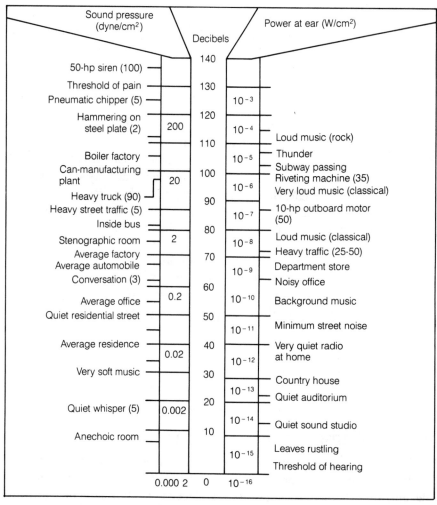

Fig. 10-4.   Relative loudness levels of common sounds.

*Table 10-8. Intensity*
*levels of various*
*musical instruments.*

| Instrument | dB intensity |
|------------|--------------|
| Piano      | 60 to 100 dB |
| Organ      | 35 to 110 dB |
| Bass drum  | 35 to 115 dB |
| Trumpet    | 55 to  95 dB |
| Violin     | 42 to  95 dB |
| Tympani    | 30 to 110 dB |
| Cymbal     | 40 to 110 dB |

While the decibel is generally used to indicate the relative volume level of ordinary sounds, there are two other units that are also used. These are the sone and the phon. Measured at a frequency of 1 kHz, 40 dB above the threshold of hearing produces a loudness of 1 sone.

The phon, a loudness unit, is based on the fact that doubling the intensity of a sound does not result in a doubling of a hearing sensation. The phon is a loudness unit that is an attempt to reflect the true loudness of a sound as perceived by the ear. The sone, on the other hand, is a unit of loudness graduated in equal steps.

Table 10-9 shows some common sounds and their loudness in phons. Table 10-10 can be used for the conversion of phons to sones or sones to phons.

*Table 10-9. Ordinary*
*sounds and their level in phons.*

| Sound | Phons |
|-------|-------|
| Threshold of hearing | 0 |
| Movement of leaves | 10−20 |
| Ticking of a clock, whispering | 20−30 |
| Soft music | 30−40 |
| Street noises | 50−60 |
| Loud voice | 60−80 |
| Orchestra playing loudly | 90 |
| Motorcycle racing its engine | 100 |
| Threshold of pain | 130 |

# Sound absorption

The sound absorption coefficient of materials varies with frequency. Table 10-11 indicates that plywood, a common building material, has a greater sound absorption coefficient at frequencies below 500 Hz, but tends to level off in the region above 2 kHz.

Table 10-10. Conversion of phons to sones or sones to phons.

| | | | | | Sones | | | | | |
|---|---|---|---|---|---|---|---|---|---|---|
| **Phons** | **0** | **+1** | **+2** | **+3** | **+4** | **+5** | **+6** | **+7** | **+8** | **+9** |
| 20 | 0.25 | 0.27 | 0.29 | 0.31 | 0.33 | 0.35 | 0.38 | 0.41 | 0.44 | 0.47 |
| 30 | 0.50 | 0.54 | 0.57 | 0.62 | 0.66 | 0.71 | 0.76 | 0.81 | 0.87 | 0.93 |
| 40 | 1.0 | 1.07 | 1.15 | 1.23 | 1.32 | 1.41 | 1.52 | 1.62 | 1.74 | 1.87 |
| 50 | 2.0 | 2.14 | 2.30 | 2.46 | 2.64 | 2.83 | 3.03 | 3.25 | 3.48 | 3.73 |
| 60 | 4.0 | 4.29 | 4.59 | 4.92 | 5.28 | 5.66 | 6.06 | 6.50 | 6.96 | 7.46 |
| 70 | 8.0 | 8.60 | 9.20 | 9.80 | 10.6 | 11.3 | 12.1 | 13.0 | 13.9 | 14.9 |
| 80 | 16.0 | 17.1 | 18.4 | 19.7 | 21.1 | 22.6 | 24.3 | 26.0 | 27.9 | 29.9 |
| 90 | 32.0 | 34.3 | 36.8 | 39.4 | 42.2 | 45.3 | 48.5 | 52.0 | 55.7 | 59.7 |
| 100 | 64.0 | 68.6 | 73.5 | 78.8 | 84.4 | 90.5 | 97.0 | 104 | 111 | 119 |
| 110 | 128 | 137 | 147 | 158 | 169 | 181 | 194 | 208 | 223 | 239 |
| 120 | 256 | 274 | 294 | 315 | 338 | 362 | 388 | 416 | 446 | 478 |

Table 10-11. Sound absorption coefficients for $^3/_{16}$-in plywood.

| Frequency (Hz) | Sound absorption coefficient |
|---|---|
| 125 | 0.35 |
| 250 | 0.25 |
| 500 | 0.20 |
| 1 000 | 0.15 |
| 2 000 | 0.05 |
| 4 000 | 0.05 |

The limits of sound absorption coefficients are 0 and 1. A substance with a coefficient of 1, or 100%, would indicate that it absorbed sound completely. It is the goal of construction materials used in anechoic chambers. A 0, or 0%, indicates that the material does not absorb sound at all.

An open window or rather that portion of it that constitutes open space can be regarded as having no reflectivity. In an echo chamber, the surfaces may have extremely high values of reflectivity. How effective a material is for the absorption of sound is indicated by its sound absorption coefficient. A cement floor has an absorption coefficient of 0.015 or 1.5%. This means that sound striking such a floor will dissipate 1.5% of the sound as heat, or slightly less, depending on how much of the sound will pass through the material. 98.5% of the sound will be reflected.

Table 10-12 indicates the effect of frequency on the sound absorption coefficients of various materials. For some materials, the absorption coefficient

*Table 10-12. Frequency versus sound absorption.*

| Material | Frequency (Hz) | | | | | |
|---|---|---|---|---|---|---|
| | 125 | 250 | 500 | 1 000 | 2 000 | 4 000 |
| Glass window | 0.35 | 0.25 | 0.18 | 0.12 | 0.07 | 0.04 |
| Lightweight drapes | 0.03 | 0.04 | 0.11 | 0.17 | 0.24 | 0.35 |
| Heavy drapes | 0.14 | 0.35 | 0.55 | 0.72 | 0.70 | 0.65 |
| Wood floor | 0.15 | 0.11 | 0.10 | 0.07 | 0.06 | 0.07 |
| Carpet (on concrete) | 0.02 | 0.06 | 0.14 | 0.37 | 0.60 | 0.65 |

*Table 10-13. Sound absorption coefficients of various materials.*

| Material | Coefficient absorption range |
|---|---|
| Linoleum on concrete floor | 0.03 to 0.08 |
| Upholstered seats | 0.05 |
| Ventilating grilles | 0.15 to 0.50 |
| Painted brick | 0.02 to 0.04 |
| Plaster on brick | 0.02 to 0.04 |
| Plaster on lath | 0.3 to 0.04 |
| Door | 0.3 to 0.05 |
| Window glass | 0.3 to 0.05 |
| Thick carpeting | 0.15 to 0.5 |
| Heavy curtains | 0.2 to 0.8 |

increases with frequency; with others it decreases; and for some it remains relatively fixed.

Absorption coefficients of building materials are often supplied as an overall range without indicating the effect of frequency, as indicated in Table 10-13.

# Reverberation time

Reverberation time is a function of room volume and sound absorption. It is directly proportional to the volume of an enclosed space, such as a recording studio or an in-home listening room and inversely proportional to the total amount of sound absorption in that enclosure (Fig. 10-5).

Reverberation time is not the same for all frequencies. The formula for calculating reverberation time, developed by Professor Wallace C. Sabine in 1895, indicates average reverberation time and does not take frequency into consideration.

$$T_{60} = \frac{0.05v}{S_a}$$

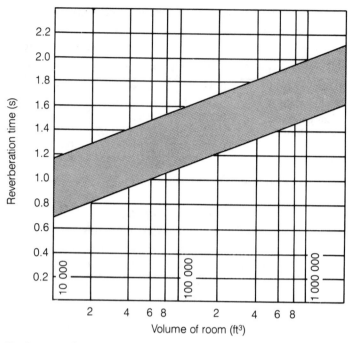

*Fig. 10-5. Dark area indicates acceptable reverberation time for rooms of different volumes.*

**Table 10-14. Sound absorption
of seats and audience at 512 Hz.**

| Item | Equivalent absorption (in sabins) |
|---|---|
| Audience, seated, units per person, depending on character of seats, etc. | 3.0−4.3 |
| Chairs, metal or wood | 0.17 |
| Pew cushions | 1.45−1.90 |
| Theater and auditorium chairs | |
|     Wood veneer seat and back | 0.25 |
|     Upholstered in leatherette | 1.6 |
|     Heavily upholstered in plush or mohair | 2.6−3.0 |
| Wood pews | 0.4 |

$T_{60}$ is the reverberation time in seconds, $v$ is the volume of the room in cubic feet and $S_a$ is the total equivalent sound absorption in sabins per square foot of surface material. Reverberation time is how long it takes reverberant sound to decrease by 60 dB, one millionth of the original sound source intensity.

#### Table 10-15. Effect of an audience on reverberation time.

| Audience (number present) | Absorption (sabins) | Reverberation time (s) |
|---|---|---|
| 0 | 1 201 | 7.3 |
| 200 | 2 011 | 4.3 |
| 400 | 2 821 | 3.1 |
| 600 | 3 631 | 2.4 |
| 800 | 4 441 | 2.0 |

#### Table 10-16. Reverberation time.

| Total distance traveled (ft) | Distance from ear to barrier (ft) | Approximate time required (s) |
|---|---|---|
| 11.2 | 5.6 | 0.01 |
| 22.4 | 11.2 | 0.02 |
| 33.6 | 16.8 | 0.03 |
| 44.8 | 22.4 | 0.04 |
| 56.0 | 28.0 | 0.05 |
| 63.2 | 31.6 | 0.06 |
| 78.4 | 39.2 | 0.07 |
| 89.6 | 44.8 | 0.08 |
| 100.8 | 50.4 | 0.09 |
| 112.0 | 56.0 | 0.1 |
| 224.0 | 112.0 | 0.2 |
| 336.0 | 168.0 | 0.3 |
| 448.0 | 224.0 | 0.4 |
| 560.0 | 280.0 | 0.5 |
| 632.0 | 316.0 | 0.6 |
| 784.0 | 392.0 | 0.7 |
| 896.0 | 448.0 | 0.8 |
| 1 008.0 | 504.0 | 0.9 |
| 1 120.0 | 560.0 | 1.0 |

Table 10-14 shows the sound absorption of seats and an audience at 512 Hz. The absorption is indicated in sabins. Table 10-15 indicates the effect of an audience on reverberation time. It takes time for sounds to reach a reflecting surface and then the ears of listeners. Table 10-16 shows the approximate time in seconds for reverberant sound to be heard.

# 11
## CHAPTER

# Filters

## Passive and active filters

Physically, a filter is a circuit, either passive or active, used for controlling the passage of a single frequency, or a group of frequencies, through a circuit. The circuit can consist of reactive elements, such as inductors and capacitors, or can include nonreactive elements such as resistors.

A passive filter (Fig. 11-1) is one without any signal gain. It will have an output that is smaller than the input. The output is usually expressed as a decimal fraction or possibly in decibels preceded by a minus sign.

An active filter (Fig. 11-2) is one that includes a device that can supply signal gain. It can have unity output or an output that is greater than its input. Signal gain can be supplied by an amplifier such as an op amp (operational amplifier).

## Filter classification

There are various ways that filters can be grouped:

- By the number of reactive elements. A filter can have one, two, or more. The simplest type of filter will have a single reactive element, either an inductor or a capacitor.
- By the number of poles. A pole includes a reactive pair, consisting of an inductor and a capacitor. A four-pole filter will have four pairs of associated reactances.
- By configuration. This is an arrangement of reactances having a resemblance to the letters T, L, or $\pi$.
- By transmission characteristic. A filter can be a low-pass, high-pass, band-pass or a band-reject type.
- By electrical characteristics such as *m*-derived, or constant-*k*.

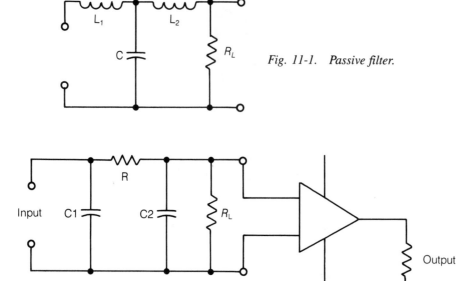

Fig. 11-1.  Passive filter.

Fig. 11-2.  Active filter using an op amp (operational amplifier).

- By their designer's names, such as Chebyshev and Butterworth.
- By the mathematical function from which it is derived, as in the case of the Bessel filter.
- By function. Filters can be named for the work they do, as power supply, noise reduction, and so on.
- By the sharpness of rolloff, specified in decibels.
- By whether they are purely electronic, using reactive elements, or are made on various piezoelectric substrates, as in the case of surface acoustic wave filters.
- By the amount of insertion loss, measured in decibels.
- By whether they are cascaded and the extent of such an arrangement.
- By whether the filter is passive or active.
- By special qualities such as comb filters, elliptic filters, iterative RC filters, digital filters, and brickwall filters.

These designations are not exclusive and are frequently combined to make the description of the filter more specific.

# Single-element low-pass RC filter

Figure 11-3 is the circuit of a filter using a single reactive element, C. The series arm uses a resistor, R, while the shunt arm consists of a capacitor, C. This filter has a very gradual rolloff beginning at the upper cutoff frequency. At this fre-

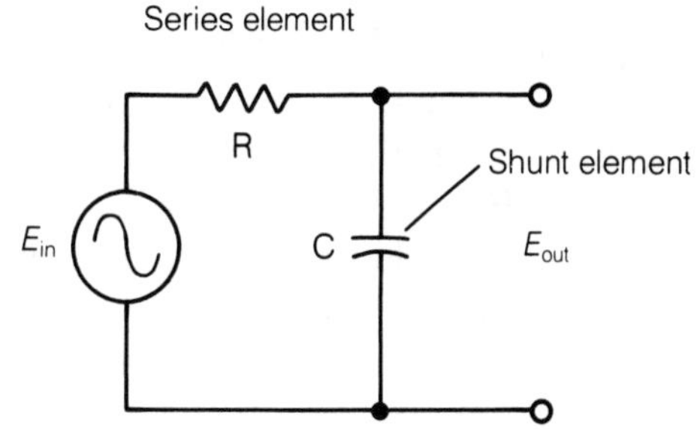

Series element

R

Shunt element

*Fig. 11-3.  Single reactive element filter.*  $E_{in}$  C  $E_{out}$

quency, the capacitive reactance is equal to the value of $R$. Because the circuit is that of a low-pass filter, there is only one cutoff frequency and it can be determined by:

$$f_c = \frac{1}{2\pi RC}$$

$$= \frac{0.159}{RC}$$

The value of 0.159 in the numerator is obtained by dividing 1 by $2\pi$.

**Example**  What is the cutoff frequency of the filter shown in Fig. 11-3 when $R = 120$ k$\Omega$ and $C = 0.002$ $\mu$F?
**Solution**  Substituting these values in the formula:

$$f_c = \frac{0.159}{(12 \times 10^4)\,(2 \times 10^{-9})}$$

$$= 0.662\,5 \text{ kHz}$$

# Filter attenuation (insertion loss)

Filter attenuation, specified in decibels, is the amount of signal loss produced at a specific frequency due to the insertion of the filter in a signal path. If this frequency is within the operative limits of the filter, it is referred to as insertion loss, although the two terms—filter attenuation and insertion loss—are sometimes used synonymously. Filter attenuation can take place outside the passband of a filter, and its amount depends on the separation of its frequency from that of the passband.

It is preferable to use inductors having a high value of $Q$, that is, a high ratio of the inductive reactance of the coil to its dc resistance. This will help minimize the insertion loss. It will also result in a smoother rolloff characteristic. The inductors can be air-core, fixed, or variable iron-core types. Cascading a number of filter sections will produce a sharper cutoff characteristic.

# Cutoff frequencies

For filters the cutoff frequencies ($f_c$) are measured at the half-power points. These are positioned at 70.7% of the peak with the real power dissipated at 50% of the maximum. Hence the points are referred to as the *half-power points*. The half-power point is the lower cutoff frequency of a high-pass filter and the upper cutoff frequency of a low-pass filter. For a bandpass filter, it consists of a pair of points—a lower and an upper cutoff.

# T, L, and $\pi$ sections

One of the ways of designating filter types is by the arrangement of the reactive elements. There are three basic configurations, as indicated in Fig. 11-4. Drawing (A) is referred to as a T section and consists of a pair of capacitors in the series

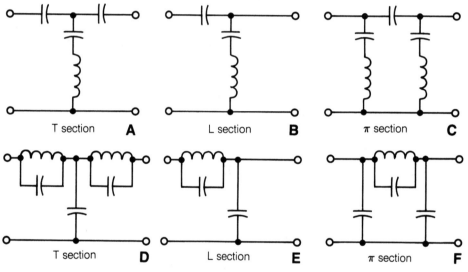

Fig. 11-4. T, L, and Pi ($\pi$) sections.

arm. The shunt arm is a series-tuned circuit. Drawing (B) is an L section, so called because it looks like an inverted L. It resembles the T section but has been changed to an L by the removal of a single series capacitor. Drawing (C) is called a $\pi$ section, or a Pi section.

Drawing (D) is a T section but is somewhat more sophisticated because it has a pair of parallel-tuned circuits in its series arm. The L section in (E) and the Pi section in (F) also use tuned circuits in their series arms. These arrangements are not the only ones because it is possible for these filters to have tuned circuits in both the series and shunt arms.

# Constant-*k* and *m*-derived filters

The division of filters into L, T, and π types is just one method of filter identification, but it has no great value other than supplying some idea of the circuit arrangement. Another technique is to refer to filters as *k* or *m* with these designations supplying information about the electrical characteristics of the filter.

For a constant-*k* filter the product of its series and parallel reactances remains constant over a limited frequency range. In terms of a formula:

$$k = X_L X_C = \frac{2\pi f L}{2\pi f C} = \frac{L}{C}$$

The *m*-derived filter is a variation of the constant-*k* and derives its name from the fact that the values of its components are multiplied by a factor, *m*. This factor is a decimal that is less than 1. The advantage of the *m*-derived filter is that it can have a sharper cutoff than constant-*k* types. The drawing in Fig. 11-5A is that of a constant-*k*, and that in (B) is an *m*-derived. The illustration is that of a basic T-type. In the *m*-derived each of the components is multiplied by *m*. Note the inclusion of an additional component in the shunt arm.

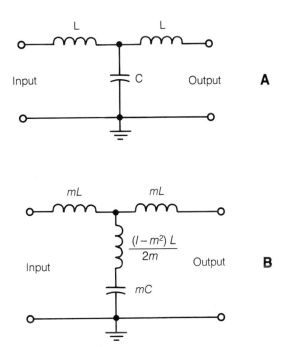

*Fig. 11-5.   Constant-k filter (A); m-derived filter (B).*

# Derivation of the value of m

The ratio of the cutoff frequency, $f_c$ to the infinite attenuation frequency, $f_\infty$, is a factor designated by the letter $m$. For a low-pass filter, $m$ can be stated as:

$$m = \sqrt{1 - \left( \frac{f^2}{f_\infty} \right)}$$

and for a high-pass filter:

$$m = \sqrt{1 - \left( \frac{f_\infty^2}{f} \right)}$$

$f_\infty$ is the infinite attenuation frequency. At some frequency, the attenuation of the filter will be so large that practically speaking it can be considered infinite.

Both $\pi$ and T-type filters can be $m$-derived. The various drawings in Fig. 11-6 show four different filters with the letter $m$ used to indicate that the filter has

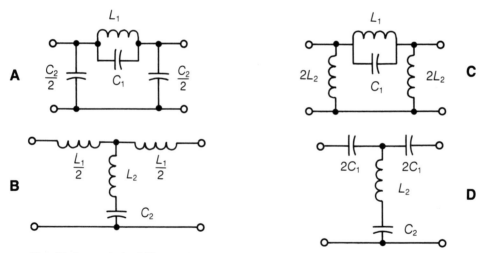

Fig. 11-6.  m-derived filters. $\pi$ section (A); T section (B); $\pi$ section (C); T section (d).

been modified. Drawings (A) and (C) are $\pi$ sections; (B) and (D) are T sections. The $m$-derived filters can be classified as either series or shunt.

# Filter functions

All filters have a basic function and can be categorized as:

High-pass
Bandpass
Band-reject (band elimination)
Low-pass

The name of the function can be combined with other designations. Thus, a filter can be identified, for example, as an *m*-derived, T-type, low-pass. Low-pass, for example, indicates the work the filter is to do, T-type supplies information about the arrangement of the reactances of the filter, and *m*-derived supplies data about the filter elements that will be used.

## High-pass filters

This is the name applied to a group of filters whose basic function is to pass high frequencies and oppose or completely block the passage of low frequencies. The simplest type consists of a single series capacitor and a shunt load resistor as shown in Fig. 11-7. This reactive element has an opposition to current flow that

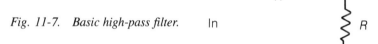

*Fig. 11-7.  Basic high-pass filter.*     In

varies inversely with frequency. The higher the frequency, the smaller the opposition, measured in ohms. Signal passage is completely or partially blocked for very low frequencies, decreasing as the frequency is increased. Another method of defining a high-pass filter is that all frequencies above a selected frequency, known as the cut-off frequency, $f_c$, are passed, all those below the cut-off frequency are frequencies that are blocked.

## High-pass constant-*k*

The arrangement of this filter is shown in Fig. 11-8. It is referred to as a *constant-k* because its series and shunt impedances, when multiplied, remain constant at all frequencies. The value of the components for this filter can be selected from:

$$C = \frac{1}{4\pi f_c R}$$

$$L = \frac{R}{4\pi f_c}$$

$$R = \sqrt{\frac{L}{C}}$$

$C$ is the series capacitance in farads, $L$ is the shunt inductance in henries, $R$ is the value of the terminating resistor in ohms, and $f_c$ is the cutoff frequency in hertz.

An operating feature of constant-$k$ filters is that they present an impedance match at one frequency only and are mismatched for all other frequencies. Another characteristic of constant-$k$ high-pass filters is that the capacitor (or capacitors) is always a series element, but the inductor (or inductors) is always positioned in parallel. These reactive elements can be arranged to form three different filter circuits, T, L, and $\pi$, as in Fig. 11-8.

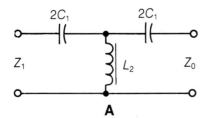

**A**

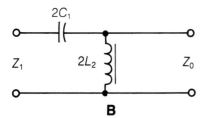

**B**

*Fig. 11-8. Types of high-pass, constant-$k$ filters. T-section (A); L section (B); $\pi$ section (C).*

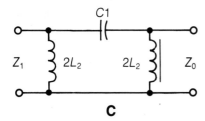

**C**

# Bandpass filter, constant-$k$

This filter is characterized by its ability to pass a selected, or continuous band of frequencies, and to reject all frequencies above or below the band. As shown in the circuit in Fig. 11-9 its basic arrangement consists of a pair of tuned circuits, one a series tuned, the other a parallel tuned. The series tuned has minimum signal rejection at its resonant frequency; the parallel tuned has maximum opposition

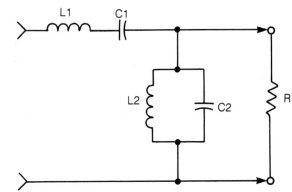

*Fig. 11-9. Bandpass filter.*

at its resonant frequency. This behavior for both circuits is designed to take place at the center of the selected frequency band. The series arm tuned circuit consists of L1 and C1; the parallel (shunt) arm tuned circuit is L2 and C2. The values of the components can be derived from:

$$C_1 = \frac{f_2 - f_1}{4\pi\, f_1\, f_2\, R}$$

$$C_2 = \frac{1}{\pi\, (f_2 - f_1)\, R}$$

$$L_1 = \frac{R}{\pi\, (f_2 - f_1)}$$

$$L_2 = \frac{(f_2 - f_1)\, R}{4\pi\, f_1\, f_2}$$

$L_1$ is the series inductor in henries; $C_1$, the series capacitor in farads, $L_2$, the parallel inductor in henries, $C_2$, the parallel capacitor in farads, $R$ is the nominal terminating resistance in ohms; $f_1$ is the lower cutoff frequency of the bandpass in hertz, and $f_2$ is the upper cutoff frequency of the bandpass in hertz. In this circuit the components L1 and C1 can be transposed without affecting the operation of the filter; similarly L2 and C2 can be transposed.

Figure 11-10 is a variation of the bandpass filter shown earlier in Fig. 11-9. The series arm now consists of two tuned circuits. L1 and C1 are series tuned circuits with this pair of circuits connected in the series arm of the filter. L2 and C2 (2 each) form a parallel-tuned circuit with this circuit connected in shunt across the filter. The inductance values of $L_1$ are to be divided by 2, and the values of capacitance of $C_1$ are to be multiplied by 2. The formulas for this filter are supplied by:

$$L_1 = \frac{Z_0}{\pi(f_2 - f_1)}$$

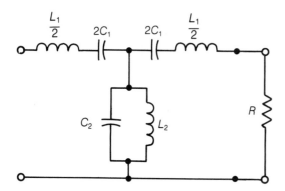

Fig. 11-10. Constant-k bandpass filter.

$$L_2 = \frac{(f_2 - f_1)}{4\pi f_1 f_2}$$

$$C_1 = \frac{(f_2 - f_1)}{4\pi f_1 f_2 Z_0}$$

$$C_2 = \frac{1}{\pi (f_2 - f_1) Z_0}$$

$$f_m = \sqrt{f_1 f_2} = \frac{1}{2\pi \sqrt{L_1 C_1}} = \frac{1}{2\pi \sqrt{L_2 C_2}}$$

$$Z_0 = \sqrt{\frac{L_1}{C_2}} = \sqrt{\frac{L_2}{C_1}}$$

$L_1$ is the inductance in henries (circuit values are to be divided by 2), $C_1$ is the capacitance in farads (circuit values are to be multiplied by 2), $L_2$ is in henries, $C_2$ is in farads, $f_1$ and $f_2$ are the lower and upper limit frequencies of the passband, in hertz, $f_m$ is the center frequency of the passband in hertz, $Z_0$ is the line impedance.

## Band-elimination filter, constant-k

This circuit shown in Fig. 11-11 works in a manner opposite that of the bandpass type. It is also referred to as a band-stop or band-rejection filter. It passes all frequencies up to a selected cutoff point, rejects or blocks the frequencies that follow, up to a second selected cutoff point, and then passes all frequencies above it.

The circuit consists of a series positioned parallel-tuned branch and a series-tuned arrangement put in parallel with the filter. The values of the components can be selected from:

$$C_1 = \frac{1}{4\pi (f_2 - f_1) R}$$

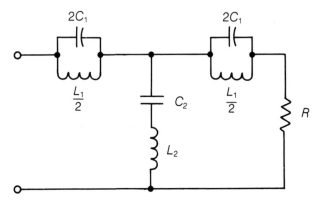

*Fig. 11-11. Band elimination filter, constant-k.*

$$C_2 = \frac{f_2 - f_1}{\pi f_1 f_2 R}$$

$$L_1 = \frac{(f_2 - f_1) R}{\pi f_1 f_2}$$

$$L_2 = \frac{R}{4\pi (f_2 - f_1)}$$

$C_1$ and $C_2$ are the capacitances in farads, $L_1$ and $L_2$ are the inductances in henries, $R$ is the resistance of the terminating resistor in ohms, $f_1$ and $f_2$ are the lower- and the upper-cutoff frequencies.

## Low-pass filter, constant-*k*

A low-pass filter (Fig. 11-12) passes all frequencies below a selected value and attenuates higher frequencies. Its action is the opposite of a high-pass filter, and its circuit arrangement is also opposite. For a low-pass constant-*k* filter:

$$C = \frac{1}{\pi f_c R}$$

$$L = \frac{R}{\pi f_c}$$

$$R = \sqrt{\frac{L}{C}}$$

$C$ is the parallel capacitor in farads, $L$ is the series inductor in henries, $R$ is the load in ohms, and $f_c$ is the cutoff frequency.

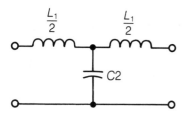

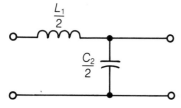

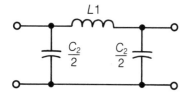

Fig. 11-12. Constant-k low-pass filters. Full section
T (A); half-section L (B); π (C).

# Cascaded filters

Filter sections can be connected in cascade, as indicated in Fig. 11-13, to obtain a
sharper rolloff. The inductors all have the same value but there is no coupling
between them, that is, the coefficient of coupling is zero. The coils must be posi-

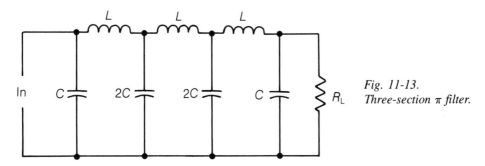

Fig. 11-13.
Three-section π filter.

tioned at right angles to each other, be widely separated, be shielded, or have
some combination of these factors. The capacitors are either $C$ or $2C$. Capacitor
values are doubled when π-sections are joined.

# Summary of filter formulas

Table 11-1 is a summary of filter formulas.

## Table 11-1. Summary of filter formulas.

Constant-$k$ low-pass

$$L_1 = \frac{Z_0}{\pi f_c}$$

$$C_2 = \frac{1}{\pi f_c Z_0}$$

$$Z_0 = \sqrt{\frac{L_1}{C_2}}$$

$$f_c = \frac{1}{\pi \sqrt{L_1 C_2}}$$

Constant-$k$ high-pass

$$L_2 = \frac{Z_0}{4 \pi f_c}$$

$$C_1 = \frac{1}{4 \pi f_c Z_0}$$

$$Z_0 = \sqrt{\frac{L_2}{C_1}}$$

$$f_c = \frac{1}{4 \pi \sqrt{L_2 C_1}}$$

Constant-$k$ bandpass

$$L_1 = \frac{Z_0}{\pi (f_2 - f_1)}$$

$$L_2 = \frac{(f_2 - f_1)}{4 \pi f_1 f_2}$$

$$C_1 = \frac{(f_2 - f_1)}{4 \pi f_1 f_2 Z_0}$$

$$C_2 = \frac{1}{\pi (f_2 - f_1) Z_0}$$

$$f_m = \sqrt{f_1 f_2} = \frac{1}{2 \pi \sqrt{L_1 C_1}} = \frac{1}{2 \pi \sqrt{L_2 C_2}}$$

*Table 11-1. Continued.*

$$Z_0 = \sqrt{\frac{L_1}{C_2}} = \sqrt{\frac{L_2}{C_1}}$$

Constant-*k* band-rejection

$$L_1 = \frac{(f_2 - f_1)\, Z_{/0}}{\pi f_1 f_2}$$

$$L_2 = \frac{Z_0}{4\pi(f_2 - f_1)}$$

$$C_1 = \frac{1}{4\pi(f_2 - f_1)\, Z_0}$$

$$C_2 = \frac{(f_2 - f_1)}{\pi(f_1 f_2 Z_0)}$$

$$f_m = \sqrt{f_1 f_2} = \frac{1}{2\pi\sqrt{L_1 C_1}} = \frac{1}{2\pi\sqrt{L_2 C_2}}$$

$$Z_0 = \sqrt{\frac{L_1}{C_2}} = \sqrt{\frac{L_2}{C_1}}$$

Series *m*-derived low-pass

$$L_1 = m\left(\frac{Z_0}{2\pi f_c}\right)$$

$$L_2 = \left(\frac{1 - m^2}{4m}\right)\left(\frac{Z_0}{2\pi f_c}\right)$$

$$C_2 = m\left(\frac{1}{\pi f_c Z_0}\right)$$

Series *m*-derived high-pass

$$L_2 = \frac{\left(\dfrac{Z_0}{4\pi f_c}\right)}{m}$$

$$C_1 = \frac{\left(\dfrac{1}{4\pi f_c Z_0}\right)}{m}$$

*Table 11-1. Continued.*

$$C_2 = \left( \dfrac{1}{\dfrac{1}{(4\pi f_c Z^0)}} \right) \left( \dfrac{1}{(4\pi f_c Z)} \right)$$

Shunt *m*-derived low-pass

$$L_1 = m \left( \dfrac{Z_0}{\pi f_c} \right)$$

$$C_1 = \left( \dfrac{1-m^2}{4m} \right) \left( \dfrac{1}{\pi f_c Z_0} \right)$$

$$C_2 = m \left( \dfrac{1}{\pi f_c Z_0} \right)$$

Shunt *m*-derived high-pass

$$L_1 = \left( \dfrac{4m}{1-m^2} \right) \left( \dfrac{Z_0}{4\pi f_c} \right)$$

$$L_2 = \dfrac{\left( \dfrac{Z_0}{4\pi f_c} \right)}{m}$$

$$C_1 = \dfrac{\left( \dfrac{1}{4\pi f_c Z_0} \right)}{m}$$

Calculation of *m*

$$m = \sqrt{1 - \left( \dfrac{f_c}{f_\infty} \right)^2}$$

$$m = \sqrt{1 \left( \dfrac{f_\infty}{f_c} \right)^2}$$

Select the formula that supplies a positive number.

# 12
## CHAPTER

# Pads and attenuators

## Pads

Pads and attenuators are made solely of resistors and are used for decreasing a variable input signal. They are inserted between a source voltage and its load, with the loss specified in decibels. As in the case of filters, the amount of signal attenuation is referred to as insertion loss.

Some of the more commonly used pad configurations consist of resistors arranged in the form of a T, H, $\pi$, L, O, bridged-T and bridged-H, as shown in Fig. 12-1. Resistor codes are alphanumeric and consist of the letter $R$ followed by a number. Identical codes indicate that the resistor values are the same. Two resistors in a diagram with both identified as R1 have the same resistance value.

The source input to a pad is usually shown at the left; the output at the right. The resistors of a pad are often selected from standard EIA values. Nonlisted values can be obtained through the use of series and parallel arrangements.

### Pad functions

A pad can be used to decrease a source voltage to keep it from overloading a following circuit. For example, a pad can be inserted between a microphone and its following amplifier when the mic is used in a very loud environment. A pad can also be used for impedance matching. Pads use fixed resistors only.

Table 12-1 supplies resistor values for three configurations, T, H and $\pi$ pads.

The $L$ pad resembles an inverted $L$, and does not have its values listed in Table 12-1. However, if the resistance of the source is greater than the resistance of the load, the value of the two arms of the pad, R1 and R2, can be calculated from:

$$R_1 = \sqrt{R_s \, (R_2 - R_L)} \qquad R_2 = \frac{R_s \times R_L}{R_1}$$

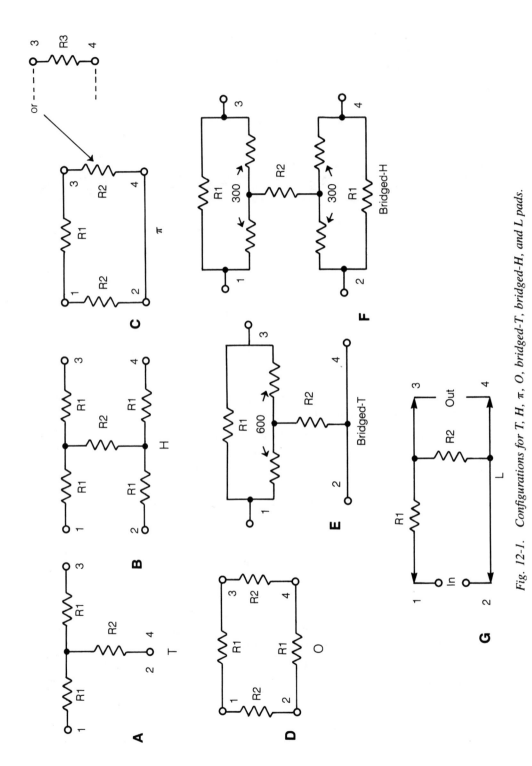

Fig. 12-1. Configurations for T, H, π, O, bridged-T, bridged-H, and L pads.

Table 12-1. Design values for pad networks.

| Loss (dB) | T pad $R_1$ | T pad $R_2$ | H pad $R_1$ | H pad $R_2$ | $\pi$ pad $R_1$ | $\pi$ pad $R_2$ |
|---|---|---|---|---|---|---|
| 0.1 | 3.58 | 50 204 | 1.79 | 50 204 | 7.20 | 100 500 |
| 0.2 | 6.82 | 26 280 | 3.41 | 26 280 | 13.70 | 57 380 |
| 0.3 | 10.32 | 17 460 | 5.16 | 17 460 | 20.55 | 34 900 |
| 0.4 | 13.79 | 13 068 | 6.90 | 13 068 | 27.50 | 26 100 |
| 0.5 | 17.20 | 10 464 | 8.60 | 10 464 | 34.40 | 20 920 |
| 0.6 | 20.9 | 8 640 | 10.45 | 8 640 | 41.7 | 17 230 |
| 0.7 | 24.2 | 7 428 | 12.1 | 7 428 | 48.5 | 14 880 |
| 0.8 | 27.5 | 6 540 | 13.75 | 6 540 | 55.05 | 13 100 |
| 0.9 | 31.02 | 5 787 | 15.51 | 5 787 | 62.3 | 11 600 |
| 1.0 | 34.5 | 5 208 | 17.25 | 5 208 | 68.6 | 10 440 |
| 2.0 | 68.8 | 2 582 | 34.4 | 2 582 | 139.4 | 5 232 |
| 3.0 | 102.7 | 1 703 | 51.3 | 1 703 | 212.5 | 3 505 |
| 4.0 | 135.8 | 1 249 | 67.9 | 1 249 | 287.5 | 2 651 |
| 5.0 | 168.1 | 987.6 | 84.1 | 987.6 | 364.5 | 2 141 |
| 6.0 | 199.3 | 803.4 | 99.7 | 803.4 | 447.5 | 1 807 |
| 7.0 | 229.7 | 685.2 | 114.8 | 685.2 | 537.0 | 1 569 |
| 8.0 | 258.4 | 567.6 | 129.2 | 567.6 | 634.2 | 1 393 |
| 9.0 | 285.8 | 487.2 | 142.9 | 487.2 | 738.9 | 1 260 |
| 10.0 | 312.0 | 421.6 | 156.0 | 421.6 | 854.1 | 1 154 |
| 11.0 | 336.1 | 367.4 | 168.1 | 367.4 | 979.8 | 1 071 |
| 12.0 | 359.1 | 321.7 | 179.5 | 321.7 | 1 119 | 1 002 |
| 13.0 | 380.5 | 282.8 | 190.3 | 282.8 | 1 273 | 946.1 |
| 14.0 | 400.4 | 249.4 | 200.2 | 249.4 | 1 443 | 899.1 |
| 15.0 | 418.8 | 220.4 | 209.4 | 220.4 | 1 632 | 859.6 |
| 16.0 | 435.8 | 195.1 | 217.9 | 195.1 | 1 847 | 826.0 |
| 17.0 | 451.5 | 172.9 | 225.7 | 172.9 | 2 083 | 797.3 |
| 18.0 | 465.8 | 152.5 | 232.9 | 152.5 | 2 344 | 772.8 |
| 19.0 | 479.0 | 136.4 | 239.5 | 136.4 | 2 670 | 751.7 |
| 20.0 | 490.4 | 121.2 | 245.2 | 121.2 | 2 970 | 733.3 |
| 22.0 | 511.7 | 95.9 | 255.9 | 95.9 | 3 753 | 703.6 |
| 24.0 | 528.8 | 76.0 | 264.4 | 76.0 | 4 737 | 680.8 |
| 26.0 | 542.7 | 60.3 | 271.4 | 60.3 | 5 985 | 663.4 |
| 28.0 | 554.1 | 47.8 | 277.0 | 47.8 | 7 550 | 649.7 |
| 30.0 | 563.0 | 37.99 | 281.6 | 37.99 | 9 500 | 639.2 |
| 32.0 | 570.6 | 30.16 | 285.3 | 30.16 | 11 930 | 630.9 |
| 34.0 | 576.5 | 23.95 | 288.3 | 23.95 | 15 000 | 624.4 |
| 36.0 | 581.1 | 18.98 | 290.6 | 18.98 | 18 960 | 619.3 |
| 38.0 | 585.1 | 15.11 | 292.5 | 15.11 | 23 820 | 615.3 |
| 40.0 | 588.1 | 12.00 | 294.1 | 12.00 | 30 000 | 612.1 |

## Insertion loss

The insertion loss of a pad is the ratio of the power input to the power output, in decibels, assuming equal impedances for the source and the load. Table 12-1 is for use where these impedances are the same. The values in the table are based on input and output impedances of 600 $\Omega$. The drawings in Fig. 12-1 show the types of pads for which the table can be used.

*Example* A simple pad is required to supply an insertion loss of 40 dB. How many resistors are needed and what is their value?

*Solution* A T pad can be used to solve this problem. The circuit is a four-terminal network as shown in the corresponding circuit diagram in Fig. 12-1. Locate 40 dB in the left-hand column of Table 12-1. To the right of this number are the values for $R_1$ and $R_2$. R1 represents two resistors, each having a value of 588.1 $\Omega$. R2 is shown as 12 $\Omega$.

*Example* It is necessary to drop the output signal voltage of a 600-$\Omega$ source by 3 dB.

*Solution* As shown in Table 12-1 a number of different pads can be used to get the same result. For an H pad, four resistors (R1) will be needed, each having a value of 51.3 $\Omega$. A single resistor (R2) of 1 703 $\Omega$ will complete the network.

The insertion loss of the L pad, the decrease in the source voltage as measured across the output, can be calculated in decibels by:

$$dB = 20 \log \left( \sqrt{\frac{R_s}{R_L}} + \sqrt{\frac{R_s}{R_L} - 1} \right)$$

dB is the amount of insertion loss, log is log to the base 10, $R_s$ is the resistance of the source, and $R_L$ is the resistance of the load.

## The *K* factor

The ratio of current, voltage, or power attenuation produced in the pad, measured in decibels, is known as the *K* factor. Table 12-2 lists the *K* factors that correspond to pad loss. The *K* factor is used in formulas for the calculation of resistor values in various pad configurations. It is applicable when the input and output impedances are equal or unequal. For equal impedances, the input and output can be transposed.

When calculating the *K* factor the result must be greater than one. If it is less than one, transpose the numerator and denominator in the ratio. Table 12-2 can be used in two ways. If the wanted insertion loss is known, listed in the column headed dB, the corresponding value of *K* is supplied in the immediately adjacent column. Either the input or output impedances of the pad can then be adjusted accordingly. If the value of *K* has been calculated, the insertion loss is then supplied in the dB column.

*Table 12-2. K factors and corresponding loss in decibels.*

| dB | K | dB | K | dB | K | dB | K |
|------|---------|------|---------|------|---------|-------|---------|
| 0.05 | 1.005 8 | 9.5 | 2.985 4 | 29.0 | 28.184 | 49.0 | 281.84 |
| 0.1 | 1.011 6 | 10.0 | 3.162 3 | 30.0 | 31.623 | 50.0 | 316.23 |
| 0.5 | 1.059 3 | 11.0 | 3.548 1 | 31.0 | 35.481 | 51.0 | 354.81 |
| 1.0 | 1.122 0 | 12.0 | 3.981 1 | 32.0 | 39.811 | 52.0 | 398.11 |
| 1.5 | 1.188 5 | 13.0 | 4.466 8 | 33.0 | 44.668 | 54.0 | 501.19 |
| 2.0 | 1.258 9 | 14.0 | 5.011 9 | 34.0 | 50.119 | 55.0 | 562.34 |
| 2.5 | 1.333 5 | 15.0 | 5.623 4 | 35.0 | 56.234 | 56.0 | 630.96 |
| 3.0 | 1.412 5 | 16.0 | 6.309 6 | 36.0 | 63.096 | 57.0 | 707.95 |
| 3.5 | 1.496 2 | 17.0 | 7.079 5 | 37.0 | 70.795 | 58.0 | 794.33 |
| 4.0 | 1.584 9 | 18.0 | 7.943 3 | 38.0 | 79.433 | 60.0 | 1 000.0 |
| 4.5 | 1.678 8 | 19.0 | 8.912 5 | 39.0 | 89.125 | 65.0 | 1 778.3 |
| 5.0 | 1.778 3 | 20.0 | 10.000 0 | 40.0 | 100.000 | 70.0 | 3 162.3 |
| 5.5 | 1.883 7 | 21.0 | 11.220 2 | 41.0 | 112.202 | 75.0 | 5 623.4 |
| 6.0 | 1.995 3 | 22.0 | 12.589 | 42.0 | 125.89 | 80.0 | 10 000 |
| 6.5 | 2.113 5 | 23.0 | 14.125 | 43.0 | 141.25 | 85.0 | 17 783 |
| 7.0 | 2.238 7 | 24.0 | 15.849 | 44.0 | 158.49 | 90.0 | 31 623 |
| 7.5 | 2.371 4 | 25.0 | 17.783 | 45.0 | 177.83 | 95.0 | 56 234 |
| 8.0 | 2.511 9 | 26.0 | 19.953 | 46.0 | 199.53 | 100.0 | $10^5$ |
| 8.5 | 2.660 7 | 27.0 | 22.387 | 47.0 | 223.87 | | |
| 9.0 | 2.818 4 | 28.0 | 25.119 | 48.0 | 251.19 | | |

Table 12-3 supplies the impedance ratio and the corresponding insertion loss in decibels. The impedance ratio can be input/output or output/input.

## T-type pad

Although Table 12-1 is convenient and helps avoid the use of formulas and arithmetic, it is limited to pads having the same input and output impedances of 600 Ω. If pads are to have identical impedances other than 600 Ω or to have unequal impedances, it becomes necessary to use formulas for the determination of the resistor values, with these formulas involving the K factor. The K factor can be used for determining the values of resistors in various configurations. One of these is the T-type pad shown earlier in Fig. 12-1A.

If the pad has equal input and output impedances the values of $R_1$, $R_2$ and $R_3$ can be calculated from:

$$R_1 \text{ and } R_2 = \left(\frac{K-1}{K+1}\right) Z$$

$$R_3 = \left(\frac{K}{K^2-1}\right) 2Z$$

Table 12-3. Impedance ratio
and corresponding loss in decibels.

| Impedance ratio $(Z_1/Z_2$ or $Z_2/Z_1)$ | Insertion loss (dB) |
|---|---|
| 1 | 0 |
| 2 | 7 |
| 3 | 8.5 |
| 4 | 11 |
| 5 | 12 |
| 6 | 13 |
| 7 | 13.5 |
| 8 | 14 |
| 9 | 14.8 |
| 10 | 15.3 |
| 20 | 18 |
| 30 | 21 |
| 40 | 22.5 |
| 50 | 23 |

An alternative method for calculating the values of $R_1$ and $R_2$ for the T pad, when input and output impedances are equal, is to use these simplified formulas in conjunction with Table 12-4.

$$R_1 = ZD$$

$$R_2 = ZE$$

Select the desired amount of attenuation by using the dB column, and then move horizontally to the numbers indicated in columns *B* to *E*. The configuration of the pad is the same as that originally illustrated in Fig. 12-1A.

A T-type pad that has unequal input and output impedances changes the formula for the determination of the resistance values. The circuit configuration remains exactly the same, but the formula for the determination of *R* values is expressed as:

$$R_1 = Z_1 \left( \frac{K^2+1}{K^2-1} \right) - 2\sqrt{Z_1 Z_2}$$

$$R_2 = Z_2 \left( \frac{K^2+1}{K^2-1} \right) - 2\sqrt{Z_1 Z_2}$$

$$R_3 = 2\sqrt{Z_1 Z_2}$$

In this instance, $Z_1$ is the larger of the two impedances.

## L pads

The L-pad configuration is illustrated in Fig. 12-1G. As shown, it consists of a single series arm and a single parallel arm. It can supply an impedance match in the direction of the series arm or in the direction of the parallel arm, but not both. If it is connected between equal impedances with impedance matching needed in the direction of the series arm, the values of the series resistor, R1, and that of the parallel resistor, R2, can be calculated from:

$$R_1 = Z \left( \frac{K-1}{K} \right)$$

$$R_2 = Z \left( \frac{1}{K-1} \right)$$

If the L pad is connected between equal impedances but the impedance match is needed in the direction of the shunt arm, the values of the two resistors, $R_1$ and $R_2$, can be calculated from:

$$R_1 = Z (K-1)$$

$$R_2 = Z \left( \frac{K}{K-1} \right)$$

The L pad can be connected between unequal impedances. If there is an impedance match needed toward the larger value, then $R_2$ can be determined by:

$$R_1 = \left( \frac{Z_1}{S} \right) \left( \frac{KS-1}{K} \right)$$

$$R_2 = \left( \frac{Z_1}{S} \right) \left( \frac{1}{K-S} \right)$$

where $S$ equals

$$\sqrt{\frac{Z_1}{Z_2}}$$

If the L pad is connected between unequal impedances and the impedance match is needed toward the smaller value, then:

$$R_1 = \left( \frac{Z_1}{S} \right) (K-S)$$

$$R_2 = \left( \frac{Z_1}{S} \right) \left( \frac{K}{KS-1} \right)$$

where $S$ equals

$$\sqrt{\frac{Z_1}{Z_2}}$$

*Table 12-4. Loss in decibels for various pads.*

| dB | A | B | C | D | E | dB | A | B | C | D | E |
|---|---|---|---|---|---|---|---|---|---|---|---|
| 0.1 | 0.988 55 | 0.011 447 | 86.360 | 0.005 756 | 86.857 | 10.0 | 0.316 23 | 0.683 77 | 0.462 48 | 0.519 49 | 0.702 73 |
| 0.2 | 0.977 24 | 0.022 763 | 42.931 | 0.011 512 | 43.426 | 11.0 | 0.281 84 | 0.718 16 | 0.392 44 | 0.560 26 | 0.612 31 |
| 0.25 | 0.971 63 | 0.028 372 | 34.247 | 0.014 390 | 34.739 | 12.0 | 0.251 19 | 0.748 81 | 0.335 45 | 0.598 48 | 0.536 21 |
| 0.3 | 0.966 05 | 0.034 046 | 28.456 | 0.017 268 | 28.947 | 12.5 | 0.237 14 | 0.762 86 | 0.310 85 | 0.616 64 | 0.502 53 |
| 0.4 | 0.954 99 | 0.045 008 | 21.219 | 0.023 022 | 21.707 | 13.0 | 0.223 87 | 0.776 13 | 0.288 45 | 0.634 16 | 0.471 37 |
| 0.5 | 0.944 06 | 0.055 939 | 16.876 | 0.028 774 | 17.362 | 14.0 | 0.199 53 | 0.800 47 | 0.249 26 | 0.667 32 | 0.415 60 |
| 0.6 | 0.933 25 | 0.066 745 | 13.982 | 0.034 525 | 14.428 | 15.0 | 0.177 83 | 0.822 17 | 0.216 29 | 0.698 04 | 0.367 27 |
| 0.7 | 0.922 57 | 0.077 429 | 11.915 | 0.040 274 | 12.395 | 16.0 | 0.158 49 | 0.841 51 | 0.188 34 | 0.726 39 | 0.325 15 |
| 0.75 | 0.917 28 | 0.082 724 | 11.088 | 0.043 147 | 11.567 | 17.0 | 0.141 25 | 0.858 75 | 0.164 49 | 0.752 46 | 0.288 26 |
| 0.8 | 0.912 01 | 0.087 989 | 10.365 | 0.046 019 | 10.842 | 17.5 | 0.133 35 | 0.866 65 | 0.153 87 | 0.764 68 | 0.271 53 |
| 0.9 | 0.901 57 | 0.098 429 | 9.159 6 | 0.051 762 | 9.633 7 | 18.0 | 0.125 89 | 0.874 11 | 0.144 02 | 0.776 37 | 0.255 84 |
| 1.0 | 0.891 25 | 0.108 75 | 8.195 5 | 0.057 501 | 8.666 7 | 19.0 | 0.112 20 | 0.887 80 | 0.126 38 | 0.798 23 | 0.227 26 |
| 1.5 | 0.841 40 | 0.158 60 | 5.305 0 | 0.086 133 | 5.761 9 | 20.0 | 0.100 000 | 0.900 00 | 0.111 111 | 0.818 18 | 0.202 02 |
| 2.0 | 0.794 33 | 0.205 67 | 3.862 1 | 0.114 62 | 4.304 8 | 21.0 | 0.089 125 | 0.910 87 | 0.097 846 | 0.836 34 | 0.179 68 |
| 2.5 | 0.749 89 | 0.250 11 | 2.998 3 | 0.142 93 | 3.426 8 | 22.0 | 0.079 433 | 0.920 57 | 0.086 287 | 0.852 82 | 0.159 87 |
| 3.0 | 0.707 95 | 0.292 05 | 2.424 0 | 0.171 00 | 2.838 5 | 22.5 | 0.074 989 | 0.925 01 | 0.081 069 | 0.860 48 | 0.150 83 |
| 3.5 | 0.668 34 | 0.331 66 | 2.015 2 | 0.198 79 | 2.415 8 | 24.0 | 0.063 096 | 0.936 90 | 0.067 345 | 0.881 30 | 0.126 70 |
| 4.0 | 0.630 96 | 0.369 04 | 1.709 7 | 0.226 27 | 2.096 6 | 25.0 | 0.056 234 | 0.943 77 | 0.059 585 | 0.893 52 | 0.112 83 |
| 4.5 | 0.595 66 | 0.404 34 | 1.473 2 | 0.253 40 | 1.846 5 | 26.0 | 0.050 119 | 0.949 88 | 0.052 763 | 0.904 55 | 0.100 49 |
| 5.0 | 0.562 34 | 0.437 66 | 1.284 9 | 0.280 13 | 1.644 8 | | | | | | |
| 6.0 | 0.501 19 | 0.498 81 | 1.004 8 | 0.332 28 | 1.338 6 | 27.0 | 0.044 668 | 0.955 33 | 0.046 757 | 0.914 48 | 0.089 515 |
| 7.0 | 0.446 68 | 0.553 32 | 0.807 28 | 0.382 47 | 1.116 0 | 27.5 | 0.042 170 | 0.957 83 | 0.044 026 | 0.919 07 | 0.084 490 |
| 7.5 | 0.421 70 | 0.578 30 | 0.729 20 | 0.406 77 | 1.025 8 | 28.0 | 0.039 811 | 0.960 19 | 0.041 461 | 0.923 43 | 0.079 748 |
| 8.0 | 0.398 11 | 0.601 89 | 0.661 43 | 0.430 51 | 0.946 17 | 30.0 | 0.031 623 | 0.968 38 | 0.032 655 | 0.938 69 | 0.063 309 |
| 9.0 | 0.354 81 | 0.645 19 | 0.549 94 | 0.476 22 | 0.811 83 | 32.0 | 0.025 119 | 0.974 88 | 0.025 766 | 0.950 99 | 0.050 269 |

Table 12-4. Continued.

| dB | A | B | C | D | E |
|---|---|---|---|---|---|
| 32.5 | 0.023 714 | 0.976 29 | 0.024 290 | 0.953 67 | 0.047 454 |
| 33.0 | 0.022 387 | 0.977 61 | 0.022 900 | 0.956 21 | 0.044 797 |
| 34.0 | 0.019 953 | 0.980 05 | 0.020 359 | 0.960 88 | 0.039 921 |
| 35.0 | 0.017 783 | 0.982 22 | 0.018 105 | 0.965 06 | 0.035 577 |
| 36.0 | 0.015 849 | 0.984 15 | 0.016 104 | 0.968 80 | 0.031 706 |
| 37.5 | 0.013 335 | 0.986 66 | 0.013 515 | 0.973 68 | 0.026 675 |
| 38.0 | 0.012 589 | 0.987 41 | 0.012 750 | 0.975 13 | 0.025 183 |
| 39.0 | 0.011 220 | 0.988 78 | 0.011 348 | 0.977 81 | 0.022 443 |
| 40.0 | 0.010 000 | 0.990 00 | 0.010 101 | 0.980 20 | 0.020 002 |
| 42.0 | 0.007 943 3 | 0.992 06 | 0.008 006 9 | 0.984 24 | 0.015 888 |
| 42.5 | 0.007 498 9 | 0.992 50 | 0.007 555 6 | 0.985 11 | 0.014 999 |
| 44.0 | 0.006 309 6 | 0.993 69 | 0.006 349 6 | 0.987 46 | 0.012 620 |
| 45.0 | 0.005 623 4 | 0.994 38 | 0.005 655 2 | 0.988 82 | 0.011 247 |
| 47.5 | 0.004 217 0 | 0.995 78 | 0.004 234 8 | 0.991 60 | 0.008 434 1 |
| 48.0 | 0.003 981 1 | 0.996 02 | 0.003 997 0 | 0.992 07 | 0.007 962 3 |
| 50.0 | 0.003 162 3 | 0.996 84 | 0.003 172 3 | 0.993 70 | 0.006 324 6 |
| 51.0 | 0.002 818 4 | 0.997 18 | 0.002 826 4 | 0.994 38 | 0.005 636 8 |
| 52.0 | 0.002 511 9 | 0.997 49 | 0.002 518 2 | 0.994 99 | 0.005 023 8 |
| 54.0 | 0.001 995 3 | 0.998 00 | 0.001 999 3 | 0.996 02 | 0.003 990 5 |
| 55.0 | 0.001 778 3 | 0.998 22 | 0.001 781 5 | 0.996 45 | 0.003 556 6 |
| 56.0 | 0.001 584 9 | 0.998 42 | 0.001 587 4 | 0.996 84 | 0.003 169 8 |
| 57.0 | 0.001 412 5 | 0.998 59 | 0.001 414 5 | 0.997 18 | 0.002 825 1 |
| 60.0 | 0.001 000 0 | 0.999 00 | 0.001 001 00 | 0.998 00 | 0.002 000 0 |
| 64.0 | 0.000 630 96 | 0.999 37 | 0.000 631 36 | 0.998 74 | 0.001 261 9 |
| 65.0 | 0.000 562 34 | 0.999 44 | 0.000 562 66 | 0.998 88 | 0.001 124 7 |
| 66.0 | 0.000 501 19 | 0.999 50 | 0.000 501 44 | 0.999 00 | 0.001 002 4 |
| 68.0 | 0.000 398 11 | 0.999 60 | 0.000 398 27 | 0.999 20 | 0.000 796 2 |
| 70.0 | 0.000 316 23 | 0.999 68 | 0.000 316 33 | 0.999 37 | 0.000 632 5 |
| 72.0 | 0.000 251 19 | 0.999 75 | 0.000 251 25 | 0.999 50 | 0.000 502 4 |
| 75.0 | 0.000 177 83 | 0.999 82 | 0.000 177 86 | 0.999 64 | 0.000 355 7 |
| 76.0 | 0.000 158 49 | 0.999 84 | 0.000 158 51 | 0.999 68 | 0.000 317 0 |
| 78.0 | 0.000 125 89 | 0.999 87 | 0.000 125 91 | 0.999 75 | 0.000 251 8 |
| 80.0 | 0.000 100 00 | 0.999 90 | 0.000 100 00 | 0.999 80 | 0.000 200 0 |
| 84.0 | 0.000 063 10 | 0.999 94 | 0.000 063 10 | 0.999 87 | 0.000 126 2 |
| 85.0 | 0.000 056 23 | 0.999 94 | 0.000 056 24 | 0.999 89 | 0.000 112 5 |
| 90.0 | 0.000 031 62 | 0.999 97 | 0.000 031 62 | 0.999 94 | 0.000 063 25 |
| 95.0 | 0.000 017 78 | 0.999 98 | 0.000 017 78 | 0.999 96 | 0.000 035 57 |
| 96.0 | 0.000 015 85 | 0.999 98 | 0.000 015 85 | 0.999 97 | 0.000 031 70 |
| 100.0 | 0.000 010 00 | 0.999 99 | 0.000 010 00 | 0.999 98 | 0.000 020 00 |

## $\pi$-type (pi) pad

The $\pi$ pad, shown in Fig. 12-1C can be inserted between equal or unequal impedances. For equal impedances, the values of the resistors can be calculated from:

$$R_1 = Z \left( \frac{K+1}{K-1} \right)$$

$$R_2 = \left( \frac{Z}{2} \right) \left( \frac{K^2-1}{K} \right)$$

If the impedances are unequal, the resistor values can be derived from:

$$R_1 = Z_1 \left( \frac{K^2-1}{K^2-2KS+1} \right)$$

$$R_2 = \left( \frac{\sqrt{Z_1 Z_2}}{2} \right) \left( \frac{K^2-1}{K} \right)$$

$$R_3 = Z_2 \left( \frac{K^2-1}{K^2-2\frac{K}{S}+1} \right)$$

where $S$ equals

$$\sqrt{\frac{Z_1}{Z_2}}$$

Note that the configuration does not change. The only modification is in the values of the series and shunt arms.

## O-type pad

The O pad uses the same formulas as the $\pi$-type pads. The O-type pad can be connected between matching or mismatched impedances. In either case, the values of the two series arm resistors should be divided by 2.

## Input and output impedances

Input and output impedances have three possibilities:

The input and output impedances are the same.
The input impedance is greater than the output impedance.
The output impedance is greater than the input impedance.

The maximum signal energy is transferred through the pad to the load when the input and output impedances match. Stated in another way, the insertion loss is minimum under these conditions. A pad can be designed to match both input and output impedances, even though these impedances aren't alike. The input impedance of the source matches the input of the pad and the output impedance of the pad matches that of the load.

For matching two impedances when the input impedance is greater than the output impedance use the formula:

$$R_1 = \sqrt{Z_1\,(Z_1 - Z_2)}$$

$$R_2 = \frac{Z_1\,Z_2}{R_1}$$

$$\text{dB loss} = 20\,\log_{10}\left(\sqrt{\frac{Z_1}{Z_2}} + \sqrt{\frac{Z_1}{Z_2} - 1}\right)$$

This formula is applicable to the unbalanced pad in Fig. 12-2.

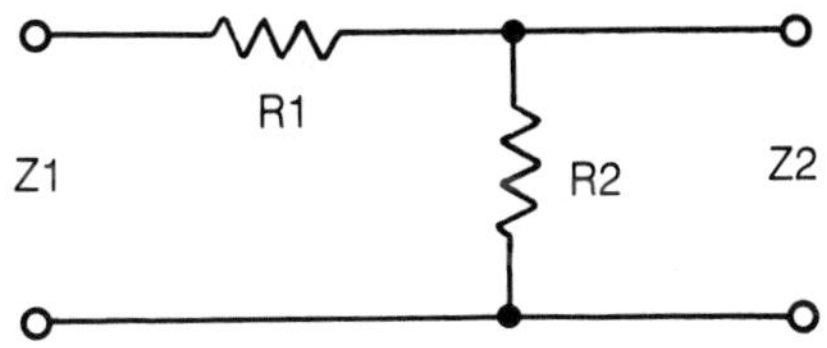

*Fig. 12-2.   Unbalanced minimum-loss pad.*

For matching two impedances when only one is to be matched, other formulas must be used. If only the larger impedance is to be matched, use a resistance $R_L$ in series with the smaller impedance so that:

$$R_L = Z_1 - Z_2$$

$$\text{dB loss} = 20\,\log_{10}\sqrt{\frac{Z_1}{Z_2}}$$

If only the smaller impedance is to be matched put a resistor $R_S$ in shunt across the larger impedance:

$$R_S = \frac{Z_1\,Z_2}{Z_1 - Z_2}$$

$$\text{dB loss} = 20\,\log_{10}\sqrt{\frac{Z_1}{Z_2}}$$

This is applicable to the balanced pad shown in Fig. 12-3.

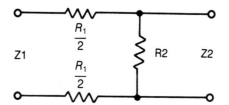

$Z1$

$Z2$

Fig. 12-3. Balanced minimum-loss pad.

## Lattice pad

The lattice pad, shown in Fig. 12-4, consists of four resistors: a series pair identified as R1 and a parallel pair R2. The parallel resistors are criss-crossed (but are joined only at their ends). It is used for connection between identical input and output impedances.

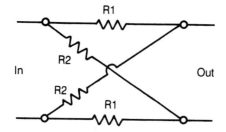

In

Out

Fig. 12-4. Lattice pad.

## Dividing pad

Pads can be used for dividing signal voltages. The dividing pad is characterized by a single input impedance but two or more matching output impedances. In the circuit of Fig. 12-5 there is a single source input and three load outputs. The value of the signal dividing resistors can be calculated from:

$$R_1 = \left(\frac{K-1}{K+1}\right) Z$$

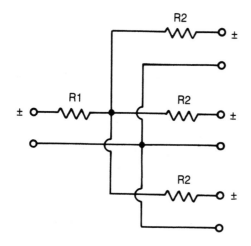

Fig. 12-5. Dividing pad.

$$R_2 = \left(\frac{K+1}{K-1}\right) Z$$

$R$ is the identical value of all the resistors, N is the number of circuits supplied with a share of the source voltage, and $Z$ is the impedance of the source.

## Bridged-T pad

The bridged pad, shown in Fig. 12-6, is a modified T pad and now has a series resistor, R1, shunting the existing series resistors whose value is $Z$. The pad is

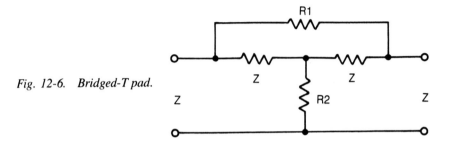

*Fig. 12-6. Bridged-T pad.*

intended for insertion between equal impedances. The values of $R_1$ and $R_2$ can be obtained from:

$$R_1 = \frac{Z}{C}$$

$$R_2 = ZC$$

The insertion loss can be selected through the use of column $C$ and the dB column listed in Table 12-4.

## Balanced-U pads

There are two types of balanced-U pads, one of which is shown in Fig. 12-7. The

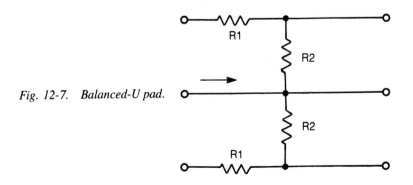

*Fig. 12-7. Balanced-U pad.*

formula for determining the resistor values is:

$$R_1 = \frac{ZB}{2}$$

$$R_2 = \frac{ZC}{2}$$

The corresponding numbers for $B$ and $C$ can be obtained from Table 12-4.

Figure 12-8 is another balanced-U pad with the series resistors R1 transposed. This pad is also for insertion between equal impedances. Although this

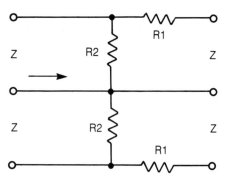

Fig. 12-8. *Alternative configuration of a balanced-U pad.*

configuration is almost the same as that in Fig. 12-7, there is some change in the values of $R_1$ and $R_2$, as indicated by their formulas:

$$R_1 = \frac{Z}{2C}$$

$$R_2 = \frac{Z}{2B}$$

These formulas also use data supplied in Table 12-4.

## Attenuators

Pads and attenuators are alike in that both can use similar configurations. The chief difference is that the resistive elements of a pad are fixed. An attenuator has at least one variable resistor. The variable resistors can be separate and independent, or they can be ganged. A dashed line between the variable resistors indicates the use of a common shaft. The variable resistors can have values that are linear or tapered. In some instances, the variable resistor is a slide type with the slide arm fastened into position at some selected point. It then functions as a pad.

The primary function of an attenuator is to produce a wanted amount of insertion loss, supplying a load with a required amount of current, voltage, or power. The amount of insertion loss can be fixed or variable but under all operating conditions, the impedance match between the source and the attenuator input and the load and attenuator output remains constant.

Like pads, attenuators are ordinarily passive devices. Because attenuators, as well as pads, work with a variety of frequencies, it is essential that their resistors be noninductive and not introduce a reactive element. Pads and attenuators have frequency limitations starting in the VHF (very high frequency) range, when connecting leads begin to exhibit reactance.

Attenuators can be designed to work between identical input and output impedances or in circuits where these impedances are different. Attenuators can not only be designed for a specific insertion loss, but also as an impedance matching network.

Aside from the introduction of one or more variable resistors, the configurations of attenuators are similar to those of pads. There are some exceptions.

### Balanced-T attenuator

An H-type attenuator configuration is shown in Fig. 12-9. The parallel resistor, $R_3$, is a variable unit and is set so it is at its resistive center. This is not necessarily

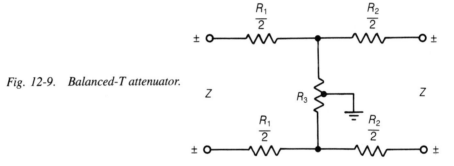

Fig. 12-9.  Balanced-T attenuator.

at the physical center of the unit. The values of the different resistors can be calculated from the same formulas used for the T pad. After determining the values of the four series resistors, divide all of them, with the exception of the variable resistor, by two. The formulas are:

$$R_1 \text{ and } R_2 = \left( \frac{K-1}{K+1} \right) Z$$

$$R_3 = \left( \frac{K}{K^2-1} \right) 2Z$$

### Bridged-T attenuator

The bridged-T attenuator in Fig. 12-10 has a pair of ganged variable resistors, R5 and R6. Their resistances have ohmic values that change inversely with respect to each other. R6 is the parallel arm. The series arm consists of two resistors, both

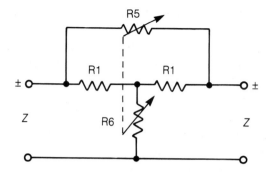

Fig. 12-10.    Unbalanced bridged-T
attenuator.

identified as R1. These are shunted by variable resistor R5. The input and output impedances, Z, are identical. The values of the components can be calculated from:

$$R_1 = Z$$

$$R_5 = (K-1)\, Z$$

$$R_6 = \left( \frac{1}{K-1} \right) Z$$

## Balanced versus unbalanced pads, attenuators, and filters

Some pads, attenuators, and filters have components in just one series leg with the other series leg having a conductor leading directly from the input to the output. Such a configuration is said to be unbalanced. A balanced one has identical parts in both series legs. The upper series leg and the lower are mirror images. The circuit shown earlier in Fig. 12-9 is a balanced configuration. The bridged T-type attenuator in Fig. 12-10 is unbalanced.

## L attenuator

The L attenuator, shown in Fig. 12-11, is the same as the L pad except that it uses a pair of variable resistors for the series and shunt arms. These resistors are

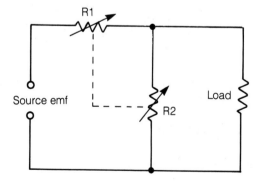

Fig. 12-11.    L attenuator.

mounted on a common shaft. The attenuator shows a constant resistance to the input emf regardless of the setting of the control knob of the two resistors. There is always an impedance match between the source and the attenuator input. The setting of the ganged resistors is also used to drop the input voltage to the amount required by the load. The total resistance of the attenuator can be expressed by:

$$R \text{ presented to the source} = R_1 + \frac{R_2 \times R_{load}}{R_2 + R_{load}}$$

The circuit is an unbalanced type. The resistance of $R_1$ and $R_2$ can be calculated by the same formulas used by the L pad.

## U attenuator

The U attenuator, shown in Fig. 12-12, gets its name from its resemblance to the letter U, placed on its side. The unit is balanced with equal value resistances, $R_1$,

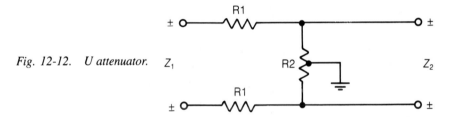

Fig. 12-12.   U attenuator.

in both series arms. The shunt resistor, $R_2$, is a variable, with its slide arm grounded at the electrical center. The values of $R_1$ and $R_2$ can be calculated from:

$$R_1 = \left(\frac{Z_1}{2S}\right) \left(\frac{KS-1}{K}\right)$$

$$R_2 = \left(\frac{Z_1}{S}\right) \left(\frac{1}{K-S}\right)$$

$R_1$ is the resistance in both series arms, $R_2$ can be a fixed or variable resistance, $Z_1$ is the input impedance, $Z_2$ is the output impedance, and $S$ is the square root of the ratio between the input and output impedances.

## Ladder attenuator

Although potentiometers are commonly used in attenuators, an alternative method is to use a rotary switch (Fig. 12-13) with fixed resistors mounted on the terminals. This is a convenient and more accurate method of selecting resistors. As the arm moves to the right, preceding resistors no longer form part of the attenuator. The attenuator is a step-loss type, and the maximum insertion loss occurs when the slide arm is on position 1. Minimum loss occurs when the slide arm is on posi-

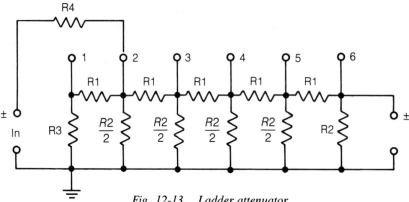

*Fig. 12-13.   Ladder attenuator.*

tion 6. This attenuator is an unbalanced type with no resistors in the lower series arm. The input and output impedances are identical. The values of the various resistors can be calculated from:

$$R_1 = \left( \frac{K_2 - 1}{2K} \right) Z$$

$$R_2 = \left( \frac{K+1}{K-1} \right) Z$$

$$R_3 = \frac{R_2 \times Z}{R_2 + Z}$$

$$R_4 = \frac{Z}{2}$$

$$Z_{in} = Z_{out}$$

## Tandem pads and attenuators

Various types of pads can be connected in tandem, such as H, L, and T. Unbalanced pads can be joined, or balanced pads can be connected. The pads can be converted to attenuators through the substitution of variable resistors.

# 13
CHAPTER

# Antennas

## Velocity factor

The velocity of a wave along a conductor, such as a transmission line, is not the same as the velocity of that wave in free space. The ratio of the two (actual velocity versus velocity in space) is known as the velocity factor. Obviously, the velocity factor must always be less than 1, and, in typical lines it varies from 0.6 to 0.97. See Table 13-1.

*Table 13-1. Velocity factors of transmission lines.*

| Type of line | Velocity factor (v) |
|---|---|
| Two-wire open line (wire with air dielectric) | 0.975 |
| Parallel tubing (air dielectric) | 0.95 |
| Coaxial line (air dielectric) | 0.85 |
| Coaxial line (solid plastic dielectric) | 0.66 |
| Two-wire line (wire with plastic dielectric) | 0.68−0.82 |
| Twisted-pair line (rubber dielectric) | 0.56−0.65 |

Table 13-2 lists the various types of antennas, supplying a description and applications.

## Coaxial cables

Coaxial cable is a type of transmission line used to connect a transceiver, a receiver, or a TV set to an antenna. The inner conductor of the cables listed in Table 13-3 is copper and so is the outside shield braid. The characteristic impedance is 50 Ω or 75 Ω. The cables in this table use a polyethylene dielectric.

*Table 13-2. Antenna types.*

| Type of antenna | Description | Application |
|---|---|---|
| **Parabolic reflector antenna** | A radiator placed at the focus of a parabola which forms a reflecting surface. Variations in the shape of the parabola provide changes in the shape of the beam produced. | Radar and satellite TV. |
| **Cosecant-squared reflector** | A reflector shaped to produce a beam pattern in which signal strength is proportional to the square of the cosecant of the angle between the horizontal and the line to the target. | Surface search by airborne radar sets. |
| **Horn antenna** | Consists of a waveguide with its mouth flared into a horn or funnel-like shape. The horn usually radiates into a reflector to provide the required beam shape. | Radar applications. |
| **End-fed hertz (Zepp)** | Half wavelength voltage-fed radiator, with signal supplies at one end by tuned, open-wire feeders. See Fig. 13-1. | For receiving and transmitting in the 16- to 30-MHz range. Most useful for multiband operation where space is limited. Use for fixed station installations. |
| **Center-fed hertz (tuned doublet or center-fed Zepp)** | A center-fed, half-wave doublet usually employing spaced feeders. Current fed on fundamental and voltage fed on all even harmonics. See Fig. 13-2. | For receiving and transmitting in the 1.6- to 30-MHz range. Can be used on any frequency if the system as a whole can be tuned to that frequency. |
| **Fuchs antenna** | Long-wire, voltage-fed radiator, an even number of quarter waves long. One end of radiator brought directly to the transmitter or tuning unit without using a transmission line. | For transmitting and receiving on any frequency where simplicity and convenience are desired. |
| **Corner reflector** | A half-wave radiator with two large metal sheets arranged so their | Used in the VHF (very high frequency) and UHF (ultra high frequency) ranges to provide directivity in the |

*Table 13-2. Continued.*

| Type of antenna | Description | Application |
|---|---|---|
| | surfaces meet at an angle whose apex lies behind the radiator. | plane which bisects the angle formed by the reflector. |
| **Marconi** | A vertical radiator approximately one-quarter wavelength long at operating frequency. One end is grounded or worked against ground. May be fed at or near base with low-impedance line. Electrical length can be increased by using loading coil in series with base or near center of radiator or by using capacitive loading at the top. The length, $L$, in feet can be computed by:<br><br>$$L = \frac{234}{f}$$<br><br>$f$ is in megahertz. $L$ is the overall length, in feet, from the top of the antenna to the point where it connects to ground or counterpoise.<br><br>    The total power dissipated in and radiated from a Marconi antenna can be calculated by<br><br>$$I^2 \, (R_r + R_g)$$<br><br>$I$ is the antenna current measured at the antenna base, $R_r$ is the radiation resistance and $R_g$ is the ground resistance. The useful radiated power is the difference between the total power consumed and the power lost in the ground resistance. See Fig. 13-3. | Widely used for medium- and low-frequency receiving and transmitting where vertical polarization is desirable. |
| **Parasitic array** | Consists of a radiator with a reflector behind and/or one | Used to develop high gain in one direction with little or no |

*Table 13-2. Continued.*

| Type of antenna | Description | Application |
|---|---|---|
| | or more directors in front. Produces a unidirectional radiation pattern. Can be either vertically or horizontally polarized. | radiation or pickup in other directions. Used on all frequencies where these characteristics are desired and space is available. |
| **Rhombic antenna** | A system consisting of four long-wire radiators arranged in the form of a diamond and fed at one end. If the corner opposite the feed point is open, response is bidirectional in a line running through these two corners. If the open end is terminated with the proper resistance, response is unidirectional in the direction of the terminated end. Gain can vary from 20 to 40 times that of a dipole, depending on the number of wavelengths in each leg. | Widely used where high gain and directivity are required. Can be used over a wide range of frequencies and is particularly useful when each leg is two or more wavelengths long on lowest frequency. Angle of radiation is lowered and vertical directivity narrowed by increasing length legs and/or increasing operating frequency. |
| **Vertical J** | A one-half wavelength vertical radiator fed at the bottom through a quarter-wave matching stub. It is omnidirectional, produces vertical polarization, and can be fed conveniently from a wide range of feed-line impedances. | Practical for use at frequencies above about 7 MHz. Normally used for fixed-frequency applications because of its extreme sensitivity to frequency changes. Efficiency falls off as frequency is raised. |
| **Coaxial antenna (Sleeve antenna)** | Vertical radiator one-half wavelength long. Upper half consists of a relatively thin radiator and the bottom half of a large diameter cylinder. Fed at the center from coaxial cable of 70 to 120 $\Omega$. | Practical for frequencies above about 7 MHz. Normally used for fixed frequency applications. Changes in frequency require that the antenna be retuned by varying length of the two halves of the radiator. Practical for operation up to about 100 MHz. |
| **Ground-plane antenna** | Omnidirectional quarter-wave vertical radiator | Practical for producing vertically polarized waves at |

*Table 13-2. Continued.*

| Type of antenna | Description | Application |
|---|---|---|
| | mounted above a horizontal reflecting surface. Its impedance is approximately 36 Ω or less. | frequencies above about 7 MHz and frequently used at frequencies as high as 300 MHz. |
| **Half rhombic (inverted V or tilted wire)** | A two-wire antenna with the legs in a vertical plane and in the shape of an inverted V. Directivity is in the plane of the legs. Feeding one end and leaving the other open results in bidirectivity. Terminating the free end with a suitable resistor produces unidirectional radiation in the direction of the termination. Gain and angle between legs depend on frequency and the number of wavelengths in each leg. | Used to provide high gain. Used where low angle of radiation is desirable. Usable over a wide frequency range. Bandwidth is greatest for terminated type. Angle of radiation is lowered as leg length and/or operating frequency is increased. |
| **Beverage antenna** | A directional long-wire horizontal antenna, two or more wavelengths long. The end nearest the distant receiving station is terminated with a 500-Ω resistor connected to a good counterpoise. The antenna, generally suspended 10 to 20 ft above ground, is nonresonant. | Used for transmitting and receiving vertically polarized waves. Often used for long-wave transoceanic broadcasts. Its input impedance is fairly constant so it can be used over a wide frequency range. Useful for frequencies between 300 kHz and 3 MHz. Highly suitable for use over dry, rocky soil. Never use over salt marshes or water. |
| **Folded dipole** | A simple center-fed dipole with a second half-wave conductor connected across its ends. Spacing between the conductors is a very small fraction of a wavelength. | Its impedance is higher than that of a simple dipole. Applications same as simple dipole. Often used in parasitic arrays to raise the feedpoint impedance to a value that can be conveniently matched to transmission line. |
| **Crow-foot antenna** | A low-frequency antenna consisting of a comparatively short vertical radiator with a | Normally used where it is impractical to erect a quarter-wave vertical |

*Table 13-2. Continued.*

| Type of antenna | Description | Application |
|---|---|---|
| | 3-wire V-shaped flat top and a counterpoise having the same shape and size as the flat top. See Fig. 13-4. | radiator. Used most frequently for reception and transmission in the 200- to 500-range. |
| **Turnstile antenna** | An omnidirectional, horizontally polarized antenna consisting of two half-wave radiators mounted at right angles to each other in the same horizontal plane. They are fed with equal currents 90° out of phase. Gain is increased by stacking. Dipoles can be single, folded, or special broadband types. | Normally used for transmission and reception of FM (frequency modulation) and television broadcast signals. |
| **Skin antennas** | Usually consist of an insulated section of the skin of an aircraft. Its radiation pattern varies with frequency, size of the radiating section, and position of the radiator on the aircraft. | Used for VHF and UHF reception and transmission in high-speed aircraft. Often used to replace fixed-wire antennas used in the 2- to 2.5-MHz range. |
| **ILAS antenna** | Localizer antennas are of several different types. One type consists of two or more square loops. Glide path is usually produced by two stacked antennas. The lower antenna is usually a horizontal loop bisected by a metal screen and supported about 6 ft off the ground. The upper antenna is a V-shaped dipole radiator with a parasitic element. Marker beacon antennas may consist of colinear dipoles or arrays. | Used to enable pilots to locate the airport and to land the plane on the desired runway when weather conditions would prohibit a landing under visual flight reference. |
| **Omni-range (VOR)** | Consists of two pairs of square-loop radiators surrounding a single square-loop radiator. | Use to provide navigation signals for aircraft in all directions from the range station. |

*Table 13-2. Continued.*

| Type of antenna | Description | Application |
|---|---|---|
| **Adcock antenna** | Consists of vertical radiators which produce bidirectional vertically polarized radiation. | Used in low-frequency radio ranges and for direction finding. |
| **Loop antennas** | A loop of wire consisting of one or more turns arranged in the shape of a square, circle, or other convenient form. It produces a bidirectional pattern along the plane of the loop. | Normally used for direction-finding applications, particularly in ships and aircraft. |
| **Stub mast** | A quarter-wave vertical radiator consisting of a metal sheath over a hardwood supporting mast. Fed with 50-$\Omega$ line with the outer conductor connected to a large metal ground surface. | Used for wide band reception and transmission of frequencies above 100 MHz. Normally used in aircraft installations. |
| **V antenna** | Bidirectional antenna made of two long-wire antennas in the form of a V and fed 180° out of phase at its apex. The V antenna is a combination of two long-wire antennas. As the length is increased, more power is concentrated near the axis of the wire. The length of each leg of a V antenna can be found by: $$L = \frac{492\,(N-0.05)}{f}$$ $N$ is the number of half wavelengths in each leg, and $f$ is the frequency in megahertz. More gain can be obtained by stacking a second V, one-half wavelength above the first. See Fig. 13-5. | Military and commercial applications. |

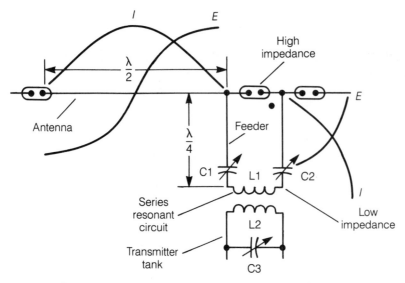

Resonant feed with odd number of quarter wavelengths

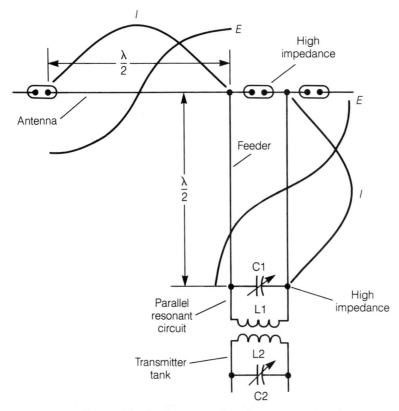

Resonant feed with even number of quarter wavelengths

*Fig. 13-1. Methods of feeding an end-fed half-wave hertz antenna.*

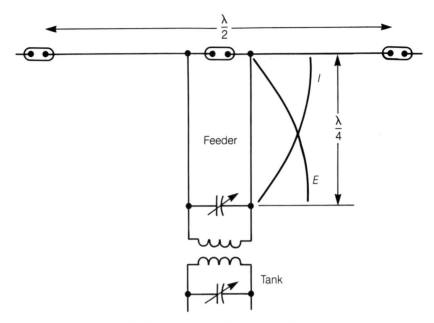

Odd number of quarter-wavelengths

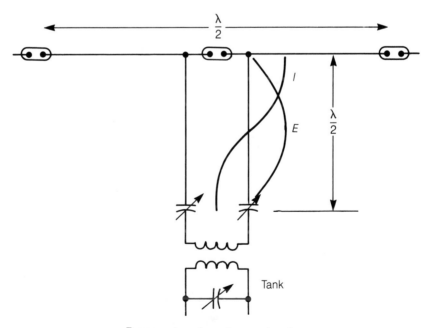

Even number of quarter-wavelengths

*Fig. 13-2. Methods of feeding a center-fed half-wave hertz antenna.*

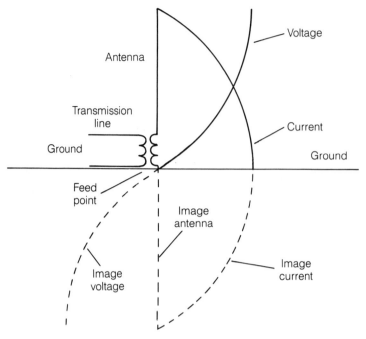

Fig. 13-3. *Current and voltage distribution of a Marconi antenna.*

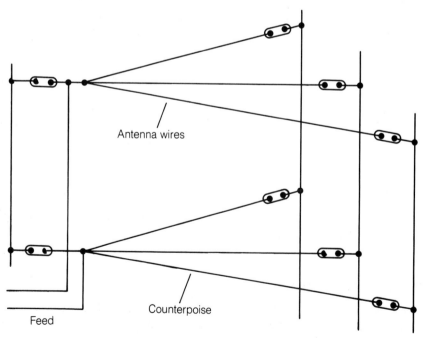

Fig. 13-4. *Arrangement of a crow-foot antenna.*

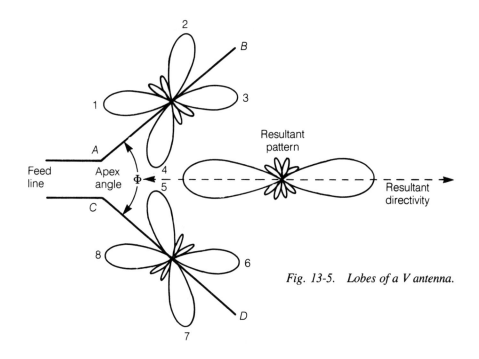

Fig. 13-5. Lobes of a V antenna.

Table 13-3. Coaxial cables for CB (citizens' band radio).

| RG/U cable type | Nominal attenuation (dB/100 ft) | | | |
|---|---|---|---|---|
| | 100 MHz | 1 000 MHz | 5 000 MHz | 10 000 MHz |
| 8/U | 1.9 | 8.0 | 27.0 | 100.0 |
| 8A/U | 1.9 | 8.0 | 27.0 | 100.0 |
| 58/U | 4.6 | 17.5 | 60.0 | 100.0 |
| 58A/U | 4.9 | 24.0 | 83.0 | 100.0 |
| 59/U | 3.4 | 12.0 | 42.0 | 100.0 |
| 59A/U | 3.4 | 12.0 | 42.0 | 100.0 |

# CB antennas

Table 13-4 lists the gain, VSWR (voltage standing-wave ratio), length, and typical weight of base station antennas.

# CB mobile antennas

Antennas for mobile CB transceivers can be trunk mounted, bumper mounted, put on a fender, on the car or truck roof. Table 13-5 supplies comparative data on a number of antennas.

## Table 13-4. Base-station antennas for CB.

| Type | Gain (dB) | VSWR | Length | Average weight (lb) |
|---|---|---|---|---|
| hor/vert 3-element | hor: 8 vert: 9.75 | 1.5:1 | 108 in boom | 28 |
| half-wave shunt-fed | 3.75 | 1.5:1 | 216 in | 10 |
| sector phased beam | 7.75 | 1.5:1 | 17.5 ft | 17 |
| vert/hor 5-element | hor: 11 vert: 12.5 | 1.5:1 | 22 ft boom | 19 |
| vertical | – | – | 234 in | 9 |
| vertical | 4 | 1.4:1 | 12 ft | 3.5 |
| vertical | 11 | 1.2:1 | 11.9 ft | 13.5 |
| vertical | 3.75 | 1:1 | 214 in | 5 |
| 3-element beam | 7.5 | 1:1 | 120 in | 15 |
| 4-element beam | 9 | 1:1 | 192 in | 20 |
| 5-element beam | 10 | 1:1 | 288 in | 25 |
| 6-element dual beam | 10.5 | 1.2:1 | 216 in | 30 |
| 8-element dual beam | 12 | 1.2:1 | 216 in | 45 |
| universal | 0 | 1.5:1 | 216 in | 2 |
| ground plane | 0 | 1.2:1 | 216 in | 5 |
| vertical | – | – | 17 ft 3 in | – |
| vertical | 1.5 | 1.5:1 | 18 ft | – |
| vertical | 9.2 | 1.1:1 | 9 ft cross boom 3 ft 1 in beam boom | 16 |
| vertical | 12.7 | 1.1:1 | 14 ft cross boom 12 ft 2 in beam boom | 30 |
| vertical | 13.9 | 1.1:1 | cross boom 20 ft beam boom | 55 |
| vertical | 8 | 1.4:1 | 12 ft boom | 7 |
| vertical | 9.5 | 1.1:1 | 18 ft boom | 18 |
| vertical | unity | 1.1:1 | 9 ft ht | 3 |
| vertical | 3 | 1.1:1 | 17 ft ht | 6 |
| vertical | 3.4 | 1.1:1 | 19 ft 10 in ht | 8 |
| vertical | 3.4 | 1.1:1 | 19 ft 10 in ht | 14 |
| attic | – | 2.1:0 | 18 in | 3 |
| dipole | – | 2.1:0 | 36 in | 6 |
| vertical | – | 2.1:0 | 18 in | 3 |

*Table 13-4. Continued.*

| Type | Gain (dB) | VSWR | Length | Average weight (lb) |
|---|---|---|---|---|
| ground plane | – | 2.1:0 | 18 in | 3 |
| vertical | – | 2.1:0 | 16 in | 10 |
| beam | 12.3 | 1.5:1 | 18 ft 8³/₄ in each element | 15 4 |
| vertical | unity | 1.5:1 | 9 ft | 4 |
| vertical | 5 | 1.5:1 | 9 ft | 7 |
| vertical | 6 | 1.2:1 | 19 ft 8 in | |
| vertical | 3.75 | 1.17:1 | 210 in | 10 |
| vertical | 10 | 1.5:1 | 204 in boom | 15 |
| vertical | 8 | 1.6:1 | 198 in boom | 11 |
| vertical | – | 1.8:1 | 108 in element 108 in radials | 4 |
| vertical | – | – | 19 ft | 12 |
| vertical | – | – | 19 ft | 11 |
| beam | – | – | 19 ft | 23 |
| vertical | – | – | 17 ft | 8 |
| vertical | – | – | 20 ft | 23 |
| vertical | 3.7 | 1.5:1 | 17 ft 3 in | – |
| beam | 8 | 1.5:1 | 224³/₄ in | 12.5 |
| beam | 8.7 | 1.5:1 | 224³/₄ in | 15 |
| beam | 9.5 | 1.5:1 | 224³/₄ in | 16.5 |
| beam | 8 | 1.5:1 | 216¹/₂ in | 14 |
| beam | 9.5 | 1.5:1 | 224³/₄ in | 20.5 |
| beam and stacking kit | 11 | 1.5:1 | 224³/₄ in | 29 |
| beam and stacking kit | 12 | 1.5:1 | 224³/₄ in | 38 |
| stacking kit | 13 | 1.5:1 | 224³/₄ in | 47 |
| conversion kit | – | – | – | 10 |
| stacking kit | – | – | – | 18 |
| stacking kit | – | – | – | 8 |
| vertical | – | 1.5:1 | 235.5 in | 7.5 |
| vertical | – | 1.5:1 | 245 in | 7.6 |
| vertical | – | – | 19 ft 8 in | 16 |
| vertical | – | – | 19 ft 8 in | 15 |
| vertical | – | – | 19 ft 10³/₄ in | 8.75 |
| beam | – | – | 18 ft 10 in boom | 8.5 |

*Table 13-4. Continued.*

| Type | Gain (dB) | VSWR | Length | Average weight (lb) |
|---|---|---|---|---|
| beam | – | – | 14 ft 1 in boom | 10.75 |
| gutter lamp | – | – | 25 in | – |
| vertical | 7.5 | 1.1:1 | – | 11 |
| vertical | 9 | 1.1:1 | – | 16 |
| vertical | 10.4 | 1.1:1 | – | 22 |
| vertical | – | 1.5:1 | 18 ft 6 in | 7.5 |
| vertical | 3.4 | 1.1:1 | 234 in | 9 |
| vertical | unity | 1.4:1 | 216 in | 7 |

*Table 13-5. CB mobile antennas.*

| Type | Gain (dB) | VSWR | Length | Average weight (lb) |
|---|---|---|---|---|
| whip, base spring | – | 1.5:1 | 108 in | 6 |
| trunk mount | – | 1.5:1 | 48 in | 1.5 |
| roof mount | – | 1.5:1 | 44 in | 1.5 |
| trunk mount | – | 1.5:1 | 48 in | 2 |
| fender mount | – | 1.5:1 | 44 in | 1.5 |
| combination CB/AM | – | 1.5:1 | 46 in | 1.5 |
| roof mount | – | 1.5:1 | 44 in | 1.5 |
| roof mount | – | 1.5:1 | 46 in | 1.5 |
| roof mount | – | 1.5:1 | 18 in | 1.25 |
| gutter mount | – | 1.5:1 | 18 in | 1.25 |
| magnetic mount | – | 1.5:1 | – | 1.25 |
| trunk mount | – | 1.5:1 | 46 in | 1.5 |
| gutter mount | – | 1.5:1 | 48 in | 1.5 |
| vertical | – | – | 96 ft | 1.5 |
| vertical | – | – | 49 in | 1.5 |
| vertical | – | – | 103 in | 1.5 |
| CB/AM | – | – | 48 in | 2 |
| vertical | – | – | 96 in | – |
| vertical | – | – | 30 in | – |
| vertical | – | – | 48 in | – |
| vertical | 1.5 | 1.5:1 | 86 in | – |

*Table 13-5. Continued.*

| Type | Gain (dB) | VSWR | Length | Average weight (lb) |
|---|---|---|---|---|
| vertical | − | 1.2:1 | − | − |
| vertical | − | 1.1:1 | 26 in | 1 |
| vertical | unity | 1.1:1 | 46 in | 2 |
| vertical | unity | 1.1:1 | 50 ft | 2 |
| vertical | unity | 1.1:1 | 46 in | 2 |
| vertical | unity | 1.1:1 | 32 in | 1 |
| vertical | unity | 1.1:1 | 59 ft | 3 |
| vertical | unity | 1.1:1 | 50 in | 2 |
| vertical | unity | 1.1:1 | 28 in | 2 |
| vertical | unity | 1.1:1 | 46 in | 1 |
| vertical | unity | 1.1:1 | 19 in | 2 |
| vertical | unity | 1.1:1 | 46 in | 3 |
| vertical | unity | 1.1:1 | 102 in | 3 |
| vertical | unity | 1.1:1 | 47 in | 3 |
| vertical | unity | 1.1:1 | $23^7/8$ in | 1.7 |
| vertical | unity | 1.1:1 | $34^5/8$ in | 3.2 |
| vertical | unity | 1.1:1 | 47 in | 3.1 |
| vertical | − | 1.1:1 | 47 in | 3.1 |
| vertical | − | 2.1:0 | 18 in | 3 |
| vertical | − | 2.1:0 | 30 in | 3 |
| vertical | − | − | 30 in | 1 |
| AM-FM/CB | − | − | 50 in | 1 |
| vertical | − | − | 102 in | 7 |
| vertical | − | 1.7:1 | 39 in | 2 |
| vertical | − | 1.7:1 | 39 in | 5 |
| vertical | − | 1.8:1 | 46 in | 4 |
| vertical | − | 1.6:1 | 102 in | 3 |
| vertical | − | 1:1 | 20 in | 3 |
| Gutter | | | | |
| vertical | − | − | 6 ft | 1.5 |
| vertical | | − | 4 ft | 0.5 |
| vertical | | − | 5 ft | 1 |
| vertical | − | − | 4 ft | 1 |
| vertical | − | − | 6 ft | 1.25 |
| vertical | unity | 1.5:1 | 63 in | − |

*Table 13-5. Continued.*

| Type | Gain (dB) | VSWR | Length | Average weight (lb) |
|------|-----------|------|--------|---------------------|
| vertical | unity | 1.5:1 | 42 in | – |
| vertical | unity | 1.5:1 | 44$^1/_2$ in | – |
| vertical | unity | 1.5:1 | 45 in | – |
| vertical | – | 1.5:1 | 8 ft 7 in | 1.75 |
| vertical | – | 1.5:1 | 45 in | 2 |
| vertical | – | 1.5:1 | 43$^3/_4$ in | 1 |
| vertical | – | 1.5:1 | 17 in | 1 |
| vertical | – | 1.5:1 | 36 in | 1 |
| vertical | – | 1.5:1 | 100 in | 3 |
| vertical | – | – | 83 ft | 2.75 |
| mobile res. | | | 29 in | 0.5 |
| mast | – | – | 54 in | 2.25 |
| mast | – | – | 54 in | 2.5 |
| vertical | – | – | – | 1 |
| vertical | – | – | – | 0.33 |
| vertical | – | – | 26$^1/_2$ in | 1.5 |
| | | | 60$^1/_2$ in | |
| vertical | – | – | 26$^1/_2$ ft | 0.75 |
| vertical | – | – | 60$^1/_2$ in | |
| vertical | – | – | – | 0.75 |
| vertical | – | – | – | 0.75 |
| vertical | – | – | – | 1 |
| AM/FM/CB | | | – | 1.5 |
| vertical | – | – | | |
| roof mount | – | – | 30 in | 0.5 |
| gutter clamp | – | – | 25 in | 1 |
| trunk groove | – | – | – | 1.25 |
| base loaded | – | – | 45$^3/_4$ in | 3 |
| trans. ant. | – | – | – | – |
| vertical | – | – | 108 in | 4.5 |
| bumper mount | – | – | 108 in | 5 |
| vertical | – | – | 102 in | 1.5 |
| vertical | – | – | 108 in | 1.5 |
| vertical | – | 1.5:1 | 8 ft | 1 |
| vertical | – | 1.5:1 | 4 ft | 0.5 |

*Table 13-5. Continued.*

| Type | Gain (dB) | VSWR | Length | Average weight (lb) |
|---|---|---|---|---|
| vertical | — | 1.5:1 | 9 ft | 1 |
| vertical | — | 1.5:1 | 36 in | 1 |
| magnetic mount | 4.1 | 1.1:25 | 32 in | — |
| vertical | 5.2 | 1.1:25 | 32 in | — |
| mtch. network | 6.4 | 1.1:0 | 10 in | — |
| vertical | — | 1.1:1 | 35 in | 0.75 |
| vertical | — | 1.1:1 | 36 in | 1 |
| vertical | — | — | 48 in | — |
| vertical | — | — | 96 in | — |
| vertical | — | — | — | — |
| vertical | — | — | 38 in | |
| vertical | — | — | 20 in | |
| vertical | — | — | 37 in | |
| vertical | — | — | 19 in | — |
| vertical | — | — | 38 in | — |
| vertical | — | — | 20 in | — |
| vertical | — | — | 3 in | 4 oz |
| vertical | — | — | 18 in | 6 oz |
| vertical | 4 | 1.1:1 | 108 in | 5 |
| vertical | unity | 1.1:1 | 46 in | 1.5 |
| vertical | unity | 1.1:1 | 45 in | 1.75 |
| vertical | unity | 1.1:1 | 18 ft | 1.5 |
| vertical | unity | 1.1:1 | 45 in | 1.5 |
| vertical | unity | 1.1:1 | 46 in | 1.5 |
| vertical | unity | 1.1:1 | 48 in | 2 |
| vertical | unity | 1.1:1 | 50 in | 2 |
| vertical | unity | 1.2:1 | 93 in | 4 |
| vertical | unity | 1.2:1 | 77 in | 4 |

# Marine antennas

Marine antennas are described in Table 13-6.

*Table 13-6. Marine antennas.*

| Type | Gain (dB) | VSWR | Length | Average weight (lb) |
|---|---|---|---|---|
| vertical | – | 1.5:1 | 97 in | 3 |
| vertical | – | – | 54 in | 2 |
| vertical | 1.5 | 1.1:1 | 19 ft 11 in | – |
| vertical | – | – | 6 ft | 4 |
| vertical | 1.5 | 1.5:1 | 79 in | – |
| vertical | – | 1.5:1 | 96³/₈ in | 2 |
| vertical | – | 1.5:1 | 12 ft | 7.5 |
| vertical | – | 1.5:1 | 18 ft 6 in | 9 |
| vertical | 4.1 | 1.1:25 | 30 in | – |
| vertical | 1.4 | 1.51 | 60 in | 2 |

# Coaxial cables for video

Various types of coaxial cable are used in video for connecting an antenna to the antenna terminals of a television receiver or for interconnecting video components. Coaxial cable is also used for connecting a dish antenna to a satellite receiver. The cable impedance is 75 $\Omega$.

Table 13-7 and Table 13-8 indicate the nominal attenuation in decibels per hundred feet of representative types of cables. Note that the attenuation, actually loss of signal along the line, increases with frequency (that is, in going from TV channel 2 to channel 83).

*Table 13-7. Nominal attenuation of coaxial cables at selected TV channels.*

| Cable | Nominal attenuation (dB/100 ft) | | | | | | | | | | |
|---|---|---|---|---|---|---|---|---|---|---|---|
| | Ch. 2 | Ch. 6 | Ch. 7 | Ch. 13 | Ch. 20 | Ch. 30 | Ch. 40 | Ch. 50 | Ch. 60 | Ch. 70 | Ch. 83 |
| **Color duct** | 2.3 | 2.7 | 3.8 | 4.2 | 6.5 | 7.0 | 7.5 | 7.8 | 8.0 | 8.4 | 9.0 |
| **Foam color duct** | 2.1 | 2.5 | 3.3 | 3.8 | 5.9 | 6.3 | 6.7 | 7.0 | 7.3 | 7.7 | 8.0 |
| **RG 59/U** | 2.6 | 3.5 | 4.9 | 5.4 | 8.3 | 8.8 | 9.2 | 9.7 | 10.3 | 11.0 | 11.9 |
| **RG 59/U foam** | 2.3 | 2.7 | 3.8 | 4.2 | 6.2 | 6.6 | 6.8 | 7.1 | 7.3 | 7.7 | 8.0 |
| **RG-6 foam** | 1.7 | 1.9 | 2.8 | 3.0 | 4.8 | 5.2 | 5.6 | 5.9 | 6.2 | 6.5 | 6.8 |
| **RG 11/U** | 1.4 | 1.7 | 2.2 | 3.2 | 5.1 | 5.3 | 5.5 | 5.7 | 6.1 | 6.2 | 6.8 |
| **RG 11/U foam** | 1.1 | 1.4 | 1.6 | 2.3 | 3.9 | 4.0 | 4.1 | 4.2 | 4.4 | 4.6 | 4.9 |
| **0.412 cable** | 0.74 | 1.0 | 1.4 | 1.5 | 2.5 | 2.6 | 2.7 | 2.9 | 3.1 | 3.3 | 3.5 |
| **0.500 cable** | 0.52 | 0.67 | 0.72 | 1.1 | 1.5 | 1.8 | 2.1 | 2.4 | 2.7 | 3.0 | 3.1 |

## Table 13-8. Nominal attenuation of coaxial cable at selected frequencies.

| Cable | Impedance | Attenuation (dB/100 ft) | | | |
|-------|-----------|--------|---------|---------|---------|
| | | 50 MHz | 144 MHz | 220 MHz | 432 MHz |
| RG-58C/U | 52.5 | 3.0 | 6.0 | 8.0 | 15.0 |
| RG-58(F) | 50 | 2.2 | 4.1 | 5.0 | 7.1 |
| RG-59B/U | 73 | 2.3 | 4.2 | 5.5 | 8.0 |
| RG-59(F) | 75 | 2.0 | 3.4 | 4.6 | 6.1 |
| RG-8A/U | 52 | 1.5 | 2.5 | 3.5 | 5.0 |
| RG-213/U | 52 | 1.5 | 2.5 | 3.5 | 5.0 |
| RG-8(F) | 50 | 1.2 | 2.2 | 2.7 | 3.9 |
| RG-11A/U | 75 | 1.55 | 2.8 | 3.7 | 5.5 |
| RG-218/U | 52 | 0.5 | 1.0 | 1.3 | 2.3 |

## Table 13-9. Return loss/VSWR.

| Return loss | VSWR | Match ratio | Reflection coefficient | Percentage reflection |
|-------------|------|-------------|------------------------|-----------------------|
| 2 dB | 8.71 | 1.26:1 | 0.79 | 79 % |
| 4 dB | 4.42 | 1.59:1 | 0.63 | 63 % |
| 6 dB | 3.01 | 1.99:1 | 0.50 | 50 % |
| 8 dB | 2.32 | 2.52:1 | 0.40 | 40 % |
| 10 dB | 1.92 | 3.16:1 | 0.32 | 32 % |
| 12 dB | 1.67 | 3.98:1 | 0.25 | 25 % |
| 14 dB | 1.50 | 5.01:1 | 0.20 | 20 % |
| 16 dB | 1.37 | 6.31:1 | 0.16 | 16 % |
| 18 dB | 1.28 | 7.94:1 | 0.13 | 13 % |
| 20 dB | 1.22 | 10.0 :1 | 0.10 | 10 % |
| 22 dB | 1.17 | 12.6 :1 | 0.079 | 7.9% |
| 24 dB | 1.13 | 15.9 :1 | 0.063 | 6.3% |
| 26 dB | 1.11 | 19.9 :1 | 0.050 | 5.0% |
| 28 dB | 1.08 | 25.1 :1 | 0.040 | 4.0% |
| 30 dB | 1.07 | 31.6 :1 | 0.032 | 3.2% |
| 32 dB | 1.05 | 39.8 :1 | 0.025 | 2.5% |
| 34 dB | 1.04 | 50.1 :1 | 0.020 | 2.0% |
| 36 dB | 1.032 | 63.1 :1 | 0.016 | 1.6% |
| 38 dB | 1.026 | 79.4 :1 | 0.013 | 1.3% |
| 40 dB | 1.020 | 100 :1 | 0.010 | 1.0% |
| 46 dB | 1.010 | 199 :1 | 0.005 | 0.5% |
| 50 dB | 1.006 | 316 :1 | 0.003 | 0.3% |
| 54 dB | 1.004 | 501 :1 | 0.002 | 0.2% |
| 60 dB | 1.002 | 1 000 :1 | 0.001 | 0.1% |

Channel Master, Div. of Avnet, Inc.

*Table 13-10. (MATV) master antenna television distribution symbols.*

| Headend equipment | | Distribution system equipment | |
|---|---|---|---|
| **Symbol** | **Description** | **Symbol** | **Description** |
| //// | Antenna | ⊿₂ | 2-way splitter |
| ⊗ | Balun | ⊿₃ | 3-way splitter |
| ⊏⊐ | Antenna joiner | ⊿₄ | 4-way splitter |
| ⊓ | Preamplifier | ⊓ | Variable-isolation wall tap |
| PS | Power supply | ⊓- | One-way line tap |
| ⊿ | Hi-Lo joiner | ⊓ | Two-way line tap |
| ⊿ | U/V joiner | ⊓ | 4-way line tap |
| ⊟ | Variable attenuator | -⊙ | "0" dB wall outlet |
| ⊟ | Fixed attenuator | ⋀⋀ | Terminator |
| -⊡ | Lo band sound carrier reducer | ⊏□⇒ | UHF/VHF band separator |
| -⊙ | Hi band sound carrier reducer | ⊏□⇒ | Matching transformer |
| -⊡⊡ | Lo band Hi "Q" trap | † | Cable adapter RG-6 |
| -⊙⊙ | Hi band Hi "Q" trap | ‡ | Cable adapter RG-11, 0.412. 0.500 to "F" |
| ⊡ | Lo band mixing unit | ‡ | Cable adapter 0.412, 0.500 entry mount with pin |
| ⊡ | Hi band mixing unit | ‡ | Cable adapter 5/8 in entry to female "F" |
| ⊿ | Converter | ⊏PS⊐ | Auxiliary power supply |
| ⊏⊐ | Broadband distribution amplifier | ▶⊐ | Line amplifier |
| ⊓ | Single-channel bandpass filter | ⊏PA⊐ | Power adder |
| ⊓ | Single-channel amplifier | ⊏PB⊐ | Voltage block |

# Standing-wave ratio

The maximum current in a transmission line, compared to the minimum current, or the maximum voltage compared to the minimum voltage, is known as the standing-wave ratio (SWR). For voltage this is sometimes shown as VSWR. SWR is an indication of the impedance mismatch between a transmission line and its load. Maximum delivery of energy takes place when the load impedance matches the impedance of the transmission line, and under these circumstances, SWR has a value of 1.

SWR is expressed by the following formula:

$$SWR = \frac{Z_1}{Z_2} \text{ or } SWR = \frac{Z_2}{Z_1}$$

$Z_1$ is the impedance of the transmission line; $Z_2$ is the terminating impedance. The larger number is always put in the numerator so as to produce a whole number.

As the SWR approaches 1, the reflection coefficient decreases, as shown in Table 13-9. The reflection coefficient, a decimal, is multiplied by 100 to produce reflection as a percentage, as indicated in columns 4 and 5 of Table 13-9.

# MATV distribution symbols

Special symbols are used in master antenna television distribution (MATV), and these are illustrated in Table 13-10.

# 14
## CHAPTER

# Video

## TV channels and frequencies

Table 14-1 supplies the channel width, and the frequencies in megahertz (MHz) of the picture and sound carriers of TV channels. The table covers VHF channels 2 to 13 (channel 1 is not used for TV) and UHF channels 14 to 83.

The bandwidth of each channel is 6 MHz regardless of frequency. In each instance, the video carrier is 1.25 MHz above the lower edge of the band while the sound carrier is 0.25 MHz lower than the high-frequency end of the channel. The separation between carriers, video and sound, is 4.5 MHz. Worldwide standards are listed in Table 14-2. The number of TV stations and sets is broken down by country in Table 14-3.

*Example*  What is the sound carrier frequency of channel 16?
*Solution*  Locate channel 16 in the left-hand column. Move to the right and under the heading of sound carrier find 487.75 MHz.

The frequencies between 806 and 890 MHz are now allocated to the land mobile services. These frequencies are designated as UHF channels 70 to 83 inclusive. Hence, channel 69 is the highest frequency UHF channel available for TV broadcasting and reception.

Television channels 2 through 13 aren't continuous bands. Channels 2 through 6, known as the low-band TV channels, extend from 54 MHz, the low-frequency end of VHF channel 2, to 88 MHz, the high frequency end of VHF channel 6. Following this is the FM band from 88 to 108 MHz. As shown in Table 14-4, this is followed by the mid band, a group of frequencies from 120 to 174 MHz. Then the regular VHF channels continue from 174 to 216 MHz followed by the super band channels from 216 to 300 MHz.

*Table 14-1. TV channels and frequencies.*

**VHF television frequencies**

| Channel no. | Frequency band (MHz) | Video carrier (MHz) | Sound carrier (MHz) |
|---|---|---|---|
| (1 not assigned) | | | |
| 2 | 54 – 60 | 55.25 | 59.75 |
| 3 | 60 – 66 | 61.25 | 65.75 |
| 4 | 66 – 72 | 67.25 | 71.75 |
| 5 | 76 – 82 | 77.25 | 81.75 |
| 6 | 82 – 88 | 83.25 | 87.75 |
| 7 | 174 – 180 | 175.25 | 179.75 |
| 8 | 180 – 186 | 181.25 | 185.75 |
| 9 | 186 – 192 | 187.25 | 191.75 |
| 10 | 192 – 198 | 193.25 | 197.75 |
| 11 | 198 – 204 | 199.25 | 203.75 |
| 12 | 204 – 210 | 205.25 | 209.75 |
| 13 | 210 – 216 | 211.25 | 215.75 |

**UHF television frequencies**

| Channel no. | Frequency band (MHz) | Video carrier (MHz) | Sound carrier (MHz) |
|---|---|---|---|
| 14 | 470 – 476 | 471.25 | 475.75 |
| 15 | 476 – 482 | 477.25 | 481.75 |
| 16 | 482 – 488 | 483.25 | 487.75 |
| 17 | 488 – 494 | 489.25 | 493.75 |
| 18 | 494 – 500 | 495.25 | 499.75 |
| 19 | 500 – 506 | 501.25 | 505.75 |
| 20 | 506 – 512 | 507.25 | 511.75 |
| 21 | 512 – 518 | 513.25 | 517.75 |
| 22 | 518 – 524 | 519.25 | 523.75 |
| 23 | 524 – 530 | 525.25 | 529.75 |
| 24 | 530 – 536 | 531.25 | 535.75 |
| 25 | 536 – 542 | 537.25 | 541.75 |
| 26 | 542 – 548 | 543.25 | 547.75 |
| 27 | 548 – 554 | 549.25 | 553.75 |
| 28 | 554 – 560 | 555.25 | 559.75 |
| 29 | 560 – 566 | 561.25 | 565.75 |
| 30 | 566 – 572 | 567.25 | 571.75 |
| 31 | 572 – 578 | 573.25 | 577.75 |
| 32 | 578 – 584 | 579.25 | 583.75 |
| 33 | 584 – 590 | 585.25 | 589.75 |
| 34 | 590 – 596 | 591.25 | 595.75 |
| 35 | 596 – 602 | 597.25 | 601.75 |
| 36 | 602 – 608 | 603.25 | 607.75 |
| 37 | 608 – 614 | 609.25 | 613.75 |
| 38 | 614 – 620 | 615.25 | 619.75 |

*Table 14-1. Continued.*

### VHF television frequencies

| Channel no. | Frequency band (MHz) | Video carrier (MHz) | Sound carrier (MHz) |
|---|---|---|---|
| 39 | 620−626 | 621.25 | 625.75 |
| 40 | 626−632 | 627.25 | 631.75 |
| 41 | 632−638 | 633.25 | 637.75 |
| 42 | 638−644 | 639.25 | 643.75 |
| 43 | 644−650 | 645.25 | 649.75 |
| 44 | 650−656 | 651.25 | 655.75 |
| 45 | 656−662 | 657.25 | 661.75 |
| 46 | 662−668 | 663.25 | 667.75 |
| 47 | 668−674 | 669.25 | 673.75 |
| 48 | 674−680 | 675.25 | 679.75 |
| 49 | 680−686 | 681.25 | 685.75 |
| 50 | 686−692 | 687.25 | 691.75 |
| 51 | 692−698 | 693.25 | 697.75 |
| 52 | 698−704 | 699.25 | 703.75 |
| 53 | 704−710 | 705.25 | 709.75 |
| 54 | 710−716 | 711.25 | 715.75 |
| 55 | 716−722 | 717.25 | 721.75 |
| 56 | 722−728 | 723.25 | 727.75 |
| 57 | 728−734 | 729.25 | 733.75 |
| 58 | 734−740 | 735.25 | 739.75 |
| 59 | 740−746 | 741.25 | 745.75 |
| 60 | 746−752 | 747.25 | 751.75 |
| 61 | 752−758 | 753.25 | 757.75 |
| 62 | 758−764 | 759.25 | 763.75 |
| 63 | 764−770 | 765.25 | 769.75 |
| 64 | 770−776 | 771.25 | 775.75 |
| 65 | 776−782 | 777.25 | 781.75 |
| 66 | 782−788 | 783.25 | 787.75 |
| 67 | 788−794 | 789.25 | 793.75 |
| 68 | 794−800 | 795.25 | 799.75 |
| 69 | 800−806 | 801.25 | 805.75 |
| 70 | 806−812 | 807.25 | 811.75 |
| 71 | 812−818 | 813.25 | 817.75 |
| 72 | 818−824 | 819.25 | 823.75 |
| 73 | 824−830 | 825.25 | 829.75 |
| 74 | 830−836 | 831.25 | 835.75 |
| 75 | 836−842 | 837.25 | 841.75 |
| 76 | 842−848 | 843.25 | 847.75 |
| 77 | 848−854 | 849.25 | 853.75 |
| 78 | 854−860 | 855.25 | 859.75 |

*Table 14-1. Continued.*

**VHF television frequencies**

| Channel no. | Frequency band (MHz) | Video carrier (MHz) | Sound carrier (MHz) |
|---|---|---|---|
| 79 | 860–866 | 861.25 | 865.75 |
| 80 | 866–872 | 867.25 | 871.75 |
| 81 | 872–878 | 873.25 | 877.75 |
| 82 | 878–884 | 879.25 | 883.75 |
| 83 | 884–890 | 885.25 | 889.75 |

*Table 14-2. Worldwide television standards.*

| Country or area | Lines/fields | System | Voltage (V) (nominal) | Frequency (Hz) |
|---|---|---|---|---|
| Afghanistan | 625/50 | PAL | 220 | 50 |
| Albania | 625/50 | SECAM | 220 | 50 |
| Algeria | 625/50 | PAL | 127–220 | 50 |
| Andorra | 625/50 | | 220 | 50 |
| Angola | 625/50 | | 220 | 50 |
| Argentina | 625/50 | PAL | 220 | 50 |
| Armenia | 625/50 | SECAM | 220 | 50 |
| Australia | 625/50 | PAL | 240 | 50 |
| Austria | 625/50 | PAL | 220 | 50 |
| Azerbaijan | 625/50 | SECAM | 220 | 50 |
| Azores | 525/60 | PAL | 220 | 50 |
| Bahamas | 525/60 | NTSC | 120 | 60 |
| Bahrain | 625/50 | PAL | 220 | 50 |
| Bangladesh | 625/50 | PAL | | |
| Barbados | 625/50 | NTSC | 120 | 50 |
| Belarus | 625/50 | SECAM | 220 | 50 |
| Belgium | 625/50 | PAL | 127–220 | 50 |
| Benin | 625/50 | | 220 | 50 |
| Bermuda | 525/60 | NTSC | 120 | 60 |
| Bolivia | 625/50 | PAL | 115–230 | 50 |
| Brazil | 525/60 | PAL | 220 | 60 |
| Bulgaria | 625/50 | SECAM | 220 | 50 |
| Burkina Faso (Upper Volta) | 625/50 | | 220 | 50 |
| Burundi | 625/50 | | 220 | 50 |
| Cameroon | 625/50 | | 127–220 | 50 |
| Canada | 525/60 | NTSC | 120–240 | 60 |
| Canary Islands | 625/50 | PAL | 127 | 50 |
| Central African Rep. | 625/50 | | 220 | 50 |
| Chad | 625/50 | | 220 | 50 |
| Chile | 525/60 | NTSC | 220 | 50 |
| China (People's Rep.) | 625/50 | PAL | 220 | 50 |

Table 14-2. Continued.

| Country or area | Lines/fields | System | Voltage (V) (nominal) | Frequency (Hz) |
|---|---|---|---|---|
| Colombia | 525/60 | NTSC | 150–240 | 60 |
| Commonwealth of Independent States (CIS) (*see invidiual countries* | 625/50 | SECAM | 220 | 50 |
| Congo Rep. | 625/50 | SECAM | 220 | 50 |
| Costa Rica | 525/60 | NTSC | 110 | 60 |
| Croatia | 625/50 | PAL | 220 | 50 |
| Cuba | 525/60 | NTSC | 120 | 60 |
| Curacao | 525/60 | NTSC | 120 | 60 |
| Cyprus | 625/50 | PAL | 220 | 50 |
| Czechoslovakia | 625/50 | SECAM | 220 | 50 |
| Denmark | 625/50 | PAL | 220 | 50 |
| Dominican Rep. | 525/60 | NTSC | 110 | 60 |
| Ecuador | 525/60 | NTSC | 120 | 60 |
| Egypt | 625/50 | SECAM | 220 | 50 |
| El Salvador | 525/60 | NTSC | 110 | 60 |
| Estonia | 625/50 | SECAM | 220 | 50 |
| Ethiopia | 625/50 | | 127 | 50 |
| Fiji | 625/50 | | 240 | 50 |
| Finland | 625/50 | PAL | 220 | 50 |
| France | 625/50 | SECAM | 115–230 | 50 |
| Gabon | 625/50 | SECAM | 127–220 | 50 |
| The Gambia | 625/50 | | | |
| Georgia | 625/50 | SECAM | 220 | 50 |
| Germany (eastern) | 625/50 | SECAM | 220 | 50 |
| Germany (western) | 625/50 | PAL | 220 | 50 |
| Ghana | 625/50 | | 230 | 50 |
| Gibraltar | 625/50 | | 230 | 50 |
| Great Britain | 625/50 | PAL | 240 | 50 |
| Greece | 625/50 | | 110–220 | 50 |
| Greenland | 525/60 | | 220 | 50 |
| Guam | 525/60 | NTSC | 110 | 60 |
| Guatemala | 525/60 | NTSC | 110–220 | 60 |
| Guinea | 625/50 | | 127–220 | 50 |
| Guyana | 625/50 | SECAM | 127 | 50 |
| Haiti | 625/50 | SECAM | 115–220 | 50 |
| Honduras | 525/60 | NTSC | 110–220 | 60 |
| Hong Kong | 625/50 | PAL | 220 | 50 |
| Hungary | 625/50 | SECAM | 220 | 50 |
| Iceland | 625/50 | PAL | 220 | 50 |
| India | 625/50 | | 230 | 50 |
| Indonesia | 625/50 | PAL | 110 | 50 |
| Iran | 625/50 | SECAM | 220 | 50 |
| Iraq | 625/50 | SECAM | 220 | 50 |
| Ireland (Rep. and No. Ire.) | 625/50 | PAL | 220 | 50 |

*Table 14-2. Continued.*

| Country or area | Lines/fields | System | Voltage (V) (nominal) | Frequency (Hz) |
|---|---|---|---|---|
| Israel | 625/50 | PAL | 230 | 50 |
| Italy | 625/50 | PAL | 127−220 | 50 |
| Ivory Coast | 625/50 | SECAM | 220 | 50 |
| Jamaica | 625/50 | PAL | 110 | 50,60 |
| Japan | 525/60 | NTSC | 100−200 | 50,60 |
| Jordan | 625/50 | PAL | 220 | 50 |
| Kazakhstan | 625/50 | SECAM | 220 | 50 |
| Kenya | 625/50 | PAL | 240 | 50 |
| Kyrgyzstan | 625/50 | SECAM | 220 | 50 |
| Korea (North) | 625/50 | | | |
| Korea (South) | 525/60 | NTSC | 100 | 60 |
| Kuwait | 625/50 | PAL | 240 | 50 |
| Latvia | 625/50 | SECAM | 220 | 50 |
| Lebanon | 625/50 | SECAM | 110−190 | 50 |
| Liberia | 625/50 | PAL | 120 | 60 |
| Libya | 625/50 | SECAM | 120 | 50 |
| Lithuania | 625/50 | SECAM | 220 | 50 |
| Luxembourg | 625/50 | SECAM | 120−208 | 50 |
| Madagascar | 625/50 | | 127−280 | 50 |
| Malawi | 625/50 | | 220 | 50 |
| Malaysia | 625/50 | PAL | 240 | 50 |
| Mali | 625/50 | | 125 | 50 |
| Malta | 625/50 | | 240 | 50 |
| Martinique | 625/50 | SECAM | 125 | 50 |
| Mauritania | 625/50 | | 220 | 50 |
| Mauritius | 625/50 | SECAM | 220 | 50 |
| Mexico | 525/60 | NTSC | 127−220 | 50,60 |
| Moldova | 625/50 | SECAM | 220 | 50 |
| Monaco | 625/50 | SECAM | 125 | 50 |
| Mongolia | 625/50 | | | |
| Morocco | 625/50 | SECAM | 115 | 50 |
| Mozambique | 625/50 | PAL | 220 | 50 |
| Netherlands | 625/50 | PAL | 220 | 50 |
| Netherlands Antilles | 525/60 | NTSC | 120−220 | 50,60 |
| New Caledonia | 625/50 | SECAM | 220 | 50 |
| New Zealand | 625/50 | PAL | 230 | 50 |
| Nicaragua | 525/60 | NTSC | 117 | 60 |
| Niger | 625/50 | | 220 | 50 |
| Nigeria | 625/50 | PAL | 220 | 50 |
| Norway | 625/50 | PAL | 230 | 50 |
| Oman | 625/50 | PAL | 220 | 50 |
| Pakistan | 625/50 | PAL | 220 | 50 |
| Panama | 525/60 | NTSC | 110 | 60 |
| Paraguay | 625/50 | PAL | 220 | 50 |

Table 14-2. Continued.

| Country or area | Lines/fields | System | Voltage (V) (nominal) | Frequency (Hz) |
|---|---|---|---|---|
| Peru | 525/60 | NTSC | 220 | 60 |
| Philippines | 525/60 | NTSC | 115 | 60 |
| Poland | 625/50 | SECAM | 220 | 50 |
| Portugal | 625/50 | | 110–220 | 50 |
| Puerto Rico | 525/60 | NTSC | 120 | 60 |
| Romania | 625/50 | | 220 | 50 |
| Russia | 625/50 | SECAM | 220 | 50 |
| Rwanda | 625/50 | | 220 | 50 |
| Samoa (Western and American) | 525/60 | NTSC | 120 | 60 |
| Saudi Arabia | 625/50 | SECAM | 120–230 | 50,60 |
| Senegal | 625/50 | SECAM | 125 | 50 |
| Sierra Leone | 625/50 | PAL | 220 | 50 |
| Singapore | 625/50 | PAL | 220 | 50 |
| Slovenia | 625/50 | PAL | 220 | 50 |
| Somalia | 625/50 | | 220 | 50 |
| South Africa | 625/50 | PAL | 220 | 50 |
| Spain | 625/50 | PAL | 127–220 | 50 |
| Sri Lanka | 625/50 | | 230 | 50 |
| St. Kitts | 525/60 | NTSC | 220 | 60 |
| Sudan | 625/50 | PAL | 220 | 50 |
| Suriname | 525/60 | NTSC | 115–127 | 50,60 |
| Swaziland | 625/50 | PAL | | |
| Sweden | 625/50 | PAL | 220 | 50 |
| Switzerland | 625/50 | PAL | 220 | 50 |
| Syria | 625/50 | SECAM | 115–220 | 50 |
| Tahiti | 625/50 | SECAM | | |
| Taiwan | 525/60 | NTSC | 100 | 60 |
| Tajikistan | 625/50 | SECAM | 220 | 50 |
| Tanzania | 625/50 | PAL | 230 | 50 |
| Thailand | 625/50 | PAL | 220 | 50 |
| Togolese Rep. | 625/50 | | 127–220 | 50 |
| Trinidad & Tobago | 525/60 | NTSC | 117 | 60 |
| Tunisia | 625/50 | SECAM | 117–220 | 50 |
| Turkey | 625/50 | PAL | 110–220 | 50 |
| Turkmenistan | 625/50 | SECAM | 220 | 50 |
| Uganda | 625/50 | PAL | 220 | 50 |
| Ukraine | 625/50 | SECAM | 220 | 50 |
| Uruguay | 625/50 | PAL | 220 | 50 |
| U.S.A. | 525/60 | NTSC | 120 | 60 |
| U.S.S.R. (see individual countries) | 625/50 | SECAM | 220 | 50 |
| Uzbekistan | 625/50 | SECAM | 220 | 50 |
| Venezuela | 525/60 | NTSC | 110–220 | 60 |
| Vietnam | 525/60 | | 120 | 50 |

Table 14-2. Continued.

| Country or area | Lines/fields | System | Voltage (V) (nominal) | Frequency (Hz) |
|---|---|---|---|---|
| Virgin Islands (U.S.) | 525/60 | NTSC | 115 | 60 |
| Western Sahara | 625/50 | | | |
| Yemen | 625/50 | | 220 | 50 |
| Yugoslavia | 625/50 | PAL | 220 | 50 |
| Zaire | 625/50 | SECAM | | |
| Zambia | 625/50 | | 230 | 50 |
| Zimbabwe | 625/50 | PAL | 220 | 50 |

Table 14-3. World TV stations and sets.
Transmitters and Sets as of September 1990.
(Low Power satellites & repeaters in parentheses.)

| Country | Stations/ transmitters | Sets black & white | Color |
|---|---|---|---|
| Afghanistan (Low power: 18) | 1 | 297 323 | 331 446 |
| Albania | 7 | 115 000 | |
| Algeria (Low power: 60) | 23 | 1 140 000 | |
| Angola | 1 | 12 300 | |
| Antiqua (Low power: 1) | 1 | 5 000 | 20 000 |
| Arab Republic of Egypt | 49 | 3 850 000 | 350 000 |
| Argentina | 35 | 5 500 000 | 500 000 |
| Australia (Low power: 763) | 312 | 96 000 | 5 011 000 |
| Austria (Low power: 512) | 904 | 351 000 | 2 331 000 |
| Bahamas | 1 | 50 000 | |
| Bahrain | 2 | 50 000 | 175 000 |
| Bangladesh (Low power: 1) | 10 | 349 091 | 89 630 |
| Barbados | 4 | 45 000 | 15 000 |
| Belgium (Low power: 12) | 38 | 422 149 | 2 852 087 |
| Belize | 2 | 12 000 | |
| Benin (Low power: 1) | 2 | 30 000 | 45 000 |
| Bermuda | 2 | 2 000 | 56 000 |
| Bolivia (Low power: 2) | 11 | 1 200 000 | 600 000 |
| Bophuthatswana | 14 | 40 000 | 55 000 |
| Brazil (Low power: 6 826) | 198 | 14 000 000 | 30 000 000 |
| British Virgin Island | 1 | 1 300 | |
| Brunei Darussalam | 4 | 3 000 | 58 500 |
| Bulgaria (Low power: 221) | 13 | 1 650 000 | 250 000 |
| Burkina Faso | 1 | 7 700 | |
| Burundi | 2 | NA | NA |
| Cameroon (Low power: 4) | 28 | 70 000 | 400 000 |
| Canada (Low power: 1 251) | 126 | 2 770 000 | 9 110 000 |

Table 14-3. Continued.

| Country | Stations/ transmitters | Sets black & white | Color |
|---|---|---|---|
| Central African Republic (Low power: 2) | 1 | 5 500 | 1 500 |
| Chad | 1 | NA | |
| Chile (Low power: 131) | 38 | 1 800 000 | 1 300 000 |
| China (Mainland) | 2 252 | 75 000 000 | |
| China (Taiwan Province) (Low power: 90) | 87 | 860 000 | 5 800 000 |
| Colombia (Low power: 84) | 36 | 2 500 000 | 500 000 |
| Congo | 1 | 2 700 | |
| Costa Rica (Low power: 22) | 28 | 150 000 | 300 000 |
| Cuba | 58 | 1 130 000 | |
| Cyprus (Low power: 65) | 9 | 89 750 | 78 000 |
| Czechoslovakia (Low power: 1 496) | 118 | 5 668 132 | |
| Denmark (Low power: 23) | 26 | 300 000 | 2 000 000 |
| Djibouti (Low power: 3) | 3 | 6 000 | 30 000 |
| Dominican Republic | 19 | 800 000 | 500 000 |
| Ecuador (Low power: 121) | 22 | 500 000 | 200 000 |
| El Salvador | 10 | 1 000 000 | 500 000 |
| Equatorial Guinea | 1 | 500 | |
| Ethiopia (Low power: 18) | 1 | 45 000 | 55 000 |
| Finland | 312 | 147 327 | 1 731 922 |
| France (Low power: 10 911) | 678 | 6 200 000 | 21 820 000 |
| French Overseas Territories (Low power: 62) | 30 | 336 500 | |
| Gabon (Low power: 5) | 1 | 5 000 | |
| Germany (Federal Republic) (Low power: 4 872) | 355 | 4 810 000 | 20 520 000 |
| Germany (Democratic Republic) (Low power: 480) | 28 | 3 903 930 | 2 234 976 |
| Ghana (Low power: 8) | 3 | 189 000 | |
| Gibraltar (Low power: 3) | 1 | 173 | 7 107 |
| Greece (Low power: 880) | 36 | 1 845 000 | 1 635 000 |
| Guatemala (Low power: 58) | 15 | 660 808 | 151 757 |
| Guinea | 3 | 30 000 | 20 000 |
| Haiti | 3 | 13 000 | |
| Honduras (Low power: 16) | 6 | 110 000 | 25 000 |
| Hong Kong (Low power: 71) | 16 | 45 000 | 1 500 000 |
| Hungary (Low power: 108) | 22 | 2 940 324 | |
| Iceland (Low power: 153) | 9 | 3 234 | 71 788 |
| India (Low power: 237) | 52 | 15 000 000 | 5 000 000 |
| Indonesia (Low power: 16) | 71 | 4 805 706 | 1 694 294 |
| Iran (Low power: 398) | 28 | 2 000 000 | |
| Iraq | 32 | 500 000 | |
| Ireland (Low power: 31) | 9 | 197 000 | 721 000 |
| Israel (Low power: 46) | 17 | 1 000 000 | 620 000 |
| Italy (Low power: 2 548) | 163 | 6 713 427 | 7 817 077 |
| Ivory Coast (Low power: 3) | 10 | 200 000 | 5 000 |
| Jamaica (Low power: 2) | 9 | 165 000 | 222 000 |
| Japan (Low power: 12 735) | 871 | 1 539 548 | 31 343 988 |

## Table 14-3. Continued.

| Country | Stations/ transmitters | Sets black & white | Color |
|---|---|---|---|
| Jordan (Low power: 21) | 9 | 250 000 | |
| Kampuchea (Low power: 5) | 1 | 30 000 | |
| Kenya | 5 | 3 833 432 | |
| Korea (South) (Low power: 504) | 127 | 2 947 200 | 4 364 510 |
| Korea (North) | 2 | 109 700 | |
| Kuwait (Low power: 2) | 18 | 8 000 | 820 000 |
| Lebanon (Low power: 5) | 10 | 200 000 | 300 000 |
| Liberia | 5 | 43 000 | 16 000 |
| Libyan Arab Republic (Low power: 2) | 12 | 235 000 | |
| Luxembourg | 4 | 20 000 | 85 000 |
| Macau (Low power: 1) | 2 | 500 | 20 000 |
| Madagascar (Low power: 30) | 2 | 40 000 | 10 000 |
| Malaysia (Low power: 14) | 51 | NA | 2 030 000 |
| Maldives | 1 | 800 | 4 356 |
| Malta | 1 | 37 059 | 102 846 |
| Mauritius (Low power: 3) | 7 | 85 000 | 25 000 |
| Mexico (Low power: 43) | 527 | 11 018 909 | 10 378 734 |
| Monaco (Low power: 1) | 47 | NA | 18 000 |
| Mongolia | 1 | 50 000 | |
| Morocco (Low power: 40) | 25 | 966 651 | 131 994 |
| Mozambique | 1 | 2 000 | 8 000 |
| Namibia (Low power: 10) | 4 | 30 000 | |
| Nepal (Low power: 1) | 2 | 10 000 | 50 000 |
| Netherlands (Low power: 15) | 19 | 5 850 000 | |
| Netherlands Antilles (Low power: 2) | 2 | 10 000 | 40 000 |
| New Zealand (Low power: 694) | 49 | 71 791 | 862 919 |
| Nicaragua (Low power: 6) | 3 | 150 000 | 60 000 |
| Niger (Low power: 8) | 11 | 20 000 | 15 000 |
| Nigeria (Low power: 1) | 72 | 3 400 000 | 600 000 |
| Norway (Low power: 2 098) | 44 | 73 971 | 1 452 850 |
| Oman (Low power: 36) | 8 | | 1 300 000 |
| Pakistan (Low power: 9) | 19 | 1 256 000 | 623 000 |
| Panama (Low power: 12) | 28 | 250 000 | 100 000 |
| Paraguay (Low power: 5) | 9 | 180 000 | 250 000 |
| Peru | 183 | 2 431 500 | |
| Philippines (Low power: 70) | 48 | 3 908 300 | 822 800 |
| Poland (Low power: 194) | 81 | 8 000 000 | 3 000 000 |
| Portugal (Low power: 34) | 25 | 1 667 675 | |
| Qatar (Low power: 5) | 14 | 50 000 | 250 000 |
| Romania (Low power: 352) | 58 | 3 912 000 | |
| St. Kitts (Low power: 3) | 1 | 2 000 | 2 500 |
| St. Vincent | 7 | 15 000 | |
| Samoa (American) (Low power: 3) | 3 | 1 000 | 6 000 |
| Saudi Arabia (Low power: 13) | 73 | 50 000 | 1 500 000 |
| Senegal (Low power: 1) | 2 | 40 000 | 10 000 |

Table 14-3. Continued.

| Country | Stations/ transmitters | Sets black & white | Color |
|---|---|---|---|
| Seychelles (Low power: 9) | 2 | 7 500 | |
| Sierra Leone | 2 | 200 000 | 10 000 |
| Singapore | 6 | 11 000 | 537 000 |
| Somali Democratic Republic | NA | | |
| South Africa (Low power: 103) | 121 | 3 100 000 | |
| Spain (Low power: 1 123) | 103 | 5 400 000 | 7 200 000 |
| Sri Lanka (Low power: 3) | 6 | 360 000 | 240 000 |
| Sudan | 2 | 90 000 | |
| Surinam (Low power: 6) | 1 | 10 000 | 30 000 |
| Swaziland (Low power: 4) | 3 | 3 469 | 11 100 |
| Sweden (Low power: 926) | 111 | 4 500 000 | |
| Switzerland (Low power: 1 230) | 128 | 2 403 655 | |
| Syria (Low power: 15) | 16 | 1 000 000 | 400 000 |
| Tanzania (Low power: 1) | 2 | 2 500 | 15 000 |
| Thailand (Low power: 24) | 51 | 2 286 400 | 1 835 800 |
| Togo (Low power: 1) | 3 | 24 000 | |
| Trinidad and Tobago (Low power: 2) | 4 | 85 000 | 250 000 |
| Tunisia (Low power: 17) | 18 | 400 000 | |
| Turkey (Low power: 858) | 195 | 5 857 000 | 3 661 000 |
| Uganda (Low power: 1) | 7 | 500 500 | |
| Union of Myanmar (Low power: 4) | 2 | 2 000 | 50 000 |
| United Arab Emirates (Low power: 4) | 18 | 350 | 145 000 |
| United Kingdom (Low power: 1 538) | 177 | 1 926 805 | 17 469 158 |
| Uruguay (Low power: 34) | 28 | 300 000 | 450 000 |
| USSR (Low power: 8 400) | 1 005 | 58 100 000 | 29 200 000 |
| Venezuela | 17 | 1 650 000 | 60 000 |
| Vietnam | 1 | 500 000 | |
| Yemen Arab Republic | NA | | |
| Yemen (People's Democratic Republic) | 5 | 30 000 | |
| Yugoslavia (Low power: 1 143) | 235 | 3 942 629 | |
| Zaire (Low power: 8) | 47 | 10 000 | 20 000 |
| Zambia | 5 | 250 000 | 80 000 |
| Zimbabwe (Low power: 2) | 10 | 140 175 | 52 325 |
| **TOTAL** | **11 237** | **320 237 893** | **350 650 964** |
| Low power repeaters | 65 057 | | |
| U.S. | 1 455 | 63 000 000 | 112 000 000 |
| **GRAND TOTAL** | **75 481** | **291 765 896** | **336 175 869** |

*Television Cable and Fact Book, Warren Publishing Inc.*

*Table 14-4. Low-band, mid-band, high-band, and super-band channels.*

| Band | Channel number | Frequency band (MHz) | Picture carrier (MHz) |
|------|:------:|:------:|:------:|
| Low | 2 | 54–60 | 55.25 |
|  | 3 | 60–66 | 61.25 |
|  | 4 | 66–72 | 67.25 |
|  | 5 | 76–82 | 77.25 |
|  | 6 | 82–88 | 83.25 |
| FM |  | 88–108 | (100.00) |
|  | A | 120–126 | 121.25 |
|  | B | 126–132 | 127.25 |
|  | C | 132–138 | 133.25 |
|  | D | 138–144 | 139.25 |
|  | E | 144–150 | 145.25 |
|  | F | 150–156 | 151.25 |
|  | G | 156–162 | 157.25 |
|  | H | 162–168 | 163.25 |
|  | I | 168–174 | 169.25 |
| High | 7 | 174–180 | 175.25 |
|  | 8 | 180–186 | 181.25 |
|  | 9 | 186–192 | 187.25 |
|  | 10 | 192–198 | 193.25 |
|  | 11 | 198–204 | 199.25 |
|  | 12 | 204–210 | 205.25 |
|  | 13 | 210–216 | 211.25 |
| Super | J | 216–222 | 217.25 |
|  | K | 222–228 | 223.25 |
|  | L | 228–234 | 229.25 |
|  | M | 234–240 | 235.25 |
|  | N | 240–246 | 241.25 |
|  | O | 246–252 | 247.25 |
|  | P | 252–258 | 253.25 |
|  | Q | 258–264 | 259.25 |
|  | R | 264–270 | 265.25 |
|  | S | 270–276 | 271.25 |
|  | T | 276–282 | 277.25 |
|  | U | 282–288 | 283.25 |
|  | V | 288–294 | 289.25 |
|  | W | 294–300 | 295.25 |

The low and high TV bands are identified by numbers and are channels 2 through 6 and 7 through 13. The mid-band channels begin with the letter *A* and extend through the letter *I*. The super-band channels start with the letter *J* and continue through the letter *W*.

In all instances, broadcast TV follows the requirements of the NTSC signal. Operators of cable TV, however, can make changes. For example, they might invert the sound and picture carrier frequencies as a means of encoding their programs.

# Standard NTSC signal for color TV transmission

The standard NTSC (National Television Standards Committee) color television signal waveform, shown in Fig. 14-1, isn't an international standard. Other nations, as listed in Table 14-2, use systems, known as PAL (phase alternation line) and SECAM (sequential and memory).

*H*, as indicated in the illustration, represents the time from the start of one horizontal line to the start of the next line. Because the horizontal sweep frequency is 15 750 Hz, the time duration of 1 line is 1/15 750 or 63.5 $\mu$s.

*V* represents the time from the start of one field to the start of the next field. Each field (Fig. 14-2) is 262$^1$/$_2$ lines, and two fields, interlaced, form a frame, or a single complete picture. Because a field requires 60 s for its completion, the time duration of a single field is 1/60 or 16 667 $\mu$s.

During the time the electron scanning beam in the picture tube retraces its path from the bottom of the screen (end of a field) to the top of the screen (start of the next field), there is a vertical blanking interval during which time the scanning beam is cut off. The time duration is equivalent to 0.05 *V* or about 14 *H* (14.5×63.5 $\mu$s) or 0.05×16 667 $\mu$s (between 820 and 920 $\mu$s approximately).

Following the sweep of the last horizontal line in a field there are six equalizing pulses (Fig. 14-3). These pulses last for a total of 3.025 *H* or 192 $\mu$s. The equalizing pulses are superimposed on the vertical blanking pulse. Figure 14-4 supplies details of the color burst signal.

Following the equalizing pulses are vertical sync pulses, and their time is equal to 3 *H*. The vertical sync pulse interval is then followed by six more equalizing pulses, and like the first six, have a similar time duration. The frequency of the equalizing pulses is 0.5 *H* or 63.5/2=31.75 $\mu$s.

Table 14-5 supplies details of the NTSC television waveform. Figure 14-5 furnishes data about the video signal following demodulation in the television receiver.

# Television interference (tvi)

The fundamental frequencies and the harmonics of other broadcast services can interfere with television signal reception. Table 14-6 (on pages 294 and 295) indi-

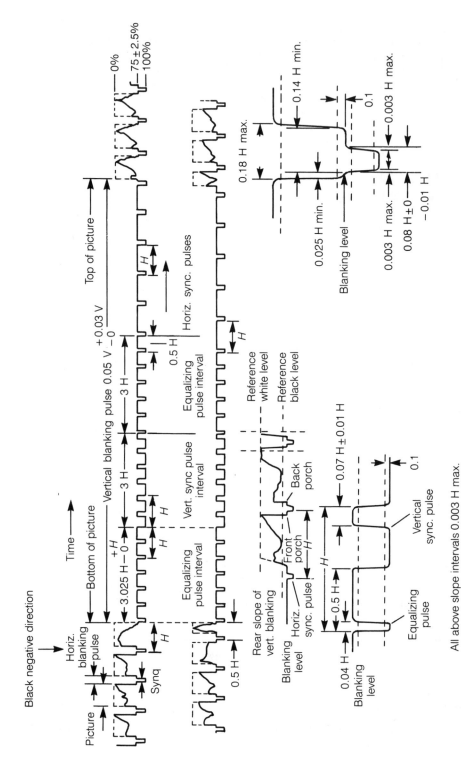

Fig. 14-1. Standard NTSC television signal.

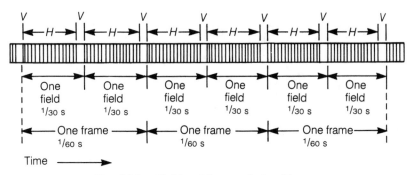

*Fig. 14-2.   Field and frame relationship.*

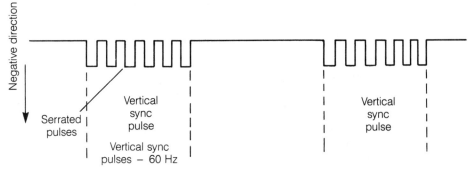

*Fig. 14-3.   The vertical sync pulse is made up of six serrated pulses.*

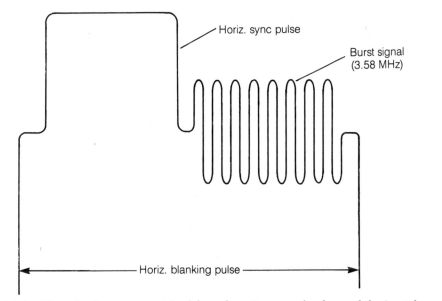

*Fig. 14-4.   The color burst is transmitted for a few microseconds after each horizontal sync pulse.*

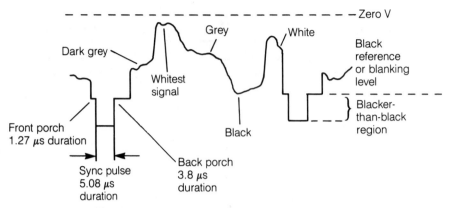

Fig. 14-5.   The video signal following demodulation.

### Table 14-5. NTSC television standards.

| Factor | Value |
|---|---|
| Field | 16.667 $\mu$s |
| Frame | 33.334 $\mu$s |
| Complete horizontal line from start of one line to beginning of next line | 63.5 $\mu$s ($^1/_{15.750}$ s) |
| Horizontal blanking | 10.16 to 11.4 $\mu$s |
| Horizontal trace without blanking time | 53.34 $\mu$s |
| Horizontal sync pulse | 5.08 to 5.68 $\mu$s |
| Vertical sync pulse interval (total of six) | 190.5 $\mu$s |
| Vertical blanking interval | 833 to 1 300 $\mu$s per field |
| Picture carrier | sound carrier − 4.5 MHz |
| Picture carrier | low end of TV channel + 1.25 MHz |
| Lower video sideband | 0.75 MHz |
| Sound carrier | upper end of TV channel minus 0.25 MHz |
| Sound carrier | minus picture carrier = 4.5 MHz |
| Channel width | 6 MHz |
| Color burst frequency | 3.579 545 MHz (Fig. 14-4) |
| Horizontal sync pulse frequency | 15.750 Hz |
| Vertical sync pulse frequency | 60 Hz |
| Type of scanning | interlaced |
| Single frame | odd field + even field |
| H | time of 1 line of picture information |

## Table 14-5. Continued.

| Factor | Value |
|---|---|
| Picture information, horizontal blanking, and horizontal sync | 93% of one field time |
| Vertical blanking, sync, and retrace period | 7% of one field or 1 270 $\mu$s |
| One field | $^1/_{30}$ s |
| One frame | $^1/_{60}$ s |
| Blanking or black reference level | 75% of carrier |
| Equalizing pulses | 0.5H or 31.75 $\mu$s |
| Front porch | 1.27 $\mu$s |
| Back porch | 3.8 $\mu$s |

cates these frequencies, their harmonics, and the TV channels that are subject to interference.

Table 14-7 (below) lists TV channels and amateur and citizens-band harmonics. Figure 14-6 illustrates the relationship of amateur band harmonics to VHF TV channels. Table 14-8 (on page 296) lists the harmonic relationship of ham bands to UHF TV channels.

## Picture tube designations

Picture tube designations are alphanumerical, using nomenclature such as 3AP1. The first number indicates the screen size, measured diagonally. The letter $P$ is

*Table 14-7. TV channels and amateur and citizens-band harmonics.*

| Channel | Frequency range (MHz) | Picture carrier frequency | CB | Harmonics | | | |
|---|---|---|---|---|---|---|---|
| | | | | 40 m | 20 m | 15 m | 10 m |
| TV IF | 40−47 | 42 | — | — | 42−43 | 42−43 | — |
| 2 | 54−60 | 55.25 | 53.9−54.8 (2nd) | 56−58.4 (8th) | 56−57.3 (4th) | — | 56−59.4 (2nd) |
| 3 | 60−66 | 61.25 | — | 63−65.7 (9th) | — | 63−64.35 (3rd) | — |
| 4 | 66−72 | 67.25 | — | 70−73 (10th) | 70−72 (5th) | — | — |
| 5 | 76−82 | 77.25 | 80.9−82.2 (3rd) | — | — | — | — |
| 6 | 82−88 | 83.25 | 82 −82.2 (3rd) | — | 84−86.4 (6th) | 84−85.8 (4th) | 84−89.1 (3rd) |

## Table 14-6. Fundamentals and harmonics of TV interference signals.

| Channel | Frequency (MHz) | CB (MHz) | Amateur bands frequency (MHz) | | | | AM | FM | FM osc. | Color osc. |
|---|---|---|---|---|---|---|---|---|---|---|
| | | 26.96 to 27.41 | 14 to 14.35 | 21 to 21.45 | 28 to 29.7 | 50 to 54 | .55 to 1.6 MHz | 88 to 108 MHz | 98 to 118 MHz | 3.53 MHz |
| TV IF | 40 to 47 | | | | | | | | | |
| TV CH. 2 | 54 to 60 | 2F | 4F | | 2F | F | | | | |
| CH. 3 | 60 to 66 | | | 3F | | | | | | |
| CH. 4 | 66 to 72 | | | | | | | | | 20F |
| CH. 5 | 76 to 82 | 3F | | | | | | | | 21F |
| CH. 6 | 82 to 88 | 3F | | 4F | 3F | | | F | | 23F |
| FM | 88–108 | 4F | | | | | | | | |
| TV CH. 7 | 174 to | | | | | | | | | 49F |

Table 14-6. Continued.

| Channel | Frequency (MHz) | CB (MHz) 26.96 to 27.41 | Amateur bands frequency (MHz) 14 to 14.35 | 21 to 21.45 | 28 to 29.7 | 50 to 54 | AM .55 to 1.6 MHz | FM 88 to 108 MHz | FM osc. 98 to 118 MHz | Color osc. 3.53 MHz |
|---|---|---|---|---|---|---|---|---|---|---|
| CH. 8 | 180 to 186 | | | | | | | | | |
| CH. 9 | 186 to 192 | 7F | | | | | | | | |
| CH. 10 | 192 to 198 | 7F | | | | | | | | |
| CH. 11 | 198 to 204 | | | | | 4F | | | 2F 99 to 102 | 57F |
| CH. 12 | 204 to 210 | | | | | 4F | | | 2F 102 to 105 | 58F |
| CH. 13 | 210 to 216 | | | | | 4F | | | 2F 105 to 108 | 60F |

All TV channels

Table 14-8. Harmonic relationships—Amateur VHF bands and UHF TV channels.

| Amateur band | Harmonic | Fundamental frequency range | Channel affected | Amateur band | Harmonic | Fundamental frequency range | Channel affected |
|---|---|---|---|---|---|---|---|
| 144 MHz | 4th | 144.0 – 144.5 | 31 | 220 MHz | 3rd | 220.00 – 220.67 | 45 |
| | | 144.5 – 146.0 | 32 | | | 220.67 – 222.67 | 46 |
| | | 146.0 – 147.5 | 33 | | | 222.67 – 224.67 | 47 |
| | | 147.5 – 148.0 | 34 | | | 224.67 – 225.00 | 48 |
| | 5th | 144.0 – 144.4 | 55 | | 4th | 220 – 221 | 82 |
| | | 144.4 – 145.6 | 56 | | | 221.0 – 222.5 | 83 |
| | | 145.6 – 146.8 | 57 | | | | |
| | | 146.8 – 148.0 | 58 | | | | |
| | 6th | 144.00 – 144.33 | 79 | 420 MHz | 2nd | 420 – 421 | 75 |
| | | 144.33 – 145.33 | 80 | | | 421 – 424 | 76 |
| | | 145.33 – 147.33 | 81 | | | 424 – 427 | 77 |
| | | 147.33 – 148.00 | 82 | | | 427 – 430 | 78 |
| | | | | | | 430 – 433 | 79 |
| | | | | | | 433 – 436 | 80 |

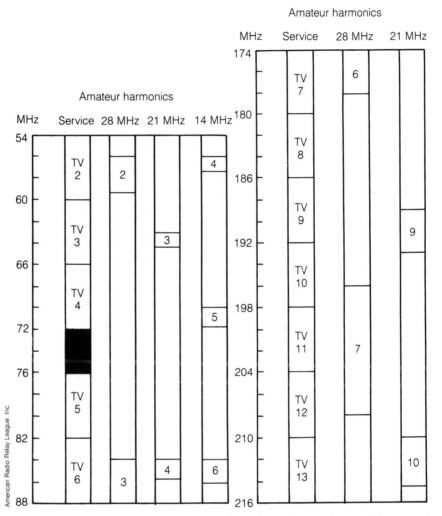

Fig. 14-6.   The relationship of amateur-band harmonics to the VHF TV channels.

the type of phosphor used and is listed in Table 14-9. A 5AP7 would be a picture tube whose screen is 5 in and uses a blue phosphor with a long persistence. The letter *A* indicates some internal tube change to distinguish it from a 5P7.

# Playing and recording time of videotape

The amount of recording and playback time of a videotape in minutes and hours is supplied in Table 14-10. The amount of time for VHS tapes is generally indicated directly on the package and is for standard play (SP). Thus, a T-60 videocassette for VHS use can play and record for one hour. If the machine has a LP (long play) and an EP (extended play) capability, that same tape can be used for two hours and three hours, respectively.

*Table 14-9. Picture tube
phosphor designations.*

| Type | Fluorescent color | Persistence |
|------|-------------------|-------------|
| P1 | green | medium |
| P2 | blue-green | long |
| P3 | yellow-green | medium |
| P4 | white | medium |
| P5 | blue | very short |
| P6 | white | medium |
| P7 | blue | long |
| P11 | blue | very short |

*Table 14-10. Maximum recording and
playback times for VHS format videocassettes.*

| VHS | Min | | | H | | |
|-----|-----|-----|-----|-----|-----|-----|
|     | SP | EP | SLP | SP | EP | SLP |
| T-20 | 20 | 40 | 60 | $1/3$ | $2/3$ | 1 |
| T-30 | 30 | 60 | 90 | $1/2$ | 1 | $1^{1}/_{2}$ |
| T-40 | 40 | 80 | 120 | $2/3$ | $1^{1}/_{3}$ | 2 |
| T-60 | 60 | 120 | 180 | 1 | 2 | 3 |
| T-80 | 80 | 160 | 240 | $1^{1}/_{3}$ | 2 | 4 |
| T-90 | 90 | 180 | 270 | $1^{1}/_{2}$ | 3 | $4^{1}/_{2}$ |
| T-100 | 100 | 200 | 300 | $1^{2}/_{3}$ | $3^{1}/_{3}$ | 5 |
| T-120 | 120 | 240 | 360 | 2 | 2 | 6 |
| T-160 | 160 | 320 | 480 | $2^{2}/_{3}$ | $5^{1}/_{3}$ | 8 |
| T-180 | 180 | 360 | 540 | 3 | 6 | 9 |

## Types of videotapes

The most commonly used videotape (other than industrial types) is VHS format
with a $1/2$-in width. Table 14-11 indicates the tapes that are available, the type of
scanning, and tape usage.

## VHS tape speeds

Tape speed is controlled by the tape speed selector in a VCR, assuming the component has such an adjustment. The three commonly used speeds are SP (standard
play). LP (long play), and EP (extended play, also known as SLP or super long
play). Table 14-12 lists the speed of VHS tapes in inches per second and millimeters per second.

Table 14-11. Videotapes and usage.

| Tape format (width) | Scan method | Mode | Where used | How used | Why used |
|---|---|---|---|---|---|
| 2-in | Quadruplex | Hi-band color Lo-band color | Broadcast TV studios—NBC CBS, ABC, PBS, local TV stations, large universities | 1 Over-the-air broadcasting 2 To make multiple copies and go many generations | Excellent color quality Very high picture definition Easily edited |
| 2-in | Helical | Monochrome and color | Schools, universities, industry | Large closed-circuit systems Master recording | Good to excellent quality picture Good copies |
| 1-in | Helical | Monochrome and color | Schools, industry professions, cable TV, broadcast TV | For anything recordable— For copies of all tape | Moderate to excellent quality picture, good editing |
| 3/4-in U-MATIC standard | Helical | Monochrome and color | Schools, industry, professions, cable TV, home use | Video cassettes | Interchangeability Good quality color 2 sound tracks (stereo sound) Easy to use |
| 1/2-in EIAJ standard | Helical | Monochrome and color | Education, industry professions, cable TV, home use | Video cartridge Teaching experimental, closed-circuit surveillance | Low cost Interchangeability Easy maintenance Simplicity, fair editing |
| 1/4-in | Helical | Monochrome and color | Education, schools, cable TV | Education, home use, on-location recording, distribution limited by nonstandardization and lack of 1/4-in equipment | Low cost Small size Light weight |

Table 14-12. VHS tape speeds.

| Mode | ips (decimals) | ips (fractions) | mm/s |
|---|---|---|---|
| Standard Play (SP) | 1.31 | $1-5/16$ | 33.27 |
| Long Play (LP) | 0.87 | $87/100$ | 22.10 |
| Extended Play (EP or super long play SP) | 0.44 | $7/16$ | 11.18 |

mm/s=millimeters per second

## Tape consumption

The amount of information to be played back or recorded on videotape depends on the amount of surface area presented to the head gap. This is determined by the tape width and the tape speed. The greater the width of the tape and the greater its speed, the more tape surface area that is scanned by a head.

Table 14-13 supplies data for various tapes. Some tapes have 1/4-in width, but these aren't widely used and aren't included. U-matic and EIAJ #1 are used in industry, and the 2-in quad is a standard broadcast tape.

Table 14-13. Tape area used per hour.

| Format | Tape width (in) | Tape speed (in/s) | ft$^2$/h |
|---|---|---|---|
| VHS standard play | 0.5 | 1.32 | 16.37 |
| long play | 0.5 | 0.66 | 8.19 |
| extended play | 0.5 | 0.45 | 5.59 |
| U-MATIC | 0.75 | 3.75 | 70.03 |
| EIAJ #1 | 0.5 | 7.5 | 93.8 |
| QUAD (TV broadcast) | 2.0 | 15.0 | 750.0 |

## Tape lengths

The designations on the package for VHS videocassettes indicates possible operating times. The chart in Table 14-14 lists the tape length in feet.

# VCR characteristics

The various types of videocassettes differ not only in physical characteristics but in electrical characteristics as well, as shown in Table 14-15.

Table 14-14.
Lengths of
VHS tapes.

| VHS | Length (ft) |
|---|---|
| T-15 | 115 |
| T-30 | 225 |
| T-45 | 335 |
| T-60 | 420 |
| T-90 | 645 |
| T-120 | 815 |
| T-180 | 1 260 |

*Table 14-15. Video tape characteristics.*

| Recorder types | Video track width (µm) | Audio track width (mm) | Drum diameter (mm) | Drum speed (rpm) | Luminance frequency (MHz) | Chroma frequency (kHz) | Cassette dimensions (mm) | Cassette volume (cm³) |
|---|---|---|---|---|---|---|---|---|
| **Consumer VCR format:** | | | | | | | | |
| VHS standard-play | 58 | 1.0 | 62 | 1 800 | 3.4–4.4 | 629 | 188×104×25 | 489 |
| VHS long-play | 35 | 1.0 | 62 | 1 800 | 3.4–4.4 | 629 | 188×104×25 | 489 |
| VR-1 000 (VX-2 000) | 48 | 0.4 | 48 | 3 600 | 3.1–4.6 | 688 | 213×146×44 | 1 368 |
| **Institutional & industrial:** | | | | | | | | |
| V-Cord II | 60 | 1.0 | 81.3 | — | 3.1–4.3 | 688 | 156×108×25 | 421 |
| V-Cord (skip-frame mode) | — | 1.0 | 81.3 | — | — | — | 156×108×25 | 421 |
| U-Matic | 85 | 0.8 | 110 | 1 800 | 3.8–5.4 | 688 | 222×140×32 | 995 |
| EIAJ open reel | 110 | 1.0 | 115.8 | — | 3.1–4.5 | 767 | — | — |

## S/N for video decks

The signal-to-noise ratios for video decks not only depends on the tape that is used, but also on the speed at which the tape is operated. As tape speed is decreased to supply longer playing or recording time, the signal-to-noise (S/N) ratio becomes poorer, as shown in Table 14-16.

## VHS connectors

Table 14-17 lists the connectors used for VHS video cassette decks.

*Table 14-16. Typical values of S/N for video cassette recorders.*

| Tape speed | S/N (dB) |
|------------|----------|
| SP | 46 |
| LP | 43 |
| EP | 40 |

*Table 14-17. VHS connectors.*

| Connection | Connector |
|------------|-----------|
| Earphone or microphone | 3.5 mm mini jack |
| Power supply terminal | 7-pin DIN |
| Camera terminal | Round 10-pin |

# SATCOM 1 (Satellite Communication Agency) reception

Certain data must be available to adjust a dish for signal reception from a particular satellite. This information includes the look angle, the tangent of that angle, the maximum height of any obstruction at a measured distance (in feet) from the dish, and the azimuth angle.

The look angle is important for it determines the maximum obstruction height. For SATCOM 1, or for any other satellite, obtain the value of the look angle. A table of trigonometric functions will supply the value of the tangent of this angle. Move the decimal point two places to the right to obtain the maximum obstruction height at a distance of 100 ft from the dish. Table 14-18 supplies information for all the states for SATCOM 1 reception.

Although the chart in Table 14-18 supplies maximum obstruction height figures for a distance of 100 ft, it is possible to determine this number for any distance. Multiply the measured distance by the tangent of the look angle for the particular state.

*Example* What is the maximum permitted obstruction height in the state of Pennsylvania for objects located at a distance of 165 ft from the dish?

*Solution* For Pennsylvania, the look angle in degrees is 16. The tangent of this angle is given as 0.29. $165 \times 0.29 = 47.85$ ft. The farther the object is from the dish, the greater the allowable obstruction size, in feet.

**Table 14-18. Dish adjustments for SATCOM 1 reception.**

| State | Look angle in degrees | Tangent of angle | Maximum height of obstruction at 100 ft | Azimuth angle (degrees) |
|---|---|---|---|---|
| Alabama | 25 | 0.47 | 47 | 244 |
| Alaska | 16 | 0.29 | 29 | 162 |
| Arizona | 43 | 0.93 | 93 | 217 |
| Arkansas | 29 | 0.55 | 55 | 239 |
| California | 43 | 0.93 | 93 | 203 |
| Colorado | 36 | 0.73 | 73 | 222 |
| Connecticut | 11 | 0.19 | 19 | 252 |
| Delaware | 14 | 0.25 | 25 | 250 |
| District of Columbia | 15 | 0.27 | 27 | 250 |
| Florida | 23 | 0.42 | 42 | 251 |
| Georgia | 24 | 0.45 | 45 | 148 |
| Hawaii | 54 | 0.99 | 99 | 129 |
| Idaho | 33 | 0.65 | 65 | 208 |
| Illinois | 24 | 0.45 | 45 | 238 |
| Indiana | 22 | 0.40 | 40 | 241 |
| Iowa | 27 | 0.51 | 51 | 233 |
| Kansas | 31 | 0.60 | 60 | 229 |
| Kentucky | 22 | 0.40 | 40 | 242 |
| Louisiana | 30 | 0.58 | 58 | 241 |
| Maine | 10 | 0.18 | 18 | 251 |
| Maryland | 15 | 0.27 | 27 | 250 |
| Massachusetts | 11 | 0.19 | 19 | 252 |
| Michigan | 20 | 0.36 | 36 | 240 |
| Minnesota | 24 | 0.45 | 45 | 229 |
| Mississippi | 29 | 0.55 | 55 | 241 |
| Missouri | 27 | 0.51 | 51 | 236 |
| Montana | 31 | 0.60 | 60 | 213 |
| Nebraska | 30 | 0.58 | 58 | 227 |
| Nevada | 41 | 0.87 | 87 | 208 |
| New Hampshire | 11 | 0.19 | 19 | 253 |
| New Jersey | 14 | 0.25 | 25 | 250 |
| New Mexico | 39 | 0.81 | 81 | 224 |
| North Carolina | 19 | 0.34 | 34 | 248 |
| North Dakota | 26 | 0.49 | 49 | 222 |
| Ohio | 18 | 0.33 | 33 | 244 |
| Oklahoma | 31 | 0.60 | 60 | 234 |
| Oregon | 38 | 0.78 | 78 | 201 |
| Pennsylvania | 16 | 0.29 | 29 | 246 |
| Rhode Island | 14 | 0.25 | 25 | 250 |
| South Carolina | 21 | 0.38 | 38 | 248 |
| South Dakota | 28 | 0.53 | 53 | 225 |
| Tennessee | 24 | 0.45 | 45 | 243 |
| Texas | 38 | 0.78 | 78 | 243 |

*Table 14-18. Continued.*

| State | Look angle in degrees | Tangent of angle | Maximum height of obstruction at 100 ft | Azimuth angle (degrees) |
|-------|----------------------|------------------|-----------------------------------------|------------------------|
| Utah | 38 | 0.78 | 78 | 214 |
| Vermont | 11 | 0.19 | 19 | 252 |
| Virginia | 18 | 0.33 | 33 | 233 |
| West Virginia | 18 | 0.33 | 33 | 227 |
| Washington | 33 | 0.65 | 65 | 200 |
| Wisconsin | 22 | 0.40 | 40 | 235 |
| Wyoming | 34 | 0.68 | 68 | 216 |

# Signal loss

The downlead from an antenna to the antenna input terminals of a TV receiver is either 75-$\Omega$ coaxial cable and, to a much lesser extent, 300-$\Omega$ transmission line. Table 14-19 shows the losses that can occur in 300-$\Omega$ line. Losses increase when the line is wet and also increase with frequency.

*Table 14-19. Possible signal losses in 300-$\Omega$ transmission line.*

Channel Master, Div. of Avnet, Inc.

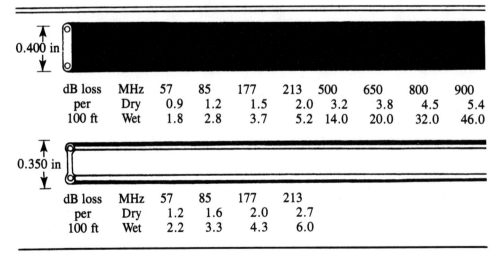

| 0.400 in | | | | | | | | | | |
|----------|-----|-----|-----|-----|-----|-----|-----|-----|-----|-----|
| dB loss | MHz | 57 | 85 | 177 | 213 | 500 | 650 | 800 | 900 | |
| per | Dry | 0.9 | 1.2 | 1.5 | 2.0 | 3.2 | 3.8 | 4.5 | 5.4 | |
| 100 ft | Wet | 1.8 | 2.8 | 3.7 | 5.2 | 14.0 | 20.0 | 32.0 | 46.0 | |

| 0.350 in | | | | | |
|----------|-----|-----|-----|-----|-----|
| dB loss | MHz | 57 | 85 | 177 | 213 |
| per | Dry | 1.2 | 1.6 | 2.0 | 2.7 |
| 100 ft | Wet | 2.2 | 3.3 | 4.3 | 6.0 |

# Signal range

Area designations for TV reception for VHF and UHF reception are shown in Table 14-20. These distances can be used to determine the type of TV antenna to select for satisfactory reception.

Table 14-20. Area designations
for TV reception.

| Area designation | For VHF | For UHF |
|---|---|---|
| Deepest fringe | 100+ mi | 60+ mi |
| Deep fringe | 100 mi | 55 mi |
| Fringe | 80 mi | 45 mi |
| Near fringe | 60 mi | 35 mi |
| Far suburban | 45 mi | 25 mi |
| Suburban | 30 mi | 15 mi |
| Urban | 20 mi | 10 mi |

Table 14-21. Dimensions of dipole elements for VHF.

| Channel number | Channel freq. (MHz) | Dipole length | Reflector length | Spacing of reflector | Director length | Spacing of director |
|---|---|---|---|---|---|---|
| 2 | 54– 60 | 8 ft 5⁵/₈ in | 8 ft 10⁷/₈ in | 2 ft 8¹/₈ in | 8 ft 1³/₄ in | 1 ft 9³/₈ in |
| 3 | 60– 66 | 7 ft 7³/₄ in | 8 ft ³/₈ in | 2 ft 3³/₄ in | 7 ft 4¹/₄ in | 1 ft 7¹/₄ in |
| 4 | 66– 72 | 6 ft 11¹/₂ in | 7 ft 3⁷/₈ in | 2 ft 2⁵/₈ in | 6 ft 8³/₈ in | 1 ft 5¹/₂ in |
| 5 | 76– 82 | 6 ft ³/₄ in | 6 ft 4¹/₂ in | 1 ft 11 in | 5 ft 10 in | 1 ft 1¹/₂ in |
| 6 | 82– 88 | 5 ft 7¹/₂ in | 5 ft 10⁵/₈ in | 1 ft 9¹/₄ in | 5 ft 4⁷/₈ in | 1 ft 2¹/₈ in |
| 7 | 174–180 | 2 ft 8 in | 2 ft 9⁵/₈ in | 10¹/₈ in | 4 ft 6³/₄ in | 6³/₄ in |
| 8 | 180–186 | 2 ft 6 in | 2 ft 8¹/₂ in | 9³/₄ in | 2 ft 5³/₄ in | 6¹/₂ in |
| 9 | 186–192 | 2 ft 9³/₄ in | 2 ft 7¹/₂ in | 9¹/₂ in | 2 ft 4³/₄ in | 6¹/₄ in |
| 10 | 192–198 | 2 ft 5 in | 2 ft 6¹/₂ in | 9¹/₄ in | 2 ft 4 in | 6¹/₈ in |
| 11 | 198–204 | 2 ft 4¹/₈ in | 2 ft 5⁵/₈ in | 8⁷/₈ in | 2 ft 3¹/₈ in | 5⁷/₈ in |
| 12 | 204–210 | 2 ft 2³/₄ in | 2 ft 4³/₄ in | 8⁵/₈ in | 2 ft 2¹/₄ in | 5³/₄ in |
| 13 | 210–216 | 2 ft 2⁵/₈ in | 2 ft 4 in | 8³/₈ in | 2 ft 1¹/₂ in | 5⁵/₈ in |

# Antenna dimensions

Television antennas are broadly tuned resonant circuits. Table 14-21 supplies the dimensions of the dipole, the lengths, and the spacing of directors and reflectors.

# Rotor cables

In some areas, TV antennas are operated by a rotor because TV stations can be in different directions. Rotor cables are either three or four conductor types of 20-gauge, 7-strand wire, covered with a vinyl jacket. Representative rotor wires and cable dimensions are shown in Fig. 14-7.

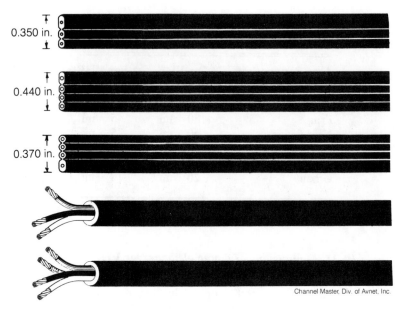

0.350 in.

0.440 in.

0.370 in.

Channel Master, Div. of Avnet, Inc.

*Fig. 14-7.   Rotor wire dimensions.*

# Microwave bands

The microwave bands extend from 1 to 300 GHz and are identified by letters, except the band from 40 to 300 GHz. This is known as the millimeter band, a reference to the wavelengths. Table 14-22 is a listing of the microwave bands.

*Table 14-22. Microwave bands.*

| Designation | Frequency (GHz) |
|-------------|-----------------|
| L band | 1.0−2.0 |
| S band | 2.0−4.0 |
| C band | 4.0−8.0 |
| X band | 8.0−12.0 |
| Ku band | 12.0−18.0 |
| K band | 18.0−27.0 |
| Ka band | 27.0−40.0 |
| Millimeter | 40.0−300.0 |

# 15
## CHAPTER

# Number systems

## Binary number system

The decimal system, using digits ranging from 0 through 9 for all number applications, is just one of a large array of number systems. Another number system, the binary, finds applications in computer and compact disc technology.

The binary system uses two digits: 0 and 1. As in the case of the decimal system, the value of the number depends on its horizontal position. Thus, in a decimal number such as 875, the digit 8 has a true value of 800, 7 has a value of 70 and 5 has a value of 5; $800+70+5=875$. The binary system is based on powers of two (Table 15-1) similar to powers of 10.

Table 15-1 supplies powers of two to $2^{60}$ and $2^{-60}$. To find the value of $2^{15}$, as an example, move down the center column headed by the letter $n$, stopping at the number 15. Move to the left and the value of $2^{15}$ is given as 32 768. To find the value of $2^{-6}$, move down the center or $n$ column and stop at the number 6. Look to the right and the value of $2^{-6}$ is supplied as 0.015 625.

Binary numbers, as in the case of decimal numbers, are generally arranged horizontally. The first binary number at the right has a value of $2^0$, the adjacent binary a value of $2^1$, etc. Thus, a number such as binary 1011 would have an equivalent decimal value of:

$$1 \qquad 0 \qquad 1 \qquad 1$$

$$(1\times2^3)+(0\times2^2)+(1\times2^1)+(1\times2^0)$$

$$8 \ + \ 0 \ + \ 2 \ + \ 1 \ = \ 11$$

Hence, binary 1011 is equivalent to decimal 11. To emphasize the number system, the subscript 2 is sometimes used to identify binary numbers, and 10 is used as a subscript for decimal numbers. In the example given: $1011_2$ equals $11_{10}$.

Table 15-1. Powers of two.

| $2^n$ | n | $2^{-n}$ |
|---:|---:|---|
| 1 | 0 | 1.0 |
| 2 | 1 | 0.5 |
| 4 | 2 | 0.25 |
| 8 | 3 | 0.125 |
| 16 | 4 | 0.062 5 |
| 32 | 5 | 0.031 25 |
| 64 | 6 | 0.015 625 |
| 128 | 7 | 0.007 812 5 |
| 256 | 8 | 0.003 906 25 |
| 512 | 9 | 0.001 953 125 |
| 1 024 | 10 | 0.000 976 562 5 |
| 2 048 | 11 | 0.000 488 281 25 |
| 4 096 | 12 | 0.000 244 140 625 |
| 8 192 | 13 | 0.000 122 070 312 5 |
| 16 384 | 14 | 0.000 061 035 156 25 |
| 32 768 | 15 | 0.000 030 517 578 125 |
| 65 536 | 16 | 0.000 015 258 789 062 5 |
| 131 072 | 17 | 0.000 007 629 394 531 25 |
| 262 144 | 18 | 0.000 003 814 697 265 625 |
| 524 288 | 19 | 0.000 001 907 348 632 812 5 |
| 1 048 576 | 20 | 0.000 000 953 674 316 406 25 |
| 2 097 152 | 21 | 0.000 000 476 837 158 203 125 |
| 4 194 304 | 22 | 0.000 000 238 418 579 101 562 5 |
| 8 388 608 | 23 | 0.000 000 119 209 289 550 781 25 |
| 16 777 216 | 24 | 0.000 000 059 604 644 775 390 625 |
| 33 554 432 | 25 | 0.000 000 029 802 322 387 695 312 5 |
| 67 108 864 | 26 | 0.000 000 014 901 161 193 847 656 25 |
| 134 217 728 | 27 | 0.000 000 007 450 580 596 923 828 125 |
| 268 435 456 | 28 | 0.000 000 003 725 290 298 461 914 062 5 |
| 536 870 912 | 29 | 0.000 000 001 862 645 149 230 957 031 25 |
| 1 073 741 824 | 30 | 0.000 000 000 931 322 574 615 478 515 625 |
| 2 147 483 648 | 31 | 0.000 000 000 465 661 287 307 739 257 812 5 |
| 4 294 967 296 | 32 | 0.000 000 000 232 830 643 653 869 628 906 25 |
| 8 589 934 592 | 33 | 0.000 000 000 116 415 321 826 934 814 453 125 |
| 17 179 869 184 | 34 | 0.000 000 000 058 207 660 913 467 407 226 562 5 |
| 34 359 738 368 | 35 | 0.000 000 000 029 103 830 456 733 703 613 281 25 |
| 68 719 476 736 | 36 | 0.000 000 000 014 551 915 228 366 851 806 640 625 |
| 137 438 953 472 | 37 | 0.000 000 000 007 275 957 614 183 425 903 320 312 5 |
| 274 877 906 944 | 38 | 0.000 000 000 003 637 978 807 091 712 951 660 156 25 |
| 549 755 813 888 | 39 | 0.000 000 000 001 818 989 403 545 856 475 830 078 125 |
| 1 099 511 627 776 | 40 | 0.000 000 000 000 909 494 701 772 928 237 915 039 062 5 |
| 2 199 023 255 552 | 41 | 0.000 000 000 000 454 747 350 886 464 118 957 519 531 25 |
| 4 398 046 511 104 | 42 | 0.000 000 000 000 227 373 675 443 232 059 478 759 765 625 |
| 8 796 093 022 208 | 43 | 0 000 000 000 000 113 686 837 721 616 029 739 379 882 812 |
| 17 592 186 044 416 | 44 | 0.000 000 000 000 056 843 418 860 808 014 869 689 941 406 |
| 35 184 372 088 832 | 45 | 0.000 000 000 000 028 421 709 430 404 007 434 844 970 703 |

*Table 15-1. Continued.*

| $2^n$ | n | $2^{-n}$ |
|---|---|---|
| 70 368 744 177 664 | 46 | 0.000 000 000 000 014 210 854 715 202 003 717 422 485 351 |
| 140 737 488 355 328 | 47 | 0.000 000 000 000 007 105 427 357 601 001 858 711 242 675 |
| 281 474 976 710 656 | 48 | 0.000 000 000 000 003 552 713 678 800 500 929 355 621 337 |
| 562 949 953 421 312 | 49 | 0.000 000 000 000 001 776 356 839 400 250 464 677 810 668 |
| 1 125 899 906 842 624 | 50 | 0.000 000 000 000 000 888 178 419 700 125 232 338 905 334 |
| 2 251 799 813 685 248 | 51 | 0.000 000 000 000 000 444 089 209 850 062 616 169 452 667 |
| 4 503 599 627 370 496 | 52 | 0.000 000 000 000 000 222 044 604 925 031 308 084 726 333 |
| 9 007 199 254 740 992 | 53 | 0.000 000 000 000 000 111 022 302 462 515 654 042 363 166 |
| 18 014 398 509 481 984 | 54 | 0.000 000 000 000 000 055 511 151 231 257 827 021 181 583 |
| 36 028 797 018 963 968 | 55 | 0.000 000 000 000 000 027 755 575 615 628 913 510 590 791 |
| 72 057 594 037 927 936 | 56 | 0.000 000 000 000 000 013 877 787 807 814 456 755 295 395 |
| 144 115 188 075 855 872 | 57 | 0.000 000 000 000 000 006 938 893 903 907 228 377 647 697 |
| 288 230 376 151 711 744 | 58 | 0.000 000 000 000 000 003 469 446 951 953 614 188 823 848 |
| 576 460 752 303 423 488 | 59 | 0.000 000 000 000 000 001 734 723 475 976 807 094 411 924 |
| 1 152 921 504 606 846 976 | 60 | 0.000 000 000 000 000 000 867 361 737 988 403 547 205 962 |

Table 15-2 lists the binary equivalents of decimal numbers ranging from 0 to 100.

*Example* What is the binary equivalent of 27?

*Solution* In Table 15-2, locate 27 in the left-hand column. Move directly across to the right and the binary equivalent is 0001 1011. The three zeros at the left of the binary can be omitted because they add nothing to the value of the number. Accordingly, decimal 27 equals binary 1 1011. Binary 1 1011 can be set up as:

$$2^4 + 2^3 + 2^2 + 2^1 + 2^0$$

$$16 + 8 + 0 + 2 + 1 = 27$$

# Decimal-to-binary conversion rules

1. Write number $n+0$ if even or $(n-1)+1$ if odd.
2. Divide even number obtained in (1) by 2.
   Write answer ($m$) below in same form:

   $$m+0 \text{ if even, } (m-1)+1 \text{ if odd.}$$

3. Continue until $m$ or $(m-1)$ becomes zero.
4. Column of ones and zeros so obtained is binary equivalent of $n$ with least significant digit at the top.

*Example* $n=327$. What is its binary equivalent?

Table 15-2. *Decimal integers to pure binary numbers.*

| Decimal integer | Binary | Decimal integer | Binary | Decimal integer | Binary | Decimal integer | Binary |
|---|---|---|---|---|---|---|---|
| 00 | 00000000 | 25 | 00011001 | 50 | 00110010 | 75 | 01001011 |
| 01 | 00000001 | 26 | 00011010 | 51 | 00110011 | 76 | 01001100 |
| 02 | 00000010 | 27 | 00011011 | 52 | 00110100 | 77 | 01001101 |
| 03 | 00000011 | 28 | 00011100 | 53 | 00110101 | 78 | 01001110 |
| 04 | 00000100 | 29 | 00011101 | 54 | 00110110 | 79 | 01001111 |
| 05 | 00000101 | 30 | 00011110 | 55 | 00110111 | 80 | 01010000 |
| 06 | 00000110 | 31 | 00011111 | 56 | 00111000 | 81 | 01010001 |
| 07 | 00000111 | 32 | 00100000 | 57 | 00111001 | 82 | 01010010 |
| 08 | 00001000 | 33 | 00100001 | 58 | 00111010 | 83 | 01010011 |
| 09 | 00001001 | 34 | 00100010 | 59 | 00111011 | 84 | 01010100 |
| 10 | 00001010 | 35 | 00100011 | 60 | 00111100 | 85 | 01010101 |
| 11 | 00001011 | 36 | 00100100 | 61 | 00111101 | 86 | 01010110 |
| 12 | 00001100 | 37 | 00100101 | 62 | 00111110 | 87 | 01010111 |
| 13 | 00001101 | 38 | 00100110 | 63 | 00111111 | 88 | 01011000 |
| 14 | 00001110 | 39 | 00100111 | 64 | 01000000 | 89 | 01011001 |
| 15 | 00001111 | 40 | 00101000 | 65 | 01000001 | 90 | 01011010 |
| 16 | 00010000 | 41 | 00101001 | 66 | 01000010 | 91 | 01011011 |
| 17 | 00010001 | 42 | 00101010 | 67 | 01000011 | 92 | 01011100 |
| 18 | 00010010 | 43 | 00101011 | 68 | 01000100 | 93 | 01011101 |
| 19 | 00010011 | 44 | 00101100 | 69 | 01000101 | 94 | 01011110 |
| 20 | 00010100 | 45 | 00101101 | 70 | 01000110 | 95 | 01011111 |
| 21 | 00010101 | 46 | 00101110 | 71 | 01000111 | 96 | 01100000 |
| 22 | 00010110 | 47 | 00100111 | 72 | 01001000 | 97 | 01100001 |
| 23 | 00010111 | 48 | 00110000 | 73 | 01001001 | 98 | 01100010 |
| 24 | 00011000 | 49 | 00110001 | 74 | 01001010 | 99 | 01100011 |
|  |  |  |  |  |  | 100 | 01100100 |

## Solution

$$326 + 1$$
$$162 + 1$$
$$80 + 1$$
$$40 + 0$$
$$20 + 0$$
$$10 + 0$$
$$4 + 1$$
$$2 + 0$$
$$0 + 1$$

Therefore, the binary equivalent of 327 is 101000111.

# BCD (Binary-coded decimals)

Binary numbers can be arranged in groups of four to correspond to decimal digits. A decimal number such as 5, for example, can be represented by binary 0101. Decimal 55 would then be 0101 0101. A setup of this kind is known as a binary-coded decimal, abbreviated as *BCD*. Note the difference between binary-coded decimals and pure binary numbers shown earlier in Table 15-2. A decimal number such as 82 in pure binary form would be 1010010, while decimal 82 in BCD notation would be 1000 0010. Table 15-3 lists BCD equivalents of decimal numbers ranging from 0 to 100.

*Table 15-3. Decimal to BCD (binary-coded decimal) notation.*

| Decimal | BCD | | Decimal | BCD | | Decimal | BCD | |
|---------|------|------|---------|------|------|---------|------|------|
| 00 | 0000 | 0000 | 30 | 0011 | 0000 | 60 | 0110 | 0000 |
| 01 | 0000 | 0001 | 31 | 0011 | 0001 | 61 | 0110 | 0001 |
| 02 | 0000 | 0010 | 32 | 0011 | 0010 | 62 | 0110 | 0010 |
| 03 | 0000 | 0011 | 33 | 0011 | 0011 | 63 | 0110 | 0011 |
| 04 | 0000 | 0100 | 34 | 0011 | 0100 | 64 | 0110 | 0100 |
| 05 | 0000 | 0101 | 35 | 0011 | 0101 | 65 | 0110 | 0101 |
| 06 | 0000 | 0110 | 36 | 0011 | 0110 | 66 | 0110 | 0110 |
| 07 | 0000 | 0111 | 37 | 0011 | 0111 | 67 | 0110 | 0111 |
| 08 | 0000 | 1000 | 38 | 0011 | 1000 | 68 | 0110 | 1000 |
| 09 | 0000 | 1001 | 39 | 0011 | 1001 | 69 | 0110 | 1001 |
| 10 | 0001 | 0000 | 40 | 0100 | 0000 | 70 | 0111 | 0000 |
| 11 | 0001 | 0001 | 41 | 0100 | 0001 | 71 | 0111 | 0001 |
| 12 | 0001 | 0010 | 42 | 0100 | 0010 | 72 | 0111 | 0010 |
| 13 | 0001 | 0011 | 43 | 0100 | 0011 | 73 | 0111 | 0011 |
| 14 | 0001 | 0100 | 44 | 0100 | 0100 | 74 | 0111 | 0100 |
| 15 | 0001 | 0101 | 45 | 0100 | 0101 | 75 | 0111 | 0101 |
| 16 | 0001 | 0110 | 46 | 0100 | 0110 | 76 | 0111 | 0110 |
| 17 | 0001 | 0111 | 47 | 0100 | 0111 | 77 | 0111 | 0111 |
| 18 | 0001 | 1000 | 48 | 0100 | 1000 | 78 | 0111 | 1000 |
| 19 | 0001 | 1001 | 49 | 0100 | 1001 | 79 | 0111 | 1001 |
| 20 | 0010 | 0000 | 50 | 0101 | 0000 | 80 | 1000 | 0000 |
| 21 | 0010 | 0001 | 51 | 0101 | 0001 | 81 | 1000 | 0001 |
| 22 | 0010 | 0010 | 52 | 0101 | 0010 | 82 | 1000 | 0010 |
| 23 | 0010 | 0011 | 53 | 0101 | 0011 | 83 | 1000 | 0011 |
| 24 | 0010 | 0100 | 54 | 0101 | 0100 | 84 | 1000 | 0100 |
| 25 | 0010 | 0101 | 55 | 0101 | 0101 | 85 | 1000 | 0101 |
| 26 | 0010 | 0110 | 56 | 0101 | 0110 | 86 | 1000 | 0110 |
| 27 | 0010 | 0111 | 57 | 0101 | 0111 | 87 | 1000 | 0111 |
| 28 | 0010 | 1000 | 58 | 0101 | 1000 | 88 | 1000 | 1000 |
| 29 | 0010 | 1001 | 59 | 0101 | 1001 | 89 | 1000 | 1001 |

*Table 15-3. Continued.*

| Decimal | BCD | | Decimal | BCD | | Decimal | BCD | |
|---|---|---|---|---|---|---|---|---|
| 90 | 1001 | 0000 | 95 | 1001 | 0101 | 100 | 0000 | 0000 |
| 91 | 1001 | 0001 | 96 | 1001 | 0110 | | | |
| 92 | 1001 | 0010 | 97 | 1001 | 0111 | | | |
| 93 | 1001 | 0011 | 98 | 1001 | 1000 | | | |
| 94 | 1001 | 0100 | 99 | 1001 | 1001 | | | |

# Bits

The words *binary digit* can be contracted to form a new word, *bit*. Unlike decimal notation, the amount of bits is sometimes used to describe a particular binary number, as shown in Table 15-4. The greater the number of bits, the larger the equivalent decimal number, assuming the leftmost binary digit to be 1. Table 15-5 shows maximum binary values and their decimal equivalents.

*Table 15-4. Binary digit designations.*

| Binary example | Designation |
|---|---|
| 0 | One bit |
| 01 | Two bits |
| 101 | Three bits |
| 1010 | Four bits |
| 10101 | Five bits |
| 100110 | Six bits |
| 1010101 | Seven bits |
| 10101011 | Eight bits |

# Voltage representation in digital form

Voltages are generally analog representations as indicated in Fig. 15-1. However, by taking a large succession of instantaneous values and converting these from decimal to binary form, you can have a digital representation of the same waveform. Table 15-6 shows how the waveform of Fig. 15-1 can be indicated digitally.

# Binary radix

Every number system has a *radix* or base. In the decimal system, the radix or base is 10, because the number of characters or symbols used in that system is 10. In

*Table 15-5. Maximum binary values and decimal equivalents.*

| Number of bits | Maximum binary value | Maximum equivalent decimal value |
|:---:|:---:|:---:|
| 1 | 1 | 1 |
| 2 | 11 | 3 |
| 3 | 111 | 7 |
| 4 | 1111 | 15 |
| 5 | 11111 | 31 |
| 6 | 111111 | 63 |
| 7 | 1111111 | 127 |
| 8 | 11111111 | 255 |
| 9 | 111111111 | 511 |
| 10 | 1111111111 | 1 023 |
| 11 | 11111111111 | 2 047 |
| 12 | 111111111111 | 4 095 |
| 13 | 1111111111111 | 8 191 |
| 14 | 11111111111111 | 16 383 |
| 15 | 111111111111111 | 32 767 |
| 16 | 1111111111111111 | 65 535 |

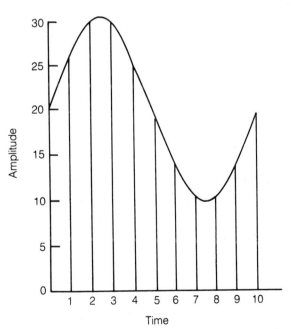

*Fig. 15-1. Method of converting analog waveform to its digital equivalent.*

*Table 15-6. Digital values of voltage wave—five-bit arrangement.*

| Time | Voltage in decimal form | Voltage in binary form |
|------|------------------------|------------------------|
| 0 | 20 | 10100 |
| 1 | 25 | 11001 |
| 2 | 28 | 11100 |
| 3 | 28 | 11100 |
| 4 | 26 | 11010 |
| 5 | 20 | 10100 |
| 6 | 14 | 01110 |
| 7 | 10 | 01010 |
| 8 | 10 | 01010 |
| 9 | 14 | 01110 |
| 10 | 20 | 10100 |

the binary system, the radix is 2, because only two symbols, 0 and 1, are used. The radix of any system is always 1 larger than the largest number used in the system. In the decimal system, the largest digit is 9, $9+1=10=$ radix or base of the decimal system. In the binary system, the largest digit is 1, $1+1=2$. Hence in the binary system, the radix is 2. The dot that separates whole numbers from fractions is the decimal point in the decimal number system; in other systems it is the radix point.

# Binary-to-decimal conversion (decoding)

An easy way to convert from binary to decimal form (known as decoding) is to set up a table such as the one shown in Fig. 15-2. Starting at the radix point and moving horizontally to the left, set up positive powers of 2, starting with $2^0$. The largest positive power of 2 should not exceed the value of the decimal. If the decimal number is 40, for example, the largest power of 2 would be $2^5=32$. $2^6$ would yield 64, which would be too large; that is, it would be larger than the decimal, 40.

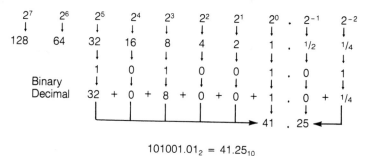

$$101001.01_2 = 41.25_{10}$$

*Fig. 15-2. Example of binary-to-decimal conversion.*

On the right side of the radix point, set up a horizontal row of negative powers of 2. The largest negative power of 2 should not exceed the value of the decimal fraction. If the decimal fraction is 0.26, then $2^{-2}$ would be the largest negative power of 2 required.

In either case, whether a positive power of 2 or a negative power of 2, no harm is done in going beyond the indicated value of the exponent. It simply means that the final answer, in binary form, will have one or more zeros at the extreme left and at the extreme right. These can be dropped since they do not contribute to the value of the binary. As a more familiar example, decimal 063 has the same value as 63. 0.780 0 is the same as 0.78. The only exception would be if you wanted to indicate an accuracy to four decimal places. Similarly, 0010101 is the same as 10101. The two zeros at the left contribute nothing. And, for a fractional binary, .010100 is the same as .0101. The two zeros at the right can be dropped.

After setting up the horizontal row of powers of 2, write the equivalent values as decimals directly below each power of 2. Add these values to obtain the required decimal value. Figure 15-2 shows how binary 101001.01 is converted to decimal 41.25. Note that this is a pure binary.

# Double-dabble method of binary-to-decimal conversion

In one method the successive values of the bits, starting with the highest-order bit, are multiplied by 2 (Fig. 15-3). The value of the next lower order bit is added in and the sum is then multiplied by 2. The process is repeated until the lowest-order bit is reached. This method is called *double dabble*, because *dabble* means to double and add.

*Example*   Convert 101010 to decimal form.

*Solution*   Set up the binary as given:

101010

$$\begin{array}{r} \times\ 2 \\ \hline 2 \end{array}$$   Multiply the MSD (most significant digit) by 2.

$$\begin{array}{r} +\ 0 \\ \hline 2 \end{array}$$   Add the value of the next lower order bit.

$$\begin{array}{r} \times\ 2 \\ \hline 4 \end{array}$$   Multiply by 2.

$$\begin{array}{r} +\ 1 \\ \hline 5 \end{array}$$   Add the value of the next lower order bit.

Convert 101111 to decimal

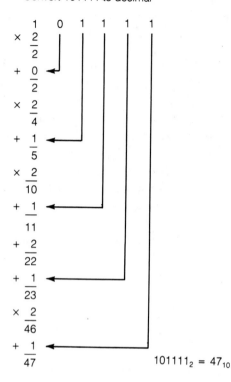

Fig. 15-3.   Conversion of a binary number to its decimal equivalent.

$$101111_2 = 47_{10}$$

$$\begin{array}{r} \times\ 2 \\ \hline 10 \end{array}$$   Multiply by 2.

$$\begin{array}{r} +\ 0 \\ \hline 10 \end{array}$$   Add the value of the next lower order bit.

$$\begin{array}{r} \times\ 2 \\ \hline 20 \end{array}$$   Multiply by 2.

$$\begin{array}{r} +\ 1 \\ \hline 21 \end{array}$$   Add the value of the next lower order bit.

$$\begin{array}{r} \times\ 2 \\ \hline 42 \end{array}$$   Multiply by 2.

$$\begin{array}{r} +\ 0 \\ \hline 42 \end{array}$$   Add value of LSD (least significant digit).

$$101010_2 = 42_{10}$$

Still another binary decoding technique is to write the binary and then to consider each binary as a coefficient of the corresponding positional decimal value, as in Table 15-7.

*Table 15-7. Example of*
*binary-to-decimal decoding of 10101110.*

| Binary | Coefficient | | Decimal value | | |
|--------|-------------|---|---------------|---|-----|
| 0 | 0 | × | 1 | = | 0 |
| 1 | 1 | × | 2 | = | 2 |
| 1 | 1 | × | 4 | = | 4 |
| 1 | 1 | × | 8 | = | 8 |
| 0 | 0 | × | 16 | = | 0 |
| 1 | 1 | × | 32 | = | 32 |
| 0 | 0 | × | 64 | = | 0 |
| 1 | 1 | × | 128 | = | 128 |
| | | | | | 174 |

$1010\ 1110_2 = 174_{10}$

# Decimal-to-binary conversion (encoding)

The conversion of decimal to binary (known as *encoding*) is the reverse of the process just described. To convert decimal to binary, start with the largest positive power of 2 that is equal to or smaller than the number.

*Example*  Find the binary equivalent of 53.
*Solution*  The nearest power of 2 is $2^5 = 32$.

1. Subtract 32 from the decimal. $53 - 32 = 21$.
2. Find the largest power of 2 that is equal to or less than 21. $2^4 = 16$.
3. Subtract 16 from 21. $21 - 16 = 5$.
4. Find the largest power of 2 that is equal to or less than 5. $2^2 = 4$. $5 - 4 = 1$.
5. The largest power of 2 that is equal to 1 is $2^0$.

Combine the terms. $2^5 + 2^4 + 2^2 + 2^0$. This can now be expanded to read

$$(1 \times 2^5) + (1 \times 2^4) + (0 \times 2^3) + (1 \times 2^2) + (0 \times 2^1) + (1 \times 2^0)$$
$$\quad 1 \qquad\quad 1 \qquad\quad 0 \qquad\quad 1 \qquad\quad 0 \qquad\quad 1$$

The binary equivalent of 53 is 110101.

## Alternative decimal-to-binary encoding

Another technique for converting decimal to binary involves repeated division by

2. Assume, for example, that you wish to change 25 to its binary equivalent. Divide 25 by 2:

$$2)\underline{25} = 12 \quad \text{with a remainder of 1}$$
$$2)\underline{12} = 6 \quad \text{with a remainder of 0}$$
$$2)\underline{6} = 3 \quad \text{with a remainder of 0}$$
$$2)\underline{3} = 1 \quad \text{with a remainder of 1}$$
$$2)\underline{1} = 0 \quad \text{with a remainder of 1}$$

The remainder shown at the top becomes the low-order digit in the binary; the remainder at the bottom is the high-order digit. Rearranging them, you have 11001. That is,

$$25_{10} = 11001_2$$

In doing the preceding conversion, you started with 25 and divided by 2. 25 divided by 2 is 12 with a remainder of 1. 12 becomes the next dividend, with 2 once again as a divisor. 2 divided into $12 = 6$. The remainder is 0, and 6 becomes the next dividend. Figure 15-4 is another example of this technique of repeated division by 2 to convert a decimal number (47) into binary form.

Convert decimal 47 to binary

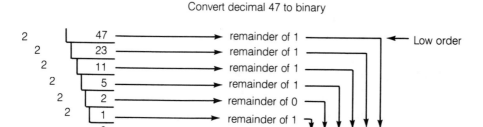

Fig. 15-4.  Conversion of a decimal to binary form.

# Hexadecimal number system

This number system uses digits 0 through 9 and letters A to F for its number applications. As with the binary and decimal systems, the value of a hexadecimal number depends on its horizontal position. Hexadecimal numbers are based on powers of 16, as shown in Table 15-8.

Hexadecimal numbers, as with binary and decimal numbers, are usually arranged horizontally. The first hexadecimal number at the right has a value of $16^0$, the adjacent hexadecimal number a value of $16^1$, etc.

In the hexadecimal system, the letter A corresponds to decimal 10, B to decimal 11, C to 12, D to 13, E to 14, and F to 15.

Table 15-8. Powers of 16.

| Power of 16 | | Decimal value | Decoding of Hex 4F3 |
|---|---|---|---|
| $16^{-1}$ | = | 0.062 5 | = 4 F 3 |
| $16^{-2}$ | = | 0.003 906 | $= (4 \times 16^2) + (15 \times 16^1) + (3 \times 16^0)$ |
| $16^{-3}$ | = | 0.000 244 | $= 4 \times 256) + (15 \times 16) + (3 \times 1)$ |
| $16^{-4}$ | = | 0.000 015 | $= 1\ 024 + 240 + 3$ |
| $16^0$ | = | 1 | $= 1\ 267$ |
| $16^1$ | = | 16 | |
| $16^2$ | = | 256 | |
| $16^3$ | = | 4 096 | |
| $16^4$ | = | 65 536 | |
| $16^5$ | = | 1 048 576 | |
| $16^6$ | = | 16 777 216 | |
| $16^7$ | = | 268 435 456 | |
| $16^8$ | = | 4 294 967 296 | |
| $16^9$ | = | 68 719 476 736 | |
| $16^{10}$ | = | 1 099 511 627 776 | |

$$4 \qquad F \qquad 3$$
$$= (4 \times 16^2) + (15 \times 16^1) + (3 \times 16^0)$$
$$= (4 \times 256) + (15 \times 16) + (3 \times 1)$$
$$= 1\ 024 + 240 + 3$$
$$= 1\ 267$$

Hexadecimal 4F3 equals decimal 1 267.

Table 15-9 lists hexadecimal equivalents for decimal numbers from 0 to 100.

## Hexadecimal-to-decimal conversions

Every hexadecimal number consists of two parts: a coefficient or multiplier and some value of 16 raised to a power. Thus, hexadecimal 5A is the same as $(5 \times 16^1) + (10 \times 16^0)$. In this hexadecimal number, 5 is a coefficient, and so is A. (A equals decimal 10.) In a hexadecimal number such as 234, these digits have a decimal value of $(2 \times 16^2) + (3 \times 16^1) + (4 \times 16^0)$. Each of the numbers 2, 3, and 4, are coefficients that are used to multiply various powers of 16. Because 234 looks like a decimal number, it can be identified as a hexadecimal number by using the subscript 16. Hence, hexadecimal 234 is more correctly written as $234_{16}$.

Table 15-10 is a convenient way of obtaining decimal equivalents of hexadecimal numbers, but some addition is involved.

*Example* What is the decimal equivalent of 5FA4?

*Solution* In Table 15-10, the first column at the left represents all possible coefficients or multipliers. In converting 5FA4 to a decimal number, work with

*Table 15-9. Decimal-to-hexadecimal conversions.*

| Decimal | Hexadecimal | Decimal | Hexadecimal | Decimal | Hexadecimal |
|---------|-------------|---------|-------------|---------|-------------|
| 0 | 0 | 35 | 23 | 70 | 46 |
| 1 | 1 | 36 | 24 | 71 | 47 |
| 2 | 2 | 37 | 25 | 72 | 48 |
| 3 | 3 | 38 | 26 | 73 | 49 |
| 4 | 4 | 39 | 27 | 74 | 4A |
| 5 | 5 | 40 | 28 | 75 | 4B |
| 6 | 6 | 41 | 29 | 76 | 4C |
| 7 | 7 | 42 | 2A | 77 | 4D |
| 8 | 8 | 43 | 2B | 78 | 4E |
| 9 | 9 | 44 | 2C | 79 | 4F |
| 10 | A | 45 | 2D | 80 | 50 |
| 11 | B | 46 | 2E | 81 | 51 |
| 12 | C | 47 | 2F | 82 | 52 |
| 13 | D | 48 | 30 | 83 | 53 |
| 14 | E | 49 | 31 | 84 | 54 |
| 15 | F | 50 | 32 | 85 | 55 |
| 16 | 10 | 51 | 33 | 86 | 56 |
| 17 | 11 | 52 | 34 | 87 | 57 |
| 18 | 12 | 53 | 35 | 88 | 58 |
| 19 | 13 | 54 | 36 | 89 | 59 |
| 20 | 14 | 55 | 37 | 90 | 5A |
| 21 | 15 | 56 | 38 | 91 | 5B |
| 22 | 16 | 57 | 39 | 92 | 5C |
| 23 | 17 | 58 | 3A | 93 | 5D |
| 24 | 18 | 59 | 3B | 94 | 5E |
| 25 | 19 | 60 | 3C | 95 | 5F |
| 26 | 1A | 61 | 3D | 96 | 60 |
| 27 | 1B | 62 | 3E | 97 | 61 |
| 28 | 1C | 63 | 3F | 98 | 62 |
| 29 | 1D | 64 | 40 | 99 | 63 |
| 30 | 1E | 65 | 41 | 100 | 64 |
| 31 | 1F | 66 | 42 | | |
| 32 | 20 | 67 | 43 | | |
| 33 | 21 | 68 | 44 | | |
| 34 | 22 | 69 | 45 | | |

each part of this hexadecimal a digit at a time, starting at the right-hand side. The first right-hand digit of 5FA4 is 4. In Table 15-10, locate the number 4 in the left-hand column. Move completely across to $16^0$, and the value of 4 is indicated as 4.

The next number of 5FA4 to consider is A. Find A in the left-hand column.

*Table 15-10. Hexadecimal-to-decimal integer conversions.*

| | $16^7$ $\times 000\ 0000$ | $16^6$ $\times 00\ 0000$ | $16^5$ $\times 0\ 0000$ | $16^4$ $\times 0000$ | $16^3$ $\times 000$ | $16^2$ $\times 00$ | $16^1$ $\times 0$ | $16^0$ $\times$ |
|---|---|---|---|---|---|---|---|---|
| 1 | 268 435 456 | 16 777 216 | 1 048 576 | 65 536 | 4 096 | 256 | 16 | 1 |
| 2 | 536 870 912 | 33 554 432 | 2 097 152 | 131 072 | 8 192 | 512 | 32 | 2 |
| 3 | 805 306 368 | 50 331 648 | 3 145 728 | 196 608 | 12 288 | 768 | 48 | 3 |
| 4 | 1 073 741 824 | 67 108 864 | 4 194 304 | 262 144 | 16 384 | 1 024 | 64 | 4 |
| 5 | 1 342 177 280 | 83 886 080 | 5 242 880 | 327 680 | 20 480 | 1 280 | 80 | 5 |
| 6 | 1 610 612 736 | 100 663 296 | 6 291 456 | 393 216 | 24 576 | 1 536 | 96 | 6 |
| 7 | 1 879 048 192 | 117 440 512 | 7 340 032 | 458 752 | 28 672 | 1 792 | 112 | 7 |
| 8 | 2 147 483 648 | 134 217 728 | 8 388 608 | 524 288 | 32 768 | 2 048 | 128 | 8 |
| 9 | 2 415 919 104 | 150 994 944 | 9 437 184 | 589 824 | 36 864 | 2 304 | 144 | 9 |
| A | 2 684 354 560 | 167 772 160 | 10 485 760 | 655 360 | 40 960 | 2 560 | 160 | 10 |
| B | 2 952 790 016 | 184 549 376 | 11 534 336 | 720 896 | 45 056 | 2 816 | 176 | 11 |
| C | 3 221 225 472 | 201 326 592 | 12 582 912 | 786 432 | 49 152 | 3 072 | 192 | 12 |
| D | 3 489 660 928 | 218 103 808 | 13 631 488 | 851 968 | 53 248 | 3 328 | 208 | 13 |
| E | 3 758 096 384 | 234 881 024 | 14 680 064 | 917 504 | 57 344 | 3 584 | 224 | 14 |
| F | 4 026 531 840 | 251 658 240 | 15 728 640 | 983 040 | 61 440 | 3 840 | 240 | 15 |

Move horizontally, but stop at the second or $16^1$ column because you are now working with the second digit of the hex number. The value of A is 160.

The third digit of 5FA4 is F in the left-hand column. Move to the third column, or the $16^2$ column. The decimal value is 3840. Finally, find the decimal equivalent of the fourth digit in 5FA4. This is the number 5. Locate 5 in the left-hand column. Move across to the right and stop in the fourth column—the $16^3$ column. The decimal equivalent here is 20 480.

Add all the decimal numbers obtained.

$$
\begin{array}{r}
4 \\
160 \\
3\ 840 \\
+\,20\ 480 \\
\hline
24\ 484
\end{array}
$$

Hexadecimal 5FA4 equals decimal 24 484. This can also be written as $5FA4_{16}$ equals $24\ 484_{10}$.

The binary, decimal, and hexadecimal number systems are just a few of those that can be used. Each of these has a different base; that is, each uses a different amount of symbols. There are only two symbols in binary—0 and 1; hence it is called a base-2 system. Decimal uses 10 symbols, and so it is a base-10 system. Hexadecimal has 16 symbols, including the digits 0 through 9, and letters A through F. The hexadecimal system is used in connection with computer printouts and also with the ASCII code, described in chapter 25.

# Base-3 system

A base-3 system is one which uses three symbols, 0, 1, and 2. Table 15-11 shows numbers in the base-3 system and their decimal equivalents. In the base-3 system,

*Table 15-11. Base-3 numbering system.*

| Base 3 | Decimal equivalent | Base 3 | Decimal equivalent |
|--------|-----------|--------|-----------|
| 000 | 0 | 022 | 8 |
| 001 | 1 | 100 | 9 |
| 002 | 2 | 101 | 10 |
| 010 | 3 | 102 | 11 |
| 011 | 4 | 110 | 12 |
| 012 | 5 | 111 | 13 |
| 020 | 6 | 112 | 14 |
| 021 | 7 | 120 | 15 |

the digit at the extreme right has the same value as in the decimal system. The center number is multiplied by 3. The number at the extreme left is multiplied by 9. Thus, a trinary number such as 221 is the same as $(2 \times 9) + (2 \times 3) + 1$ or $18 + 6 + 1 = 25$. Hence, 221 in trinary form is equivalent to 25 in decimal form.

# Base-4 system

In the quaternary, or base 4, system there are four symbols. These symbols are 0, 1, 2, and 3. Table 15-12 shows quaternary numbers and their decimal equivalents. In the quaternary system, the rightmost digit has the same value as the same digit in the same position in the decimal system. However, that digit cannot have a value greater than 3. The center digit is equivalent to that digit multiplied by 4. The leftmost digit is equivalent to that digit multiplied by 16. For example, 322 in quaternary is equal to $(3 \times 16) + (2 \times 4) + (2 \times 1) = 48 + 8 + 2 = 58$. Hence, quaternary 322 = decimal 58.

# Base-5 system

A base-5 system makes use of five symbols and these are 0, 1, 2, 3, and 4. Table 15-13 shows base five numbers (also called the quinary system) and their decimal equivalents. As in the preceding number systems, the rightmost digit is directly equivalent in decimal. The center digit is to be multiplied by 5 and the left-hand digit by 25. Thus, quinary 214 is equivalent to $(25 \times 2) + (5 \times 1) + (1 \times 4) = 59$. This number, 59, is in decimal form.

*Table 15-12. Base-4 numbering system.*

| Quaternary | Decimal | Quaternary | Decimal |
|------------|---------|------------|---------|
| 000 | 00 | 031 | 13 |
| 001 | 01 | 032 | 14 |
| 002 | 02 | 033 | 15 |
| 003 | 03 | 100 | 16 |
| 010 | 04 | 101 | 17 |
| 011 | 05 | 102 | 18 |
| 012 | 06 | 103 | 19 |
| 013 | 07 | 110 | 20 |
| 020 | 08 | 111 | 21 |
| 021 | 09 | 112 | 22 |
| 022 | 10 | 113 | 23 |
| 023 | 11 | 120 | 24 |
| 030 | 12 | 121 | 25 |

*Table 15-13. Base-5 numbering system.*

| Base 5 | Decimal equivalent | Base 5 | Decimal equivalent |
|--------|--------------------|--------|--------------------|
| 000 | 000 | 013 | 008 |
| 001 | 001 | 014 | 009 |
| 002 | 002 | 020 | 010 |
| 003 | 003 | 021 | 011 |
| 004 | 004 | 022 | 012 |
| 010 | 005 | 023 | 013 |
| 011 | 006 | 024 | 014 |
| 012 | 007 | 030 | 015 |

# Octal number system

The octal number system uses 8 symbols, 0, 1, 2, 3, 4, 5, 6, and 7. Just as you can have powers of 2, or 10, or 16, so too can you have powers of 8. These are shown in Table 15-14.

*Example*   Convert $254_8$ to its decimal equivalent.

*Solution*   The subscript, 8, in connection with 254 indicates that the number is in the octal system.

$$254_8 = (2 \times 8^2) + (5 \times 8^1) + (4 \times 8^0)$$
$$= (2 \times 64) + (5 \times 8) + (4 \times 1)$$
$$= 128 + 40 + 4$$
$$= 172_{10}$$

### Table 15-14. Powers of 8.

| Power | | Value | Power | | Value |
|-------|---|-------|-------|---|-------|
| $8^0$ | = | 1 | | | |
| $8^1$ | = | 8 | $8^{-1}$ | = | 0.125 |
| $8^2$ | = | 64 | $8^{-2}$ | = | 0.015 6 |
| $8^3$ | = | 512 | $8^{-3}$ | = | 0.001 95 |
| $8^4$ | = | 4 096 | $8^{-4}$ | = | 0.000 244 |
| $8^5$ | = | 32 768 | | | |
| $8^6$ | = | 262 144 | | | |

Table 15-15 is a listing of decimal and octal equivalents from decimal 0 to decimal 100. Table 15-16 is a listing of decimal exponential values in octal, ranging from $10^0$ to $10^{10}$. As in the decimal system, various mathematical constants are

### Table 15-15. Decimal-to-octal conversion.

| Decimal | Octal | Decimal | Octal | Decimal | Octal | Decimal | Octal |
|---------|-------|---------|-------|---------|-------|---------|-------|
| 00 | 00 | 26 | 32 | 51 | 63 | 76 | 114 |
| 01 | 01 | 27 | 33 | 52 | 64 | 77 | 115 |
| 02 | 02 | 28 | 34 | 53 | 65 | 78 | 116 |
| 03 | 03 | 29 | 35 | 54 | 66 | 79 | 117 |
| 04 | 04 | 30 | 36 | 55 | 67 | 80 | 120 |
| 05 | 05 | 31 | 37 | 56 | 70 | 81 | 121 |
| 06 | 06 | 32 | 40 | 57 | 71 | 82 | 122 |
| 07 | 07 | 33 | 41 | 58 | 72 | 83 | 123 |
| 08 | 10 | 34 | 42 | 59 | 73 | 84 | 124 |
| 09 | 11 | 35 | 43 | 60 | 74 | 85 | 125 |
| 10 | 12 | 36 | 44 | 61 | 75 | 86 | 126 |
| 11 | 13 | 37 | 45 | 62 | 76 | 87 | 127 |
| 12 | 14 | 38 | 46 | 63 | 77 | 88 | 130 |
| 13 | 15 | 39 | 47 | 64 | 100 | 89 | 131 |
| 14 | 16 | 40 | 50 | 65 | 101 | 90 | 132 |
| 15 | 17 | 41 | 51 | 66 | 102 | 91 | 133 |
| 16 | 20 | 42 | 52 | 67 | 103 | 92 | 134 |
| 17 | 21 | 43 | 53 | 68 | 104 | 93 | 135 |
| 18 | 22 | 44 | 54 | 69 | 105 | 94 | 136 |
| 19 | 23 | 45 | 55 | 70 | 106 | 95 | 137 |
| 20 | 24 | 46 | 56 | 71 | 107 | 96 | 140 |
| 21 | 25 | 47 | 57 | 72 | 110 | 97 | 141 |
| 22 | 26 | 48 | 60 | 73 | 111 | 98 | 142 |
| 23 | 27 | 49 | 61 | 74 | 112 | 99 | 143 |
| 24 | 30 | 50 | 62 | 75 | 113 | 100 | 144 |
| 25 | 31 | | | | | | |

## Table 15-16. Decimal exponential values in octal.

| Decimal | Octal 11 9 8-8 | 8 6 8-8 | 5 3 8-8 | 2 0 8-8 |
|---|---|---|---|---|
| $10^0$ | 000 | 000 | 000 | 000 |
| $10^1$ | 000 | 000 | 000 | 012 |
| $10^2$ | 000 | 000 | 000 | 144 |
| $10^3$ | 000 | 000 | 0 001 | 750 |
| $10^4$ | 000 | 000 | 023 | 420 |
| $10^5$ | 000 | 000 | 303 | 240 |
| $10^6$ | 000 | 003 | 641 | 100 |
| $10^7$ | 000 | 046 | 113 | 200 |
| $10^8$ | 000 | 575 | 360 | 400 |
| $10^9$ | 007 | 346 | 545 | 000 |
| $10^{10}$ | 112 | 402 | 762 | 000 |

used in the octal system (Table 15-16). It isn't necessary (although it is sometimes helpful) to go through the decimal system in converting from one system to another. Table 15-17 is a short listing of octal to hexadecimal conversions. Table

## Table 15-17. Octal-to-hexadecimal conversion.

| Octal | Hexadecimal | Octal | Hexadecimal |
|---|---|---|---|
| 00 | 0 | 10 | 8 |
| 01 | 1 | 11 | 9 |
| 02 | 2 | 12 | A |
| 03 | 3 | 13 | B |
| 04 | 4 | 14 | C |
| 05 | 5 | 15 | D |
| 06 | 6 | 16 | E |
| 07 | 7 | 17 | F |

15-18 consists of the conversion of octal fractions to decimal fractions.

One of the basic differences in the various numbering systems is in the place values. In the decimal system, the rightmost number is in the units column, the next number to its left is in the 10s column, and the one to the left of that in the 100s column. Table 15-19 lists place values for the decimal system, the duodecimal, octal, and binary systems.

*Table 15-18. Conversion of octal fractions to decimal fractions.*

| Octal | Decimal | Octal | Decimal |
|---|---|---|---|
| 0.000000 | 0.000000 | 0.000025 | 0.000080 |
| 0.000001 | 0.000003 | 0.000026 | 0.000083 |
| 0.000002 | 0.000007 | 0.000027 | 0.000087 |
| 0.000003 | 0.000011 | 0.000030 | 0.000091 |
| 0.000004 | 0.000015 | 0.000031 | 0.000095 |
| 0.000005 | 0.000019 | 0.000032 | 0.000099 |
| 0.000006 | 0.000022 | 0.000033 | 0.000102 |
| 0.000007 | 0.000026 | 0.000034 | 0.000106 |
| 0.000010 | 0.000030 | 0.000035 | 0.000110 |
| 0.000011 | 0.000034 | 0.000036 | 0.000114 |
| 0.000012 | 0.000038 | 0.000037 | 0.000118 |
| 0.000013 | 0.000041 | 0.000040 | 0.000122 |
| 0.000014 | 0.000045 | 0.000041 | 0.000125 |
| 0.000015 | 0.000049 | 0.000042 | 0.000129 |
| 0.000016 | 0.000053 | 0.000043 | 0.000133 |
| 0.000017 | 0.000057 | 0.000044 | 0.000137 |
| 0.000020 | 0.000061 | 0.000045 | 0.000141 |
| 0.000021 | 0.000064 | 0.000046 | 0.000144 |
| 0.000022 | 0.000068 | 0.000047 | 0.000148 |
| 0.000023 | 0.000072 | 0.000050 | 0.000152 |
| 0.000024 | 0.000076 | 0.000051 | 0.000156 |

*Table 15-19. Place values in various number systems.*

| Decimal | | Duodecimal | | Octal | | Binary | | | | |
|---|---|---|---|---|---|---|---|---|---|---|
| tens | ones | twelves | ones | eights | ones | sixteens | eights | fours | twos | ones |
| | 0 | | 0 | | 0 | | | | | 0 |
| | 1 | | 1 | | 1 | | | | | 1 |
| | 2 | | 2 | | 2 | | | | 1 | 0 |
| | 3 | | 3 | | 3 | | | | 1 | 1 |
| | 4 | | 4 | | 4 | | | 1 | 0 | 0 |
| | 5 | | 5 | | 5 | | | 1 | 0 | 1 |
| | 6 | | 6 | | 6 | | | 1 | 1 | 0 |
| | 7 | | 7 | | 7 | | | 1 | 1 | 1 |

*Table 15-19. Continued.*

| Decimal | | Duodecimal | | Octal | | Binary | | | | |
|---|---|---|---|---|---|---|---|---|---|---|
| | 8 | | 8 | 1 | 0 | | 1 | 0 | 0 | 0 |
| | 9 | | 9 | 1 | 1 | | 1 | 0 | 0 | 1 |
| 1 | 0 | | A | 1 | 2 | | 1 | 0 | 1 | 0 |
| 1 | 1 | | B | 1 | 3 | | 1 | 0 | 1 | 1 |
| 1 | 2 | 1 | 0 | 1 | 4 | | 1 | 1 | 0 | 0 |
| 1 | 3 | 1 | 1 | 1 | 5 | | 1 | 1 | 0 | 1 |
| 1 | 4 | 1 | 2 | 1 | 6 | | 1 | 1 | 1 | 0 |
| 1 | 5 | 1 | 3 | 1 | 7 | | 1 | 1 | 1 | 1 |
| 1 | 6 | 1 | 4 | 2 | 0 | 1 | 0 | 0 | 0 | 0 |
| 1 | 7 | 1 | 5 | 2 | 1 | 1 | 0 | 0 | 0 | 1 |
| 1 | 8 | 1 | 6 | 2 | 2 | 1 | 0 | 0 | 1 | 0 |
| 1 | 9 | 1 | 7 | 2 | 3 | 1 | 0 | 0 | 1 | 1 |
| 2 | 0 | 1 | 8 | 2 | 4 | 1 | 0 | 1 | 0 | 0 |
| 2 | 1 | 1 | 9 | 2 | 5 | 1 | 0 | 1 | 0 | 1 |
| 2 | 2 | 1 | A | 2 | 6 | 1 | 0 | 1 | 1 | 0 |
| 2 | 3 | 1 | B | 2 | 7 | 1 | 0 | 1 | 1 | 1 |
| 2 | 4 | 2 | 0 | 3 | 0 | 1 | 1 | 0 | 0 | 0 |

# 16
CHAPTER

# Electronic logic

All number systems are concerned with quantities. Although letters are used in algebra as an aid in problem solving, the letters must ultimately be converted to numeric values. All number systems and algebra are related. They have been developed as specialized tools; it is not only possible to convert from one number system to another, it is frequently necessary to do so.

The mathematical operations performed by a computer are handled by electronic circuits. Number systems do not tell us anything about these circuits. They enable you to input data and get certain results expressed in mathematical form. To describe the circuits and how they work would require that you draw the diagram and supply an explanation.

The simplest kind of circuit is the on−off switch, whose diagram is shown in Fig. 16-1. When the switch arm is up, the circuit is open; when it is down, it is closed. This is a common type of switch used for turning lights on and off in a home, a mechanical type of switch with an extremely low operating speed. A computer switch is electronic and is capable of moving from on, to off, to on millions of times in a second.

## Boolean algebra

There are many other kinds of switches in computers, and although it is easy to draw one of them, the problem becomes complex when a large number of switches are involved, especially since the switches are interdependent. A mathematical system, devised by George Boole in 1854, can be used to describe these circuits and to indicate the way they interact, without the need for drawing elaborate and involved diagrams. In ordinary algebra, letters are used to represent quantities. In Boolean algebra, letters are used to represent switches. Boolean algebra can be used advantageously for the design of computer switching circuits.

# Switching-circuit arrangements

Two or more switches arranged as shown in Fig. 16-2 form a serial (or series) circuit. The switches are in series, and any electrical pulse that moves through one of the switches will also move through the other, assuming both are closed. Figure 16-2 shows just two switches in series, but any number of switches can be connected.

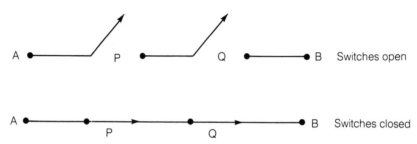

*Fig. 16-2.  Both switches (P AND Q) must be closed before an electrical pulse can travel from A to B.*

## The AND condition

If an electrical pulse is to travel from point A to point B, it must pass through a conducting wire from A to switch P, through switch P, along the conducting wire to switch Q, through switch Q, and finally to point B. This is possible only if switch P *and* switch Q are closed, as shown in the lower drawing in Fig. 16-2. This is referred to as an *AND* circuit. An AND circuit can consist of two or more series-connected switches. The word *series* refers to the fact that the pulse travels serially through the switches, that is, through the first switch (P) and then through the second (Q). AND circuits are series circuits.

## The OR condition

In the circuit arrangement of Fig. 16-3, it is not necessary that both switches be closed. If switch P is closed, then an electrical pulse has a path of travel from point A to point B. Figure 16-4 shows the two possible switching arrangements for the passage of pulses from A to B. Because either switch P *or* Q can be used to supply a connecting path for the electrical pulse, the arrangement is called an *OR* circuit. OR circuits are parallel connected (also known as shunt connected) and can consist of two or more switches.

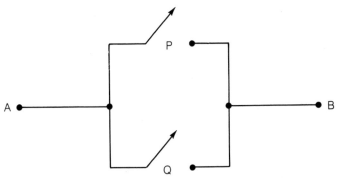

*Fig. 16-3. Either switch P OR Q must be closed for an electrical pulse to travel from A to B.*

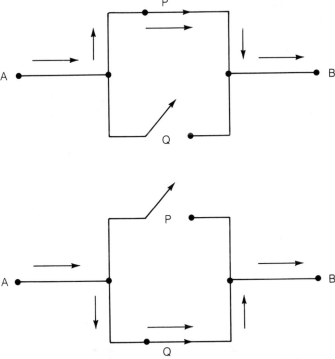

*Fig. 16-4. Electrical pulses can travel through switch P OR switch Q, depending on which is closed. The arrows indicate the path of travel of the pulses.*

## The AND−OR condition

Finally, switches can be arranged in series—parallel, as in Fig. 16-5. Again, this represents a simplification, because circuit arrangements are usually much more complex. Various combinations of series and parallel switches can take many different forms.

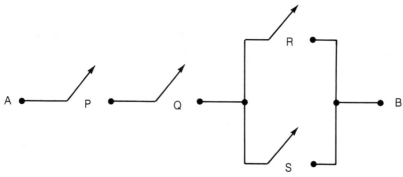

Fig. 16-5. For an electrical pulse to travel from A to B, both series switches must be closed and either one of the parallel switches must also be closed.

The switching arrangement in Fig. 16-5 is an *AND−OR* circuit, because switches P AND Q must be closed and switch R OR S must be closed for an electrical pulse to travel from A to B.

### The 0 or 1 state

When a single switch is open, no electrical pulses can pass through it. This open-circuit condition is called the *0 state*. A closed switch is represented by the *1 state*. These two conditions are shown in Fig. 16-6.

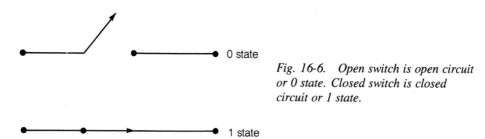

Fig. 16-6. Open switch is open circuit or 0 state. Closed switch is closed circuit or 1 state.

### Gates

Switches, or switching systems, are sometimes known as *gates*. An open gate allows a signal to pass; a closed gate means the signal cannot go through. An open gate corresponds to a closed switch; a closed gate corresponds to an open switch. Gates may be mechanical, electromechanical, or electronic. Lighting switches used in the home are *mechanical switches*. Operating switches on computer equipment panels can also be mechanical. These switches require some sort of human finger action to operate.

An *electromechanical switch* is one that is closed or opened by a flow of current through some switch-activating device. A relay is an electromechanical switch. When an electrical current flows through a coil, the coil is made into an electromagnet capable of attracting and holding a switch blade. When the current stops flowing, a spring pulls the switch back into its original position.

An *electronic switch* can be either a vacuum tube, a gas-filled tube, or a semiconductor. Semiconductors include diodes and transistors, similar to those used in home radio and television receivers. Diodes (two-element semiconductors) and transistors (three elements or more) can be used as electronic switches or gates. They have the advantage of small size, light weight, no moving parts, and the ability to operate at tremendously high speeds.

## Operation symbols

Just as symbols are used in ordinary algebra to represent some quantity, so are symbols used in connection with switches to indicate some kind of operation. What looks like an upside-down vee ($\wedge$) represents the word *and*. The word *or* is indicated by $\vee$. Negation is shown by a line drawn over the symbol negated, and equivalence by three parallel lines ($\equiv$).

To see how these operation symbols are used, refer once again to Fig. 16-2. Here are two series gates. Both gates must be open—that is, both switches must be closed—before an electrical pulse (an electron current) can move from A to B. Using symbolic terminology, you could describe this circuit as $P \wedge Q$.

In the parallel circuit shown in Fig. 16-3, is another pair of gates. In this switching arrangement, an electrical pulse can travel from A to B either through gate P or through gate Q, depending upon which of the gates is open or closed. To describe the fact that an electrical pulse could travel through either P or Q, you can state $P \vee Q$.

The circuits of Figs. 16-2 and 16-3 can be combined to form the series–parallel circuit of Fig. 16-5. The series group and the parallel group can now be regarded as subcircuits, with the combination referred to as a *total circuit*. When more than one subcircuit is present, the letters representing the gates are ordinarily enclosed in parentheses. For an electrical pulse to travel from A or B in the series–parallel circuit of Fig. 16-5, $(P \wedge Q) \wedge (R \vee S)$. This means that P and Q gates must both be opened as well as either R or S.

## Circuit simplification

The fewer the number of circuits in a computer, the lower is its manufacturing cost and the less chance there is of *downtime*, time during which the computer is inoperative for repair. Boolean algebra can be used as a tool for circuit simplification. As an example, examine the series–parallel arrangement of Fig. 16-7. Here a pair of series gates, P and Q, are in parallel with another pair of series gates, P and R. An electrical impulse can travel from A to B either through the upper

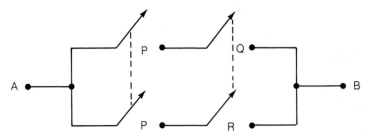

Fig. 16-7.   *In this circuit, a pair of series gates (P AND Q) is in parallel with another pair of series gates (P AND R).*

series gates or through the lower series gates. The Boolean statement is:

$$(P \wedge Q) \vee (P \wedge R)$$

In this Boolean statement, which can be read as P and Q or P and R, P can be factored, for it appears in both subcircuits. If you factor P:

$$P \wedge (Q \vee R)$$

Read this statement as P and Q or R. If read aloud, pauses are often used where parenthesis indicate groupings. In the simplified statement there is just one gate, P. You can then draw the gating circuit as shown in Fig. 16-8.

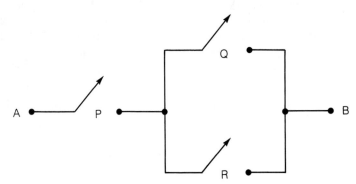

Fig. 16-8.   *Simplification of the circuit of Fig. 16-7. Gate P is now common to both parallel gates, Q and R.*

In the computer either the circuit of Fig. 16-7 or that shown in Fig. 16-8 can be used. The two circuits can be made to perform equivalent functions; hence, the Boolean statements describing these gates are also equivalent. To show this equivalency, you can write

$$(P \wedge Q) \vee (P \wedge R) \equiv P \wedge (Q \vee R)$$

The gating circuits shown in the previous illustrations are fundamental types. A more elaborate arrangement is shown in Fig. 16-9. Here there are a total of 12 gates in a series—parallel arrangement. Switches A, B, and C are in series, but

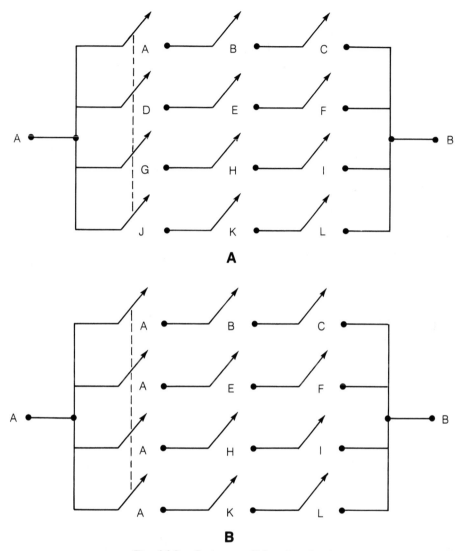

Fig. 16-9.  *Series-parallel gating circuit.*

this series subcircuit is in parallel with switches D, E, and F. In turn, D, E, and F form a series subcircuit in parallel with switches G, H, I, J, K, and L.

An electrical impulse travelling toward the gates from point A has four possible paths to follow. It can move through gates A, B, and C, provided these gates are open. Or it can move through gates D, E, and F, or through G, H, and I or, finally, through gates J, K, and L. The only requirement is that one of these subcircuits should consist of three open gates. Set this up as a Boolean expression this way:

$$(A \wedge B \wedge C) \vee (D \wedge E \wedge F) \vee (G \wedge H \wedge I) \vee (J \wedge K \wedge L)$$

If you examine gates A, D, G, and J, you will see that each of these is a series switch for its own particular subcircuit. Since they all perform identical subcircuit gating functions, we can identify them with the same letter (A), as shown in Fig. 16-9B. The Boolean expression is the same except that some letters have been changed:

$$(A \wedge B \wedge C) \vee (A \wedge E \wedge F) \vee (A \wedge H \wedge I) \vee (A \wedge K \wedge L)$$

The reason for the change of letter becomes obvious when you examine the expression. Because the letter A is common to all the subcircuits—that is, the letter A appears in all the subcircuit descriptions—it can be factored, and the expression can be simplified into this form:

$$A \wedge ((B \wedge C) \vee (E \wedge F) \vee (H \wedge I) \vee (K \wedge L))$$

The interpretation of this expression is that an electrical impulse can flow from point A to point B by passing through switches

1. A and B and C; or
2. A and E and F; or
3. A and H and I; or
4. A and K and L.

Because you have now factored the original Boolean statement, you can set up the original and the factored form as an equivalency statement:

$$(A \wedge B \wedge C) \vee (A \wedge E \wedge F) \vee (A \wedge H \wedge I) \vee (A \wedge K \wedge L)$$
$$\equiv A \wedge ((B \wedge C) \vee (E \wedge F) \vee (H \wedge I) \vee (K \wedge L))$$

Because the two Boolean statements are equivalent, read either the same way:

A and B and C OR A and E and F OR A and H and I OR A and K and L

## The OR symbols

The OR symbol is frequently represented by what looks like the letter vee ( $\vee$ ) but sometimes a plus sign is used in its place. Boolean OR terms can be written as

$$A \vee B$$
$$A + B$$

Both terms are read as A OR B.

## Interpreting Boolean statements

The symbol $\wedge$ or the word *and* always means a series circuit. A statement such as A $\wedge$ B $\wedge$ C means you have three switches in series, three because you have three letters identifying the switches, and in series because the AND symbol is used to associate the letters.

The symbol ∨ or the word *or* always means a parallel circuit. A statement such as A ∨ B means we have two switches, two because you have two identifying letters, and parallel because you have used the OR symbol, ∨. OR and AND symbols can be used in a variety of combinations to supply series–parallel circuits.

Return to the earlier Boolean statement obtained by factoring:

$$A \wedge ((B \wedge C) \vee (E \wedge F) \vee (H \wedge I) \vee (K \wedge L))$$

To interpret this Boolean statement, follow these steps.

1. There are four terms in parentheses. This means there are at least four subcircuits.
2. There is one term not in parentheses. This single term indicates the presence of another subcircuit. We now have a total of $1 + 4 = 5$ subcircuits.
3. Examine the number of letters in each subcircuit term in parentheses. In the example given, there are two for each subcircuit. This means that each such subcircuit contains two gates.
4. Examine the symbol connecting the letters inside the parentheses. In the example given, the symbol is ∧. This AND symbol means that the subcircuit is a series circuit. Since the same symbol is used in each subcircuit, they are all series circuits.
5. Finally, examine the letter not enclosed in parentheses. It is a single letter and so represents a single gate. It is joined to the other subcircuits by the AND symbol, ∧, and so is in series with them.

From the information given, it is now possible to draw the simplified gating circuit. It is shown in Fig. 16-10. The advantage of the arrangement in Fig. 16-10 is that it is simpler than the original circuit of Fig. 16-9. Three gates have been eliminated. However, there is still another advantage. Go back to Fig. 16-9 for a moment and imagine an electrical pulse travelling from point A over to the four switches marked A in each of the subcircuits. Assume the top switch, A, is closed. The pulse can continue along through B and C, but only if B and C are also closed. If switches K and L are closed instead, the pulse has no way of getting through to point B, even though three switches are closed. In Fig. 16-10, if switch A is closed, then if any of the series combinations such as B, C, or E, F, or H, I, or K, L are also closed, the pulse will reach B. Switch A is now common—that is, it is a path used by all the other switches.

*Example* Assume that you are given the following Boolean expression and are required (1) to draw the original circuit and (2) to factor the Boolean expression to supply a simplified circuit. The advantage of Boolean algebra is that the

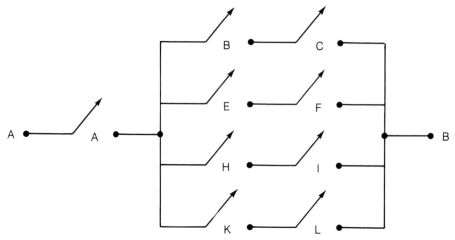

*Fig. 16-10. Simplification of the circuit of Fig. 16-9.*

circuits need not be drawn or constructed initially. All the work can be done, and then the final circuit can be obtained from the results of simplifying the original Boolean statement. The original circuit can be drawn, if desired, to supply a visual comparison between the two circuits. For very complex gating circuits, it is often convenient to avoid drawing the original circuit and to concentrate on obtaining the simplified expression:

$$(A \wedge C \wedge D) \vee (A \wedge B \wedge D) \vee (A \wedge B \wedge E)$$

**Solution**    This Boolean statement reveals a number of facts about the circuit. It consists of three subcircuits, and each of these subcircuits is a series arrangement because of the AND sign between each of the letters. The three subcircuits are connected in parallel because of the OR sign existing between each of the expressions in parentheses. Finally, one of the switches, A, can be factored out of each subcircuit; and another switch, B, can be factored out of two of them.

Starting with switch A, and factoring A,

$$(A \wedge C \wedge D) \vee (A \wedge B \wedge D) \vee (A \wedge B \wedge E)$$
$$\equiv A \wedge ((C \wedge D) \vee (B \wedge D) \vee (B \wedge E))$$

Now factor B out of the remaining two statements and get

$$A \wedge ((C \wedge D) \vee (B \wedge (D \vee E)))$$

To compare the two circuits, examine the original statement and, following its instructions, draw the circuit. It will appear as Fig. 16-11A. Then, using the final Boolean statement obtained, draw the circuit as in Fig. 16-11B.

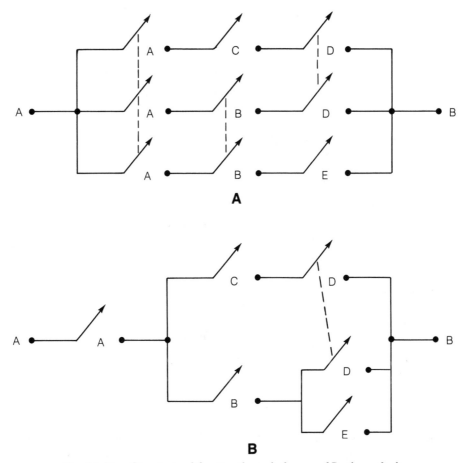

*Fig. 16-11. Circuit simplification through the use of Boolean algebra.*

## Truth table (OR operation)

Assume a simple switching circuit such as that indicated by the two switches in Fig. 16-12. Because the switches are in parallel, it is an OR circuit, because a pulse can travel to the output (point C) if switch A is closed OR if switch B is closed. If, as indicated earlier, an open switch can be represented by 0 and a closed switch by 1, it becomes possible to set up a table showing the effect of opening and closing the switches. If both switches are open (0), there is no output pulse. The condition at point C is equivalent to 0. However, if either switch A or B (1 or 1) is closed, the condition at the output, point C is 1.

To understand this more completely, imagine that both switches, A and B, control a light bulb at point C. If a lighted bulb is represented by 1, then an unlit bulb is equivalent to 0. It is possible to have light if both A and B are closed, OR if

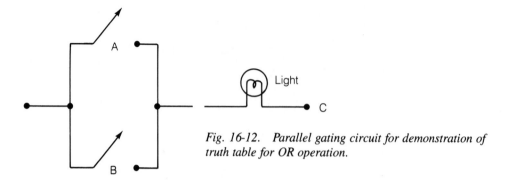

Fig. 16-12. *Parallel gating circuit for demonstration of truth table for OR operation.*

either are closed. An arrangement of all possible switching conditions and the results is known as a *truth table* for the OR operation (Table 16-1).

The truth table for the OR operation tells you, in its first row, that if switches A and B are open, the output is 0. The remaining three rows indicate the other

| A | + | B | = | C |
|---|---|---|---|---|
| 0 | | 0 | | 0 |
| 0 | | 1 | | 1 |
| 1 | | 0 | | 1 |
| 1 | | 1 | | 1 |

*Table 16-1.*
*OR truth*
*table.*

switching possibilities. If switch B is closed (row 2) or if switch A is closed (row 3) or if both switches A or B are closed (row 4), then a 1 condition will exist at point C. The symbol + shown in the table is equivalent to the OR symbol. The statement across the top of the table reads A OR B equals C.

## Truth table (AND operation)

In Fig. 16-13, there is another switching circuit consisting of two switches, A and B, connected in series. If a pulse is to reach point C, both switches must be closed. To operate a light bulb at point C, both switches must be in their closed position before the light can go on. An on light at C can be represented by 1; an off light by 0. A truth table for the AND operation can be set up as in Table 16-2.

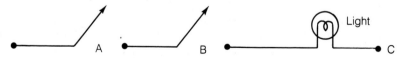

Fig. 16-13. *Series gating circuit for demonstration of truth table for AND operation.*

**340**  *Chapter 16*

| A ∧ B = C | | |
|---|---|---|
| 0 | 0 | 0 |
| 0 | 1 | 0 |
| 1 | 0 | 0 |
| 1 | 1 | 1 |

*Table 16-2.*
*AND truth*
*table.*

The truth table shows that there is no output at C, unless both switches A and B are closed.

## AND function addition table

In a series arrangement as in Fig. 16-13, both switches A AND B must be closed for a pulse to be present at point C. You can adapt the truth table for the AND operation and develop an AND function addition table. The AND symbol, earlier represented by ∧, is sometimes represented in Boolean algebra as a dot, somewhat resembling a radix point, but placed halfway up between the letters. Instead of writing A ∧ B, you can write A·B. Both representations mean A AND B. Sometimes the dot, known as a connective, is omitted, and the letters are placed immediately adjacent. AB means A AND B. It is important not to confuse this with ordinary algebraic operations in which AB means A multiplied by B. For AND operation, the statement can be made in four different forms:

$$A \wedge B = C$$

$$A \cdot B = C$$

$$AB = C$$

$$A \times B = C$$

The terms *series* and *parallel* are quite common, but are not always used. The word *shunt* is synonymous with parallel. The word OR is often used in place of parallel, and the word AND in place of series. A set of switches placed in parallel is said to be ORed; those put in series are said to be ANDed.

## Gating symbols

Drawing switches working in gating functions can become tedious, especially if a large number of switches is involved. Simplified symbols are used for both OR and AND gates and are shown in Fig. 16-14. Drawing (A) is that of an OR gate and consists of two switches A and B, in parallel. Drawing (B) is that of an AND gate comprising two switches in series. Note that the shapes of these symbols are

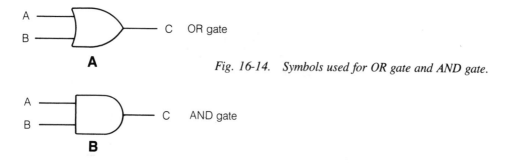

Fig. 16-14. Symbols used for OR gate and AND gate.

different. In the OR symbol, all three sides are curved. The AND symbol has one side that is straight.

# Logic

Arithmetic is such an interwoven part of living it is sometimes difficult to analyze the automatic thinking that goes into the setup of a problem in arithmetic. Problems using arithmetic, algebraic, or Boolean symbols are derived from statements consisting of words. If you want to calculate the area of a room, you measure its length and its width. You can state: length times width equals area. But this statement is not in a form that will allow you to do any calculations. It is only when you use the algebraic process of allowing literals to represent length and width that you are able to set up a formula that will permit arithmetic. The statement length times width equals area is then expressed as

$$A = L \times W$$

These letters represent values, and when the values are substituted for the letters, the final step in arithmetic can be taken.

The statement, "I have a car and I also have the key." could be considered an AND statement. The car is essential, but so is the key. One is useless without the other. An OR statement allows an alternative; an AND statement does not.

The pulse output of an AND or OR statement is 1 or positive. You can also have a negative condition. A negative AND output is called *NOT AND* and is sometimes referred to as *NAND*. A negative OR output is called *NOT OR*, but is often abbreviated *NOR*.

# The NOR function

An OR function can be represented by a pair of switches in parallel, as in circuit (A) in Fig. 16-15. To operate, either switch A or switch B must be in the 1 position—that is, they must be closed. According to the OR function addition table, either $1+0=1$, or $0+1=1$, or $1+1=1$, another way of saying that switch A or switch B (or both) must be closed for a pulse to pass through. A condition in which both

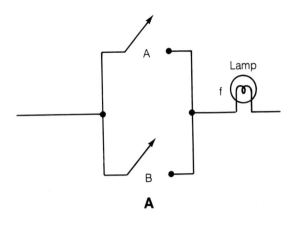

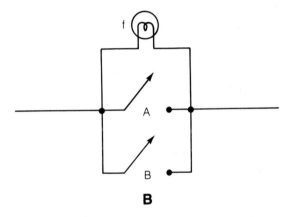

*Fig. 16-15. In drawing (A), either switch A OR B must be closed for an electrical pulse to light the lamp at point f. In drawings (B) and (C), all switches must be open for the lamp to light.*

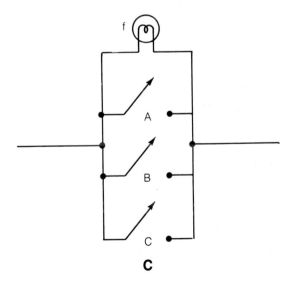

switches must be open to produce a given result is called a NOR function—that is, the behavior of the circuit is exactly opposite that of an OR arrangement.

To see how this is possible, examine drawing (B) in Fig. 16-15. Here you have the same pair of switches, but now there is a lamp, $f$, shunted across the switches. An electrical pulse will not go through the switches if they are open, but will go through $f$. You can represent an unlighted lamp by 0, and a lighted lamp by 1. Using 0 for an open switch and 1 for a closed switch, we would have

$$0 \wedge 0 = 1$$

Compare this with the results obtained in the OR function addition table. That table states that $0 \wedge 0 = 0$.

Although Figs. 16-15A and B show only two switches, you could also have three or more. However, no electrical pulse will pass through component $f$, a lamp in this case, unless all the switches in Fig. 16-15C are open.

Table 16-3 is a truth table for the NOR operation based on the three-switch circuit of Fig. 16-15C.

| A | v | B | v | C | = | f |
|---|---|---|---|---|---|---|
| 0 | | 0 | | 0 | | 1 |
| 0 | | 0 | | 1 | | 0 |
| 0 | | 1 | | 0 | | 0 |
| 1 | | 0 | | 0 | | 0 |
| 0 | | 1 | | 1 | | 0 |
| 1 | | 0 | | 1 | | 0 |
| 1 | | 1 | | 0 | | 0 |
| 1 | | 1 | | 1 | | 0 |

*Table 16-3. NOR truth table.*

The symbol for a NOR circuit resembles that for an OR circuit, but is modified as shown in Fig. 16-16. Algebraically, you can describe this circuit as

$$f = \overline{A + B + C}$$

The horizontal line drawn above the letters on the right side is called an overline or vinculum. The terms overlined are said to be negated. If A is positive, then you can express negated A as $\overline{A}$. $\overline{A}$ is usually read as *Not A*. If $A = 1$, then $\overline{A} = 0$. Note that A and $\overline{A}$ are opposites.

# The NOT operation

It is possible to arrange two switches in series or in parallel, such that one switch is open and the other is closed. If the open switch is 0 and the closed switch is 1, then you can also designate these two switches as A and not A, or A and $\overline{A}$. Fig-

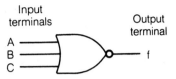

*Fig. 16-16.* *Symbol for NOR operation.*

ure 16-17 shows the possible settings of these switches. In drawing Fig. 16-17A the first switch is closed and is designated as A. The second switch, connected in series with the first, is not closed, and because it is open is in the 0 condition. Its condition is also exactly opposite that of the first switch, and so you can designate it as $\overline{A}$.

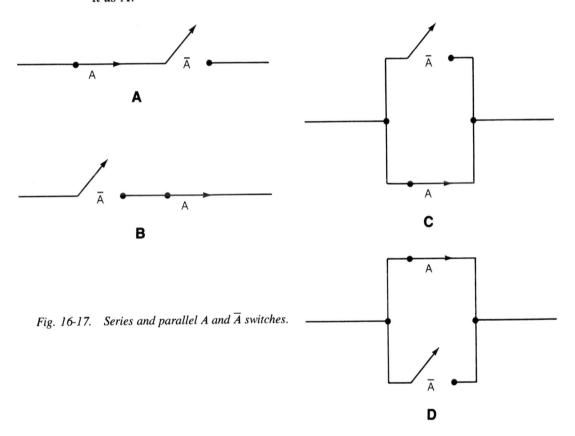

*Fig. 16-17.* *Series and parallel A and $\overline{A}$ switches.*

In drawing B, the two switches have changed roles. The first switch is now open and becomes $\overline{A}$. The second switch is closed and is identified as A. Drawings C and D show A and $\overline{A}$ for a parallel arrangement.

Because the NOT circuit is A or not A, the truth table (Table 16-4) is very simple. Figure 16-18 shows the symbol for NOT operation. The circle on a NOR or NAND gate represents the NOT operation.

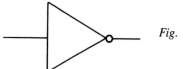

| **A** | **Ā** |
|-------|-------|
| 1 | 0 |
| 0 | 1 |

*Table 16-4.*
*NOT truth table.*

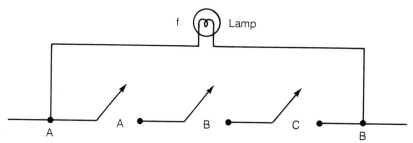

*Fig. 16-18. Symbol for NOT operation.*

# The NAND function

Figure 16-19 shows the circuit used to represent the NAND function. It consists of three switches in series, shunted by a lamp, *f*. The lamp can light if any one of the switches is open, or if they are all open. That is, *f*=1 when A and B and C are 0. Because *f* and A and B and C are opposites, the logical equation for the circuit is

*Fig. 16-19. Circuit used to represent the NAND function.*

$$f=\overline{ABC}$$

You can set up a truth table (Table 16-5) for the three switches shown in Fig. 16-19.

If you examine the first row of the truth table, you will see that *f* is on, or is 1, when switches A, B, and C are open or 0. In the last row, when all switches are closed, or 1, then no pulse can reach *f*, because the electrical impulse will take the path presented by the closed switches. Under these conditions, *f*=0. Figure 16-20 shows the symbol used for the NAND operation.

Note that this is not the same as $\overline{A} \wedge \overline{B} \wedge \overline{C}$ because this would only be 1 when all inputs were 0.

| A ∧ B ∧ C = f | | | |
| --- | --- | --- | --- |
| 0 | 0 | 0 | 1 |
| 0 | 0 | 1 | 1 |
| 0 | 1 | 0 | 1 |
| 0 | 1 | 1 | 1 |
| 1 | 0 | 0 | 1 |
| 1 | 0 | 1 | 1 |
| 1 | 1 | 0 | 1 |
| 1 | 1 | 1 | 1 |

*Table 16-5. NAND truth table.*

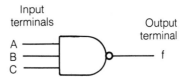

*Fig. 16-20. Symbol for NAND circuit.*

# Laws of Boolean algebra

Just as there are laws governing the manipulation of literals in algebra, there are laws in Boolean algebra. There are three such laws. One is the associative law, one is the commutative law, and one is the distributive law.

### Associative law

Letters connected by like signs may be bracketed in any manner. This means letters all separated by the AND symbol may be rebracketed. A similar statement holds for the OR symbol.

$$A \lor B \lor C = (A \lor B) \lor C = A \lor (B \lor C)$$

An inspection of this statement reveals that the circuit consists of three switches in parallel. The switches can be arranged in any order, provided they always remain in parallel. If you will refer to the diagram of Fig. 16-21, the associative law indicates that any parallel-connected switch may be interchanged with any other parallel-connected switch.

The associative law can also be stated as

$$A \land B \land C = (A \land B) \land C = A \land (B \land C)$$

This statement refers to three switches connected in series.

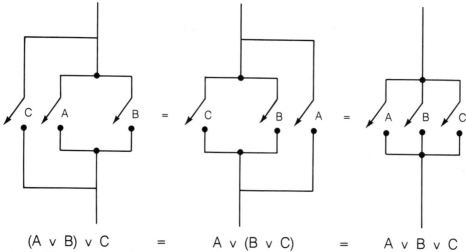

$$(A \lor B) \lor C \quad = \quad A \lor (B \lor C) \quad = \quad A \lor B \lor C$$

Fig. 16-21. *Associative law permits switches in OR condition to be arranged in any parallel combination.*

The law is *not* applicable in Boolean statements using different connectives. It would not, for example, be used for a statement such as A ∨ B ∧ C.

## Commutative law

The commutative law states that letters may be arranged (without grouping) in any desired manner, provided the same connective is used between letters. Thus:

$$A \land B \land C = B \land C \land A = C \land A \land B$$

This is shown in Fig. 16-22.

Both the associative and commutative laws can be used for a portion of a large Boolean expression, provided that portion is connected by similar signs. In a series of statements such as A ∧ B ∧ C ∨ D ∧ E ∨ F, the first three variables, A, B, and C can be grouped or commuted. When no parentheses are included, AND functions take precedence over OR functions.

## Distributive law

Factoring an ordinary algebraic expression is a way of simplifying an algebraic statement by showing that it is the product of simpler terms. Conversely, an ordinary algebraic statement can be expanded by multiplying pairs of terms. In Boolean algebra, an analogous procedure is used and is known as the distributive law. The distributive law is used only for the AND operation.

Consider the Boolean statement:

$$(A \land B) \lor (A \land C)$$

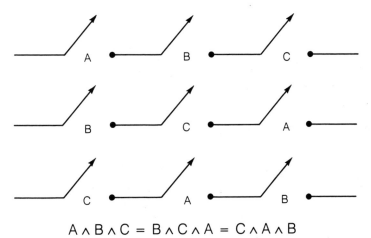

$$A \wedge B \wedge C = B \wedge C \wedge A = C \wedge A \wedge B$$

*Fig. 16-22.   Communitative law permits switches in series to be rearranged.*

You can also write this as:

$$AB + AC$$

In either case, however, the statement can be rearranged into $A(B+C)$. Similarly, a group such as $ABC + ABD$ can be rearranged into $AB(C+D)$.

If a group of terms is joined by the same connective and has a common variable, the variable can be factored, but only for the AND operation:

$$AB + BC + BD = B(A + C + D)$$

Note that

$$(A+B)(A+C)(A+D) \text{ does not equal } A+BCD$$

## Complements

Gates or switches can be designed so that they open or close simultaneously, or in phase. If a pair of switches operates in phase, they can be assigned the same letter. Customarily, however, some gates open in the computer at the moment other gates close, depending on the signal that reaches them. If a single signal opens one gate while closing another, or closes one gate while opening another, the two gates operate out of phase or complement each other. If an open gate equals 1, a closed gate equals 0. 1 and 0 are complements. If the open gate is identified as A, the closed gate is $\overline{A}$ (not A).

$$A = 1$$

$$\overline{A} = 0$$

# Identities

Boolean expressions such as $A=1$ and $\overline{A}=0$ are called identities. In ordinary algebra they would be referred to as equations.

Various other identities are possible. An expression such as

$$A+\overline{A}=1$$

can be read as A OR NOT A=1. The presence of the connective OR indicates the statement refers to a parallel circuit, such as the one shown earlier in Fig. 16-3. The Boolean expression $A+A=1$ means that the two switches are actuated by a single signal, the switches operate out of phase, and an electrical pulse will always be present at the output (point B in Fig. 16-3). In the design of computer circuitry, it is much easier and faster to write $A+A=1$ than to draw a parallel circuit with one open and one closed switch.

An expression such as:

$$AB+AC$$

can be factored to:

$$A(B+C)$$

Although the two statements are equivalent, they represent different circuit arrangements. But, although the circuits may be physically different (one might have more switches than the other), they both perform identical functions. In more complex Boolean statements, there can be a number of possible ways of writing equivalent statements. However, these can all be expressed by a single letter. Thus:

$$D=A(B+C)$$
$$=AB+AC$$

The complement of this statement is:

$$D=\overline{A(B+C)}$$

The vinculum shown above the term at the right has a double function. It is a grouping symbol and also indicates some function to be performed. In the statement, the vinculum indicates that all the literals on the right-hand side are to be treated as a group.

# Summary of logic diagrams and truth tables

Table 16-6 is a summary of logic diagrams and truth tables for AND/OR gates.

For simplification, this chapter uses switches for examples. The gates, AND, OR, NOR, and NAND are actually special electronic circuits that combine high and low digital pulses in the way necessary to have the outputs shown by the truth tables.

*Table 16-6. Summary of logic diagrams and truth tables for AND/OR gates.*

| AND | | OR | | Truth Tables | | |
|---|---|---|---|---|---|---|
| Gate | Equation | Gate | Equation | A | B | C |
| (AND gate) | $A \wedge B$ | (NOR gate) | $\overline{A} \vee \overline{B}$ | 0 0 0<br>0 1 0<br>1 0 0<br>1 1 1 | | |
| (NAND gate) | $\overline{A \wedge B}$ | (OR gate, inputs inverted) | $\overline{A} \vee \overline{B}$ | 0 0 1<br>0 1 1<br>1 0 1<br>1 1 0 | | |
| (AND, inputs inverted) | $\overline{A} \wedge \overline{B}$ | (OR gate) | $A \vee B$ | 0 0 0<br>0 1 1<br>1 0 1<br>1 1 1 | | |
| (NAND, inputs inverted) | $\overline{A} \wedge \overline{B}$ | (NOR gate) | $\overline{A \vee B}$ | 0 0 1<br>0 1 0<br>1 0 0<br>1 1 0 | | |
| (AND, A inverted) | $\overline{A} \wedge B$ | (NOR, B inverted) | $\overline{A \vee \overline{B}}$ | 0 0 0<br>0 1 1<br>1 0 0<br>1 1 0 | | |
| (AND, B inverted) | $A \wedge \overline{B}$ | (OR, A inverted) | $\overline{A} \vee B$ | 0 0 0<br>0 1 0<br>1 0 1<br>1 1 0 | | |
| (NAND, A inverted) | $\overline{\overline{A} \wedge B}$ | (OR, B inverted) | $A \vee \overline{B}$ | 0 0 1<br>0 1 0<br>1 0 1<br>1 1 1 | | |
| (NAND, B inverted) | $\overline{A \wedge \overline{B}}$ | (OR, A inverted) | $\overline{A} \vee B$ | 0 0 1<br>0 1 1<br>1 0 0<br>1 1 1 | | |

# 17

### CHAPTER

# Solid state

## Transistor alpha and beta

The current-amplification factor of a transistor can be expressed in terms of either alpha ($\alpha$) or beta ($\beta$). The ratio between a change in collector current for a change in emitter current is called alpha, and because alpha is less than unity, it is given as a decimal.

When emitter current changes, not only collector current changes but base current as well. The ratio of a change in collector current to a change in base current is also used as a measure of the amplification of a transistor and is represented by beta ($\beta$).

The relationship between alpha and beta is shown in these formulas:

$$\beta = \frac{\alpha}{1-\alpha}$$

$$\alpha = 1 - \left( \frac{1}{\beta+1} \right)$$

Because alpha is a decimal, it sometimes makes a problem awkward to handle. For this reason, it is sometimes more convenient to give current amplification as beta. Because beta is always in whole numbers it becomes convenient when comparing the current gains of different transistors. Table 17-1 supplies common values of alpha and corresponding values for beta. The range of alpha is from 0.5 to 0.996 4; that of beta is from 1 to 270.

*Example*   The (alpha) current gain of a transistor is 0.973 0. What is its current gain in terms of beta?

*Solution*   Table 17-1 shows that an alpha of 0.973 0 corresponds to a beta of 36.

*Table 17-1. Alpha versus Beta.*

| β | α | β | α | β | α |
|---|---|---|---|---|---|
| 1 | 0.500 0 | 41 | 0.976 2 | 81 | 0.987 8 |
| 2 | 0.666 6 | 42 | 0.976 7 | 82 | 0.988 0 |
| 3 | 0.750 0 | 43 | 0.977 3 | 83 | 0.988 1 |
| 4 | 0.800 0 | 44 | 0.977 8 | 84 | 0.988 2 |
| 5 | 0.833 3 | 45 | 0.978 2 | 85 | 0.988 4 |
| 6 | 0.857 1 | 46 | 0.978 6 | 86 | 0.988 5 |
| 7 | 0.875 0 | 47 | 0.979 2 | 87 | 0.988 6 |
| 8 | 0.888 9 | 48 | 0.979 6 | 88 | 0.988 8 |
| 9 | 0.900 0 | 49 | 0.980 0 | 89 | 0.988 9 |
| 10 | 0.909 1 | 50 | 0.980 4 | 90 | 0.989 0 |
| 11 | 0.916 7 | 51 | 0.980 8 | 91 | 0.989 1 |
| 12 | 0.923 1 | 52 | 0.981 1 | 92 | 0.989 2 |
| 13 | 0.928 6 | 53 | 0.981 5 | 93 | 0.989 4 |
| 14 | 0.933 3 | 54 | 0.981 8 | 94 | 0.989 5 |
| 15 | 0.937 5 | 55 | 0.982 1 | 95 | 0.989 6 |
| 16 | 0.941 2 | 56 | 0.982 5 | 96 | 0.989 7 |
| 17 | 0.944 4 | 57 | 0.982 8 | 97 | 0.989 8 |
| 18 | 0.947 4 | 58 | 0.983 1 | 98 | 0.989 8 |
| 19 | 0.950 0 | 59 | 0.983 3 | 99 | 0.990 0 |
| 20 | 0.952 4 | 60 | 0.983 6 | 100 | 0.990 1 |
| 21 | 0.954 5 | 61 | 0.983 9 | 101 | 0.990 9 |
| 22 | 0.956 5 | 62 | 0.984 1 | 120 | 0.990 7 |
| 23 | 0.958 3 | 63 | 0.984 4 | 125 | 0.992 1 |
| 24 | 0.960 0 | 64 | 0.984 6 | 130 | 0.993 1 |
| 25 | 0.961 5 | 65 | 0.984 8 | 140 | 0.993 2 |
| 26 | 0.963 0 | 66 | 0.985 1 | 150 | 0.993 3 |
| 27 | 0.964 3 | 67 | 0.985 3 | 160 | 0.993 8 |
| 28 | 0.965 5 | 68 | 0.985 5 | 170 | 0.994 2 |
| 29 | 0.966 7 | 69 | 0.985 7 | 180 | 0.994 5 |
| 30 | 0.967 7 | 70 | 0.985 9 | 190 | 0.994 8 |
| 31 | 0.968 8 | 71 | 0.986 1 | 200 | 0.995 2 |
| 32 | 0.969 7 | 72 | 0.986 3 | 210 | 0.995 4 |
| 33 | 0.970 6 | 73 | 0.986 5 | 220 | 0.995 6 |
| 34 | 0.971 4 | 74 | 0.986 7 | 230 | 0.995 8 |
| 35 | 0.972 2 | 75 | 0.986 8 | 240 | 0.996 0 |
| 36 | 0.973 0 | 76 | 0.987 0 | 250 | 0.996 2 |
| 37 | 0.973 7 | 77 | 0.987 2 | 260 | 0.996 3 |
| 38 | 0.974 4 | 78 | 0.987 3 | 270 | 0.996 4 |
| 39 | 0.975 0 | 79 | 0.987 5 | | |
| 40 | 0.975 6 | 80 | 0.987.7 | | |

*Example* What is the alpha of a transistor if its beta value is 76?

*Solution* Locate 76 in the column marked beta. To the right find the value of beta in terms of alpha, 0.987 0.

The values of alpha and beta can also be calculated using formulas. Obviously, you cannot change alpha without affecting beta.

*Example* A transistor has an alpha of 95%. What is the value of beta?

*Solution*

$$\beta = \frac{\alpha}{1-\alpha}$$

$$\beta = \frac{0.95}{1-0.95}$$

$$\beta = \frac{0.95}{0.05} = 19$$

*Example* A transistor has a beta of 45. What is the value of alpha?

*Solution*

$$\beta = \frac{\alpha}{1-\alpha}$$

$$45 = \frac{\alpha}{1-\alpha}$$

Multiply both sides of this equation by $1-\alpha$:

$$45(1-\alpha) = \alpha$$

$$45 - 45\alpha = \alpha$$

Transpose the $-45\alpha$ to the right-hand side:

$$45 = 45\alpha + \alpha \text{ or } 45 = 46\alpha$$

Dividing both sides of the equation by 46:

$$45/46 = \alpha \text{ or } \alpha = 0.978 \text{ or } 97.8\%$$

The current gain, alpha, of the common-base arrangement can also be expressed as:

$$\alpha = \frac{\Delta I_c}{\Delta I_e} \text{ (with } V_c \text{ constant)}$$

and the current gain of the common-emitter configuration, beta, can appear as:

$$\beta = \frac{\Delta I_c}{\Delta I_b} \text{ (with } V_c \text{ constant)}$$

# Power gain

The power gain is a comparison between the output power and the input power and is written as:

$$A_p = \frac{\Delta P_o}{\Delta P_i}$$

# Input resistance

The input resistance of a transistor can be determined by:

$$R_i = \frac{\Delta V_i}{\Delta I_i}$$

$R_i$ is the input resistance. The Greek letter delta indicates that the value is the quantity of change measured in the variable. $V_i$ is the voltage drop as measured between the base and emitter and $I_i$ is the base to emitter current, usually in microamperes.

The circuit in Fig. 17-1 is the technique used for measuring input resistance. The collector-emitter is open circuited. The amount of dc forward voltage, repre-

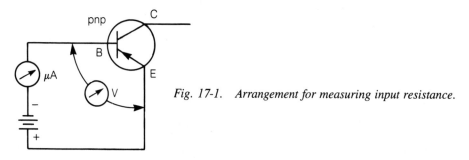

Fig. 17-1.  *Arrangement for measuring input resistance.*

sented by the battery, is the forward bias with its amount determined by the location of the operating point on the transistor load line. This value is used because it will determine the point at which the transistor will operate. Table 17-2 shows expected results when making resistance measurements.

# Output resistance

The output resistance is determined by:

*Table 17-2. Method for testing transistors.*

| Meter connection (polarity) | | | Resistance (×100 scale) | |
|---|---|---|---|---|
| **Emitter** | **Base** | **Collector** | **pnp type** | **npn type** |
| positive | negative | (none) | low | high |
| negative | positive | (none) | high | low |
| (none) | positive | negative | high | low |
| (none) | negative | positive | low | high |
| negative | (none) | positive | high | mid |
| positive | (none) | negative | mid | high |
| positive | to emitter | negative | high | |
| positive | to collector | negative | low | |
| negative | to emitter | positive | | high |
| negative | to collector | positive | | low |

$$R_o = \frac{\Delta V_o}{\Delta I_o}$$

$R_o$ is the output resistance, $V_o$ is the output (base to collector) voltage, and $I_o$ is the output current base to collector. Figure 17-2 shows the test method used for finding the value of the output resistance. As indicated in the drawing the base floats (remains disconnected). For transistors preceding an output transistor currents are

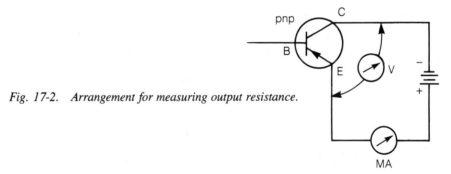

*Fig. 17-2. Arrangement for measuring output resistance.*

measured in microamperes and milliamperes. If the transistor is used as an output type, its current will be in the order of milliamperes and amperes.

Table 17-3 shows the resistances for various circuit configurations. These results, and the formulas listed earlier, are only for triode pnp and npn transistors. Table 17-4 lists their characteristics.

# Transistor functions

Numerous transistors exist. They can be categorized by function, by type, by an assigned number. A listing by function is broad based, as indicated in Table 17-5.

*Table 17-3. Transistor resistances.*

| Circuit | Input resistance | Output resistance |
|---|---|---|
| Common base | $R_e + R_b$ | $R_c + R_b$ |
| Common emitter | $R_b + R_e$ | $R_c + R_e$ |
| Common collector | $R_b + R_c$ | $R_e + R_c$ |

*Table 17-4. Basic transistor circuit characteristics.*

| Circuit schematic | Common base | Common emitter | Common collector |
|---|---|---|---|
| | | | |
| Power gain | Yes | Yes (highest) | Yes |
| Voltage gain | Yes (approx. same CE) | Yes | No (less than unity) |
| Current gain | No (less than unity) | Yes | Yes |
| Input impedance | Lowest (approx. 50 Ω) | Intermediate (approx. 1 kΩ) | Highest (approx. 300 kΩ) |
| Output impedance | Highest (approx. 1 MΩ) | Intermediate (approx. 50 kΩ) | Lowest (approx. 300 Ω) |
| Phase inversion | No | Yes | No |

# Transistor types

Numerous types have been developed since the introduction of the original transistor. Some of these had a brief popularity and then disappeared. Newer types sometimes consisted of older units that were revised and assigned a new name. The list in Table 17-6 consists of a mix of the old and the new, with some no longer available. In some instances transistors were developed to meet a special need, and disappeared when the need faded away.

Table 17-7 supplies the characteristics of some of the various transistor types. Figure 17-3 shows symbols for a limited number of transistors.

*Table 17-5. Transistor types according to function.*

**Functional transistor types**

Small input-signal transistors
Power transistors
Multifunction transistors
Special purpose transistors
High-frequency transistors
Audio transistors
Switching transistors

*Table 17-6. Transistor types.*

| | |
|---|---|
| Triode npn | Unijunction |
| Triode pnp | Bipolar |
| Tetrode | Intrinsic |
| Surface-barrier | Hook |
| Diffused | Multiheaded |
| Micro-alloy | Alloy junction |
| Junction | MOSFET |
| Field-effect (FET) | MNOS |
| Power | IGFET |
| JFET | VMOS |
| CMOS (COS/MOS) | Avalanche |
| Mesfet | GaAsFET |
| NMOS | |

*Table 17-7. Transistor types and their characteristics.*

| Type | Characteristics |
|---|---|
| Field-effect | Referred to as FET. Junction transistor. Highly sensitive. Capable of handling large signals. Used in receiver front ends, in logic circuits, in computer memory, and as amplifiers for electret and condenser microphones. Three-terminal device. Voltage operated. Terminals consist of a source, gate, and a drain. Can be used as an amplifier or switch. Does not load source input. |
| GaAsFET | Ga is the chemical symbol for gallium; As for arsenic. FET is an abbreviation for field-effect transistor. Gallium arsenide is the doping material. This transistor behaves like a vacuum tube since it is a voltage, not a current amplifier. Its linearity is good and it has stable impedance. Widely used in LNA (low-noise amplifiers) associated with satellite TV receivers. |
| JFET | Abbreviation for junction field-effect transistor. Available with either an n- or a p-channel, with two different symbols for these types. |
| MOSFET | Abbreviations for metal-oxide, field-effect transistor. Made of either n- or p-type semiconductor material. An N MOSFET is an npn arrangement; a P MOSFET |

*Table 17-7. Continued.*

| Type | Characteristics |
|---|---|
| | is a pnp. Unit has high-input impedance. Two types available, enhancement and enhancement-depletion. The enhancement MOSFET has two n types of semiconductor material in a p-type substrate. The gate-to-source voltage must have a minimum drain current of 10 $\mu$A. At voltages below this minimum, the MOSFET is cut off. |
| CMOS | Also known as COS/MOS. Utilizes a p-channel metal-oxide and an n-channel metal-oxide semiconductor. Known as a complementary symmetry MOS or COS/MOS. Consists of a pair of independent transistors, manufactured in a series arrangement, separated by a guard band. The unit is basically a FET. Uses little power during non-operating time. Applications include logic circuits in calculators and microprocessors. |
| MESFET | Do not confuse with MOSFET. This unit is a metal semiconductor FET using silicon on sapphire. Highly efficient and can function up to 3 GHz. Enhancement MESFET is identified as E-MESFET and depletion-mode type as D-MESFET. |
| IGFET | Insulated-gate FET with its gate electrically insulated from its channel by a layer of silicon dioxide. |
| NMOS | A negative channel metal-oxide semiconductor FET. |
| VMOS | Also called a vertical metal-oxide semiconductor, a vertical MOS, or a V-groove MOS. Works as a high-power capability FET. Built on a positive n substrate that functions as the drain. Characterized by a V-shaped groove etched through its epitaxial layer. Surface of its groove is covered with silicon dioxide. Source and gate connections are by aluminum metalization. |
| Avalanche | Consists of a pnp or npn modified for a specific function. When in the avalanche mode a large current passes through the transistor. Avalanche point is reached by reverse biasing. Can work as a switch with a small voltage controlling a large current flow. Behavior is similar to that of a zener diode. |
| Unijunction | Sometimes referred to as a UJT. Design is similar to double-based diode. Functions as a voltage-controlled switch or as an oscillator. Oscillator range is from very low ac, close to dc, to 1 MHz. High sensitivity. |

*Table 17-7. Continued.*

| Type | Characteristics |
|------|-----------------|
| | Current turns on when switching mode is less than 0.5 $\mu$A to 12 $\mu$A. |
| pnp | A commonly used three-element type. Can have widely varying electrical characteristics and so is available via numerous part numbers. Unit consists of a positive emitter, negative base and positive collector. Input between emitter and base is positive biased; output between collector and base is negative biased. Unit is a three-layer, bipolar semiconductor. Used in radio receivers, audio amplifiers and in oscillator, logic, and switching circuits. |
| npn | Three-layer bipolar semiconductor. Biased with base positive with respect to the emitter. Collector is positive with respect to the emitter. Except for differences in polarity is similar to the pnp type. Used in radio receivers, audio amplifiers, oscillator, logic, and switching circuits. |
| Tetrode | A three-element transistor with an additional lead attached to the base. Designed to function as npn or pnp types. Forward and reverse biased in same way as triode npn or pnp. |
| Surface-barrier | Designed for high-frequency use. Separation between emitter and collector as little as 0.000 1 in. Reduced power-handling capability. Emitter and collector electrodes made of indium. Symbol is same as that of pnp transistor. |
| Surface-controlled avalanche | Avalanche breakdown voltage controlled by an external field. Will function at frequencies up to 10 GHz. |
| Diffused | Somewhat similar to the surface-barrier transistor. In the surface barrier, the indium is a plated electrode. The diffused transistor depends on the melting of the indium to narrow the base. For both types of transistors the symbol is identical and is that of a pnp transistor. |
| Diffused-emitter-and-base | Sometimes known as a double-diffused transistor. Produced by gaseous treatment using p- and n-type impurities forming pn junctions. |
| Diffused-emitter-and-collector | A member of the diffusion family of transistors. Both emitter and collector are produced by diffusion. |
| Diffused-junction | Emitter and collector formed by a diffusion of an |

*Table 17-7. Continued.*

| Type | Characteristics |
|------|-----------------|
| | impurity into the semiconductor material. The technique does not involve heating of the impurity. |
| Diffused mesa | Collector-base junction formed by diffusion. Emitter-base junction formed by gaseous diffusion or by an evaporated metal strip. The term mesa describes the result of etching emitter and base regions. |
| Diffused metal oxide | Referred to as a DMOS transistor. A specialized diffusion technique for the production of discrete FETs for improvement of gain in the UHF region and for supplying better frequency response. |
| Micro-alloy | There are different methods used in putting the emitter and collector electrodes onto a surface-barrier transistor. They can be plated on or they can be alloyed as a very thin film of metal on an n-type base. This produces a unit known as a micro-alloy transistor. An alloy consists of two substances that are melted or diffused into each other. The micro-alloy transistor is closely related to the diffused type. |
| Intrinsic | Also known as intrinsic-region transistors. Unlike the surface-barrier and diffused transistors in which the base is as thin as possible, the intrinsic transistor may increase the base thickness and yet have an extremely high alpha cutoff frequency. The intrinsic region consists of a semiconductor free of any doping impurity. The value of the intrinsic region is that it separates the collector and the base. The large separation between collector and base reduces the capacitance between them, raising the alpha cutoff frequency. Available in two forms: pnip and npin. |
| Hook | This unit is a pnpn type. the collector has a slice of p-type semiconductor sandwiched in between itself and the base. This transistor is unusual in that it has an alpha greater than 1. Unlike the usual grounded-emitter transistor arrangement, the hook transistor has an output that is in phase with its input. |
| Multiheaded | Unit contains two transistors in one package. They are completely independent of each other, can be used singly or in combination. Each multiheaded unit has six leads. |
| Tandem | The tandem transistor is connected internally. The emitter of the first is wired to the base of the following transistor. In some tandem transistors, the |

*Table 17-7. Continued.*

| **Type** | **Characteristics** |
|---|---|
| | collector is the wired element, but all sorts of combinations are possible. |
| Power | Power transistors are the final link between the signal and the load. Power transistors are no different than ordinary transistors except they are designed to handle large values of current. In ordinary amplifying transistors the currents are in microamperes and milliamperes. In power transistors, collector currents can be 10 A or more. A heat sink is an important part of the power transistor assembly. Instead of having three leads, power transistors usually have just two. The leads are for the base and emitter. The collector is mechanically and electrically connected to the shell. |

*Fig. 17-3. Transistor symbols.*

*Fig. 17-3. Continued.*

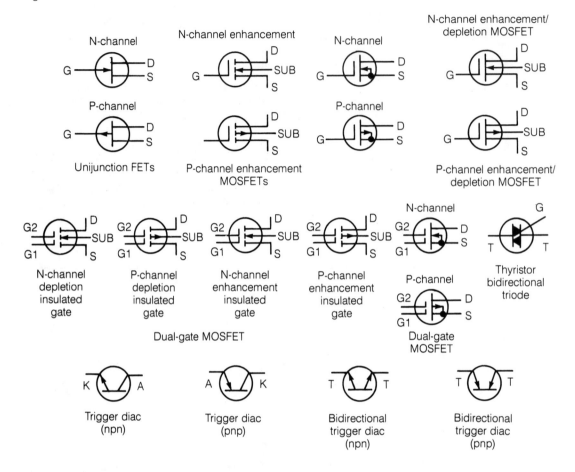

## Semiconductor rectifiers

Semiconductors are used for the rectification of power-line voltages, as demodulators of signal voltages, as current control devices, as switching devices, as dc voltage regulators (Fig. 17-4), as signal clampers, and in other applications. Table 17-8 lists the most common types and indicates some of their characteristics. Figure 17-5 shows the symbols for various solid-state diodes.

## Chips and integrated circuits

Although the words *chip* and *integrated circuit* are often used interchangeably they are not synonymous. A chip consists of a very small section of semiconductor material. The housing for that material is the IC (integrated circuit). Thousands of circuits consisting of transistors and diodes can be positioned on the chip,

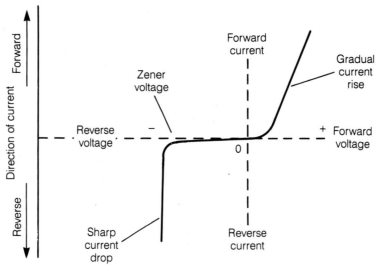

*Fig. 17-4.  Zener diode current flow.*

*Table 17-8. Semiconductor rectifiers.*

| Type | Characteristics |
|------|-----------------|
| Zener diode | At a particular value of reverse bias—the avalanche point—the current through the diode increases substantially. This behavior is shown in Fig. 17-4. The zener is used to regulate the output voltage of a power supply. |
| Power supply diodes | Semiconductor diodes are used in various types of power supplies, including half-wave, full-wave, and bridge. |
| Pin diodes | The name pin is obtained from the construction of this unit. It consists of p-type semiconductor and n-type sandwiching an intrinsic (I) semiconductor. The characteristics are determined by the thickness of the I layer. At low frequencies, the pin diode works as a junction diode and, at high frequencies, as a resistor. The frequency at which the crossover takes place is: $$f = 0.5 \times t$$ $t$ is supplied in specification sheets; $f$ is the transition frequency. |
| Varistor | The varistor consists of a pair of diodes connected so that the cathode of one is joined to the anode of the other. It is used as a shunt to protect a transistor collector against positive or negative voltage spikes. The resistance of the varistor depends on voltage, current, and polarity. Like |

*Table 17-8. Continued.*

| Type | Characteristics |
|---|---|
| | transistors, varistors are silicon types. Also known as a MOV (metal-oxide varistor). |
| Tunnel diode | Also known as Esaki diodes (after the inventor). An unusual characteristic is its negative resistance. As a result, its current flow decreases when the applied voltage increases. The unit finds application as an oscillator or amplifier. A pnp diode, it is capable of working at microwave frequencies. |
| Varactor diode | Also known as a varicap, a voltage-sensitive capacitor, or voltage-variable capacitor. Its junction, when reverse biased, works as a capacitance controlled by the reverse voltage. It is commonly used for electronic tuning of TV receivers, but it can also work as an electronic switch. |
| Silicon-controlled rectifier | Also known as an SCR or thyristor. Unit has three terminals—a cathode, an anode, and a gate. It will not conduct, even when forward biased, until its gate is pulsed. The SCR is a four-layer pnpn device. Conduction continues with the control signal removed until the anode supply is reduced, its polarity reversed, or removed completely. It works as an electrically controlled switch for dc loads. |
| Triac | The triac is a bidirectional rectifier and consists of two SCRs in shunt. It is also known as a bidirectional triode thyristor. The structure is npnpn and can be triggered into either forward or reverse conduction. The component functions as an electrically controlled switch. The connecting terminals are known as main terminal 1, main terminal 2, and gate. Alternatively, these terminals are sometimes identified as A1 and A2. With a dc voltage applied to the two main terminals the component works as an SCR. SCRs are listed as low-, medium-, and high-current types. |
| Clamping diodes | The clamp consists of a pair of diodes. The purpose of the clamp is to keep the signal voltage from rising above a predetermined level. The clamping circuit is passive, so there are some losses, which can be overcome by a following amplifier. |
| Thermistors | Manufactured by sintering mixtures of metal-oxide powders. High-temperature, nonlinear negative or positive coefficient of resistance. Thermistor is an acronym coined from *thermally sensitive resistor*, with the resistance varying inversely with temperature. |

*Table 17-8. Continued.*

| Type | Characteristics |
|------|-----------------|
| Photodiodes | A pn semiconductor diode. Light increases its reverse leakage current, that is, its resistance. Manufactured so light can reach its junction, which is larger than those in ordinary diodes. Diodes are usually packaged so that light cannot reach its elements. |
| LASCR | Light-activated SCR. pnpn type with incident light supplying gate current. Functions as a photo switch and is triggered into conduction when light falls on its base-collector junction. Light sensitivity can be controlled by a positive signal applied to its gate. |
| Light-emitting diode | Abbreviated as LED. Consists of a pn, which emits light when forward biased. Light ranges from infrared to violet. Most popular light colors are red, yellow, and green. Made of a semiconductor material such as gallium arsenide (GaAs) or lead telluride (PbTe). Light is emitted from the junction of the diode. |

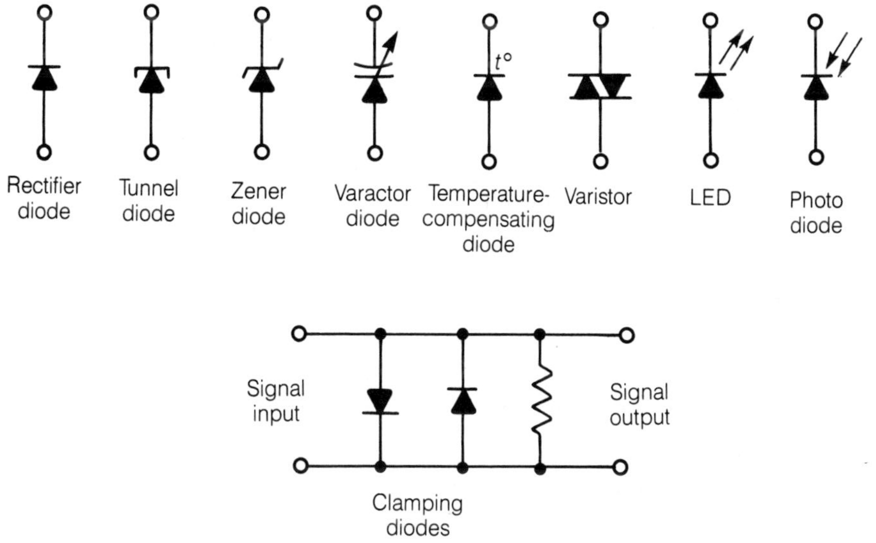

Rectifier diode   Tunnel diode   Zener diode   Varactor diode   Temperature-compensating diode   Varistor   LED   Photo diode

Signal input    Clamping diodes    Signal output

*Fig. 17-5. Symbols for solid-state diodes.*

whose area is referred to as real estate. IC sizes are small and typically measure about $1/8$ to $1/4$ in$^2$.

The chip consists of a semiconductor substrate and can be active or inactive. An active substrate is one that is an element of the diodes and transistors it supports. An inactive substrate has no connections to the semiconductor components.

The substrate is a silicon wafer, and when completed with diodes and transistors is the chip. The encased chip is an IC. There are two basic functions of chips, processing and storage. Processing chips are active and perform control functions or do the arithmetic. Storage chips hold information or instructions. There are two formats for storage chips, RAM (random-access memory) and ROM (read-only memory). An IC can contain both processing and storage functions.

There are a number of different types of ICs but there is no standardization. They can be classified by function, by size (that is, by the number of active components) and by speed of function. These are listed in Table 17-9.

*Table 17-9. Classification of ICs.*

| Type | Characteristics |
|------|-----------------|
| Small scale | An IC (integrated circuit) with fewer than 100 solid-state components. Abbreviated as *SSI*. |
| Medium scale | An IC with more than 100 but fewer than 1 000 solid-state components. Abbreviated as *MSI*. |
| Large scale | An IC containing between 1 000 and 5 000 solid-state components. Abbreviated as *LSI*. |
| Very large scale | Characterized by using a very small section of real estate but containing more than 5 000 solid-state components. Abbreviated as *VLSI*. |
| Super large scale | An IC that contains 100 000 or more solid-state diodes and transistors. Abbreviated *SLSI*. |
| Very high speed | IC that utilizes CMOS/SOS (complementary metal-oxide semiconductor, silicon-on-sapphire) chips. |
| Optoelectronic | Contains a semiconductor laser or photodiode plus an FET IC. This unit is a laser diode whose light output is several milliwatts. Capable of high speeds in the gigahertz region. Used as a light transmitter, modulator, or repeater in optical communications, as a low-noise laser source or sensor. |

*Table 17-10. IC package identification.*

| Type | Carrier | Characteristics |
|------|---------|-----------------|
| SOIC | small-outline IC | Widely used. Has gull-wing outline. |
| PLCC | plastic-lead chip carrier | Leads in J shape. |
| LCCC | leadless ceramic chip carrier | No leads. Metallic contacts on bottom surface of the package. |

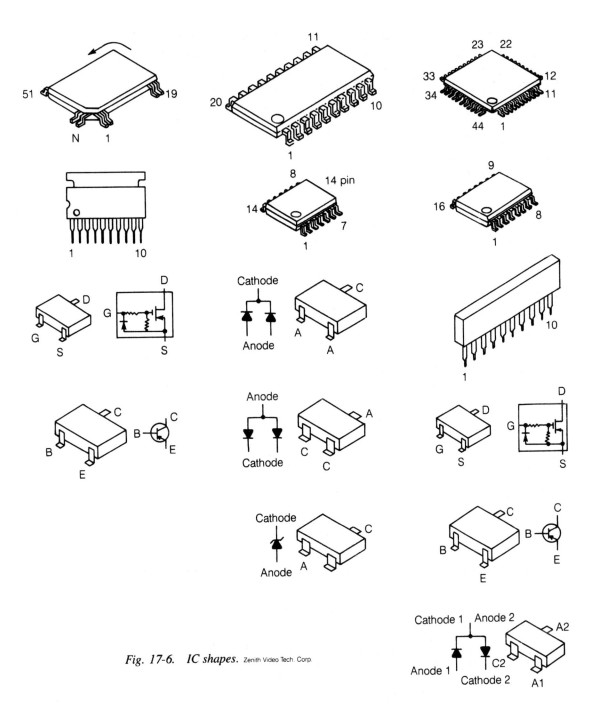

*Fig. 17-6.* **IC shapes.** Zenith Video Tech. Corp.

# IC packaging

ICs are mounted on PC (printed circuit) boards and commonly the board will have a number of them. ICs can be plug-in types or may be surface mounted. Identification of a particular IC is by part number positioned on one of its package surfaces or by the arrangement of its leads. These leads, connected to the internally positioned chip, can be located on one or more sides of the IC. Commonly the IC can have from as few as three pins to as many as 40, but some have as many as 244. Table 17-10 lists package identifications. Figure 17-6 shows some of the various IC packages.

# 18
## CHAPTER

# Wire, cable, and connectors

## Identifying wire

Wire can be categorized by:

gauge number
type of insulation
whether it is stranded or solid
by the components it connects
by voltage

Low-voltage wire is for any voltage that is 30 V or less. Table 18-1 lists the various kinds of wire.

## Wire gauge

The diameter of a wire is indicated by its gauge number, the lower the number, the thicker the wire. There are different wire gauges, including AWG (American Wire Gauge) and B&S (Brown and Sharpe), which are used for copper wire; others are used for brass, iron and sheet metal. AWG is the standard used for wire in electronics. Its gauge numbers range from the thickest wire, gauge 0000, to gauge 40. Gauge 0000 has a thickness of 0.460 in, and gauge 40 has a thickness of 0.003 1 in. The most commonly used wire gauges in electronics are 16 to 22. Wire used in electronics usually has an even gauge number (16, 18, 20), but wire with odd gauge numbers is available. Table 18-2 is for both single solid and stranded wire. Table 18-3 has a more extensive gauge range but covers solid wire only.

*Table 18-1. Wire types.*

| Type | Characteristics |
|------|-----------------|
| Bell wire | Known at one time as annunciator wire (Fig. 18-1). Consists of a single solid conductor insulated with a double layer of cotton. Each layer is wound in opposite directions and impregnated with paraffin. Sometimes called pushback wire because the insulated covering can easily be pushed back. |
| Single wire line | Almost like bell wire except for its insulation, often plastic, available in different colors (Fig. 18-2A). The insulation isn't pushback but can be removed with a single-edge razor blade, diagonal cutter, or wire stripper. |
| Twisted pair | Consists of two intertwined insulated conductors (Fig. 18-2B). Conductors for interconnecting electronic components are parallel and do not use the twisted arrangement. |
| Litzendraht | Better known as *Litz*. Each strand is insulated from the others. The effect is an increase in total surface area plus a reduction in ac resistance. The individual strands are joined at their ends. Effective up to about 1 MHz. Removing the insulation from each individual strand presents connection problems. Sometimes used in phono cartridges, but not often because of its cost. |
| Tinned | Tinned surface is a thin layer of tin or solder. Desirable because it makes soldering much easier. |
| Lamp cord | Also known as zip cord because its individual conductors can be "zipped" apart. Commonly used for connecting speakers. Made of seven strands of wire whose cross-sectional area is the equivalent of 18- or 20-gauge wire. Lamp cord can be polarized (not all lamp cords are). Polarized cords can be identified by a ridge above one of the pairs, or it can be covered with a silver- or copper-colored plastic (Fig. 18-3). |

Fig. 18-1.   Bell wire.

**A**

**B**

Fig. 18-2.   Single wire line (A); twisted pair (B).

Fig. 18-3.   Lamp cord.

### Table 18-2. Single and stranded wire gauges.

| Size, approx. AWG—American wire gauge | Number of strands | Strand diameter, nominal (in) | Strand area (circular mils) | Conductor diameter, average (in) | dc resistance (Ω/1 000 ft) |
|---|---|---|---|---|---|
| 32 | 7 | 0.003 1 | 67 | 0.010 | 183.0 |
| 30 | 7 | 0.004 0 | 112 | 0.013 | 109.9 |
| 28 | 7 | 0.005 0 | 175 | 0.016 | 70.4 |
| 26 | 1 | 0.015 9 | 253 | — | 46.3 |
| 26 | 7 | 0.006 3 | 278 | 0.020 | 44.3 |
| 24 | 1 | 0.020 1 | 404 | — | 28.3 |
| 24 | 7 | 0.008 0 | 448 | 0.025 | 27.5 |
| 24 | 19 | 0.005 0 | 475 | 0.026 | 25.7 |
| 22 | 1 | 0.025 3 | 640 | — | 17.9 |
| 22 | 7 | 0.010 0 | 700 | 0.031 | 17.6 |
| 22 | 19 | 0.006 3 | 754 | 0.032 | 16.3 |
| 20 | 1 | 0.032 0 | 1 024 | — | 11.2 |
| 20 | 7 | 0.012 6 | 1 111 | 0.038 | 10.9 |
| 20 | 10 | 0.010 0 | 1 000 | 0.038 | 12.3 |
| 20 | 19 | 0.008 0 | 1 216 | 0.041 | 10.1 |
| 18 | 1 | 0.040 3 | 1 624 | — | 7.05 |
| 18 | 7 | 0.015 9 | 1 770 | 0.048 | 6.89 |
| 18 | 16 | 0.010 0 | 1 600 | 0.048 | 7.69 |
| 18 | 19 | 0.010 0 | 1 900 | 0.051 | 6.48 |
| 16 | 1 | 0.050 8 | 2 581 | — | 4.43 |
| 16 | 19 | 0.011 3 | 2 426 | 0.058 | 5.02 |
| 16 | 26 | 0.010 0 | 2 600 | 0.061 | 4.73 |
| 14 | 1 | 0.064 1 | 4 109 | — | 2.79 |
| 14 | 19 | 0.014 2 | 3 831 | 0.072 | 3.18 |
| 14 | 41 | 0.010 0 | 4 100 | 0.076 | 3.00 |
| 12 | 1 | 0.080 8 | 6 529 | — | 1.76 |
| 12 | 19 | 0.017 9 | 6 088 | 0.090 | 2.00 |
| 12 | 65 | 0.010 0 | 6 500 | 0.096 | 1.89 |
| 10 | 104 | 0.010 0 | 10 380 | 0.121 | 1.16 |

The cross-sectional area of wire approximately doubles for every three gauge numbers (Table 18-3). At the same time, the current carrying capacity is doubled. Every wire size becomes 12.3% greater in diameter as the wire gauge number is decreased by one. Wire gauges are based on this computation:

Wire gauge diameter (mils) × 1.123 = diameter of next larger size.

| AWG | Diameter (mils, $d$) | Area (circular mils, $d^2$) | $\Omega/1\,000$ ft at 20°C, or 68°F |
|---|---|---|---|
| 0000 | 460.00 | 211 600 | 0.049 01 |
| 000 | 409.64 | 167 805 | 0.061 80 |
| 00 | 364.80 | 133 079 | 0.077 93 |
| 0 | 324.86 | 105 534 | 0.098 27 |
| 1 | 289.30 | 83 694 | 0.123 9 |
| 2 | 257.63 | 66 373 | 0.156 3 |
| 3 | 229.42 | 52 634 | 0.197 0 |
| 4 | 204.31 | 41 743 | 0.248 5 |
| 5 | 181.94 | 33 102 | 0.313 3 |
| 6 | 162.02 | 26 250 | 0.395 1 |
| 7 | 144.28 | 20 817 | 0.498 2 |
| 8 | 129.49 | 16 768 | 0.628 2 |
| 9 | 114.43 | 13 094 | 0.792 1 |
| 10 | 101.89 | 10 382 | 0.998 9 |
| 11 | 90.742 | 8 234.1 | 1.260 |
| 12 | 80.808 | 6 529.9 | 1.588 |
| 13 | 71.961 | 5 178.4 | 2.003 |
| 14 | 64.084 | 4 106.8 | 2.525 |
| 15 | 57.068 | 3 256.8 | 3.184 |
| 16 | 50.820 | 2 582.7 | 4.016 |
| 17 | 45.257 | 2 048.2 | 5.064 |
| 18 | 40.303 | 1 624.3 | 6.385 |
| 19 | 35.890 | 1 288.1 | 8.051 |
| 20 | 31.961 | 1 021.5 | 10.15 |
| 21 | 28.465 | 810.10 | 12.80 |
| 22 | 25.347 | 642.47 | 16.14 |
| 23 | 22.571 | 509.45 | 20.36 |
| 24 | 20.100 | 404.01 | 25.67 |
| 25 | 17.900 | 320.41 | 32.37 |
| 26 | 15.940 | 254.08 | 40.81 |
| 27 | 14.195 | 201.50 | 51.47 |
| 28 | 12.641 | 159.79 | 64.90 |
| 29 | 11.257 | 126.72 | 81.83 |
| 30 | 10.025 | 100.50 | 103.2 |
| 31 | 8.928 | 79.71 | 130.1 |
| 32 | 7.950 | 63.20 | 164.1 |
| 33 | 7.080 | 50.13 | 206.9 |
| 34 | 6.305 | 39.75 | 260.9 |
| 35 | 5.615 | 31.53 | 329.0 |
| 36 | 5.000 | 25.00 | 414.8 |

Table 18-3. Continued.

| AWG | Diameter (mils, $d$) | Area (circular mils, $d^2$) | $\Omega$/1 000 ft at 20 °C, or 68 °F |
|---|---|---|---|
| 37 | 4.453 | 19.83 | 523.1 |
| 38 | 3.965 | 15.72 | 059.6 |
| 39 | 3.531 | 12.47 | 831.8 |
| 40 | 3.145 | 9.89 | 1049 |

# Fusing currents

Table 18-4 gives the fusing currents in amperes for five commonly used types of wires. The current $I$ in amperes at which a wire will melt can be calculated from:

$$I = Kd^{3/2}$$

$d$ is the wire diameter in inches and $K$ is a constant that depends on the metal concerned. A wide variety of factors influence the rate of heat loss and these figures must be considered as approximations.

# dc wire resistance

The dc resistance of a wire—its opposition to the flow of a direct current—is generally measured per 1 000 ft at 68 °F (20 °C). As temperature increases, so does resistance. The resistance of copper wire becomes greater for each increase in gauge number. Knowing the resistance of 1 000 ft of any wire, it is possible to calculate the resistance of 1 000 ft of wire for the next larger gauge number.

The resistance of 1 000 ft of 20-gauge wire is 10.15 $\Omega$. Multiplying this by 1.26 supplies a resistance of 12.789 $\Omega$ for 21-gauge wire. Conversely, to find the resistance of 1 000 ft of the next lower gauge number, divide by 1.26. The resistance of smaller lengths can be calculated by multiplying or dividing by a factor such as 10, 100 or 1 000. Thus, the resistance of 300 ft of 30-gauge stranded wire is $(300/1\ 000) \times 103.2 = 30.96\ \Omega$. $103.2\ \Omega$ is the resistance of 1 000 ft of this wire. The resistance of a wire is proportional to its length in feet and inversely proportional to its cross-sectional area.

# Stranded wire

Stranded wire, commonly consisting of seven strands of bare copper wire twirled to form the equivalent of a single solid conductor, is used extensively in electronics. Assume a wire of seven strands of 32-gauge wire (identified as 7/32 wire), each of which has a diameter of 3.1 mils. Each wire will have a circular mil area of $3.1 \times 3.1 = 9.61$ circular mils. As there are seven wires the total area is $7 \times 9.61 = 67.27$ circular mils.

## Table 18-4. Fusing currents of wires.

| AWG B&S* gauge | $d$ (in) | Copper K=10 244 | Aluminum K=7 585 | German silver K=5 320 | Iron K=3 148 | Tin K=1 642 |
|---|---|---|---|---|---|---|
| 40 | 0.003 1 | 1.77 | 1.31 | 0.90 | 0.54 | 0.28 |
| 38 | 0.003 9 | 2.50 | 1.85 | 1.27 | 0.77 | 0.40 |
| 36 | 0.005 0 | 3.62 | 2.68 | 1.85 | 1.11 | 0.58 |
| 34 | 0.006 3 | 5.12 | 3.79 | 2.61 | 1.57 | 0.82 |
| 32 | 0.007 9 | 7.19 | 5.32 | 3.67 | 2.21 | 1.15 |
| 30 | 0.010 0 | 10.2 | 7.58 | 5.23 | 3.15 | 1.64 |
| 28 | 0.012 6 | 14.4 | 10.7 | 7.39 | 4.45 | 2.32 |
| 26 | 0.015 9 | 20.5 | 15.2 | 10.5 | 6.31 | 3.29 |
| 24 | 0.020 1 | 29.2 | 21.6 | 14.9 | 8.97 | 4.68 |
| 22 | 0.025 3 | 41.2 | 30.5 | 21.0 | 12.7 | 6.61 |
| 20 | 0.031 9 | 58.4 | 43.2 | 29.8 | 17.9 | 9.36 |
| 19 | 0.035 9 | 69.7 | 51.6 | 35.5 | 21.4 | 11.2 |
| 18 | 0.040 3 | 82.9 | 61.4 | 42.3 | 25.5 | 13.3 |
| 17 | 0.045 2 | 98.4 | 72.9 | 50.2 | 30.2 | 15.8 |
| 16 | 0.050 8 | 117 | 86.8 | 59.9 | 36.0 | 18.8 |
| 15 | 0.057 1 | 140 | 103 | 71.4 | 43.0 | 22.4 |
| 14 | 0.064 1 | 166 | 123 | 84.9 | 51.1 | 26.6 |
| 13 | 0.071 9 | 197 | 146 | 101 | 60.7 | 31.7 |
| 12 | 0.080 8 | 235 | 174 | 120 | 72.3 | 37.7 |
| 11 | 0.090 7 | 280 | 207 | 143 | 86.0 | 44.9 |
| 10 | 0.101 9 | 333 | 247 | 170 | 102 | 53.4 |
| 9 | 0.114 4 | 396 | 293 | 202 | 122 | 63.5 |
| 8 | 0.128 5 | 472 | 349 | 241 | 145 | 75.6 |
| 7 | 0.144 3 | 561 | 416 | 287 | 173 | 90.0 |
| 6 | 0.162 0 | 668 | 495 | 341 | 205 | 107 |

*Brown and Sharpe.

# ac resistance

The resistance of a wire to the flow of dc through it can be measured with an ohm-meter. At zero frequency (dc) the entire volume of the wire is used for the passage of current. This assumes pure dc, that is, dc with no variations.

Skin effect, also known as ac resistance, is the tendency of an alternating current to flow near or on the surface of a conductor. At a frequency as low as 1 kHz, the current flows through the outside skin to a depth of 0.09 cm. At 1 MHz, skin depth is only 0.007 cm. Stranded wire is often used in preference to solid conductors for radio-frequency currents, as the surface area is much larger than that of a comparable solid wire. Skin effect is equivalent to a reduction of the volume of copper through which a current flows.

# Connecting single conductors

Joining single conductors is usually a two-step operation. The initial step is the formation of a good mechanical connection followed by soldering. Table 18-5 shows various connection techniques.

*Table 18-5. Wire connection methods.*

| Type | Method |
|---|---|
| Wire splice | Twist one wire around the other as in drawing (A) in Fig. 18-4. Continue wrapping until the two wires form a mechanical joint. |
| Tap | Strip wire anywhere along its length. Strip second wire for about 1/2-in and wrap first wire as in (B) in Fig. 18-4. |
| Western Union splice | Wrap one wire around the other as in (C) in Fig. 18-4. |
| Staggered splice | Wires are wrapped as in drawing (D) in Fig. 18-4. This technique keeps wires from shorting. |
| Rattail joint | The two exposed wire ends are tightly twisted, (drawing E) in Fig. 18-4. |

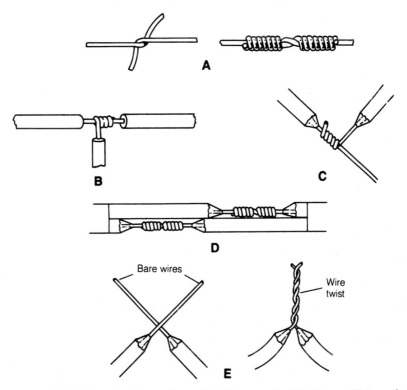

Fig. 18-4.  Methods of joining wires: wire splice (A); tap strip (B); Western Union splice (C); staggered splice (D); rattail joint (E).

# Cables

These consist of two or more wires that are grouped. Table 18-6 describes various types.

*Table 18-6. Cables.*

| Type | Method |
|---|---|
| Flat | Sometimes called ribbon cable (Fig. 18-5). Can be color coded. Can consist of two or more wires insulated from each other. End wire may be color coded as a reference. The word *flat* refers only to the fact that the adjoining wires form a flat surface. Various other types of flat cable include the mainspring (Fig. 18-6A), the rollup (drawing B) and the accordion (drawing C). |
| Multiconductor | Sometimes used when one of the wires, covered with white insulation (Fig. 18-7), carries signals; the other wire, the "hot" lead, supplies a dc voltage; and the black lead supplies a common, ground, or negative lead. The cable can be round or flat. Any group of one or more wires, plus shield braid, is known as multiconductor cable. The wires can be color coded red, black, or white; two of the wires can be white and one black; or they can be no particular color, as in coaxial cable. White or red is generally considered the plus, positive or "hot" lead; black the minus, negative, ground or "cold" lead. A cold lead is a ground lead and can be used as a voltage reference point. The color coding of multiconductor cable is the opposite of polarity used for electrical wiring. In house wiring, the black lead is the hot, or positive; the white lead is ground. |
| VCR | Those used by VHS recorders are sometimes called VHS cables. |
| Rotor | The problem of having a TV antenna face broadside to various TV stations is overcome by using a rotary antenna. Rotor cables can be three or four conductors, using 20-gauge wire. The wire is stranded and is covered by black vinyl insulation. The cables can be flat or round (Fig. 18-8). The wires in the round cable are color coded, with one black and the other wire or wires white. The flat cable isn't color coded, but one of the wires has a larger layer of insulation. The wire consists of seven strands, 20 gauge. |

*Table 18-6. Continued.*

| Type | Method |
|------|--------|
| Double receptacle connector cable | Permits the operation of two components simultaneously from a single battery. Both components must have the same dc voltage requirements but can have different current needs. |
| Extension cord for vehicles | This cable is intended for plugging into the cigarette lighter of a vehicle. The cord must be sufficiently long to permit components to be operated at some distance from the car. Typical cord length is 25 ft. Used for operating a portable video camera. |
| Dubbing | Used for transferring video and audio signals from one VCR to another. The cable is twinned or joined (Fig. 18-9). A dedicated cable, used only for dubbing, is twinned for the greater part of its length, but the individual conductors are separated a few inches from their connectors (plugs). |

Fig. 18-5.  *Flat ribbon cable.*

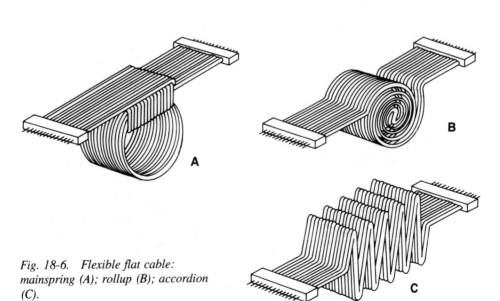

Fig. 18-6.  *Flexible flat cable: mainspring (A); rollup (B); accordion (C).*

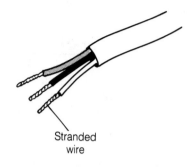

Fig. 18-7.   Multiconductor cable.

Stranded
wire

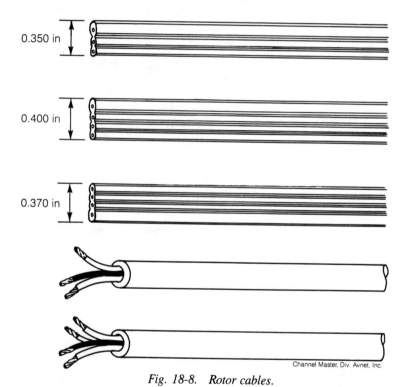

0.350 in

0.400 in

0.370 in

Channel Master, Div. Avnet, Inc.

Fig. 18-8.   Rotor cables.

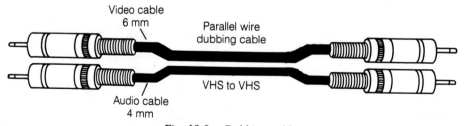

Video cable
6 mm

Parallel wire
dubbing cable

VHS to VHS

Audio cable
4 mm

Fig. 18-9.   Dubbing cable.

# Transmission line

Also called download, transmission cable connects an antenna to the input terminals on a receiver. There are various types of transmission line used for audio, radio, and television. These include single-wire line, twin lead, and coaxial cable.

# Velocity factor

The velocity of a wave, such as an RF (radio frequency) modulated composite video signal, is dependent on the medium through which it travels. It varies from one type of transmission line to another. The ratio of the actual velocity of the wave to its velocity in space is the velocity factor. In space, the velocity or speed of a radio wave is the same as that of the speed of light, approximately 186 000 mi/s. The velocity of radio waves in transmission lines is always less than this. Table 18-7 lists the velocity factor for radio waves in different types of transmission lines.

*Table 18-7. Transmission line velocity.*

| Type of transmission line | Velocity factor (V) |
|---|---|
| Coaxial line (air dielectric) | 0.85 |
| Coaxial line (solid plastic dielectric) | 0.66 |
| Two-wire line (wire with plastic dielectric) | 0.68−0.82 |
| Twisted-pair line (rubber dielectric) | 0.56−0.65 |

# Twin lead

Also known as two-wire line, or 300-Ω line, it is used primarily as the transmission line or download from an FM or TV antenna. Short lengths connected to the antenna input terminals of a TV set can be used as an interference signal trap.

The impedance is 300 Ω, regardless of length, but the greater the length, the larger the signal loss. Each of the two conductors in twin-lead cable (Fig. 18-10) is made of stranded wire, generally seven strands of 28-gauge wire, with the conductors jointly forming the equivalent of 20-gauge solid wire.

Twin lead cable is passive and introduces signal losses, expressed in decibels per 100 ft. The loss increases with frequency and the condition of the line, whether dry or wet. This loss increases with age as the plastic begins to weather from outside use. In the home, twin lead is sometimes painted over, increasing deterioration and capacitance effect.

*Fig. 18-10. Twin-lead transmission line equipped with open spade lugs.*

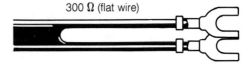

300 Ω (flat wire)

# Shielded twin lead

Unshielded twin lead can pick up electrical noise; this is overcome by using shielded twin lead. It consists of a pair of conductors made up of seven strands of bare, twisted copper wire. The wire is insulated with plastic and then covered with a foil sheathing. Running along the foil is a bare wire that can be used as a separate ground lead. The twin conductors, the ground wire and the metallic sheath are covered with a layer of plastic, such as Mylar.

Shielded twin lead (Fig. 18-11) is thicker, heavier, is more difficult to handle than unshielded twin lead, and costs more. It may be necessary to use in areas having high electrical interference.

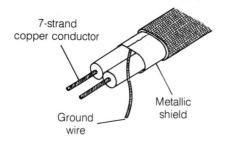

7-strand copper conductor

Ground wire

Metallic shield

Fig. 18-11.   Shielded twin lead.

# Coaxial cable

There are two types of video coaxial cable, balanced and unbalanced. Unbalanced coaxial (coax) is more widely used.

# Unbalanced coax

This cable (Fig. 18-12) consists of a central conductor of solid or stranded wire and is made of pure copper. It is the hot lead, and is surrounded by a layer of insulating material made of polyethylene plastic. Around it is a jacket of wire mesh or braid, or sometimes just a wire wrap in the form of a continuous loop. The braid

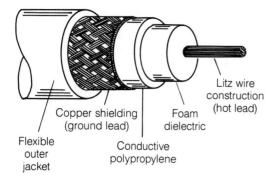

Flexible outer jacket

Copper shielding (ground lead)

Conductive polypropylene

Foam dielectric

Litz wire construction (hot lead)

Fig. 18-12.   Unbalanced coaxial cable.

works as a conducting lead—the ground or cold lead. The braid and the center lead form the two conductors required for the transfer of a signal.

## Unbalanced coaxial cable types

Not all unbalanced coaxial cables are alike. That in Fig. 18-13A uses a single layer of braid. The coax in drawing (B) has a layer of foil beneath the braid. The coax in drawing (C) uses Litz wire to help overcome skin effect and has a silver-coated double layer of shield braid.

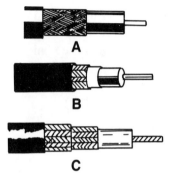

*Fig. 18-13. Unbalanced coaxial cable. Single layer of braid (A); layer of foil beneath braid (B); silver-coated double braid (C).*

## Dual coaxial cable

Sometimes two signals must be transferred independently and in that case dual coax is available (Fig. 18-14). Each of the two cables is unbalanced and the cables use a common protective outer covering.

*Fig. 18-14. Dual coaxial cable.*

## Messenger cable

For in-home satellite TV it is not only necessary to supply coax for bringing the signal indoors but also to use wiring to supply dc operating voltages for the dish motor, the LNA (low-noise amplifier), the downconverter, and the solid-state components mounted on the dish structure. Unbalanced coax is used to handle the signal and a single conductor can deliver the dc voltage, using messenger cable (Fig. 18-15). The single wire is positive for dc, and the shield braid works as the minus lead. The braid also works as the negative lead for the coax.

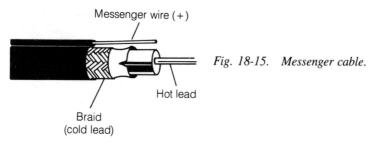

Messenger wire (+)

*Fig. 18-15.  Messenger cable.*

Hot lead

Braid
(cold lead)

# Hardline

This is a type of unbalanced coax using a solid outside metallic shield instead of flexible braid. Aside from this physical characteristic, hardline works the same as ordinary coax. It is used only for special applications, not including TV.

# Siamese cable

Siamese cable (Fig. 18-16) consists of coax plus three additional wires, two of which are color coded white, the other black. These conductors and the cable are independent but are housed in a single jacket.

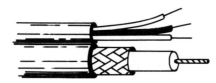

*Fig. 18-16.  Siamese cable.*

The black wire, sometimes called a drain, is the ground lead, so this arrangement does not use the shield braid as a common wire. One of the white wires is used for supplying a dc voltage to the low-noise amplifier and also to a downconverter mounted on the dish structure of a satellite TV system. The remaining white wire is used for supplying a dc voltage for a VTO (voltage-tuned oscillator). Two of the white wires deliver dc; one must supply a constant voltage, usually 22 V, to the LNA and the downconverter. The VTO supply is 8 V and is used intermittently.

# Balanced coax

Balanced coax consists of two independent conductors throughout the length of the cable (Fig. 18-17). The two conductors are used for carrying signals and the braid, usually grounded, serves as a shield against electrical noise and extraneous signals.

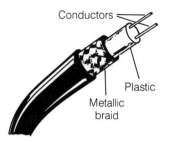

Fig. 18-17. *Balanced coaxial cable.*

Conductors

Plastic

Metallic
braid

# Coaxial cable identification

Coaxial cable is identified by type number, with each type having different physical and electrical characteristics. For television work, the most commonly used cable is RG-59U. The center conductor is usually 22 gauge. Heavier wire gauges are used for commercial installations such as the coax installed by cable TV companies. Table 18-8 lists commonly used shielded cables.

# Signal loss

Signal loss twin lead depends on whether the cable is wet or dry (Table 18-9). Signal loss in coaxial cable is a function of its length. This is dc resistance loss. The dc resistance loss for a 50-MHz signal and one for 55 MHz will be about the same. Another type of loss is frequency selective. The bypassing action is more effective at the high-frequency end. For audio frequencies the high-frequency loss will be insignificant. For video, which has an upper limit of 4.2 MHz, the loss at this end will result in degradation of fine picture detail. The amount of loss depends on dc resistance loss; cable capacitance, usually supplied in picofarads; signal frequency; and losses in the dielectric. Table 18-10 lists the attenuation of various types of coax for specific TV channels.

# Audio cables

Although audio cables (Fig. 18-18) are constructed along the lines of coaxial cable, they are seldom referred to as such. Audio cables do not include extra wires such as Siamese or messenger cables.

# Microphone cables

These can be balanced or unbalanced. Lower grades have a center conductor covered with thermoplastic insulation. The shield is a spiral wrap of wire. If a mic is to be fixed in position, mounted permanently, cables having a diameter of $1/4$ to $5/16$ in can be used, but this thickness makes the cable stiffer. If the mic is to be carried the cable should be flexible with a thickness of about $3/16$ in.

*Table 18-8. Coaxial cables and characteristics.*

| Type | Jacket O.D., inch (mm) | Jacket type | Shield | Dielectric O.D. & type, inch (mm) | Center conductor (mm) | pF/ft (pF/m) | Nom. imp. (Ω) | Nominal attenuation (dB) | |
|---|---|---|---|---|---|---|---|---|---|
| | | | | | | | | 1 MHz | 100 ft |
| 59/U | 0.242 (6.15) | Black vinyl | Bare copper 95% coverage | 0.146PE (3.71) | 22 Ga (0.643) copperweld | 21 (68.9) | 73 | 100 200 400 | 3.4 4.9 7.1 |
| 59/U | 0.242 (6.15) | Black vinyl | Bare copper 80% coverage | 0.146PE (3.71) | 22 Ga (0.643) copperweld | 21 (68.0) | 73 | 100 200 400 | 3.4 4.9 7.1 |
| 59/U stranded center | 0.242 (6.15) | Black vinyl | Bare copper 95% coverage | 0.146FPE (3.71) | 22 Ga 7×30 (0.76) bare copper | 17.37 (56.8) | 75 | 100 200 400 | 3.0 4.4 6.5 |
| 59/U cable for "F" 59 connectors | 0.242 (6.15) | Black | Bonded aluminum +60% Alum. braid 100% coverage | 0.146FPE (3.71) | 22 Ga (0.81) copperweld | 17.3 (56.8) | 75 | 50 100 200 500 900 | 1.8 2.6 3.8 6.2 8.4 |
| 62A/U | 0.242 (6.15) | Black | Bare copper 95% coverage | 0.146SSPE (3.71) | 22 Ga (0.81) | 13.5 (44.3) | 93 | 100 200 400 900 | 3.1 4.4 6.3 11.0 |

Table 18-8. Continued.

| Type | Jacket O.D., inch (mm) | Jacket type | Shield | Dielectric O.D. & type, inch (mm) | Center conductor (mm) | pF/ft (pF/m) | Nom. imp. (Ω) | Nominal attenuation (dB) 1 MHz | 100 ft |
|---|---|---|---|---|---|---|---|---|---|
| Video double braid | 0.304 (7.72) | Non-contaminating vinyl | Tinned copper double braid 98% coverage | 0.200PE (5.08) | 20 Ga (0.813) bare copper | 21 (68.9) | 75 | 01<br>10<br>1<br>4.5<br>10<br>100 | 0.06<br>0.08<br>0.25<br>0.45<br>0.78<br>2.70 |
| 213/U | 0.405 (10.29) | Black vinyl | Bare copper 97% coverage | 0.285PE (7.24) | 13 Ga (2.17) bare copper | 30.8 (101) | 50 | 100<br>200<br>400<br>900 | 2.0<br>3.0<br>4.7<br>7.8 |
| 214/U | 0.425 (10.80) | Black jacket | 2 Silver coated copper 98% coverage | 0.285PE (7.24) | 13 Ga (2.26) silver coated copper | 30.8 (101) | 50 | 100<br>200<br>400<br>900 | 2.0<br>3.0<br>4.7<br>7.8 |
| Dual RG/59 coax cable | 0.242 × 0.505 | Black PVC | Bare copper 95% coverage | 0.146PE (3.71) | 22 Ga (0.570) copperweld | 20.5 (67.3) | 75 | 100<br>200<br>400 | 3.5<br>5.1<br>7.5 |

## Table 18-9. Signal loss for twin lead.

| MHz | 57 | 85 | 177 | 213 | 500 | 650 | 800 | 900 |
|---|---|---|---|---|---|---|---|---|
| dB loss per 100 ft Dry | 0.9 | 1.2 | 1.5 | 2.0 | 3.2 | 3.8 | 4.5 | 5.4 |
| Wet | 1.8 | 2.8 | 3.7 | 5.2 | 14.0 | 20.0 | 32.0 | 46.0 |

| MHz | 57 | 85 | 177 | 213 |
|---|---|---|---|---|
| dB loss per 100 ft Dry | 1.2 | 1.6 | 2.0 | 2.7 |
| Wet | 2.2 | 3.3 | 4.3 | 6.0 |

## Table 18-10. Coaxial cable attenuation.

| MHz | 57 | 85 | 177 | 213 | 500 | 650 | 800 | 900 |
|---|---|---|---|---|---|---|---|---|
| dB loss per 100 ft Dry | 2.1 | 2.5 | 3.6 | 3.8 | 5.9 | 6.8 | 7.7 | 8.0 |
| Wet | 2.1 | 2.5 | 3.6 | 3.8 | 5.9 | 6.8 | 7.7 | 8.0 |

### Cable characteristics
### Nominal attenuation (dB/100 ft)

| Cable | Ch. 2 | Ch. 6 | Ch. 7 | Ch. 13 | Ch. 20 | Ch. 30 | Ch. 40 | Ch. 50 | Ch. 60 |
|---|---|---|---|---|---|---|---|---|---|
| Color duct | 2.3 | 2.7 | 3.8 | 4.2 | 6.5 | 7.0 | 7.5 | 7.8 | 8.0 |
| Foam color duct | 2.1 | 2.5 | 3.3 | 3.8 | 5.9 | 6.3 | 6.7 | 7.0 | 7.3 |
| RG-59/U | 2.6 | 3.5 | 4.9 | 5.4 | 8.3 | 8.8 | 9.2 | 9.7 | 10.3 |
| RG-59/U foam | 2.3 | 2.7 | 3.8 | 4.2 | 6.2 | 6.6 | 6.8 | 7.1 | 7.3 |
| RG-6 foam | 1.7 | 1.9 | 2.8 | 3.0 | 4.8 | 5.2 | 5.6 | 5.9 | 6.2 |
| RG-11/U | 1.4 | 1.7 | 2.2 | 3.2 | 5.1 | 5.3 | 5.5 | 5.7 | 6.1 |
| RG-11/U foam | 1.1 | 1.4 | 1.6 | 2.3 | 3.9 | 4.0 | 4.1 | 4.2 | 4.4 |
| 0.412 cable | 0.74 | 1.0 | 1.4 | 1.5 | 2.5 | 2.6 | 2.7 | 2.9 | 3.1 |
| 0.500 cable | 0.52 | 0.67 | 0.72 | 1.1 | 1.5 | 1.8 | 2.1 | 2.4 | 2.7 |

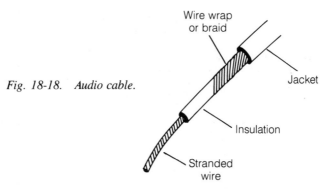

Fig. 18-18.   Audio cable.

# Adapter cable

This is an accessory cable that permits the connection of one type of terminal to another. An example would be connecting a VCR to a video camera when these have incompatible terminals. The fact that a cable plug will fit into a jack does not necessarily indicate compatibility.

# Impedance of coaxial cable

Referred to as characteristic or surge impedance, this specification is indicated in ohms and is represented by the letter Z. It is the distributed inductance and capacitance of the cable. The impedance is stated by:

$$Z = \sqrt{L/C}$$

Z is the impedance in ohms, L the inductance in henries, and C the capacitance in farads. A cable having an impedance of 75 Ω has the same impedance for a length of 10 ft, 300 ft, or any other length.

The characteristic (surge) impedance is directly dependent on the physical dimensions as indicated in Fig. 18-19. A is the outside diameter of the inner conductor (the hot lead), and B is the inside diameter of the shield (the cold lead).

$$Z = 138 \log_{10}(B/A)$$

Z is the impedance in ohms, A and B can be any measurement units, provided they are in the same units.

Fig. 18-19.   *Physical characteristics of coaxial cable used for the calculation of surge impedance.*

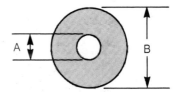

The dc resistance of coaxial cable can be calculated from:

$$R_o = P_o L/s$$

$R_o$ is the resistance in ohms; $P_o$ is the coefficient of resistance for the kind of material used as the conductor; $L$ is the length of the conductor in meters, and $s$ is the cross-sectional area of the conductor in square meters.

There is a formula that takes both frequency and physical characteristics into consideration.

$$R_s = (4.16)10^{-8} \times \sqrt{f\left(\frac{1}{A} + \frac{1}{B}\right)}$$

$R_s$ is the calculated value of resistance in ohms per meter at the skin effect frequency, $f$, and $A$ and $B$ are the dimensions of the cable conductors as shown earlier in Fig. 18-19.

# Impedance of twin lead

The characteristic impedance of two-wire transmission line can be calculated from:

$$Z = \frac{275}{\sqrt{K}} \log_{10} \frac{b}{a}$$

$K$ is the dielectric constant of the material between the parallel conductors. Its value is 1 for air. When polyethylene is used, the value of $K$ is 0.675. The impedance usually ranges from 50 to 300 $\Omega$, but for TV it is generally 300 $\Omega$. In the formula above, $b$ is the spacing center to center of the conductors and $a$ is their diameter (Fig. 18-20).

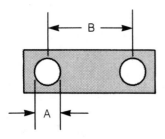

Fig. 18-20.   Physical characteristics for determining the impedance of twin lead.

# Voltage standing-wave ratio

Abbreviated as the VSWR of a transmission line, it is the difference between the maximum and minimum voltage along that line. Ideally, the best VSWR is unity, a ratio of 1 to 1. VSWR is also an indication of the loss of signal energy along a transmission line due to reflected signal energy from the load to the source. Table 18-11 shows various values of VSWR.

*Table 18-11. Return signal loss/VSWR.*

| VSWR | Reflection coefficient | Percentage reflection | VSWR | Reflection coefficient | Percentage reflection |
|------|------------------------|-----------------------|------|------------------------|-----------------------|
| 8.71 | 0.790 | 79% | 1.110 | 0.050 | 5.0% |
| 4.42 | 0.630 | 63% | 1.080 | 0.040 | 4.0% |
| 3.01 | 0.500 | 50% | 1.070 | 0.032 | 3.2% |
| 2.32 | 0.400 | 40% | 1.050 | 0.025 | 2.5% |
| 1.92 | 0.320 | 32% | 1.040 | 0.020 | 2.0% |
| 1.67 | 0.250 | 25% | 1.032 | 0.016 | 1.6% |
| 1.50 | 0.200 | 20% | 1.026 | 0.013 | 1.3% |
| 1.37 | 0.160 | 16% | 1.020 | 0.010 | 1.0% |
| 1.28 | 0.130 | 13% | 1.010 | 0.005 | 0.5% |
| 1.22 | 0.100 | 10% | 1.006 | 0.003 | 0.3% |
| 1.17 | 0.079 | 7.9% | 1.004 | 0.002 | 0.2% |
| 1.13 | 0.063 | 6.3% | 1.002 | 0.001 | 0.1% |

# Connectors

There are many types of connectors used for joining a wire to some component. In some instances, the connector is nothing more than a screw. A pair of these are used for connecting twin lead to a TV receiver. Although screw connectors are simple, they are not as convenient as plugs that can be removed or reinserted easily and rapidly. Table 18-12 is a listing of various plug types. Figure 18-21 shows various types of banana plugs.

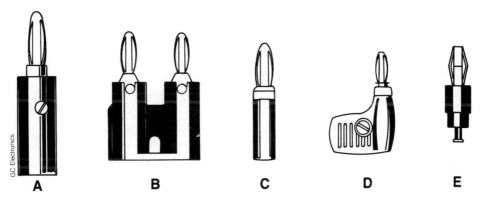

*Fig. 18-21. Banana plugs. Single conductor (A); twin (B); midget (C); right angle (D); sub-miniature (E).*

*Table 18-12. Plug types.*

| Type | Characteristics |
|------|-----------------|
| Banana | Used for connecting single, unshielded wires. They mate with jacks having a bore of 0.161 in. The joining of a wire to a banana plug can be a solderless or a soldered connection. |
| |     Banana plugs are equipped with four-leaved phosphor bronze or beryllium copper-nickel-plated spring tips. The springs are compressed to fit the opening of the mating jack. Because of the difference in the dimensions of the plug and jack, the plug makes a force fit, but the spring leaves of the tip permit easy removal. The various banana plugs appear in Fig. 18-21. |
| Dual banana | When a pair of wires is required, a dual banana plug can be used with polarity indicated on the top and side. The mating jack must be 3/4 in on centers. |
| Midget banana | Used where economy of space is essential. |
| Right angle banana | Used for connecting probes to a test instrument. Like other plugs in this family, it has standard-size tip with a diameter ranging from 0.170 to 0.180 in. |
| Subminiature | Available with a tip diameter that is only 0.120 in. Used for printed circuit connections. The mating jack should have a diameter of 0.104 in. |

# Tip plugs

Tip plugs have solid metallic tips made according to standard industry specifications of 0.080-in diameter. Because there is no spring action, as in the case of banana plugs, the tips of these plugs must be machined to very close tolerances. The connecting wire to the solid tip plug can be soldered to the plug or the connection can be solderless.

The tip plug shown in Fig. 18-22A is solderless, with the connecting wire tightened securely under the knurled collar. The tip plug in drawing (B) differs from the others, as its tip is hollow. However, its tip diameter is the same as that of other tip plugs. The tip plug in drawing (C) is a right-angle type. The plastic handle consists of a pair of mating halves that can be assembled with one machine screw and nut. The right-angle tip plug is popularly used for meter test leads. Tip plugs are also available as dual units (D). This is solderless, and plugs into a corresponding pair of jacks having a spacing of 3/4 in on centers.

# Mini plug

The mini plug (Fig. 18-23) accepts wire using shield braid. The tip of the plug makes contact with the center conductor of the wire, while the cylindrical metal

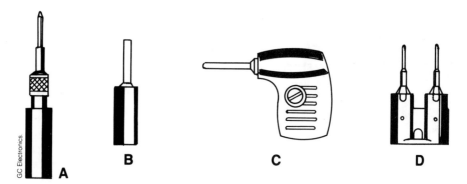

*Fig. 18-22. Tip plugs: solderless type (A); hollow tip (B); right angle (C); double plug (D).*

*Fig. 18-23. Mini plug.*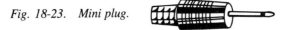

extension (also called a barrel) of the plug makes contact with the shield braid. The wire has some resemblance to unbalanced coax, but in this case the shield braid is not coaxial with the central hot lead. In coaxial cable the insulation around the hot lead is such that the central conductor is equidistant from its surrounding shield braid. This is not true with ordinary shielded wire used in audio applications.

When a barrel is internally threaded, as some of them are, it is identified by its diameter and the number of threads per inch. In its specs, a barrel could be $5/8 \times 27$ TPI ($5/8$-in diameter; 27 threads per inch).

# Phone plug

The phone plug is a larger version of the mini plug and is sometimes used as the terminating element for cables joining audio components. The phone plug looks just like its smaller version, the mini. It is a two- or three-element type and has a barrel that is $1^{1}/4$ in (31.8 mm) long. Microphone cables use either of these two plugs for connection to VCRs (Fig. 18-24).

Both plugs, the mini and the phone, can be either monophonic or stereophonic. The phone plug is so called because it was originally used as a connector for telephone switchboards. Some of these plugs have an insulated casing, and others use metal, which not only adds strength to the plug, but works as a shield as well. It is also possible to obtain these plugs in right-angle form.

# Phono plug (RCA plug)

The names for these two different plugs, phone and phono, are so similar that they are often confused. A phono plug, also called an RCA plug or a pin plug (Fig. 18-25), is commonly used as a connector for audio components.

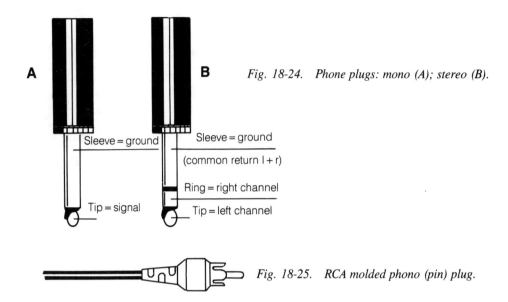

A  B    *Fig. 18-24.  Phone plugs: mono (A); stereo (B).*

Sleeve = ground

Tip = signal

Sleeve = ground

(common return l + r)

Ring = right channel

Tip = left channel

*Fig. 18-25.  RCA molded phono (pin) plug.*

# Plug adapters

Ordinarily a mini plug would mate with a mini jack, a phone plug with a phone jack and a phono plug with a phono jack. A problem arises, though, if a cable is equipped with one type of plug, and the component to which it must deliver a signal has a completely different type of jack. There are various solutions. The existing plug can be removed, the jack can be changed, or you can try to get a cable with the right kind of plug. The easiest (and fastest) technique is to use an adapter.

The three plugs shown in Fig. 18-26—the phone plug, the phono plug (it's less confusing to call it an RCA plug), and the mini plug—have adapters available so they can be made to join with other types of jacks. As is indicated at the top of the drawing, a phone plug that would ordinarily require a phone jack can be made to mate with an RCA jack. Along the same lines, an RCA plug can join with a phone jack and a mini plug with a phone jack. Adapters for other types of plugs are also available. This does not mean a cable with the right kind of plugs isn't desirable—it is, but a cable with one kind of plug at one end and a different type on the other end isn't always easy to get. A possible alternative is to customize your own, starting with bulk cable and mounting the specific plugs required.

Figure 18-27 shows RCA plugs mounted on twinned cable. The cables are joined for almost their complete length, with separation near the connectors. If further separation is required, the cables can be pulled apart. Twin cables have several advantages, the installation is neater, it is easier to trace the connection, and it avoids the exasperation of losing one cable of a pair.

The outer body of the plug is sometimes color coded as an aid in observing polarity. If the plugs are the same color, the symbol (+) might sometimes be

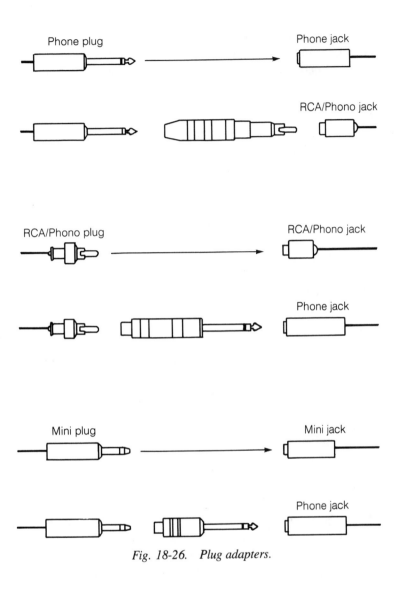

Fig. 18-26.   Plug adapters.

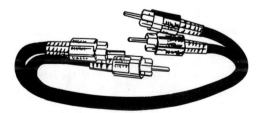

Fig. 18-27.   RCA connectors for a hi-fi system.

stamped on a corresponding pair of input and output plugs. Color coded plugs, often using black for the negative lead and red for the positive lead, permit quick identification. Stamped polarity symbols often are difficult to read.

# Belling and Lee plug

The plug in Fig. 18-28 is a push-in connector, attached to coaxial cable. It is a standard type used in Europe, Africa, the Middle East, and Great Britain.

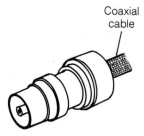

Coaxial
cable

Fig. 18-28.   Belling and Lee plug.

# Motorola plug

This plug, shown in Fig. 18-29, is used to connect video components to a corresponding Motorola jack, used in a wall plate with connections to a master antenna. This is part of a television system sometimes used in high-rise apartment houses.

Fig. 18-29.   Motorola plug.

# F plug

The F plug (Fig. 18-30) is widely used, and is available in a wide assortment. It can be straight or have a right-angle bend to keep a cable close to equipment. It is used with unbalanced coaxial cable, which can be used for the downlead from an antenna or for interconnecting video components.

The F plug is on cables that connect a VCR to a TV set. It also is found between switchers and the equipment to which the switchers are attached, on accessory devices such as baluns (impedance-matching transformers), on band separators, on signal splitters, and so on.

F connectors can be threaded or push-on. The advantage of the threaded F plugs is that the connection is tighter and more secure. They are less likely to con-

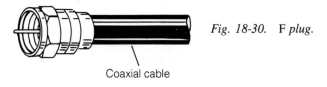

Fig. 18-30.   F plug.

Coaxial cable

tribute to cable intermittents, often difficult to find during servicing. However, if a cable must be disconnected often, using a screw-on connector can be a nuisance.

# BNC plug

The purpose of a plug and jack is to ensure a good transfer of a signal or a dc voltage. The plug and its mating jack must make a good fit. To ensure the fit, some plugs are equipped with rotatable rings that are threaded so the plug can be screwed into place. Other plugs depend on good manufacturing tolerances so the plug and its jack can make a force fit. A compromise between these two extremes is the BNC plug shown in Fig. 18-31. Comparable to a twist-on bayonet bulb, this plug requires a partial turn to lock the plug into position. It has the advantage of permitting a quick connection, plus one that is positive in its action.

*Fig. 18-31. BNC plug. It has an outside diameter of 0.25 in and is used with coaxial cable.*

# T connector

It is sometimes necessary to make a connection to coaxial cable somewhere along its length. For this a special connector is needed (Fig. 18-32). Although generally referred to as a plug, it is a hybrid, composed of two jacks and a plug. Two ends of coaxial cable with F-type male connectors are plugged into the sides, while an F-type jack is connected to the remaining port at the center.

*Fig. 18-32. T connector.*

# PL-259 plug

Another type of plug is the PL-259, also known as a UHF plug, used with RG-59/U coaxial cable. This is a standard UHF connector often supplied with a screw-on lock. It is also available as a solderless push-on type, with the connector screwed onto its connecting cable.

# Test instrument connectors

Plugs are used for making connections to test instruments such as a volt-ohm-milliameter (VOM) (Fig. 18-33). The leads can be terminated in a pair of insulated alligator clips or in a pair of needle-point test prods. The plugs used with test leads are often pin or banana plugs. The plugs can be straight or right-angle units.

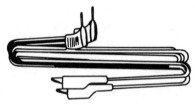

*Fig. 18-33.   Test leads are terminated in insulated alligator clips at one end and pin plugs or banana jacks at the other.*

## Cable connectors

Different components can have identical jacks, in which case the cable used to connect them will have identical plugs, one at each end of the cable. However, jacks on one component are often different from those on another, especially if the components are made by different manufacturers. As a result, a cable may have different plugs at each end.

In Fig. 18-34, drawing (A) shows a cable with a pair of F plugs indicating that each of the components to be connected is equipped with F jacks. This cable can be used to interface a pair of video components. Drawing (B) is a cable using RCA plugs on both ends. This type of connector is frequently used to interconnect audio components. The cable is audio coax but more frequently called audio cable. In (C), the cable has an F plug at one end and a Motorola plug at the other. The Motorola plug connects to a wall plate wired to a master antenna, and the F plug is used to connect the coaxial cable to a television set. Drawing (D) shows a cable with a pair of BNC plugs. Drawing (E) shows a pair of F plugs, but one is a right-angle type. A right-angle plug is useful if the cable is to go straight up or down from the component to which it is connected. Drawing (F) shows a paired (twinned) set of plugs, RCA to RCA, an arrangement commonly used for connecting high-fidelity components. Drawing (G) illustrates an F plug to RCA plug cable, and (H) shows an RCA plug connected to a 3.5 mm mini plug.

## DIN (Deutsche Industrie Normenausschuss) connector

All the plugs previously described involve the use of a single cable. Since a component may have a number of jacks, a relatively large collection of cables may be required, especially if a number of components are involved. As a result, the area behind a rack housing those components can look like a wiring maze. Not only is it unsightly, but it can lead to making wrong connections and also makes it difficult to locate and trace individual cables.

One solution is the use of a DIN cable for interconnecting audio components. Another is flat cable of the type widely used for connecting computer units, consisting of four or more wires so that multiple connections can be made with a sin-

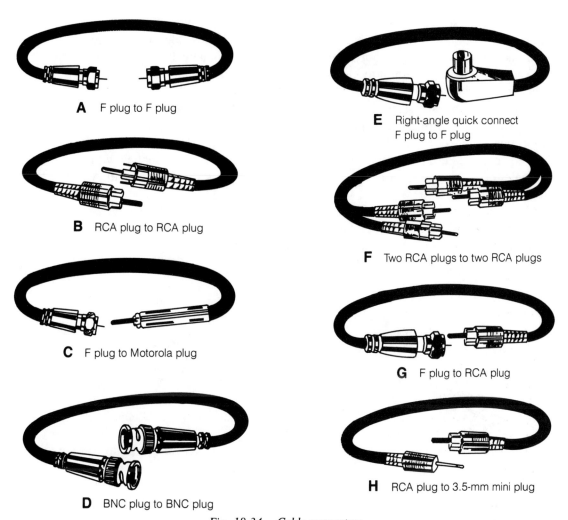

**A**   F plug to F plug

**B**   RCA plug to RCA plug

**C**   F plug to Motorola plug

**D**   BNC plug to BNC plug

**E**   Right-angle quick connect
F plug to F plug

**F**   Two RCA plugs to two RCA plugs

**G**   F plug to RCA plug

**H**   RCA plug to 3.5-mm mini plug

*Fig. 18-34.   Cable connectors.*

gle cable. The flat cable has had very limited application for audio components, but the DIN connector is used more often. DIN cables usually have three or five wires. Figure 18-35 shows a DIN plug. Five-pin DIN connectors can be used for connecting cassette decks to a receiver or integrated amplifier.

The ports (jacks) for the input and output of a tape deck are often identified in different ways. It is sometimes easy to transpose the connectors. Because the DIN

*Fig. 18-35.   DIN connector.*

connector has a keyway, the connection becomes certain and wiring errors are eliminated. Not only does the DIN arrangement make it impossible to transpose input and output connections, but polarity is followed automatically. With separate cables you must be careful to identify which line is left channel and which is right channel when signal polarity is important. The difficulty with a DIN cable is that manufacturers can assign different functions to the various wires inside the cable. Thus, the fact that a DIN plug will fit into a DIN jack is no guarantee of connection success. Fortunately, manufacturers often supply DIN cables with their equipment. Further, components such as tape decks are often supplied with two types of ports: one for a DIN connector and other ports for the use of RCA plugs.

Three- and five-pin DIN connectors are common, but they can range to as many as eight pins (Fig. 18-36). Not all DIN pin arrangements are the same. Note

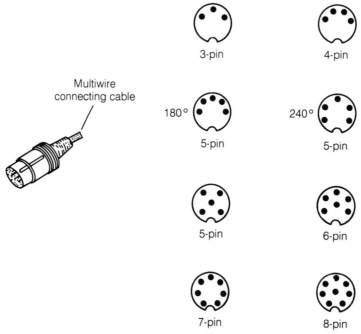

Fig. 18-36. Multiple-pin DIN connectors.

in the drawing the differences in the five-pin types. DIN plugs may also be straight or right angle. The right-angle DIN plug permits the cable to form an angle of 90° with the equipment to which it is connected. DIN connectors look alike, and a three-pin DIN can have the same outward appearance as an eight-pin DIN. To determine the type, count the number of pins.

# XLR plug

XLR plugs, sometimes called Cannon plugs (Fig. 18-37) are used with micro-

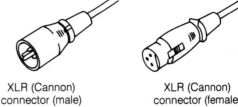

*Fig. 18-37.  Male and female XLR connectors.*

XLR (Cannon)
connector (male)

XLR (Cannon)
connector (female)

phone cables and, as shown in Fig. 18-38, are three- and five-contact keyed contacts, so it is possible to insert the plug in its jack in only one way.

XLR plugs and DIN connectors are functionally and physically different. The DIN connector uses wire pairs to carry signals to and from a component. The XLR is intended only for carrying signals from a mic to some component, such as a mixer.

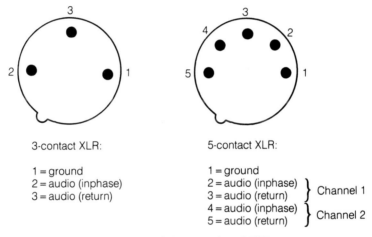

3-contact XLR:

1 = ground
2 = audio (inphase)
3 = audio (return)

5-contact XLR:

1 = ground
2 = audio (inphase) ⎫
3 = audio (return) ⎬ Channel 1
4 = audio (inphase) ⎫
5 = audio (return) ⎬ Channel 2

*Fig. 18-38.   3- and 5-contact keyed XLR contacts.*

The numbering of the pin positions on the plug or jack (sometimes called a socket) usually is clockwise when the plug or jack is held as shown in Fig. 18-39. (Sometimes, though, the XLR pins are numbered counterclockwise.) There is no standardization as to the way in which the wires of the cable are connected to the terminal points of the plug or jack. Drawing (A) shows the connections for balanced cable, and drawing (B) illustrates how an unbalanced cable can be connected. As there are only two connections for unbalanced cable the center, or hot, lead is wired to terminal 3. The shield, which also works as a signal conductor, is wired to terminal 2. Note that terminal 1 is unused. For a three-pin XLR, use pin 2 as the grounded shield connection, with either pin 1 or pin 3 for the hot lead. Drawing (C) shows the alternate connections for unbalanced cable.

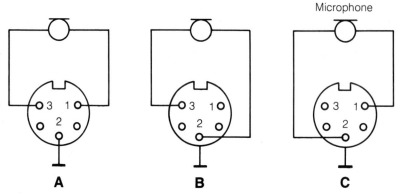

Fig. 18-39. Alternative XLR 3-pin connections.

Figure 18-40 shows the connections for a three-pin XLR. Note that the terminals are not numbered. Drawing (A) is an end view of the plug and the jack, and (B) shows the wiring arrangement for unbalanced cable. Drawing (C) indicates the connections for balanced coaxial cable, and (D) illustrates an unbalanced to a balanced output, using a center-tapped transformer. Terminals 2 and 3 of the output side of the transformer are the signal leads; the center tap connection is grounded.

Although there is no agreement on XLR connections, a standard for the three-pin XLR does exist and it is IEC 268-12. Using this standard, pin 1 is common, pin 2 is positive, and pin 3 is negative.

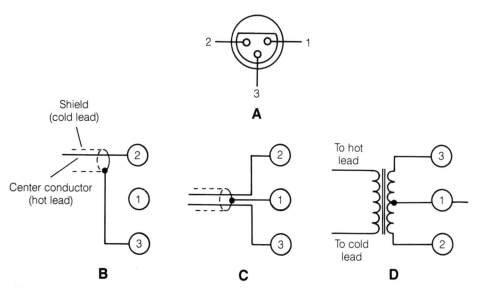

Fig. 18-40. 3-pin XLR connector: end view of plug or jack (A); unbalanced connection (B); balanced connection (C); unbalanced-to-balanced connection (D).

# Video camera connectors

Connectors for video cameras are often 10-pin types but some are also fourteen pins. There is no standardization for such connectors used to form a link between the camera and VCRs. The fact that a connector will fit does not automatically mean it is the correct one, as manufacturers may assign different circuits to the wiring. The connectors in Fig. 18-41 are quick-fit. Some video camera manufacturers supply interfacing cables to permit their cameras to be used with just about any VCR. Typical cable length is about 5 ft, but longer extension cables are available. Professional video cameras can use connectors having as many as fifty pins.

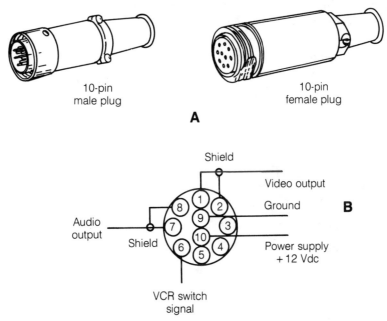

*Fig. 18-41. 10-pin male and female VCR/camera connectors (A) and wiring diagram (B).*

# Modem connections

Computer data can be delivered via telephone lines using a modem, an acronym from the words *modulate* and *demodulate*. Modulation is associated with signal transmission; demodulation with signal reception. There are two basic techniques: the acoustic and the direct mode. Figure 18-42 is an illustration of the use of the acoustic mode. The telephone handset is removed from its cradle and is positioned in the modem. The modem is connected to the computer via a 4-ft (1 220 mm) cable. The other end of this cable fits into a jack at the rear of the modem. The cable is keyed so it can fit one way. The telephone is connected to a telephone outlet and this can be done by a modular jack or a standard phone jack.

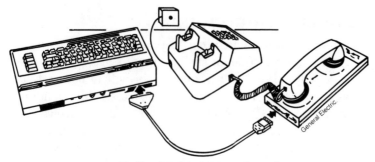

*Fig. 18-42.   Modem in acoustic mode.*

In the direct mode, the telephone handset remains in its cradle (Fig. 18-43). The modem is plugged into the telephone line, and the cable connection between the computer keyboard and the modem is the same as that shown in Fig. 18-42. In order for your computer to communicate over the telephone line you must first load a communications program into your computer. To place the modem on line, run the program.

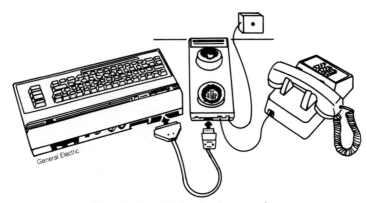

*Fig. 18-43.   Modem in direct mode.*

# Printed circuit board connections

The conductors etched into a board can terminate in a series of connectors. These connectors can be joined to a jack, which can then mate with a plug having corresponding pins.

# Jacks

Making a connection from one component to another involves plugs and jacks; the plugs are mounted on cables, the jacks on the equipment and sometimes on cables as well.

# Plugs and jacks for printed circuit boards

Printed circuit (PC) boards can make use of plugs and jacks specifically designed for them. The miniature horizontal jack can accept a tip plug of 0.080-in diameter in either end, or from the top or bottom. This jack has a width of 0.400 in and a depth of 0.156 in. The contact is silver-plated beryllium copper and has a current capacity of 5 A maximum.

# Banana and tip jacks

These jacks are mounted on the chassis of components or on the rear apron, although some, such as jacks for microphones or headphones, are positioned on the front for user convenience. A typical banana jack is shown in Fig. 18-44A. The threaded jack is held in place by a hex nut, and the internal connection is made to a soldering lug positioned between the hex nut and the front of the jack. In banana jacks the bore must be such that it will accommodate all standard banana plugs. The bore is smooth. As the banana plug is inserted in its jack the four-spring metal leaves of the plug are compressed, forcing good contact between the plug and the jack. Figure 18-44B shows the jack for a tip plug.

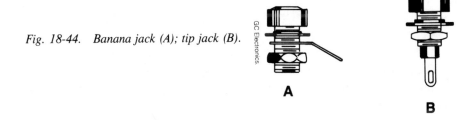

*Fig. 18-44.   Banana jack (A); tip jack (B).*

**A**

**B**

# Jack for unbalanced connector

Figure 18-45 shows the details for an unbalanced phone jack. It has two connecting elements. One is the tip connector that makes contact with the tip of the plug and is the connection for the hot lead. The other is the sleeve connector and this element of the jack contacts the sleeve of the plug, which, in turn, is attached to the cable shield braid. Inside the component on which the jack is mounted, the sleeve connector is automatically grounded to the chassis. A contact on the tip connector is wired to a circuit using a small length of shielded cable. The shield of that cable is also grounded to the chassis.

# Balanced connectors

Figure 18-46 shows the arrangement for two-wire, balanced or stereo connections

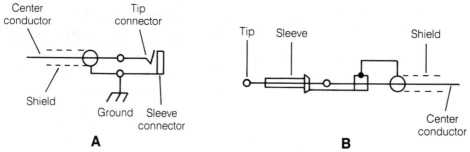

Fig. 18-45. *Connection details for a phone jack (A) and plug (B).*

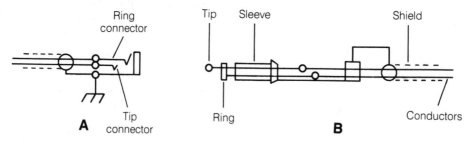

Fig. 18-46. *Jack (A) and plug (B) connections for balanced cable.*

for a jack. This jack has separate contacts for the tip and the ring of the mating plug. The metal frame of the jack makes contact with the long sleeve of the plug and is the ground connection. The tip and the ring are insulated from each other and from the shield of the plug. The tip and the ring, however, are connected to the two hot lead conductors in the plug.

## Shorting and nonshorting jacks

Jacks can be categorized in various ways; but one is whether the jack is a shorting or nonshorting type. Figure 18-47A illustrates a nonshorting jack, and (B) shows a shorting jack. In the nonshorting jack, the hot lead connector remains "floating," that is, it remains open. In the shorting type, the two upper terminals make contact. This contact is broken when the plug is inserted. The shorting arrangement can put a resistive load across the hot connection, killing any possible unwanted signal pickup, such as a hum or noise voltage, by the hot lead. The jack in (B) is sometimes called a closed-circuit jack.

## Inline and open-frame jacks

Jacks can be mounted on cables. Known as inline jacks, they are completely shielded, protecting the wiring of the cables to which they are attached. Jacks

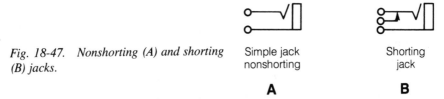

*Fig. 18-47. Nonshorting (A) and shorting (B) jacks.*

Simple jack
nonshorting

**A**

Shorting
jack

**B**

used on components such as receivers and amplifiers are open frame, so called because they have no need of a surrounding shell like that used by inline jacks. Inline jacks use a small machine screw for making connections; open-frame jacks have soldering terminals. Another type of jack used to replace the open-frame type is the enclosed jack. Electrically, the connections are the same as those used with the open-frame type but the metal enclosure of the jack protects the jack against unwanted signal pickup.

## Jack adapters

Because of the diversity of plugs and jacks, adapters are needed. Thus a component equipped with an RCA jack can be connected to some other component using a BNC jack. Many other variations are possible. Adapters can be used for jacks and some of these appear in Fig. 18-48. Drawing (A) shows an F-jack to F-jack

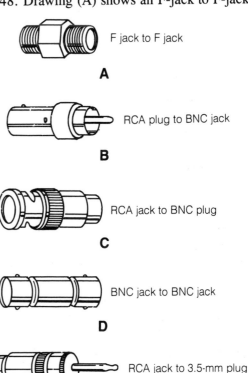

F jack to F jack

**A**

RCA plug to BNC jack

**B**

*Fig. 18-48. Jack adapters: F jack to F jack (A); RCA plug to BNC jack (B); RCA jack to BNC plug (C); BNC jack to BNC jack (D); RCA jack to 3.5-mm plug (E).*

RCA jack to BNC plug

**C**

BNC jack to BNC jack

**D**

RCA jack to 3.5-mm plug

**E**

adapter. An adapter can be a combined plug-jack, as in (B). Here the input is a BNC jack; the output is an RCA plug. This adapter could be used, for example, if a connecting cable was equipped with a BNC plug. It would be inserted into the BNC jack portion of the adapter. The end of the plug equipped with the RCA plug could then be inserted into an RCA jack.

Drawing (C) indicates connections that are exactly the opposite of those in (B). Here the RCA jack can accommodate a cable equipped with an RCA plug. The BNC plug output of the adapter can then be inserted into a BNC jack. Drawing (D) is a BNC jack to BNC jack adapter. An adapter of this kind could be used when a cable equipped with BNC plugs is to be extended. Finally, in drawing (E) there is an RCA jack to a 3.5-mm plug adapter. This adapter could be used with a cable equipped with an RCA plug. This plug could be inserted into the RCA jack end of the adapter, and the 3.5-mm plug could be inserted into a corresponding jack.

# 19
CHAPTER

# Summary of electronics formulas

## dc circuits

### Ohm's law for dc

$$I = E/R = P/E = \sqrt{P/R}$$

$$R = E/I = P/I^2 = E^2/P$$

$$E = IR = P/I = \sqrt{PR}$$

$$P = EI = E^2/R = I^2R$$

Ohm's law

$I$ is in amperes

$R$ is in ohms

$E$ is in volts

$P$ is in watts

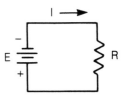

### n resistors in series

$$R_t = R_1 + R_2 + \ldots + R_n$$

### Three resistors in series

$$R_t = R_1 + R_2 + R_3$$

$$\theta = 0°$$

$R$ is in ohms, or identical multiples

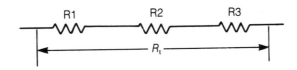

# n resistors in parallel

$$\frac{1}{R_t} = \frac{1}{R_1} + \frac{1}{R_2} + \dots + \frac{1}{R_n}$$

$$R_t = \frac{1}{\dfrac{1}{R_1} + \dfrac{1}{R_2} + \dots + \dfrac{1}{R_n}}$$

# Two resistors in parallel

$$R_t = \frac{R_1 \times R_2}{R_1 + R_2}$$

$$\frac{1}{R_t} = \frac{1}{R_1} + \frac{1}{R_2}$$

$$R_t = \frac{R_1}{2} \text{ or } \frac{R_2}{2} \text{ (when both resistors are identical)}$$

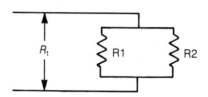

# Three resistors in parallel

$$\frac{1}{R_t} = \frac{1}{R_1} + \frac{1}{R_2} + \frac{1}{R_3}$$

$$R_t = \frac{1}{\dfrac{1}{R_1} + \dfrac{1}{R_2} + \dfrac{1}{R_3}}$$

$$R_t = \frac{R_1}{1 + \dfrac{R_1}{R_2} + \dfrac{R_1}{R_3} + \dfrac{R_1}{R_4}}$$

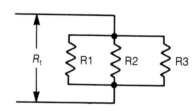

# Resistors in series-parallel (example)

$$R_t = R_1 + \frac{1}{\dfrac{1}{R_2} + \dfrac{1}{R_3} + \dfrac{1}{R_4}}$$

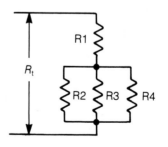

$$R_t = R_1 + \frac{R_2 \times R_3}{R_2 + R_3} + R_4 + R_5 + \frac{R_6 \times R_7}{R_6 + R_7}$$

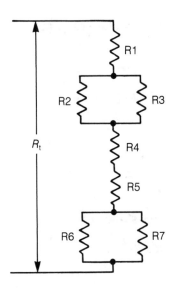

### dc voltage drop

$$e = E \times \frac{R_1}{R_1 + R_2}$$

$E$ is the dc source voltage
R1 and R2 are series resistors
$e$ is voltage across R1

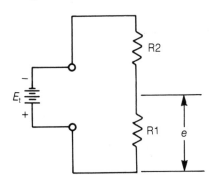

### Voltage divider

$$E_t = I(R_1 + R_2)$$
$$E_t = E_1 + E_2$$

## Parallel voltage drops

$$E = E_1 = E_2$$

$$I_1 = \frac{I_2 \times R_2}{R_1}$$

$$R_1 = \frac{I_2 \times R_2}{I_1}$$

$$I_2 = \frac{I_1 \times R_1}{R_2}$$

$$R_2 = \frac{I_1 \times R_1}{I_2}$$

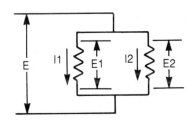

## Conductances in parallel

$G_t = G_1 + G_2 + G_3 \ldots$

$E = I/G$

$I = E/G$

$G = \dfrac{1}{R}$

$R = \dfrac{1}{G}$

Ohm's law for conductance

$G$ is in siemens (mhos)

$E$ is in volts

$I$ is in amperes

$R$ is in ohms

## Resistance of a conductor

$R = K\dfrac{l}{D^2}$

$R$ is in ohms

$K$ is resistance of a mil-foot (specific resistance)

$l$ is length in feet and

$D$ is diameter in mils

## Power

$P = I^2 \times R$

$P = E^2/R$

$P = E \times I$

$kW = \dfrac{E \times I}{1\ 000}$

Kilowatts = watts/1 000

Watts = kilowatts × 1 000

Watt-hours = watts × hours

Kilowatt-hours = kilowatts × hours

## Current

$I = \sqrt{P/R}$

$I = P/E$

$I^2 = P/R$

## Voltage

$E = \sqrt{P \times R}$

$E^2 = P \times R$

$E = P/I$

## Percentage voltage regulation

$$\text{Voltage regulation} = \frac{E_{nl} - E_{fl}}{E_{fl}} \times 100$$

$E_{nl}$ is the no-load voltage
$E_{fl}$ is the full-load voltage

## Resistance

$R = P/I^2$
$R = E^2/P$

## Shunt law

$I_1 \times R_1 = I_2 \times R_2$

$I_1$ and $I_2$ are currents in amperes
$R_1$ and $R_2$ are resistances in ohms

$$I_1 = \frac{I_2 \times R_2}{R_1}$$

$$I_2 = \frac{I_1 \times R_1}{R_2}$$

$$R_1 = \frac{I_2 \times R_2}{I_1}$$

$$R_2 = \frac{I_1 \times R_1}{I_2}$$

## Kirchhoff's current law

$I_1 + I_2 = I_3 + I_4 + I_5$

$I_1 + I_2 - I_3 - I_4 - I_5 = 0$

J = junction

The sum of the currents flowing to and from a junction is zero

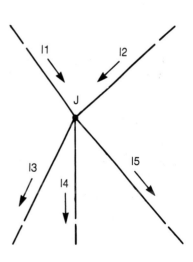

## Kirchhoff's voltage law

$E_1+E_2+E_3=0$

The sum of the voltage drops in a closed loop is equal to zero

# Time constants

### Time constant for series RL

$t=L/R$

$t$ is the time (in seconds) for current to reach 63.2% of peak
$L$ is in henries
$R$ is in ohms

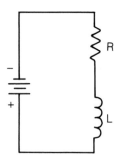

### Time constant for series RC

$t=R\times C$

$t$ is the time (in seconds) for the voltage across the capacitor to reach 63.2% of peak
$R$ is in ohms
$C$ is in farads

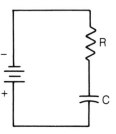

# Meters and measuring circuits

$I_s+R_s=I_m\times R_m$

$I_s$ is shunt current
$I_m$ is meter current
Both $I_s$ and $I_m$ must be in same current units.
$R_m$ is meter resistance
$R_s$ is the shunt resistance
Both $R_s$ and $R_m$ must be in same resistance unit

### Meter multiplier resistance

$R=R_m(n-1)$

$R$ is value of multiplier resistance
$R_m$ is meter resistance
n is the multiplication factor

## Meter sensitivity

$$M_s = \frac{R_m + R_1}{E}$$

$M_s$ is meter sensitivity in ohms per volt
$R_m$ is meter resistance
$R_1$ is multiplier resistance
$E$ is full-scale reading in volts

## Wheatstone bridge

$$R_x = \frac{R_3 \times R_2}{R_1}$$

$R_x$ is unknown resistance value
$R_1$, $R_2$, and $R_3$ are resistance values of
the bridge elements
G is a galvanometer

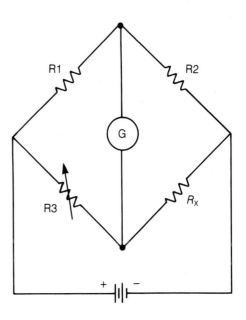

## Inductance bridge

$$L_x = R_1 \times R_2 \times C_1$$

$R_1$ and $R_2$ are in ohms
$C_1$ is in farads
$L_x$ is in henries

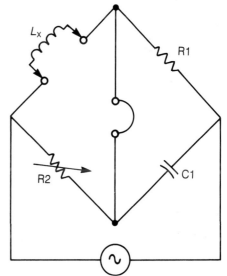

## Capacitance bridge

$$C_x = C_1 \times \frac{R_1}{R_2}$$

The unknown capacitance, $C_x$ and standard capacitance, $C_1$, are in $\mu F$
Fixed resistance $R_1$ and variable resistance $R_2$ are in ohms

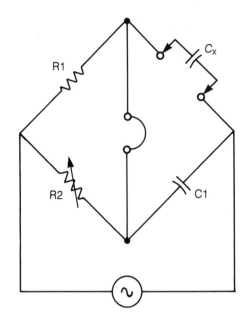

## Slide-wire bridge

$$R_X = \frac{L_2}{L_1 - L_2} \times R_1$$

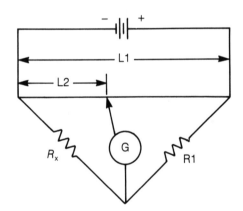

# ac circuits
## Ohm's law for ac

$$E_C \quad = I \times X_c$$

$$E_L \quad = I \times X_L$$

$$E_R \quad = I \times R$$

$$E_{source} = I \times Z$$

Peak ac power in resistive circuit

$$P_{peak} = \frac{E^2_{peak}}{R}$$

Impedance

$$Z = E/I$$

Current

$$I = E/Z$$

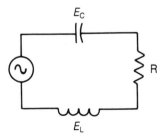

### Single capacitor

$Z = X_c$

$\theta = -90°$

### n capacitors in series

$$\frac{1}{C_t} = \frac{1}{C_1} + \frac{1}{C_2} + \ldots + \frac{1}{C_n}$$

$$C_t = \frac{1}{\dfrac{1}{C_1} + \dfrac{1}{C_2} + \ldots + \dfrac{1}{C_n}}$$

$$Z = \frac{1}{2\pi f C_t}$$

$\theta = -90°$

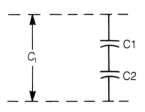

### Two capacitors in series

$$C_t = \frac{C_1 \times C_2}{C_1 + C_2}$$

All capacitance units must be in farads
or in identical submultiples

$$Z = \frac{1}{2\pi f}\left(\frac{C_1 + C_2}{C_1 C_2}\right)$$

$\theta = -90°$

### n capacitors in parallel

$$C_t = C_1 + C_2 + \ldots + C_n$$

$$Z = \frac{1}{2\pi f C_t}$$

$\theta = -90°$

**Three capacitors in parallel**

$C_t = C_1 + C_2 + C_3 \ldots$

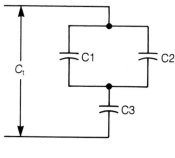

$$Z = \cfrac{1}{\cfrac{1}{X_{C1}} + \cfrac{1}{X_{C2}} + \cfrac{1}{X_{C3}}}$$

$\theta = -90°$

**Capacitors in series-parallel (example)**

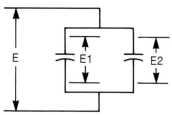

$$C_t = \frac{(C_1 + C_2)\,(C_3)}{C_1 + C_2 + C_3}$$

**Parallel capacitor voltages**

$E = E_1 = E_2$

# Inductors

**Single inductance**

$Z = X_L$

$\theta = +90°$

**n inductors in series (no coupling)**

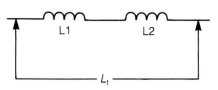

$L_t = L_1 + L_2 + \ldots + L_3$

$Z = 2\pi f L_t$

$\theta = +90°$

All inductance units must be in henries or identical submultiples

## Two inductors in series (no coupling)

$L_t = L_1 + L_2$

$Z = X_{L1} + X_{L2}$

$\theta = +90°$

## Inductors in series (aiding)

$L_t = L_1 + L_2 + 2M$

$M$ is mutual inductance, in henries

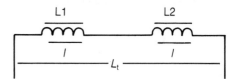

## Inductors in series (opposing)

$L_t = L_1 + L_2 - 2M$

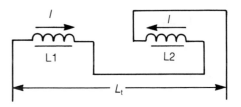

## n inductors in parallel (no coupling)

$$\frac{1}{L_t} = \frac{1}{L_1} + \frac{1}{L_2} + \ldots + \frac{1}{L_n}$$

$$L_t = \cfrac{1}{\cfrac{1}{L_1} + \cfrac{1}{L_2} + \ldots + \cfrac{1}{L_n}}$$

$Z = 2\pi f L_t$

$\theta = +90°$

## Two inductors in parallel (no coupling)

$$L_t = \frac{L_1 \times L_2}{L_1 + L_2}$$

$$Z = 2\pi f \left( \frac{L_1 \times L_2}{L_1 + L_2} \right)$$

$$\theta = +90°$$

# Combined impedance

Resistance and inductance in series

$$Z = \sqrt{R^2 + X_L{}^2}$$

$$\theta = \arctan \frac{X_L}{R}$$

Resistance and capacitance in series

$$Z = \sqrt{R^2 + X_C{}^2}$$

$$\theta = \arctan \frac{X_C}{R}$$

Inductance and capacitance in series

$$Z = X_L - X_C$$

$\theta = -90°$ when $X_L < X_C$

$\quad = 0°$ when $X_L = X_C$

$\quad = +90°$ when $X_L > X_C$

Resistance, inductance, and capacitance in series when $X_L > X_C$

$$Z = \sqrt{R^2 + (X_L - X_C)^2}$$

$$\theta = \arctan \frac{X_L - X_C}{R}$$

Two resistances in parallel

$$Z = \frac{R_1 R_2}{R_1 + R_2}$$

$$\theta = 0°$$

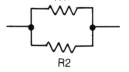

Three or more resistors in parallel

$$Z = \frac{1}{\dfrac{1}{R_1} + \dfrac{1}{R_2} + \dfrac{1}{R_3} \ \cdots}$$

$$\theta = 0°$$

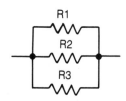

Inductance and resistance in parallel

$$Z = \frac{RX_L}{\sqrt{R^2 + X_L^2}}$$

$$\theta = \arctan \frac{R}{X_L}$$

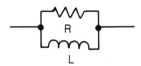

Capacitance and resistance in parallel

$$Z = \frac{RX_C}{\sqrt{R^2 + X_C^2}}$$

$$\theta = -\arctan \frac{R}{X_C}$$

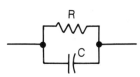

Inductance and capacitance in parallel

$$Z = \frac{X_L X_C}{X_L - X_C}$$

$$\theta = 0° \text{ when } X_L = X_C$$

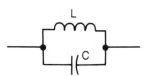

Inductance, resistance, and capacitance in parallel

$$Z = \frac{RX_L X_C}{\sqrt{X_L^2 X_C^2 + R^2 (X_L - X_C)^2}}$$

$$\theta = \arctan \frac{RX_C - RX_L}{X_L X_C}$$

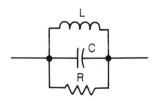

Inductance and series resistance in parallel with capacitance

$$Z = X_C \sqrt{\frac{R^2 + X_L^2}{R^2 + (X_L - X_C)^2}}$$

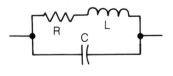

$$\theta = \arctan\left(\frac{X_L(X_C - X_L) - R^2}{RX_C}\right)$$

Capacitance and series resistance in parallel with inductance and series resistance

$$Z = \sqrt{\frac{(R_L^2 + X_L^2)(R_C^2 + X_C^2)}{(R_L + R_C)^2 + (X_L - X_C)^2}}$$

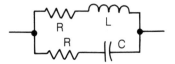

$$\theta = \arctan \frac{X_L(R_C^2 + X_C^2) - X_C\,(R_L^2 + X_L^2)}{R_L(R_C^2 + X_C^2) + R_C\,(R_L^2 + X_L^2)}$$

## Q or figure of merit

a coil

$$Q = \frac{X_L}{R_L}$$

a single capacitor

$$Q = \frac{X_C}{R_C}$$

## Wavelength and frequency of a sine wave

$$\lambda = \frac{300\ 000}{f}$$
$\lambda$ is wavelength in meters
$f$ is frequency in kilohertz

$$\lambda = \frac{30}{f}$$
$\lambda$ is wavelength in centimeters
$f$ is frequency in gigahertz

$$\lambda = \frac{984\ 000}{f}$$
$\lambda$ is wavelength in feet
$f$ is frequency in kilohertz

$$\lambda = \frac{984}{f}$$
$\lambda$ is wavelength in feet
$f$ is frequency in megahertz

$$\lambda = \frac{11.8}{f}$$

$\lambda$ is wavelength in inches
$f$ is frequency in gigahertz

$$f = \frac{3 \times 10^5}{\lambda} \text{ (kHz)}$$

$\lambda$ = wavelength in meters

$$f = \frac{3 \times 10^4}{\lambda} \text{ (MHz)}$$

$\lambda$ = wavelength in centimeters

$\mathring{A} = 3.937 \times 10^{-9}$ inch

$\mathring{A}$ is 1 angstrom unit

$\mathring{A} = 1 \times 10^{-10}$ meter

$\mathring{A} = 1 \times 10^{-4}$ micron

The micron was formerly referred to as a micrometer

1 micron $= 3.937 \times 10^{-5}$ in
1 micron $= 1 \times 10^{-6}$ m
1 micron $= 1 \times 10^4$ $\mathring{A}$

## Period of a sine wave

$t = 1/f$

$t$ is time in seconds
$f$ is frequency in hertz

## Sine wave voltages or currents

Average value of voltage or current sine wave

$$E_{av} = 0.637 \times E_{peak}$$

Peak value of voltage or current sine wave

$$E_{peak} = \frac{E_{av}}{0.637}$$

Peak value of voltage or current sine wave

$$E_{peak} = 1.57 \times E_{av}$$

Peak-to-peak value of voltage or current sine wave

$$E_{p\text{-}p} = 2 \times E_{peak}$$

Peak value of voltage or current sine wave

$E_p = E_{p-p}/2$

Average value of voltage or current sine wave

$E_{av} = E_{p-p} \times 0.318\ 5$

Average value of voltage or current sine wave

$E_{av} = (E_{p-p}/2) \times 0.637$

Average value of voltage or current sine wave

$E_{av} = E_{p-p} \times 0.5 \times 0.637$

Average value of voltage or current sine wave

$E_{av} = \dfrac{E_{p-p} \times 1}{2} \times 0.637 = E_{p-p} \times 1/2 \times 0.637$

Peak-to-peak value of voltage or current sine wave

$E_{p-p} = \dfrac{E_{av}}{0.318\ 5}$

rms (effective) value of voltage or current sine wave

$I_{eff} = 0.707 \times I_{peak}$

rms value of voltage or current sine wave

$I_{eff} = \dfrac{I_{peak}}{1.414}$

Peak value of voltage or current sine wave

$I_{peak} = \dfrac{I_{eff}}{0.707}$

Peak-to-peak value of voltage or current sine wave

$E_{p-p} = 2.828 \times E_{eff}$

rms value of voltage or current sine wave

$E_{eff} = \dfrac{E_{p-p}}{2.828} = \dfrac{E_{peak}}{\sqrt{2}} = \dfrac{E_{peak}}{1.414} = 0.707\ E_{peak}$

## Susceptance

$$B = \frac{X}{R^2 + X^2}$$

$$B = 1/X$$

$B$ = susceptance in siemans (mhos)
$R$ = resistive ohms
$X$ = reactive ohms

## Apparent power

$$P = E \times I$$

$P$ = apparent power in watts
$E$ = voltage in volts
$I$ = current in amperes

## True power

$$P = EI \cos \phi = EI \times pf$$

$$pf = P/EI = \cos \phi$$

$pf$ = power factor

$$\cos \phi = \text{true power/apparent power}$$

$$E = I \times Z$$

$$E = \frac{P}{I \times \cos \phi}$$

$$E = \sqrt{\frac{P \times Z}{\cos \phi}}$$

$$Z = \frac{E}{I}$$

$$Z = \frac{E^2 \times \cos \phi}{P}$$

$$Z = \frac{P}{I^2 \times \cos \phi}$$

$$pf = \frac{R}{Z}$$

$$pf = \frac{W}{VA}$$

$$P = \frac{E^2 \times \cos \phi}{Z}$$

$$P = I^2 \times R$$

$$P = E \times I \times \cos \phi$$

$$P = I^2 \times Z \times \cos \phi$$

$$I = \sqrt{\frac{P}{Z \times \cos \phi}}$$

$$I = \sqrt{\frac{P}{R}}$$

$$I = \frac{E}{Z}$$

$$I = \frac{P}{E \times \cos \phi}$$

Average ac power in resistive circuit

$$P_{av} = I^2_{av} \times R$$

Peak ac power in resistive circuit

$$P_{peak} = I^2_{peak} \times R$$

Average ac power in resistive circuit

$$P_{av} = \frac{E^2_{av}}{R}$$

## ac voltage

ac voltage across a capacitor or coil

$$E = I \times X$$

Source voltage

$$E_{source} = \sqrt{E_R^2 + E_X^2}$$

Applied voltage in RC circuit

$$E_{source} = I \times \sqrt{R^2 + X^2}$$

Voltages in series RC circuits

$$E = I \times Z = \text{source voltage}$$
$$I = E/Z$$
$$Z = E/I$$

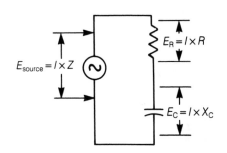

Applied voltage in RL circuit

$$E_{source} = I \times \sqrt{R^2 + X_L^2}$$

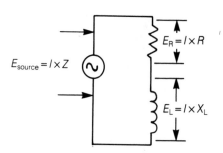

ac voltage across a coil

$$E_L = I_L \times X_L$$

Current through a coil

$$I_L = \frac{E_L}{X_L}$$

## Reactance

Inductive reactance

$$X_L = 2\pi f L = 6.28\, fL$$

$\pi = 3.141\ 59$ approximately
$f$ is frequency in Hz
$C$ is capacitance in farads
$L$ is inductance in henries

Capacitive reactance

$$X_C = \frac{1}{2\pi f C}$$

$$= \frac{1}{6.28\, fC}$$

$$X_L = \frac{E_L}{I_L}$$

Voltages in series RL circuits

$E = I \times Z =$ source voltage
$I = E/Z$
$Z = E/I$

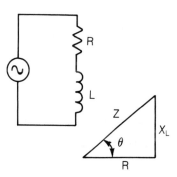

Phase angle of series RL circuit

$$\phi = \arctan \frac{X_L}{R}$$

Phase angle of series RC circuit

$$\phi = \arctan \frac{X_C}{R}$$

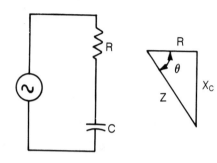

In series circuits where the phase angle and any two of the Z, R, and X components are known, the unknown component can be determined from:

$$Z = \frac{R}{\cos \phi}$$

$$Z = \frac{X}{\sin \phi}$$

$$R = Z \cos \phi$$

$$X = Z \sin \phi$$

## Resonance

Impedance at resonance: $X_L = X_C$

Resonant frequency

$$f_0 = \frac{1}{2 \pi \sqrt{LC}}$$

$f_o$ is the frequency in hertz
$L$ is in henries
$C$ is in farads

Series resonance

Z (at any frequency) $= R + j(X_L - X_C)$
Z (at resonance) $= R$

Parallel resonance

$$Z_{max} \text{ (at resonance)} = \frac{X_L X_C}{R} = \frac{X_L^2}{R}$$

$$(X_L = X_C) = Q X_L = \frac{L}{RC}$$

## Filters

$R$ = ohms
$L$ = henries
$C$ = farads
$(1\ \mu\mathrm{F} = 10^{-6}\ \mathrm{F})$
$T$ = time constant (seconds)
$f_\mathrm{r}$ = resonant frequency (hertz)
$\omega$ = $2\pi f$
$2\pi$ = 6.28

$$\frac{1}{2\pi} = 0.159\,2$$

### Low pass RC

$$\frac{E_{\mathrm{out}}}{E_{\mathrm{in}}} = \frac{1}{\sqrt{1+\omega^2 T^2}}$$

$$T = RC$$

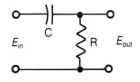

### High pass RC

$$\frac{E_{\mathrm{out}}}{E_{\mathrm{in}}} = \frac{1}{\sqrt{1+\dfrac{1}{\omega^2 T^2}}}$$

$$T = RC$$

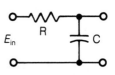

### Low pass RL

$$\frac{E_{\mathrm{out}}}{E_{\mathrm{in}}} = \frac{1}{\sqrt{1+\omega^2 T^2}}$$

$$T = \frac{L}{R}$$

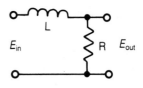

### High pass RL

$$\frac{E_{\mathrm{out}}}{E_{\mathrm{in}}} = \frac{1}{\sqrt{1+\dfrac{1}{\omega^2 T^2}}}$$

$$T = \frac{L}{R}$$

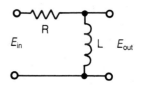

## Low pass LC

$$\frac{E_{out}}{E_{in}} = \frac{1}{\omega^2 LC - 1} = \frac{1}{\frac{f^2}{f_0^2} - 1}$$

$$f_0 = \frac{0.159\,2}{\sqrt{LC}}$$

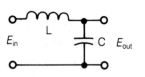

## High pass LC

$$\frac{E_{out}}{E_{in}} = \frac{1}{\frac{1}{\omega^2 LC} - 1} = \frac{1}{\frac{f_0^2}{f^2}}$$

$$f_0 = \frac{0.159\,2}{\sqrt{LC}}$$

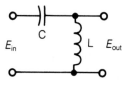

# Transformers

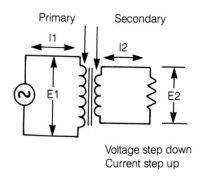

Voltage step down
Current step up

For all transformer turns ratios:

$P = E \times I$

$E_1 \times I_1 = E_2 \times I_2$

(disregarding transformer losses)

Transformer primary turns

$$N_1 = \frac{E_1 \times N_2}{E_2}$$

Transformer secondary turns

$$N_2 = \frac{E_2 \times N_1}{E_1}$$

Transformer primary current

$$I_1 = \frac{I_2 \times N_2}{N_1}$$

**Transformer secondary current**

$$I_2 = \frac{I_1 \times N_1}{N_2}$$

**Transformer primary turns**

$$N_1 = \frac{I_2 \times N_2}{I_1}$$

**Transformer secondary turns**

$$N_2 = \frac{I_1 \times N_1}{I_2}$$

**Voltage**

$$E_1 = \frac{E_2 \times I_2}{I_1}$$

$$E_2 = \frac{E_1 \times I_1}{I_2}$$

**Current**

$$I_1 = \frac{E_2 \times I_2}{E_1}$$

$$I_2 = \frac{E_1 \times I_1}{E_2}$$

**Transformer primary voltage**

$$E_1 = \frac{E_2}{T_r}$$

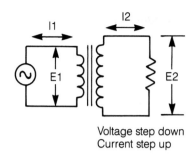

Voltage step down
Current step up

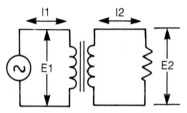

Primary current times primary
voltage = secondary current
times secondary voltage

Small magnetizing
current flows in
primary winding

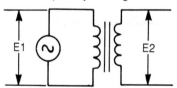

Secondary is open circuited
No current flows
No power transferred from
primary to secondary
Voltage exists across secondary

$T_r$=turns ratio

Transformer secondary voltage

$E_2 = E_1 \times T_r$

Transformer primary voltage

$E_1 = \dfrac{E_2 \times N_1}{N_2}$

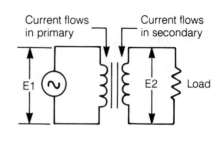

Transformer secondary voltage

$E_2 = \dfrac{E_1 \times N_2}{N_1}$

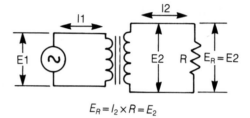

$E_R = I_2 \times R = E_2$

Turns ratio

$T_r = \sqrt{\dfrac{Z_2}{Z_1}} = \dfrac{N_2}{N_1}$    $Z_2$ and $Z_1$ are secondary and primary impedances in ohms
$N_1$ = primary turns
$N_2$ = secondary turns

Hysteresis loss

$P = kVf\,B^x_{max}$

$P$ = power loss, watts
$k$ = a constant for a given specimen
$V$ = volume of iron, meters$^3$
$f$ = frequency, hertz
$B$ = flux density, tesla
x = an index between 1.5 and 2.3, often taken as 2

Eddy current loss

$P = kt^2f^2B^2$

$P$ = power loss, watts
$k$ = a constant for a given specimen
$t$ = thickness of laminations, meters
$f$ = frequency, hertz
$B$ = flux density, tesla

Maximum ac magnetizing force

$$H = \frac{NI\sqrt{2}}{l}$$

$H$ = magnetizing force, amperes per meter
$N$ = number of turns in magnetizing coil
$I$ = rms current, amperes
$l$ = magnetic path length, meters

## Incremental permeability

$$\mu_\Delta = \left( \frac{10^9 l}{8\pi^2 N_\Delta{}^2 fA} \right) Z$$

$$Z = \sqrt{R^2 + \omega^2 L^2}$$

$$\omega = 2\pi f = 6.28f$$

$$\theta = \tan^{-1}\left( \frac{R}{\omega L} \right)$$

$\mu_\Delta$ = relative incremental permeability
$\theta$ = phase angle of complex permeability, degrees
$R$ = resistance, ohms
$\omega$ = angular frequency, radians per second
$l$ = magnetic path length, meters
$N_\Delta$ = number of turns
$f$ = frequency, hertz
$L$ = inductance, henries
$A$ = cross-sectional area, square meters

Conversion factors

One ampere-turn per meter $= 4\pi \times 10^{-3}$ Oe
1 Wb/m$^2$ $= 10^4$ G
1 Wb/m$^2$ $= 1$ T
1 W/kg $= 0.454$ W/lb
1 W/lb $= 2.2$ W/kg

Transformer and magnet formulas

$$H = \frac{NI}{l}$$

$H$ = magnetizing force, amperes per meter
$N$ = number of turns in magnetizing coil
$I$ = current, amperes
$l$ = magnetic path length, meters

## Permeability

$$B = \mu H$$

$B$ = flux density, tesla
$\mu$ = permeability
$H$ = magnetizing force, amperes per meter

## Resistivity

$$\varrho = \frac{RA}{l}$$

$\varrho$ = resistivity, ohm meters

$R$ = resistance, ohms
$A$ = area, square meters
$l$ = length, meters

## Transformer equation

$$E = 4.44 \, Nf\phi$$

$E$ = induced voltage
$N$ = number of turns
$f$ = frequency, hertz
$\phi$ = magnetic flux, webers

## Area

$$A = l \times w$$

$l$ = length
$w$ = width

## Volume

$$V = l \times w \times d$$

$V$ = volume in cubic units
$A$ = area in square units
$d$ = depth

# Generators

## Generator frequency

$$f = \frac{\text{rpm} \times p}{120}$$

$f$ = frequency in hertz
$p$ = number of poles
rpm = revolutions per minute

Rotating speed of a generator

$$\text{rpm} = \frac{120 \times f}{p}$$

Number of poles of generator

$$p = \frac{120 \times f}{\text{rpm}}$$

Star-wound ac generator

$$E_{\text{line}} = 1.732 \times E_{2\text{-phase}}$$

Power in single-phase reactive circuit

$$P = E \times I \times pf \qquad\qquad pf = \text{power factor}$$

Power in three-phase reactive circuit

$$P = E \times I \times pf \times 1.732$$

Current in three-phase reactive circuit

$$I = \frac{P}{E \times pf \times 1.732}$$

Voltage in three-phase reactive circuit

$$E = \frac{P}{I \times pf \times 1.732}$$

Current (single-phase)

$$I = \frac{746 \times \text{hp}}{E \times \eta \times pf} \qquad\qquad \eta = \text{efficiency}$$

Current (three-phase)

$$I = \frac{746 \times \text{hp}}{E \times \eta \times pf \times 1.732} \qquad\qquad \text{hp} = \text{horsepower}$$

Single-phase kva

$$\text{kva} = \frac{E \times I}{1\,000} \qquad\qquad \text{kva} = \text{kilovolt-amperes}$$

Three-phase kva

$$\text{kva} = \frac{E \times I \times 1.732}{1\ 000}$$

Two- or three-wire distribution system

$$E_{\text{load}} = E_{\text{source}} \pm \text{feeder drops}$$

# Illumination

lumens = candlepower × 12.57
candlepower = lumens/12.57

# Cross-sectional area

### Square mils

Square mils = (any side)²

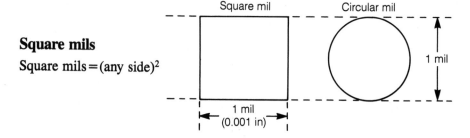

### Circular mils

Circular mils = $d \times d = d^2$
Circular mil area = Square mils/0.785 4
Square mil area = Circular mil area × 0.785 4

# Specific resistance

$R = \varrho \dfrac{l}{A}$ at 20 °C for drawn copper

### Mil foot

Wire standard consisting of wire whose length is 1 ft and whose diameter is 1 mil.

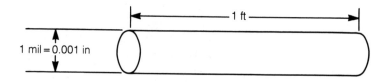

# Efficiency

Output = Input × Efficiency

(efficiency expressed as a decimal)

$$\text{Efficiency} = \frac{\text{Output}}{\text{Input}}$$

Output and input in watts

$$\text{Input} = \frac{\text{Output}}{\text{Efficiency}}$$

Percentage efficiency

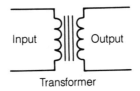

Input    Output

Transformer

$$\% \text{ Efficiency} = \frac{\text{Output}}{\text{Input}} \times 100$$

# Temperature coefficiency of resistance

$$R_t = R_0(1 + \alpha T)$$

$R_0$ = resistance at $0\,°\text{C}$
$\alpha$ = temperature coefficient of resistance of copper wire at $0\,°\text{C}$, or
0.004 27
$T$ = temperature in $°\text{C}$

# Heating effect of a current

$$H = 0.057\ 168 \times R$$

$H$ is heat in calories per second
$R$ is resistance in ohms

# Temperature

$$F = (C \times 9/5) + 32$$
$$C = (F - 32) \times 5/9$$
$$\eta = P_{out}/P_{in}$$

$F$ is temperature in $°\text{F}$
$C$ is temperature in $°\text{C}$
$\eta$ is efficiency
$P_{out}$ is output power
$P_{in}$ is input power
Multiply answer by 100 to obtain
efficiency in terms of percentage
rather than a decimal

# Horsepower

1 hp = 746 W = 550 ft lb/sec      hp is horsepower

1 W = 1/746 hp = 0.001 341 hp

$$kW = \frac{hp \times 746}{10^3}$$

$$hp = \frac{No.\ of\ ft \cdot lbs/min}{33\,000}$$

$$hp = \frac{E \times I}{746}$$

$E$ = voltage in volts

$I$ = current in amperes

Ampere-hours

Ampere-hours = amperes × hours

Current (dc)

$$I = \frac{746 \times hp}{E \times \eta}$$

$\eta$ = efficiency

Current (ac)

$$I = \frac{746 \times hp}{E \times \eta \times pf \times 1.732}$$

# Bandwidth

$$\Delta = \frac{f_r}{Q} = \frac{R}{2\pi L}$$

# Transistor characteristics

$$\alpha = \frac{\Delta i_c}{\Delta i_e}$$

$\alpha$ is current gain

$i_c$ is collector current

$i_e$ is emitter current

### Resistance gain

$$r_g = \frac{r_o}{r_i}$$

$r_g$ is resistance gain

$r_o$ is output resistance

$r_i$ is input resistance

### Voltage gain

$$v_g = \frac{e_o}{e_i} = \frac{i_c r_o}{i_e r_i} = \frac{r_o}{r_i} \alpha$$

$e_o$ is output voltage

$e_i$ is input voltage

### Power gain

$$P_g = \alpha^2 \frac{r_o}{r_i}$$

### Base-current amplification factor

$$\beta = \frac{\Delta i_c}{\Delta i_b}$$

$i_c$ is collector current

$\beta$ is base current amplication factor (beta)

# 20

## CHAPTER

# Electronics symbols and abbreviations

## Symbols

The symbols shown in Table 20-1 are those that are commonly used, although there are some variations.

## Abbreviations

Technical abbreviations shown in Table 20-2 appear in electronics text material, in circuit diagrams, in reports, and in magazine articles. There is no industry-wide style standard. The best that can be hoped for is that a selected style will be consistent throughout a book, report, or article. Abbreviations might be in lower-case, capital letters, or some combination of both. Abbreviations might or might not have periods. Periods are desirable where the abbreviation could be mistaken for a word.

## Notation

Components can be identified alphanumerically. Figure 20-1 shows the three different types of notation that are used. The numbers following the letters are used for identification only and should not be regarded as multipliers.

## Schematic symbols

Schematic symbols (Fig. 20-2) are an electronic shorthand way of depicting a component.

*Table 20-1. Electronic units, abbreviations, and symbols.*

| Electronic unit | Unit abbreviation | Component symbol |
|---|---|---|
| Capacitance | F=farad<br>μF=microfarad<br>pF=picofarad | Symbol $C$ |
| Frequency | Hz=hertz (cycles per second)<br>kHz=kilohertz<br>MHz=megahertz | Symbol $f$ |
| Inductance | H=henry<br>mH=millihenry<br>μH=microhenry | Symbol $L$ |
| Resistance | Ω=ohm<br>KΩ=kilohms<br>MΩ=megohms | Symbol $R$ |
| Time | s=second<br>ms=millisecond<br>μs=microsecond<br>ns=nanosecond | Symbol $t$ |
| Current | A=ampere<br>mA=milliampere<br>μA=microampere | Symbol $I$ or $i$ |
| Voltage | V=volt<br>mV=millivolt<br>μV=microvolt<br>kV=kilovolt | Symbol $E$ or $V$ |
| Power | W=watt<br>mW=milliwatt<br>μW=microwatt<br>kW=kilowatt<br>MW=megawatt | Symbol $W$ |

*Table 20-2. Electronic abbreviations.*

| Abbreviation | Term |
|---|---|
| a | alpha; ammeter |
| A | ammeter; ampere(s); area |
| Å | Angstrom unit |
| abc | automatic base compensation |
| ac | alternating current or voltage |
| acc | automatic chroma control |
| a/d | analog/digital |
| adf | automatic direction finder |
| adj | adjacent; adjustment |

*Table 20-2. Continued.*

| Abbreviation | Term |
|---|---|
| adp | automatic data processing |
| Aesop-1 | Addressable Extension System with Over-night Program |
| AF | audio frequencies |
| afc | automatic frequency control |
| aft | audio-frequency transformer; automatic fine tuning |
| agc | automatic gain control |
| Ah | ampere-hour |
| alc | automatic level control |
| algol | alogrithmic language |
| alu | arithmetic logic unit |
| AM | amplitude modulation |
| amp | ampere(s); amplifier |
| amp-hr | ampere hour |
| AND | gate |
| anl | automatic noise limiter |
| anr | automatic noise reduction |
| ant | antenna |
| antilog | antilogarithm |
| apc | automatic phase control |
| apld | automatic program locating device |
| arm | armature |
| ASA | American Standards Association |
| ASCII | American Standard Code for Information Interchange |
| att | attenuator; attenuation |
| atto | $10^{-18}$ |
| autotrans | autotransformer |
| aux | auxiliary |
| av | average |
| avc | automatic volume control |
| ave | automatic volume expansion |
| AWG | American Wire Gauge |
| b | bel; beta, magnetic flux density |
| b or base | base (transistor) |
| *B* | magnetic flux density; susceptance |
| bal | balance |
| bal mod | balanced modulator |
| balun | balanced-to-unbalanced transformer |
| BASIC | beginner's all-purpose symbolic instruction |
| batt | battery |
| bcd | binary coded decimal |
| bci | broadcast interference |
| bcl | broadcast listener |
| bd | baud code |
| bfo | beat-frequency oscillator |
| bin | binary |
| bit(s) | binary digit(s) |

*Table 20-2. Continued.*

| Abbreviation | Term |
|---|---|
| bo | Barkhausen oscillation |
| bp | bandpass |
| bps | bits per second |
| bto | blocking-tube oscillator |
| Btu | British thermal unit |
| b/w | black and white |
| c | centi (one-hundredth; $10^{-2}$); collector (transistor) |
| C, cap | capacitor (or capacitance) |
| °C | Celsius (degrees) |
| calib | calibrate |
| cath foll | cathode follower |
| catv | community antenna television |
| CB | citizens' band |
| cc | cotton covered (wire) |
| ccd | charge coupled device |
| cctv | closed circuit television |
| ccw | counterclockwise |
| cemf | counter electromotive force |
| cent | centering |
| centi (or c) | $10^{-2}$ |
| cgs or CGS | centimeter-gram-second |
| ch | choke |
| chan | channel |
| chg | charge |
| chrome | chromium |
| cir mil(s) | circular mil(s) |
| ckt brkr | circuit breaker |
| cm | centimeters |
| cm³ | cubic centimeters |
| CMOS | COS/MOS transistor |
| CMOS/SOS | complementary metal-oxide semiconductory silicon or sapphire transistor |
| coax | coaxial cable |
| COBOL | common business-oriented language |
| colog | cologarithm |
| com | common |
| cond | conductor |
| conn | connection |
| cont | control |
| conv | convergence; converter |
| cos | cosine |
| cosh | hyperbolic cosine |
| cot | cotangent |
| counter emf; cemf | counter electromotive force |
| cps | cycles per second (hertz); characters per second |

## Table 20-2. Continued.

| Abbreviation | Term |
|---|---|
| CPU | central processing unit |
| c-r | cathode-ray |
| cro | cathode-ray oscilloscope |
| $CrO_2$ | chromium dioxide |
| crt | cathode-ray tube |
| csc; cosc | cosecant |
| ct | center tap |
| cu | control unit |
| cw | clockwise; continuous wave |
| d | diode; deci (one-tenth; $10^{-1}$); diameter |
| d/a | digital/analog |
| dB | decibel |
| dBf | signal reference to power |
| dblr | doubler |
| dBmV | decibels referred to 1 mW across 600 $\Omega$ |
| DBS | direct broadcast satellite |
| dc | direct current or voltage |
| dc rest | dc restorer |
| dcc | double cotton covered (wire) |
| deci | $10^{-1}$ |
| defl | deflection |
| deg | degrees (angle); degrees (temperature) |
| deka | ten |
| demod | demodulator |
| det | detector |
| dielec | dielectric |
| diff | differentiator |
| DIN | Deutsche Industrie Normenausschuss (German Industrial Standards) |
| disch | discharge |
| discrim | discriminator |
| dk | deka (ten) |
| dkm | decameters |
| dm | decimeters |
| dp | data processing |
| dpcm | differential pulse code modulation |
| DPDT | double-pole, double-throw (switch) |
| DPST | double-pole, single-throw (switch) |
| dsc | double silk covered (wire) |
| dx | distance |
| dyn | dynamic, dyne |
| e | emitter (transistor); voltage; electronic charge |
| EBCDIC | Extended Binary Coded Decimals Interchange Code |
| ec | enamel covered |
| eco | electron-coupled oscillator |

*Table 20-2. Continued.*

| Abbreviation | Term |
|---|---|
| EDP | Electronic Data Processing |
| eff | effective (rms) |
| ehf | extremely high frequencies (30 to 300 GHz) |
| EIA | Electronic Industries Association |
| EIAJ | Electronic Industries Association of Japan |
| elec | electric; electrolytic |
| elect | electrode |
| elf | extremely low frequencies |
| emf | electromotive force (voltage) |
| emu | electromagnetic unit(s) |
| enam | enameled (wire) |
| encl | enclosure |
| EP | extended play |
| eprom | erasable programmable read-only memory |
| eq | equalization |
| equiv | equivalent |
| erase hd | erase head |
| erf | error function |
| erp | effective radiated power |
| esu | electrostatic unit(s) |
| eV | electron volt(s) |
| ext | external or extension |
| f | femto ($10^{-15}$) |
| f or freq | frequency |
| °F | Fahrenheit (degrees) |
| F | farad(s) |
| $\mathcal{F}$ | magnetomotive force |
| fA | femtoampere; $10^{-15}$ A |
| fax | facsimile |
| $f_c$ | cutoff frequency |
| FCC | Federal Communications Commission |
| FeCr | ferrichrome |
| femto | $10^{-15}$ |
| FET | field-effect transistor |
| ff | fast forward |
| FHR | fixed-head recorder |
| FM | frequency modulation |
| foll | follower(-ing) |
| fone | headphones; earphones (see also phone) |
| Fortran | formula translation (language) |
| FS | Fourier series |
| G | conductance |
| g | gram; grid (in tube diagrams); conductance |
| Gb | gilberts |
| gca | ground-controlled approach |

*Table 20-2. Continued.*

| Abbreviation | Term |
|---|---|
| gdo | grid-dip oscillator |
| gen | generator |
| GHz | gigahertz (kilomegahertz) |
| giga | $10^9$ |
| $G_m$ | mutual conductance |
| GMT | Greenwich mean time |
| gnd | ground |
| H | henry; henries |
| h | hour |
| ham | radio amateur operator |
| hd | head |
| hecto | $10^2$ (one hundred) |
| hex | hexadecimal |
| hf | high frequency (3 000 to 30 000 kHz); high filter |
| hi fi | high fidelity |
| hi pot | high potential |
| hm | hectometers |
| hor or horiz | horizontal |
| hp | horsepower |
| htr (in diagrams) | heater |
| hv | high voltage |
| hvr | home video recorder |
| Hz | hertz (cycles per second) |
| i | current (instantaneous value) |
| ic | integrated circuit |
| icw | interrupted continuous waves |
| IEC | International Electrotechnical Commission |
| i-f or i.f. | intermediate frequency |
| i-f or i.f.t. | intermediate-frequency transformer |
| IGFET | Insulated Gate Field-Effect Transistor |
| IHF | Institute of High Fidelity |
| ils | instrument landing system |
| im | intermodulation; intermodulation distortion |
| in | inch; input |
| int | integrator |
| i/o | input/output |
| ips | inches per second |
| IR | voltage; voltage drop |
| j | imaginary number |
| J | joule |
| jb | junction box |
| jct | junction |
| JFET | junction field effect transistor |
| K | numerical constant; coupling coefficient |

*Table 20-2. Continued.*

| Abbreviation | Term |
|---|---|
| k | kilo; thousand; $10^3$ |
| *k* | dielectric constant |
| kg | kilogram |
| kHz | kilohertz |
| kilo | $10^3$ |
| km | kilometers |
| kV | kilovolt(s) |
| kVA | kilovolt-amperes |
| kVAR | reactive kilovolt amperes |
| kW | kilowatt |
| kWh | kilowatt-hours |
| L | coil; inductor; inductance; load |
| laser | light amplification by stimulated emission of radiation |
| LC | inductance-capacitance |
| LCD | liquid crystal diode |
| LED | light emitting diode |
| lf | low frequency (30 to 300 kHz) |
| lin | linearity; linear |
| lm | limiter |
| ln; $\log_e$ | Napierian logarithm |
| lna | low-noise amplifier |
| log; $\log_{10}$ | common logarithm |
| $\log^{-1}$ | antilogarithm |
| lp | long play |
| ls | limit switch |
| lsd | least significant digit |
| lsi | large scale integration |
| LVR | longitudinal video recording |
| m | milli ($10^{-3}$); one-thousandth; meter |
| M | mutual inductance |
| $\mu$ | micro ($10^{-6}$); one-millionth; amplication factor; permeability |
| $\mu$A | microampere |
| mA | milliampere |
| mag | magnetic |
| matv | master antenna television |
| max | maximum |
| md | mean deviation |
| mds | multipoint distribution service |
| meg | megohm |
| mega | million; $10^6$ |
| mem | memory |
| MESFET | metal semiconductor field-effect transistor |
| MeV | million electron volts |
| $\mu$f | microfarad |
| mf | medium frequencies (300 to 3 000 kHz) |

*Table 20-2. Continued.*

| Abbreviation | Term |
|---|---|
| mfb | motional feedback |
| mH | millihenry(ies) |
| MHz | megahertz |
| $\mu$H | microhenry(ies) |
| mic or mike | microphone |
| micro | one-millionth; $10^{-6}$ |
| micromicro | one-millionth of a millionth; $10^{-12}$ (see also pico) |
| milli | $10^{-3}$ (one-thousandth) |
| min | minimum |
| mks or MKS | meter-kilogram-second |
| mm | millimeters; moving magnet |
| mmf | magnetomotive force |
| MNOS | metal nitric oxide semiconductor |
| mod | modulation; modulator; modulus |
| modem | modulator-demodulator |
| mol | maximum output level |
| mon | monitor |
| mono | monophonic; monochrome; monaural |
| MOS | metal oxide semiconductor |
| MOSFET | metal oxide semiconductor field effect transistor |
| MOST | metal oxide semiconductor transistor |
| mpx | multiplex |
| $\mu$s | microsecond |
| ms | millisecond |
| msd | most significant digit |
| mtv | music television |
| mult | multiplier |
| $\mu$V | microvolt |
| mv | multivibrator |
| mV | millivolt |
| MVA | megavolt-ampere |
| mvb | multivibrator |
| mvc | manual volume control |
| $\mu$V/m | microvolts per meter |
| $\mu$W | microwatt |
| MW | megawatt |
| mW | milliwatt |
| mWh | megawatt-hour |
| my | myria; (ten thousand; $10^4$) |
| mym | myriameters |
| n | nano; $10^{-9}$ |
| N | number of turns |
| NAB | National Assoc. of Broadcasters |
| NAND | NOT AND gate |
| nano or n | $10^{-9}$ |
| nbfm | narrow-band FM |

*Table 20-2. Continued.*

| Abbreviation | Term |
|---|---|
| nc | normally-closed (switch or relay); neutralizing capacitor; no connection |
| Ne | neon |
| neg | negative; minus |
| net | network |
| nf | negative feedback |
| NI | ampere turns |
| NMOS | negative channel metal oxide semiconductor |
| no | normally open (switch or relay) |
| NOR | NOT OR gate |
| Np | neper |
| npn or NPN | negative-positive-negative (transistor) |
| NTSC | National Television Standards Committee |
| n type | semiconductor with excess of negative carriers |
| nvr | no voltage release |
| OD | outside diameter |
| Oe | oersted |
| $\Omega$/V | ohms per volt |
| omni | omnidirectional |
| OR | or gate |
| osc | oscillator |
| out | output |
| p | pole; pico ($10^{-12}$) |
| P | power |
| pa | public address; power amplifier |
| PAL | phase alternation line |
| pam | pulse amplitude modulation |
| pb | playback |
| pc | photocell |
| pcm | pulse code modulation |
| pd | potential difference |
| pdm | pulse duration modulation |
| perm | permanent |
| pf | power factor |
| pF | picofarad |
| phone(s) (or fone) | headphones; earphones |
| photo mult | photomultiplier |
| pi | $\pi$; approximately 3.141 6 |
| pico | formerly designated as micromicro $10^{-12}$ |
| pix | picture |
| pixel | picture element |
| pl | pilot lamp |
| pL/1 | programming language/one |
| pll | phase locked loop |
| pm | permanent magnet (speaker); phase modulation; pulse modulation |

*Table 20-2. Continued.*

| Abbreviation | Term |
|---|---|
| pn | diode or transistor junction |
| pnp or PNP | positive-negative-positive (transistor) |
| pos | positive; plus |
| pot | potentiometer; potential |
| pp | peak-to-peak; pushpull |
| ppi | plan-position indicator (radar) |
| ppm | parts per million; pulse position modulation |
| pps | pulses per second |
| preamp | preamplifier |
| prf | pulse repetition frequency |
| pri | primary |
| PROM | programmable read-only memory |
| psi | pounds per square inch |
| pt | phototube |
| ptm | pulse time modulation |
| p type | semiconductor with excess of positive carriers |
| ptv | projection television |
| pwr | power |
| Q | reactance-resistance ratio; transistor; electric charge; Q factor; Q signal |
| quad | quadrature, quadraphonic (quadriphonic) |
| Qube | two-way cable tv |
| R | resistor |
| $R$ | resistance |
| r | radius |
| $\mathcal{R}$ | reluctance |
| rad | radian |
| RAM | random access memory |
| RC | resistance-capacitance |
| r-c | radio control |
| rcdg | recording |
| rcdr | recorder |
| rcvr | receiver |
| re | reference |
| rec | record |
| rect | rectifier |
| reg | regulator |
| regen | regeneration |
| rev | reverse |
| reverb | reverberation |
| rf | radio frequency |
| rfc | radio-frequency choke |
| rft | radio-frequency transformer |
| rgb or RGB | red, green, and blue |
| rheo | rheostat |

*Table 20-2. Continued.*

| Abbreviation | Term |
|---|---|
| RIAA | Record Industry Association of America |
| RL | resistance-inductance |
| RLC | resistance-inductance-capacitance |
| rms | root-mean-square; effective |
| ROM | read-only memory |
| rpm | revolutions per minute |
| ry | relay |
| ry, nc | relay, normally closed |
| ry, no | relay, normally open |
| s | second(s) |
| s or sw | switch |
| scc | single-cotton covered (wire) |
| sce | single cotton enameled (wire) |
| scope | oscilloscope |
| scr | silicon-controlled rectifier |
| sec | secant; secondary; second |
| SECAM | sequential and memory |
| sech | hyperbolic secant |
| sels | selsyn |
| sep | separator |
| sg | screen grid |
| shf | super-high frequency (3 000 to 300 000 MHz) |
| sig | signal |
| sin | sine |
| sinh | hyperbolic sine |
| sld | solenoid |
| SLP | super long play |
| S/N | signal-to-noise ratio |
| sp | single-pole |
| SP | standard play |
| SPDT | single-pole, double-throw |
| SPDTDB | single-pole, double-throw, double-break |
| SPDTNCDB | single-pole, double-throw, normally-closed, double-break |
| SPDTNO | single-pole, double-throw, normally-open |
| SPDTNODB | single-pole, double-throw, normally-open, double-break |
| spec(s) | specifications |
| spkr | loudspeaker |
| spl | sound pressure level |
| SPSTNC | single-pole, single-throw, normally-closed |
| SPSTNO | single-pole, single-throw, normally-open |
| sq | square |
| ssb | single sideband |
| ssc | single silk covered (wire) |
| ssdd | specific signal display device |
| stereo | stereophonic |
| strobe | stroboscope |

*Table 20-2. Continued.*

| Abbreviation | Term |
|---|---|
| stv | subscription TV |
| sup | suppressor |
| superhet | superheterodyne |
| sw | switch; short wave |
| swl | short-wave listener |
| swr | standing-wave ratio |
| sync | synchronization; synchronous |
| t | transformer; trimmer capacitor; tera; $10^{12}$; time |
| tacho | tachometer |
| tan | tangent |
| tanh | hyperbolic tangent |
| teleg | telegraph; telegram |
| tera | $10^{12}$ |
| term | terminal |
| thd | total harmonic distortion |
| TIM | transient intermodulation distortion |
| Tocom | two-way cable TV |
| tr | transmit-receive; turns ratio; transient response |
| trans | transformer |
| trf | tuned-radio frequency |
| trig | trigger |
| tsf | telegraphie sans fil (wireless telegraphy) |
| tsi | time slot interchange |
| TV | television |
| tvi | television interference |
| uhf | ultra-high frequencies (300 to 3 000 MHz) |
| UL | Underwriters' Laboratories, Inc. |
| V | volt(s); transistor; voltmeter; volume; velocity of sound |
| VA | voltamperes (apparent power) voltage amplifier |
| Vac, Vdc | volts ac, dc |
| VAR | variable; reactive volt-amperes; varistor |
| vc | voice coil |
| vcr | video cassette recorder |
| vdt | video display terminal |
| vers | versed sine |
| vert | vertical |
| vfo | variable frequency oscillator |
| vhf | very high frequencies (30 to 300 MHz) |
| vid | video |
| vir | vertical interval reference |
| vlf | very low frequencies (below 30 kHz) |
| vlsi | very large scale integration |
| V/M | volts per meter |
| VMOS | vertical metal-oxide semiconductor |
| vol | volume |

*Table 20-2. Continued.*

| Abbreviation | Term |
|---|---|
| vom | volt-ohm-milliammeter; volt-ohmmeter |
| VSWR | voltage standing wave ratio |
| vt or v | tube |
| vtf | vertical tracking force |
| vtvm | vacuum-tube voltmeter |
| vu | volume unit |
| W | watt(s) |
| Wh | watt-hour |
| wpm | words per minute |
| wrms | weighted rms |
| WVdc | working volts, dc |
| X | reactance |
| $X_C$ | capacitive reactance |
| $X_L$ | inductive reactance |
| xformer | transformer |
| xmit | transmitter |
| xmission | transmission |
| xtal | crystal |
| Y | admittance |
| Z | impedance; characteristic impedance |

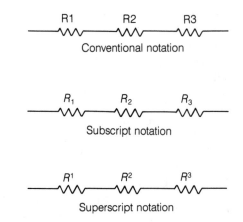

*Fig. 20-1.  Types of notation used in electronic circuits.*

## Semiconductor devices

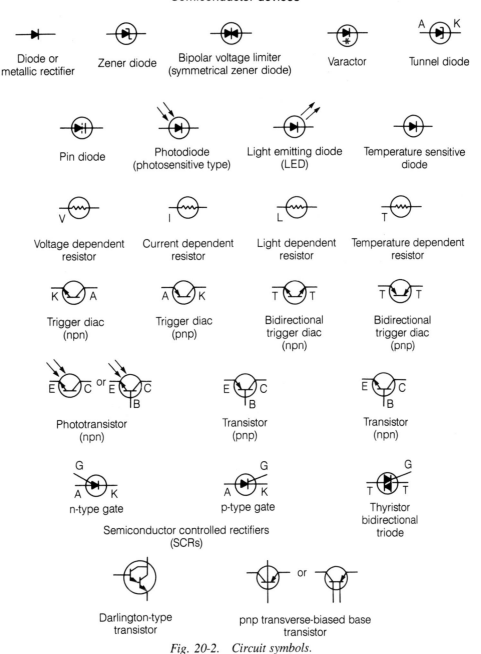

Fig. 20-2.   Circuit symbols.

## 7-segment LED indicator

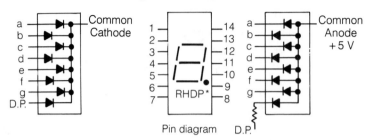

*Decimal point (D.P.) available for right hand, left hand, or universal—must specify.

## Computer symbols

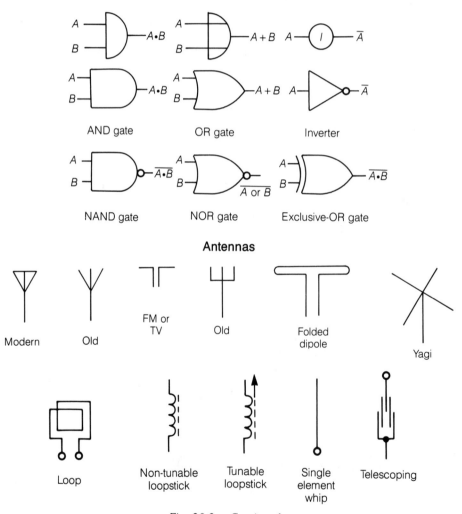

*Fig. 20-2. Continued.*

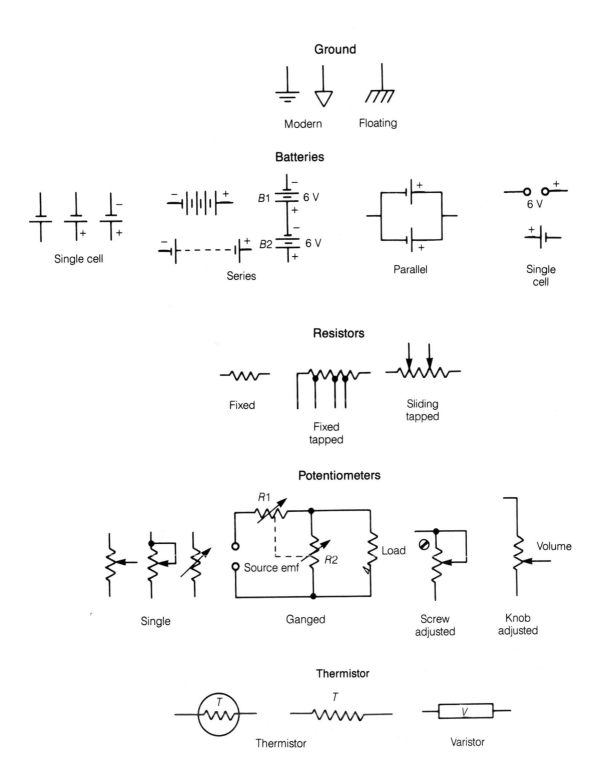

# Ground

Modern    Floating

# Batteries

Single cell

$B1$ 6 V
$B2$ 6 V

Series

Parallel

6 V

Single cell

# Resistors

Fixed

Fixed tapped

Sliding tapped

# Potentiometers

Single

$R1$
Source emf
$R2$
Load

Ganged

Screw adjusted

Volume

Knob adjusted

# Thermistor

$T$

Thermistor

$T$

$V$

Varistor

*Fig. 20-2. Continued.*

## Fixed capacitors

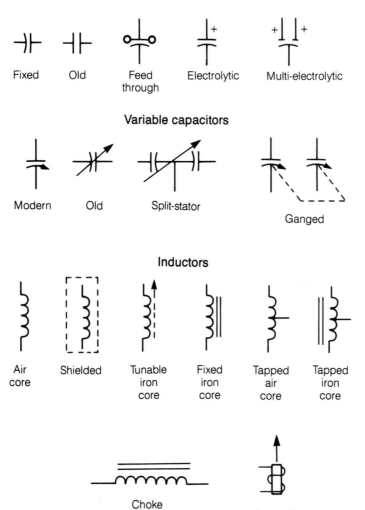

Fixed    Old    Feed through    Electrolytic    Multi-electrolytic

## Variable capacitors

Modern    Old    Split-stator

Ganged

## Inductors

Air core    Shielded    Tunable iron core    Fixed iron core    Tapped air core    Tapped iron core

Choke

Solenoid

## Transformers

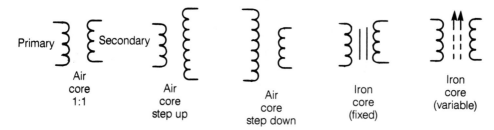

Primary    Secondary

Air core 1:1    Air core step up    Air core step down    Iron core (fixed)    Iron core (variable)

*Fig. 20-2.    Continued.*

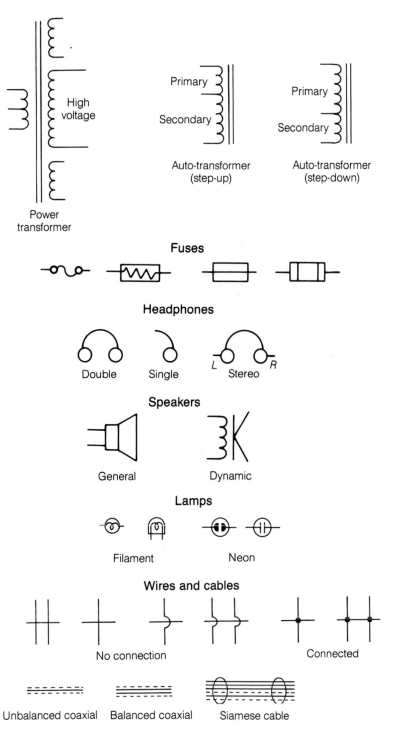

*Fig. 20-2. Continued.*

## Plugs

Two-conductor phone   Three-conductor phone   Phono   Phono jack   117 Vac Grounded ac line plug   117 Vac Ungrounded ac line plug

## Jacks

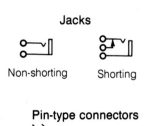

Non-shorting   Shorting

## Pin-type connectors

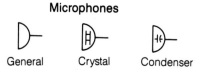

Male   Female

## Microphones

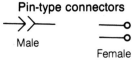

General   Crystal   Condenser

## Switches

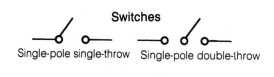

Single-pole single-throw   Single-pole double-throw

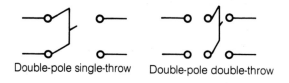

Double-pole single-throw   Double-pole double-throw

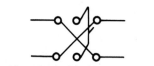

Double-pole double-throw reversing

*Fig. 20-2. Continued.*

## Wafer switches

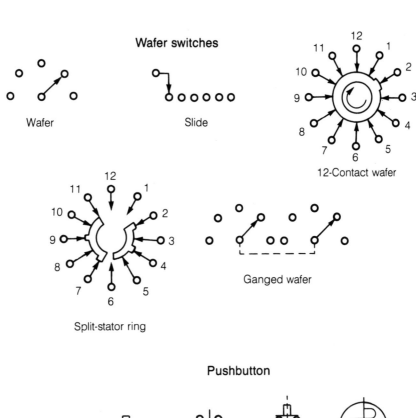

Wafer

Slide

12-Contact wafer

Split-stator ring

Ganged wafer

## Pushbutton

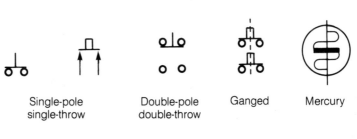

Single-pole
single-throw

Double-pole
double-throw

Ganged

Mercury

## Circle symbols

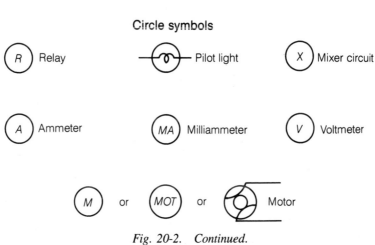

R  Relay

Pilot light

X  Mixer circuit

A  Ammeter

MA  Milliammeter

V  Voltmeter

M  or  MOT  or  Motor

*Fig. 20-2.  Continued.*

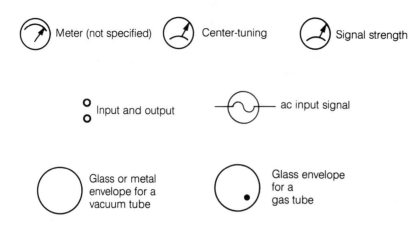

Meter (not specified)

Center-tuning

Signal strength

Input and output

ac input signal

Glass or metal envelope for a vacuum tube

Glass envelope for a gas tube

## Relays

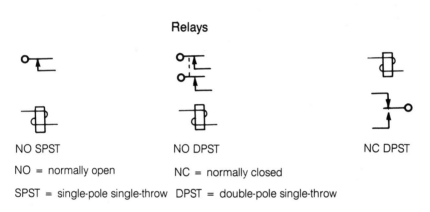

NO SPST

NO DPST

NC DPST

NO = normally open

NC = normally closed

SPST = single-pole single-throw   DPST = double-pole single-throw

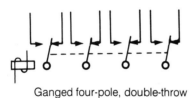

Ganged four-pole, double-throw

## Phono cartridges

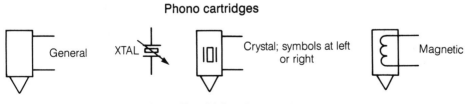

General

XTAL

Crystal; symbols at left or right

Magnetic

*Fig. 20-2.   Continued.*

## OP amps

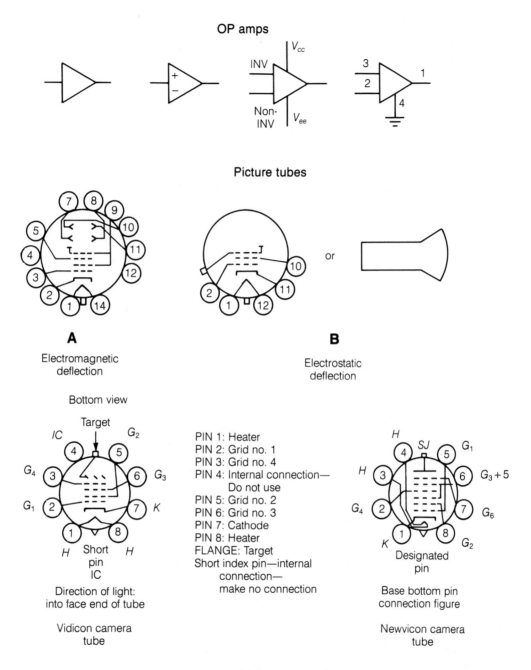

## Picture tubes

**A**

Electromagnetic deflection

**B**

Electrostatic deflection

or

**Bottom view**

Target

IC          G₂

G₄          G₃

G₁          K

H    Short    H
     pin
     IC

Direction of light:
into face end of tube

Vidicon camera
tube

PIN 1: Heater
PIN 2: Grid no. 1
PIN 3: Grid no. 4
PIN 4: Internal connection—
    Do not use
PIN 5: Grid no. 2
PIN 6: Grid no. 3
PIN 7: Cathode
PIN 8: Heater
FLANGE: Target
Short index pin—internal
    connection—
    make no connection

H          SJ          G₁

H                      G₃ + 5

G₄                     G₆

K                      G₂

Designated
pin

Base bottom pin
connection figure

Newvicon camera
tube

*Fig. 20-2.   Continued.*

# System flowchart symbols

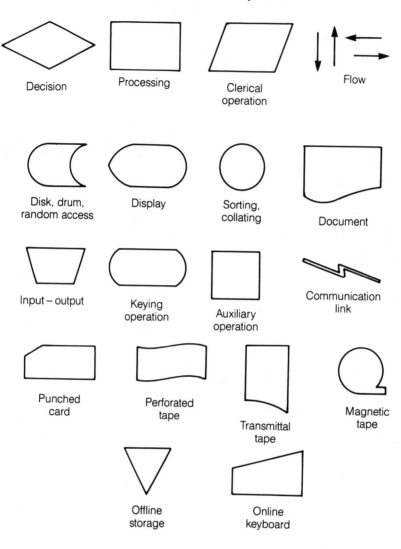

Decision

Processing

Clerical operation

Flow

Disk, drum, random access

Display

Sorting, collating

Document

Input – output

Keying operation

Auxiliary operation

Communication link

Punched card

Perforated tape

Transmittal tape

Magnetic tape

Offline storage

Online keyboard

*Fig. 20-2.    Continued.*

# 21
CHAPTER

# Properties of matter

## Heat and energy

Table 21-1 shows the relationship between heat and its equivalent in work or energy.

*Table 21-1. Heat versus energy.*

| Amount of heat | Equivalent amount of Work or energy |
|---|---|
| 1 calorie | 4.186 J |
| 1 Btu | 778 lb·ft |
| 0.239 calorie | 1 J |

## Metals

Table 21-2 lists metallic elements and their physical properties. Metals are good conductors of electricity and heat.

## Density

Density is a physical description of the mass of a substance per unit volume. Table 21-3 lists the densities of solids and liquids.

## Crystals

A crystal is a body formed by the solidification of a chemical element, a compound, or a mixture. It has a regularly repeating internal arrangement of its atoms. Table 21-4 provides information on crystals.

*Table 21-2. Metallic elements.*

| Element | Symbol | Color | Atomic weight | Specific gravity or density | Specific heat | Melting point (°C) | Coefficient of linear expansion |
|---|---|---|---|---|---|---|---|
| Aluminum | Al | Tin-white | 27.1 | 2.67 | 0.214 0 | 657 | 0.000 023 1 |
| Antimony | Sb | Bluish-white | 120.2 | 6.71–6.86 | 0.050 8 | 630 | 0.000 010 5 |
| Arsenic | As | Steel-gray | 75.0 | 5.72 | 0.081 | 450 | 0.000 005 5 |
| Barium | Ba | Pinkish-gray | 137.4 | 3.8 | 0.068 | 850 | – |
| Beryllium | Be | Silver-white | 9.1 | 1.9 | 0.582 0 | – | – |
| Bismuth | Bi | Pinkish-white | 208.0 | 9.823 | 0.030 5 | 268 | 0.000 001 4 |
| Bromine | Br | – | 79.6 | – | – | – | – |
| Cadmium | Cd | Tin-white | 112.4 | 8.546–8.667 | 0.054 8 | 322 | 0.000 002 7 |
| Calcium | Ca | Yellow | 40.1 | 1.578 | 0.170 0 | 800 | 0.000 026 9 |
| Cerium | Ce | Gray | 140.2 | 7.64 | 0.044 8 | 623 | – |
| Cesium | Cs | Silver-white | 132.8 | 1.9 | 0.048 | 27 | – |
| Chromium | Cr | Gray | 52.0 | 6.81–7.3 | 0.120 0 | 1 700 | – |
| Cobalt | Co | Grayish-white | 59.0 | 8.5–8.7 | 0.107 0 | 1 490 | 0.000 012 3 |
| Columbium (see Niobium) | Cb | | | | | | |
| Copper | Cu | Red | 63.6 | 8.92–8.95 | 0.095 2 | 1 100 | 0.000 016 7 |
| Erbium | E | – | 166.0 | – | – | – | – |
| Gadolinium | Gd | – | 156.0 | – | – | – | – |
| Gallium | Ga | Bluish-white | 69.9 | 5.9 | 0.079 | 30 | – |
| Germanium | Ge | Bluish-white | 72.5 | 5.5 | 0.074 | 900 | 0.000 016 7 |
| Gold | Au | Yellow | 197.2 | 19.265 | 0.032 4 | 1 065 | 0.000 013 6 |
| Iridium | Ir | Steel-white | 193.1 | 22.38 | 0.032 6 | 2 250 | 0.000 006 5 |
| Indium | In | White | 114.8 | 7.42 | 0.057 0 | 176 | 0.000 041 7 |
| Iron | Fe | Silver-white | 55.9 | 7.84 | 0.114 0 | 1 550 | 0.000 011 6 |
| Lanthanum | La | Gray | 139.0 | 6.163 | 0.044 9 | 826 | – |
| Lead | Pb | Bluish-white | 207.1 | 11.254–11.38 | 0.031 4 | 328 | 0.000 002 7 |
| Lithium | Li | Silver-white | 7.02 | 0.589–0.598 | 0.941 0 | 180 | – |
| Magnesium | Mg | Silver-white | 24.3 | 1.75 | 0.250 0 | 632 | 0.000 026 9 |
| Manganese | Mn | Reddish-gray | 55.0 | 8.0 | 0.122 0 | 1 245 | – |

Table 21-2. Continued.

| Element | Symbol | Color | Atomic weight | Specific gravity or density | Specific heat | Melting point (°C) | Coefficient of linear expansion |
|---|---|---|---|---|---|---|---|
| Mercury | Hg | Bluish-white | 200.0 | 13.594 | 0.031 9 | −40 | 0.000 061 0 |
| Molybdenum | Mo | Silver-white | 96.0 | 8.6 | 0.072 2 | 2 450 | — |
| Neodymium | Nd | — | 143.6 | 7.0 | — | 840 | — |
| Nickel | Ni | — | 58.7 | 8.9 | 0.108 0 | 450 | 0.000 012 7 |
| Niobium | Nb | Steel-gray | 93.5 | 12.1 | 0.071 | 1 950 | — |
| Osmium | Os | Bluish-white | 190.9 | 22.5 | 0.031 1 | 2 500 | 0.000 006 5 |
| Palladium | Pd | Tin-white | 106.7 | 11.4 | 0.059 3 | 1 549 | 0.000 011 7 |
| Platinum | Pt | — | 195.2 | 21.5 | 0.032 4 | 1 780 | 0.000 008 9 |
| Potassium | K | Silver-white | 39.10 | 0.875 | 0.166 0 | 60 | 0.000 084 1 |
| Praseodymium | Pr | — | 140.5 | 6.5 | — | 940 | — |
| Radium | Ra | — | 225.0 | — | — | — | — |
| Rhenium | Re | Silver-white | 186.2 | 21.04 | — | 3 178 3 182 | — |
| Rhodium | Rh | Tin-white | 102.9 | 12.1 | 0.058 0 | 2 000 | 0.000 008 5 |
| Rubidium | Rb | Silver-white | 85.5 | 1.52 | 0.077 | 38.5 | 0.000 071 |
| Ruthenium | Ru | — | 101.7 | 12.261 | 0.061 1 | 2 400 | 0.000 009 6 |
| Samarium | Sm | — | 150.3 | 7.7 | — | 1 350 | — |
| Scandium | Sc | — | 44.1 | — | — | — | — |
| Silver | Ag | White | 107.9 | 10.4–10.57 | 0.056 0 | 962 | 0.000 019 2 |
| Sodium | Na | Silver-white | 23.0 | 0.98 | 0.293 | 96 | 0.000 071 |
| Strontium | Sr | Yellow | 87.6 | 2.5 | — | 800 | — |
| Tantalum | Ta | Black | 181.6 | 16.8 | 0.036 5 | 2 910 | 0.000 007 9 |
| Tellurium | Te | — | 127.5 | 6.25 | 0.049 | 452 | 0.000 016 7 |
| Terbium | Tb | — | 160 | — | — | — | — |
| Thallium | Tl | Bluish-white | 204.0 | 11.8 | 0.033 5 | 303 | 0.000 030 2 |
| Thorium | Th | Gray | 232.4 | 11.2 | 0.027 6 | 1 690 | — |
| Thullium | Tm | — | 171 | — | — | — | — |
| Tin | Sn | White | 119.0 | 7.293 | 0.055 9 | 232 | 0.000 020 3 |

Table 21-2. Continued.

| Element | Symbol | Color | Atomic weight | Specific gravity or density | Specific heat | Melting point (°C) | Coefficient of linear expansion |
|---------|--------|-------|---------------|-----------------------------|---------------|--------------------|---------------------------------|
| Titanium | Ti | Dark gray | 48.1 | 3.6 | 0.13 | 1 800 | — |
| Tungsten | W | Light gray | 184.0 | 19.129 | 0.033 4 | 3 000 | — |
| Uranium | U | Grayish-white | 238.5 | 18.33 | 0.027 7 | 1 500 | — |
| Vanadium | V | Whitish-gray | 51.1 | 5.9 | 0.125 | 1 680 | — |
| Ytterbium | Yb | — | 173.0 | — | — | — | — |
| Yttrium | Yt | Gray | 89.0 | 3.80 | — | — | — |
| Zinc | Zn | Bluish-white | 65.4 | 7.1 | 0.093 5 | 419 | 0.000 027 4 |
| Zirconium | Zr | Gray | 90.6 | 4.15 | 0.066 2 | 1 300 | — |

## Table 21-3. Densities of solids and liquids.

| Substance | Density g/cm³ | lb/ft³ |
|---|---|---|
| | $g/cm^3$ | $lb/ft^3$ |
| Aluminum | 2.58 | 1.61.1 |
| Copper | 8.9 | 555.4 |
| Gold | 19.3 | 1 205.0 |
| Ice | 0.916 7 | 57.2 |
| Iron | 7.87 | 491.3 |
| Lead | 11.0 | 686.7 |
| Mercury | 13.596 | 848.7 |
| Nickel | 8.80 | 549.4 |
| Platinum | 21.50 | 1 342.2 |
| Sea water | 1.025 | 64.0 |
| Silver | 10.5 | 655.5 |
| Tin | 7.18 | 448 |
| Tungsten | 18.6 | 1 161.2 |
| Uranium | 18.7 | 1 167.4 |
| Water | 1.000 | 62.4 |
| Zinc | 7.19 | 448.6 |

## Table 21-4. Crystal data.

| Crystal | Symbols |
|---|---|
| Bornite | $3Cu_2S_3Fe_sS_3$ |
| Carborundum | SiC |
| Cassiterite (tinstone) | $SnO^2$ |
| Copper pyrites | $Cu_2S_2FeS_2$ |
| Galena | PbS |
| Graphite | C |
| Hertzite | PbS |
| Iron pyrites | $FeS_2$ |
| Malachite | $CuCo_2CuH_2O$ |
| Molybdenite | $M_0S_2$ |
| Silicon | Si |
| Tellurium | Te |
| Zincite | ZnO |

**Crystal combinations**
Carborundum with steel
Iron pyrites with silicon
Carborundum with silicon
Galena with tellurium
Tellurium with zincite
Copper pyrites with tellurium

*Table 21-5. Potentials of thermocouples.*

| Substance | Point | °C | °F | emf (mV) Platinum-rhodium couple | emf (mV) Copper-constantan couple |
|---|---|---|---|---|---|
| Tin | melting | 231.9 | 449.0 | 1.706 | 11.009 |
| Cadmium | melting | 320.9 | 609.6 | 2.503 | 16.083 |
| Zinc | melting | 419.4 | 786.9 | 3.430 | |
| Sulphur | boiling | 444.5 | 920.1 | 3.672 | |
| Aluminum | melting | 658.7 | 1 217.6 | 5.827 | |
| Silver | melting | 960.2 | 1 760.3 | 9.111 | |
| Copper | melting | 1 082.8 | 1 981.0 | 10.534 | |
| Nickel | melting | 1 452.6 | 2 646.6 | 14.973 | |
| Platinum | melting | 1 755 | 3 191 | 18.608 | |

# Thermocouples

A thermocouple consists of two dissimilar metals with one at a higher temperature than the other. A voltage is produced under these circumstances, with the amount of current generated proportional to the temperature difference. Table 21-5 shows voltages produced by thermocouple materials with the cold junction kept at 0 °C. In Table 21-5, the column at the extreme left shows one element of the thermocouple; the other element is either platinum-rhodium or copper-constantan.

# 22
CHAPTER

# Conversions

## Frequency to wavelength conversion (kilohertz to meters)

The relationship between the frequency of a wave ($f$) in hertz and its wavelength, in meters, is supplied by the formula:

$$\lambda = \frac{300\,000\,000}{f}$$

300 000 000 is the velocity of light (and of radio waves) in space, and is a constant. $\lambda$ = the wavelength in meters.

Table 22-1 supplies the abbreviations and descriptions of waves whose frequency extends from 30 Hz to 3 000 GHz.

The velocity of light is in meters per second. Although it is frequently rounded off to 300 000 000, its more probable value is 299 820 000 m/s, the value used in Table 22-2. When the frequency is in kilohertz:

$$\lambda = \frac{299\,820}{f}$$

When the frequency is in megahertz:

$$\lambda = \frac{299.82}{f}$$

Wavelength and frequency have an inverse relationship. As frequency increases, wavelength decreases. Conversely, a decrease in frequency means an increase in wavelength. In terms of formulas: $\lambda$ equals 299 820 000 divided by $f$, or $f$ equals 299 820 000 divided by $\lambda$. Consequently, the columns indicated in Table 22-2 are interchangeable.

Table 22-1. Frequency bands.

| Frequency | Designation | Abbreviation |
|---|---|---|
| 30 to 300 Hz | extremely low frequencies | ELF |
| 20 Hz to 20 kHz | sound frequencies | AF |
| 20 to 30 kHz | very low frequency | VLF |
| 30 to 300 kHz | low frequency | LF |
| 300 to 3 000 kHz | medium frequency | MF |
| 3 000 to 30 000 kHz | high frequency | HF |
| 30 to 300 MHz | very high frequency | VHF |
| 300 to 3 000 MHz | ultra high frequency | UHF |
| 3 000 to 30 000 MHz | super high frequency | SHF |
| 30 000 to 300 000 MHz (30 to 300 GHz) | extremely high frequency | EHF |
| 300 to 3 000 GHz | no designation | |

Table 22-2. Kilohertz to meters or meters to kilohertz.

| kHz | m | kHz | m | kHz | m | kHz | m | kHz | m |
|---|---|---|---|---|---|---|---|---|---|
| 1 | 299 820 | 170 | 1 764 | 420 | 713.9 | 670 | 447.5 | 920 | 325.9 |
| 2 | 149 910 | 180 | 1 666 | 430 | 697.3 | 680 | 440.9 | 930 | 322.4 |
| 3 | 99 940 | 190 | 1 578 | 440 | 681.4 | 690 | 434.5 | 940 | 319.0 |
| 4 | 79 955 | 200 | 1 499 | 450 | 666.3 | 700 | 428.3 | 950 | 315.6 |
| 5 | 59 964 | 210 | 1 428 | 460 | 651.8 | 710 | 422.3 | 960 | 312.3 |
| 6 | 49 970 | 220 | 1 363 | 470 | 637.9 | 720 | 416.4 | 970 | 309.1 |
| 7 | 42 831 | 230 | 1 304 | 480 | 624.6 | 730 | 410.7 | 980 | 305.9 |
| 8 | 37 478 | 240 | 1 249 | 490 | 611.9 | 740 | 405.2 | 990 | 302.8 |
| 9 | 33 313 | 250 | 1 199 | 500 | 599.6 | 750 | 399.8 | 1 000 | 299.8 |
| 10 | 29 982 | 260 | 1 153 | 510 | 587.9 | 760 | 394.5 | 1 010 | 296.9 |
| 20 | 14 991 | 270 | 1 110 | 520 | 570.6 | 770 | 389.4 | 1 020 | 293.9 |
| 30 | 9 994 | 280 | 1 071 | 530 | 565.7 | 780 | 384.4 | 1 030 | 291.1 |
| 40 | 7 495 | 290 | 1 034 | 540 | 552.2 | 790 | 379.5 | 1 040 | 288.3 |
| 50 | 5 996 | 300 | 999.4 | 550 | 545.1 | 800 | 374.8 | 1 050 | 285.5 |
| 60 | 4 997 | 310 | 967.2 | 560 | 535.4 | 810 | 370.1 | 1 060 | 282.8 |
| 70 | 4 283 | 320 | 936.9 | 570 | 526.0 | 820 | 365.6 | 1 070 | 280.2 |
| 80 | 3 748 | 330 | 908.5 | 580 | 516.9 | 830 | 361.2 | 1 080 | 277.6 |
| 90 | 3 331 | 340 | 881.8 | 590 | 508.2 | 840 | 356.9 | 1 090 | 275.1 |
| 100 | 2 998 | 350 | 856.6 | 600 | 499.7 | 850 | 352.7 | 1 100 | 272.6 |
| 110 | 2 726 | 360 | 832.8 | 610 | 491.5 | 860 | 348.6 | 1 110 | 270.1 |
| 120 | 2 499 | 370 | 810.3 | 620 | 483.6 | 870 | 344.6 | 1 120 | 267.7 |
| 130 | 2 306 | 380 | 789.0 | 630 | 475.9 | 880 | 340.7 | 1 130 | 265.3 |
| 140 | 2 142 | 390 | 768.8 | 640 | 468.5 | 890 | 336.9 | 1 140 | 263.0 |
| 150 | 1 999 | 400 | 749.6 | 650 | 461.3 | 900 | 333.1 | 1 150 | 260.7 |
| 160 | 1 874 | 410 | 731.3 | 660 | 454.3 | 910 | 329.5 | 1 160 | 258.5 |

*Table 22-2. Continued.*

| kHz | m | kHz | m | kHz | m | kHz | m | kHz | m |
|---|---|---|---|---|---|---|---|---|---|
| 1 170 | 256.3 | 1 590 | 188.6 | 2 010 | 149.2 | 2 430 | 123.4 | 2 850 | 105.2 |
| 1 180 | 254.1 | 1 600 | 187.4 | 2 020 | 148.4 | 2 440 | 122.9 | 2 860 | 104.8 |
| 1 190 | 251.9 | 1 610 | 186.2 | 2 030 | 147.7 | 2 450 | 122.4 | 2 870 | 104.5 |
| 1 200 | 249.9 | 1 620 | 185.1 | 2 040 | 147.0 | 2 460 | 121.9 | 2 880 | 104.1 |
| 1 210 | 247.8 | 1 630 | 183.9 | 2 050 | 146.3 | 2 470 | 121.4 | 2 890 | 103.7 |
| 1 220 | 245.8 | 1 640 | 182.8 | 2 060 | 145.5 | 2 480 | 120.9 | 2 900 | 103.4 |
| 1 230 | 243.8 | 1 650 | 181.7 | 2 070 | 144.8 | 2 490 | 120.4 | 2 910 | 103.0 |
| 1 240 | 241.8 | 1 660 | 180.6 | 2 080 | 144.1 | 2 500 | 119.9 | 2 920 | 102.7 |
| 1 250 | 239.9 | 1 670 | 179.5 | 2 090 | 143.5 | 2 510 | 119.5 | 2 930 | 102.3 |
| 1 260 | 238.0 | 1 680 | 178.5 | 2 100 | 142.8 | 2 520 | 119.0 | 2 940 | 102.0 |
| 1 270 | 236.1 | 1 690 | 177.4 | 2 110 | 142.1 | 2 530 | 118.5 | 2 950 | 101.6 |
| 1 280 | 234.2 | 1 700 | 176.4 | 2 120 | 141.4 | 2 540 | 117.1 | 2 960 | 101.3 |
| 1 290 | 232.4 | 1 710 | 175.3 | 2 130 | 140.8 | 2 550 | 117.6 | 2 970 | 100.9 |
| 1 300 | 230.6 | 1 720 | 174.3 | 2 140 | 140.1 | 2 560 | 117.1 | 2 980 | 100.6 |
| 1 310 | 228.9 | 1 730 | 173.3 | 2 150 | 139.5 | 2 570 | 116.7 | 2 990 | 100.3 |
| 1 320 | 227.1 | 1 740 | 172.3 | 2 160 | 138.8 | 2 580 | 116.2 | 3 000 | 99.94 |
| 1 330 | 225.4 | 1 750 | 171.3 | 2 170 | 138.2 | 2 590 | 115.8 | 3 010 | 99.61 |
| 1 340 | 223.7 | 1 760 | 170.4 | 2 180 | 137.5 | 2 600 | 115.3 | 3 020 | 99.28 |
| 1 350 | 222.1 | 1 770 | 169.4 | 2 190 | 136.9 | 2 610 | 114.9 | 3 030 | 98.95 |
| 1 360 | 220.5 | 1 780 | 168.4 | 2 200 | 136.3 | 2 620 | 114.4 | 3 040 | 98.63 |
| 1 370 | 218.8 | 1 790 | 167.5 | 2 210 | 135.7 | 2 630 | 114.0 | 3 050 | 98.30 |
| 1 380 | 217.3 | 1 800 | 166.6 | 2 220 | 135.1 | 2 640 | 113.6 | 3 060 | 97.98 |
| 1 390 | 215.7 | 1 810 | 165.6 | 2 230 | 134.4 | 2 650 | 113.1 | 3 070 | 97.66 |
| 1 400 | 214.2 | 1 820 | 164.7 | 2 240 | 133.8 | 2 660 | 112.7 | 3 080 | 97.34 |
| 1 410 | 212.6 | 1 830 | 163.8 | 2 250 | 133.3 | 2 670 | 112.3 | 3 090 | 97.03 |
| 1 420 | 211.1 | 1 840 | 162.9 | 2 260 | 132.7 | 2 680 | 111.9 | 3 100 | 96.72 |
| 1 430 | 209.7 | 1 850 | 162.1 | 2 270 | 132.1 | 2 690 | 111.5 | 3 110 | 96.41 |
| 1 440 | 208.2 | 1 860 | 161.2 | 2 280 | 131.5 | 2 700 | 111.0 | 3 120 | 96.10 |
| 1 450 | 206.8 | 1 870 | 160.3 | 2 290 | 130.9 | 2 710 | 110.6 | 3 130 | 95.79 |
| 1 460 | 205.4 | 1 880 | 159.5 | 2 300 | 130.4 | 2 720 | 110.2 | 3 140 | 95.48 |
| 1 470 | 204.0 | 1 890 | 158.6 | 2 310 | 129.8 | 2 730 | 109.8 | 3 150 | 95.18 |
| 1 480 | 202.6 | 1 900 | 157.8 | 2 320 | 129.2 | 2 740 | 109.4 | 3 160 | 94.88 |
| 1 490 | 201.2 | 1 910 | 157.0 | 2 330 | 128.7 | 2 750 | 109.0 | 3 170 | 94.58 |
| 1 500 | 199.9 | 1 920 | 156.2 | 2 340 | 128.1 | 2 760 | 108.6 | 3 180 | 94.28 |
| 1 510 | 198.6 | 1 930 | 155.3 | 2 350 | 127.6 | 2 770 | 108.2 | 3 190 | 93.99 |
| 1 520 | 197.3 | 1 940 | 154.5 | 2 360 | 127.0 | 2 780 | 107.8 | 3 200 | 93.69 |
| 1 530 | 196.0 | 1 950 | 153.8 | 2 370 | 126.5 | 2 790 | 107.5 | 3 210 | 93.40 |
| 1 540 | 194.7 | 1 960 | 153.0 | 2 380 | 126.0 | 2 800 | 107.1 | 3 220 | 93.11 |
| 1 550 | 193.4 | 1 970 | 152.2 | 2 390 | 125.4 | 2 810 | 106.7 | 3 230 | 92.82 |
| 1 560 | 192.2 | 1 980 | 151.4 | 2 400 | 124.9 | 2 820 | 106.3 | 3 240 | 92.54 |
| 1 570 | 191.0 | 1 990 | 150.7 | 2 410 | 124.4 | 2 830 | 105.9 | 3 250 | 92.25 |
| 1 580 | 189.8 | 2 000 | 149.9 | 2 420 | 123.9 | 2 840 | 105.6 | 3 260 | 91.97 |

*Table 22-2. Continued.*

| kHz | m | kHz | m | kHz | m | kHz | m | kHz | m |
|---|---|---|---|---|---|---|---|---|---|
| 3 270 | 91.69 | 3 690 | 81.25 | 4 110 | 72.95 | 4 530 | 66.19 | 4 950 | 60.57 |
| 3 280 | 91.41 | 3 700 | 81.03 | | | 4 540 | 66.04 | 4 960 | 60.45 |
| 3 290 | 91.13 | 3 710 | 80.81 | 4 120 | 72.77 | 4 550 | 65.89 | | |
| 3 300 | 90.85 | | | 4 130 | 72.60 | 4 560 | 65.75 | 4 970 | 60.33 |
| 3 310 | 90.58 | 3 720 | 80.60 | 4 140 | 72.42 | | | 4 980 | 60.20 |
| | | 3 730 | 80.38 | 4 150 | 72.25 | 4 570 | 65.61 | 4 990 | 60.08 |
| 3 320 | 90.31 | 3 740 | 80.17 | 4 160 | 72.07 | 4 580 | 65.46 | 5 000 | 59.96 |
| 3 330 | 90.04 | 3 750 | 79.95 | | | 4 590 | 65.32 | 5 010 | 59.84 |
| 3 340 | 89.97 | 3 760 | 79.74 | 4 170 | 71.90 | 4 600 | 65.18 | | |
| 3 350 | 89.50 | | | 4 180 | 71.73 | 4 610 | 65.04 | 5 020 | 59.73 |
| 3 360 | 89.23 | 3 770 | 79.53 | 4 190 | 71.56 | | | 5 030 | 59.61 |
| | | 3 780 | 79.32 | 4 200 | 71.39 | 4 620 | 64.90 | 5 040 | 59.49 |
| 3 370 | 88.97 | 3 790 | 79.11 | 4 210 | 71.22 | 4 630 | 64.76 | 5 050 | 59.37 |
| 3 380 | 88.70 | 3 800 | 78.90 | | | 4 640 | 64.62 | 5 060 | 59.25 |
| 3 390 | 88.44 | 3 810 | 78.69 | 4 220 | 71.05 | 4 650 | 64.48 | | |
| 3 400 | 88.18 | | | 4 230 | 70.88 | 4 660 | 64.34 | 5 070 | 59.14 |
| 3 410 | 87.92 | 3 820 | 78.49 | 4 240 | 70.71 | | | 5 080 | 59.02 |
| | | 3 830 | 78.28 | 4 250 | 70.55 | 4 670 | 64.20 | 5 090 | 58.90 |
| 3 420 | 87.67 | 3 840 | 78.08 | 4 260 | 70.38 | 4 680 | 64.06 | 5 100 | 58.79 |
| 3 430 | 87.41 | 3 850 | 77.88 | | | 4 690 | 63.93 | 5 110 | 58.67 |
| 3 440 | 87.16 | 3 860 | 77.67 | 4 270 | 70.22 | 4 700 | 63.79 | | |
| 3 450 | 86.90 | | | 4 280 | 70.05 | 4 710 | 63.66 | 5 120 | 58.56 |
| 3 460 | 86.65 | 3 870 | 77.47 | 4 290 | 69.89 | | | 5 130 | 58.44 |
| | | 3 880 | 77.27 | 4 300 | 69.73 | 4 720 | 63.52 | 5 140 | 58.33 |
| 3 470 | 86.40 | 3 890 | 77.07 | 4 310 | 69.56 | 4 730 | 63.39 | 5 150 | 58.22 |
| 3 480 | 86.16 | 3 900 | 76.88 | | | 4 740 | 63.25 | 5 160 | 58.10 |
| 3 490 | 85.91 | 3 910 | 76.68 | 4 320 | 69.40 | 4 750 | 63.12 | | |
| 3 500 | 85.66 | | | 4 330 | 69.24 | 4 760 | 62.99 | 5 170 | 57.99 |
| 3 510 | 85.42 | 3 920 | 76.48 | 4 340 | 69.08 | | | 5 180 | 57.88 |
| | | 3 930 | 76.29 | 4 350 | 68.92 | 4 770 | 62.86 | 5 190 | 57.77 |
| 3 520 | 85.18 | 3 940 | 76.10 | 4 360 | 68.77 | 4 780 | 62.72 | 5 200 | 57.66 |
| 3 530 | 84.93 | 3 950 | 75.90 | | | 4 790 | 62.59 | 5 210 | 57.55 |
| 3 540 | 84.69 | 3 960 | 75.71 | 4 370 | 68.61 | 4 800 | 62.46 | | |
| 3 550 | 84.46 | | | 4 380 | 68.45 | 4 810 | 62.33 | 5 220 | 57.44 |
| 3 560 | 84.22 | 3 970 | 75.52 | 4 390 | 68.30 | | | 5 230 | 57.33 |
| | | 3 980 | 75.33 | 4 400 | 68.14 | 4 820 | 62.20 | 5 240 | 57.22 |
| 3 570 | 83.98 | 3 990 | 75.14 | 4 410 | 67.99 | 4 830 | 62.07 | 5 250 | 57.11 |
| 3 580 | 83.75 | 4 000 | 74.96 | | | 4 840 | 61.95 | 5 260 | 57.00 |
| 3 590 | 83.52 | 4 010 | 74.77 | 4 420 | 67.83 | 4 850 | 61.82 | | |
| 3 600 | 83.28 | | | 4 430 | 67.68 | 4 860 | 61.69 | 5 270 | 56.89 |
| 3 610 | 83.05 | 4 020 | 74.58 | 4 440 | 67.53 | | | 5 280 | 56.78 |
| | | 4 030 | 74.40 | 4 450 | 67.38 | 4 870 | 61.56 | 5 290 | 56.68 |
| 3 620 | 82.82 | 4 040 | 74.21 | 4 460 | 67.22 | 4 880 | 61.44 | 5 300 | 56.57 |
| 3 630 | 82.60 | 4 050 | 74.03 | | | 4 890 | 61.31 | 5 310 | 56.46 |
| 3 640 | 82.37 | 4 060 | 73.85 | 4 470 | 67.07 | 4 900 | 61.19 | | |
| 3 650 | 82.14 | | | 4 480 | 66.92 | 4 910 | 61.06 | 5 320 | 56.36 |
| 3 660 | 81.92 | 4 070 | 73.67 | 4 490 | 66.78 | | | 5 330 | 56.25 |
| | | 4 080 | 73.49 | 4 500 | 66.63 | 4 920 | 60.94 | 5 340 | 56.15 |
| 3 670 | 81.69 | 4 090 | 73.31 | 4 510 | 66.48 | 4 930 | 60.82 | 5 350 | 56.04 |
| 3 680 | 81.47 | 4 100 | 73.13 | 4 520 | 66.33 | 4 940 | 60.69 | 5 360 | 55.94 |

*Table 22-2. Continued.*

| kHz | m | kHz | m | kHz | m | kHz | m | kHz | m |
|---|---|---|---|---|---|---|---|---|---|
| 5 370 | 55.83 | 5 790 | 51.78 | 6 210 | 48.28 | 6 630 | 45.22 | 7 050 | 42.53 |
| 5 380 | 55.73 | 5 800 | 51.69 | | | 6 640 | 45.15 | 7 060 | 42.47 |
| 5 390 | 55.63 | 5 810 | 51.60 | 6 220 | 48.20 | 6 650 | 45.09 | | |
| 5 400 | 55.52 | | | 6 230 | 48.13 | 6 660 | 45.02 | 7 070 | 42.41 |
| 5 410 | 55.42 | 5 820 | 51.52 | 6 240 | 48.05 | | | 7 080 | 42.35 |
| | | 5 830 | 51.43 | 6 250 | 49.97 | 6 670 | 44.95 | 7 090 | 42.29 |
| 5 420 | 55.32 | 5 840 | 51.34 | 6 260 | 47.89 | 6 680 | 44.88 | 7 100 | 42.23 |
| 5 430 | 55.22 | 5 850 | 51.25 | | | 6 690 | 44.82 | 7 110 | 42.17 |
| 5 440 | 55.11 | 5 860 | 51.16 | 6 270 | 47.82 | 6 700 | 44.75 | | |
| 5 450 | 55.01 | | | 6 280 | 47.74 | 6 710 | 44.68 | 7 120 | 42.11 |
| 5 460 | 54.91 | 5 870 | 51.08 | 6 290 | 47.67 | | | 7 130 | 42.05 |
| | | 5 880 | 50.99 | 6 300 | 47.59 | 6 720 | 44.62 | 7 140 | 41.99 |
| 5 470 | 54.81 | 5 890 | 50.90 | 6 310 | 47.52 | 6 730 | 44.55 | 7 150 | 41.93 |
| 5 480 | 54.71 | 5 900 | 50.82 | | | 6 740 | 44.48 | 7 160 | 41.87 |
| 5 490 | 54.61 | 5 910 | 50.73 | 6 320 | 47.44 | 6 750 | 44.42 | | |
| 5 500 | 54.51 | | | 6 330 | 47.36 | 6 760 | 44.35 | 7 170 | 41.82 |
| 5 510 | 54.41 | 5 920 | 50.65 | 6 340 | 47.29 | | | 7 180 | 41.76 |
| | | 5 930 | 50.56 | 6 350 | 47.22 | 6 770 | 44.29 | 7 190 | 41.70 |
| 5 520 | 54.32 | 5 940 | 50.47 | 6 360 | 47.14 | 6 780 | 44.22 | 7 200 | 41.64 |
| 5 530 | 54.22 | 5 950 | 50.39 | | | 6 790 | 44.16 | 7 210 | 41.58 |
| 5 540 | 54.12 | 5 960 | 50.31 | 6 370 | 47.07 | 6 800 | 44.09 | | |
| 5 550 | 54.02 | | | 6 380 | 46.99 | 6 810 | 44.03 | 7 220 | 41.53 |
| 5 560 | 53.92 | 5 970 | 50.22 | 6 390 | 46.92 | | | 7 230 | 41.47 |
| | | 5 980 | 50.14 | 6 400 | 46.85 | 6 820 | 43.96 | 7 240 | 41.41 |
| 5 570 | 53.83 | 5 990 | 50.05 | 6 410 | 46.77 | 6 830 | 43.90 | 7 250 | 41.35 |
| 5 580 | 53.73 | 6 000 | 49.97 | | | 6 840 | 43.83 | 7 260 | 41.30 |
| 5 590 | 53.64 | 6 010 | 49.89 | 6 420 | 46.70 | 6 850 | 43.77 | | |
| 5 600 | 53.54 | | | 6 430 | 46.63 | 6 860 | 43.71 | 7 270 | 41.24 |
| 5 610 | 53.44 | 6 020 | 49.80 | 6 440 | 46.56 | | | 7 280 | 41.18 |
| | | 6 030 | 49.72 | 6 450 | 46.48 | 6 870 | 43.64 | 7 290 | 41.13 |
| 5 620 | 53.35 | 6 040 | 49.64 | 6 460 | 46.41 | 6 880 | 43.58 | 7 300 | 41.07 |
| 5 630 | 53.25 | 6 050 | 49.56 | | | 6 890 | 43.42 | 7 310 | 41.02 |
| 5 640 | 53.16 | 6 060 | 49.48 | 6 470 | 46.34 | 6 900 | 43.45 | | |
| 5 650 | 53.07 | | | 6 480 | 46.27 | 6 910 | 43.39 | 7 320 | 40.96 |
| 5 660 | 52.97 | 6 070 | 49.39 | 6 490 | 46.20 | | | 7 330 | 40.90 |
| | | 6 080 | 49.31 | 6 500 | 46.13 | 6 920 | 43.33 | 7 340 | 40.85 |
| 5 670 | 52.88 | 6 090 | 49.23 | 6 510 | 46.06 | 6 930 | 43.26 | 7 350 | 40.79 |
| 5 680 | 52.79 | 6 100 | 49.15 | | | 6 940 | 43.20 | 7 360 | 40.74 |
| 5 690 | 52.69 | 6 110 | 49.07 | 6 520 | 45.98 | 6 950 | 43.14 | | |
| 5 700 | 52.60 | | | 6 530 | 45.91 | 6 960 | 43.08 | 7 370 | 40.68 |
| 5 710 | 52.51 | 6 120 | 48.99 | 6 540 | 45.84 | | | 7 380 | 40.63 |
| | | 6 130 | 48.91 | 6 550 | 45.77 | 6 970 | 43.02 | 7 390 | 40.57 |
| 5 720 | 52.42 | 6 140 | 48.83 | 6 560 | 45.70 | 6 980 | 42.95 | 7 400 | 40.52 |
| 5 730 | 52.32 | 6 150 | 48.75 | | | 6 990 | 42.89 | 7 410 | 40.46 |
| 5 740 | 52.23 | 6 160 | 48.67 | 6 570 | 45.63 | 7 000 | 42.83 | | |
| 5 750 | 52.14 | | | 6 580 | 45.57 | 7 010 | 42.77 | 7 420 | 40.41 |
| 5 760 | 52.05 | 6 170 | 48.59 | 6 590 | 45.50 | | | 7 430 | 40.35 |
| | | 6 180 | 48.51 | 6 600 | 45.43 | 7 020 | 42.71 | 7 440 | 40.30 |
| 5 770 | 51.96 | 6 190 | 48.44 | 6 610 | 45.36 | 7 030 | 42.65 | 7 450 | 40.24 |
| 5 780 | 51.87 | 6 200 | 48.36 | 6 620 | 45.29 | 7 040 | 42.59 | 7 460 | 40.10 |

Table 22-2. Continued.

| kHz | m | kHz | m | kHz | m | kHz | m | kHz | m |
|---|---|---|---|---|---|---|---|---|---|
| 7 470 | 40.14 | 7 890 | 38.00 | 8 310 | 36.08 | 8 730 | 34.34 | 9 150 | 32.77 |
| 7 480 | 40.08 | 7 900 | 37.95 | 8 320 | 36.04 | 8 740 | 34.30 | 9 160 | 32.73 |
| 7 490 | 40.03 | 7 910 | 37.90 | 8 330 | 35.99 | 8 750 | 34.27 | 9 170 | 32.70 |
| 7 500 | 39.98 | 7 920 | 37.86 | 8 340 | 35.95 | 8 760 | 34.23 | 9 180 | 32.66 |
| 7 510 | 39.92 | 7 930 | 37.81 | 8 350 | 35.91 | 8 770 | 34.19 | 9 190 | 32.62 |
| 7 520 | 39.87 | 7 940 | 37.76 | 8 360 | 35.86 | 8 780 | 34.15 | 9 200 | 32.59 |
| 7 530 | 39.82 | 7 950 | 37.71 | 8 370 | 35.82 | 8 790 | 34.11 | 9 210 | 32.55 |
| 7 540 | 39.76 | 7 960 | 37.67 | 8 380 | 35.78 | 8 800 | 34.07 | 9 220 | 32.52 |
| 7 550 | 39.71 | 7 970 | 37.62 | 8 390 | 35.74 | 8 810 | 34.03 | 9 230 | 32.48 |
| 7 560 | 39.66 | 7 980 | 37.57 | 8 400 | 35.69 | 8 820 | 33.99 | 9 240 | 32.45 |
| 7 570 | 39.61 | 7 990 | 37.52 | 8 410 | 35.65 | 8 830 | 33.95 | 9 250 | 32.41 |
| 7 580 | 39.55 | 8 000 | 37.48 | 8 420 | 35.61 | 8 840 | 33.92 | 9 260 | 32.38 |
| 7 590 | 39.50 | 8 010 | 37.43 | 8 430 | 35.57 | 8 850 | 33.88 | 9 270 | 32.34 |
| 7 600 | 39.45 | 8 020 | 37.38 | 8 440 | 35.52 | 8 860 | 33.84 | 9 280 | 32.31 |
| 7 610 | 39.40 | 8 030 | 37.34 | 8 450 | 35.48 | 8 870 | 33.80 | 9 290 | 32.27 |
| 7 620 | 39.35 | 8 040 | 37.29 | 8 460 | 35.44 | 8 880 | 33.76 | 9 300 | 32.24 |
| 7 630 | 39.29 | 8 050 | 37.24 | 8 470 | 35.40 | 8 890 | 33.73 | 9 310 | 32.20 |
| 7 640 | 39.24 | 8 060 | 37.20 | 8 480 | 35.36 | 8 900 | 33.69 | 9 320 | 32.17 |
| 7 650 | 39.19 | 8 070 | 37.15 | 8 490 | 35.31 | 8 910 | 33.65 | 9 330 | 32.14 |
| 7 660 | 39.14 | 8 080 | 37.11 | 8 500 | 35.27 | 8 920 | 33.61 | 9 340 | 32.10 |
| 7 670 | 39.09 | 8 090 | 37.06 | 8 510 | 35.23 | 8 930 | 33.57 | 9 350 | 32.07 |
| 7 680 | 39.04 | 8 100 | 37.01 | 8 520 | 35.19 | 8 940 | 33.54 | 9 360 | 32.03 |
| 7 690 | 38.99 | 8 110 | 36.97 | 8 530 | 35.15 | 8 950 | 33.13 | 9 370 | 32.00 |
| 7 700 | 38.94 | 8 120 | 36.92 | 8 540 | 35.11 | 8 960 | 33.46 | 9 380 | 31.96 |
| 7 710 | 38.89 | 8 130 | 36.88 | 8 550 | 35.07 | 8 970 | 33.42 | 9 390 | 31.93 |
| 7 720 | 38.84 | 8 140 | 36.83 | 8 560 | 35.03 | 8 980 | 33.39 | 9 400 | 31.90 |
| 7 730 | 38.79 | 8 150 | 36.79 | 8 570 | 34.98 | 8 990 | 33.35 | 9 410 | 31.86 |
| 7 740 | 38.74 | 8 160 | 36.74 | 8 580 | 34.94 | 9 000 | 33.31 | 9 420 | 31.83 |
| 7 750 | 38.69 | 8 170 | 36.70 | 8 590 | 34.90 | 9 010 | 33.28 | 9 430 | 31.79 |
| 7 760 | 38.64 | 8 180 | 36.65 | 8 600 | 34.86 | 9 020 | 33.24 | 9 440 | 31.76 |
| 7 770 | 38.59 | 8 190 | 36.61 | 8 610 | 34.82 | 9 030 | 33.20 | 9 450 | 31.73 |
| 7 780 | 38.54 | 8 200 | 36.56 | 8 620 | 34.78 | 9 040 | 33.17 | 9 460 | 31.69 |
| 7 790 | 38.49 | 8 210 | 36.52 | 8 630 | 34.74 | 9 050 | 33.13 | 9 470 | 31.66 |
| 7 800 | 38.44 | 8 220 | 36.47 | 8 640 | 34.70 | 9 060 | 33.09 | 9 480 | 31.63 |
| 7 810 | 38.39 | 8 230 | 36.43 | 8 650 | 34.66 | 9 070 | 33.06 | 9 490 | 31.59 |
| 7 820 | 38.34 | 8 240 | 36.39 | 8 660 | 34.62 | 9 080 | 33.02 | 9 500 | 31.56 |
| 7 830 | 38.29 | 8 250 | 36.34 | 8 670 | 34.58 | 9 090 | 32.98 | 9 510 | 31.53 |
| 7 840 | 38.24 | 8 260 | 36.30 | 8 680 | 34.54 | 9 100 | 32.95 | 9 520 | 31.49 |
| 7 850 | 38.19 | 8 270 | 36.25 | 8 690 | 34.50 | 9 110 | 32.91 | 9 530 | 31.46 |
| 7 860 | 38.15 | 8 280 | 36.21 | 8 700 | 34.46 | 9 120 | 32.88 | 9 540 | 31.43 |
| 7 870 | 38.10 | 8 290 | 36.17 | 8 710 | 34.42 | 9 130 | 32.84 | 9 550 | 31.39 |
| 7 880 | 38.05 | 8 300 | 36.12 | 8 720 | 34.38 | 9 140 | 32.80 | 9 560 | 31.36 |

*Table 22-2. Continued.*

| kHz | m | kHz | m | kHz | m | kHz | m | kHz | m |
|---|---|---|---|---|---|---|---|---|---|
| 9 570 | 31.33 | 9 670 | 31.01 | 9 770 | 30.69 | 9 870 | 30.38 | 9 970 | 30.07 |
| 9 580 | 31.30 | 9 680 | 30.97 | 9 780 | 30.66 | 9 880 | 30.35 | 9 980 | 30.04 |
| 9 590 | 31.26 | 9 690 | 30.94 | 9 790 | 30.63 | 9 890 | 30.32 | 9 990 | 30.01 |
| 9 600 | 31.23 | 9 700 | 30.91 | 9 800 | 30.59 | 9 900 | 30.28 | 10 000 | 29.98 |
| 9 610 | 31.20 | 9 710 | 30.88 | 9 810 | 30.56 | 9 910 | 30.25 | | |
| 9 620 | 31.17 | 9 720 | 30.85 | 9 820 | 30.53 | 9 920 | 30.22 | | |
| 9 630 | 31.13 | 9 730 | 30.81 | 9 830 | 30.50 | 9 930 | 30.19 | | |
| 9 640 | 31.10 | 9 740 | 30.78 | 9 840 | 30.47 | 9 940 | 30.16 | | |
| 9 650 | 31.07 | 9 750 | 30.75 | 9 850 | 30.44 | 9 950 | 30.13 | | |
| 9 660 | 31.04 | 9 760 | 30.72 | 9 860 | 30.41 | 9 960 | 30.10 | | |

*Example*   A radio wave has a frequency of 500 kHz. What is its wavelength in meters?

*Solution*   Find the number 500 in Table 22-2. Move horizontally and you will see the corresponding wavelength of 599.6 m.

*Example*   One of the bands of a short-wave receiver covers the range from 3 000 to 5 000 kHz. What wavelength range does this include?

*Solution*   Table 22-2 shows that 3 000 kHz corresponds to 99.94 m and that 5 000 kHz corresponds to 59.96 m. Thus, this particular band is from approximately 60 to 100 m.

*Example*   What is the length, in feet, of a wave having a frequency of 4 280 kHz?

*Solution*   Locate this frequency (4 280 kHz) in the kHz column. You will note it corresponds to a wavelength of 70.05 m. However, the problem calls for the answer in feet. Consult Table 22-3 and you will see that 70 m is 229.66 ft.

# Meters to feet

The distance from the start to the finish of a single cycle of a wave, called the wavelength, is usually specified in meters. Table 22-3 supplies data on the conversion of meters to feet. The table can be extended by moving the decimal point. Move the decimal point an equal number of places in the same direction in both columns. Thus, a wavelength of 18 m corresponds to 59.055 ft. And 180 m corresponds to 590.5 ft, etc. Use Table 22-4 for converting feet to meters.
columns. Thus, a wavelength of 18 m corresponds to 59 055 ft. And 180 m corresponds to 590.5 ft, etc. Use Table 22-4 for converting feet to meters.

*Example*   What is the length, in feet, of a wave that is 36 m long?
*Solution*   Table 22-3 shows that 36 m equals 118.08 ft ($36 \times 3.28$).

### Table 22-3. Meters to feet.

| m | ft | m | ft | m | ft | m | ft |
|---|-----|---|-----|---|------|---|------|
| 1 | 3.280 8 | 26 | 85.302 | 51 | 167.32 | 76 | 249.34 |
| 2 | 6.561 7 | 27 | 88.583 | 52 | 170.60 | 77 | 256.62 |
| 3 | 9.842 5 | 28 | 91.863 | 53 | 173.88 | 78 | 255.90 |
| 4 | 13.123 | 29 | 95.144 | 54 | 177.16 | 79 | 259.19 |
| 5 | 16.404 | 30 | 98.425 | 55 | 180.45 | 80 | 262.47 |
| 6 | 19.685 | 31 | 101.71 | 56 | 183.73 | 81 | 265.75 |
| 7 | 22.966 | 32 | 104.99 | 57 | 187.01 | 82 | 269.03 |
| 8 | 26.247 | 33 | 108.27 | 58 | 190.29 | 83 | 272.31 |
| 9 | 29.527 | 34 | 111.55 | 59 | 193.57 | 84 | 275.59 |
| 10 | 32.808 | 35 | 114.83 | 60 | 196.85 | 85 | 278.87 |
| 11 | 36.089 | 36 | 118.08 | 61 | 200.13 | 86 | 282.15 |
| 12 | 39.370 | 37 | 121.39 | 62 | 203.41 | 87 | 285.43 |
| 13 | 42.651 | 38 | 124.67 | 63 | 206.69 | 88 | 288.71 |
| 14 | 45.932 | 39 | 127.95 | 64 | 209.97 | 89 | 291.99 |
| 15 | 49.212 | 40 | 131.23 | 65 | 213.25 | 90 | 295.27 |
| 16 | 52.493 | 41 | 134.51 | 66 | 216.53 | 91 | 298.56 |
| 17 | 55.774 | 42 | 137.80 | 67 | 219.82 | 92 | 301.84 |
| 18 | 59.055 | 43 | 141.08 | 68 | 223.10 | 93 | 305.12 |
| 19 | 62.336 | 44 | 144.36 | 69 | 226.38 | 94 | 308.40 |
| 20 | 65.617 | 45 | 147.64 | 70 | 229.66 | 95 | 311.68 |
| 21 | 68.897 | 46 | 150.92 | 71 | 232.94 | 96 | 314.96 |
| 22 | 72.178 | 47 | 154.20 | 72 | 236.22 | 97 | 318.23 |
| 23 | 75.459 | 48 | 157.48 | 73 | 239.50 | 98 | 321.52 |
| 24 | 78.740 | 49 | 160.76 | 74 | 242.78 | 99 | 324.80 |
| 25 | 82.021 | 50 | 164.04 | 75 | 246.06 | 100 | 328.08 |

**Example**   What is the frequency of a wave whose wavelength is 20 ft?

**Solution**   Table 22-3 shows 19.68 ft equals 6 m. For greater accuracy, using Table 22-4 (1 ft equals 0.304 800 6 m), 20 ft×0.304 8 equals 6.096 m.

Table 22-2 does not have a value for 6 m but it does have an amount that is very close. This is 60.08 m, corresponding to a frequency of 4 990 kHz. Frequency varies inversely with wavelength and so you must multiply this answer by 10. 4 990 kHz becomes 49 990 kHz. For greater accuracy use:

$$f = 299\,820\,000/\lambda$$

$$= \frac{299\,820\,000}{6.09}$$

$$= 49\,231\,527 \text{ Hz}$$

$$= 49\,231 \text{ kHz}$$

Table 22-4. Feet to meters.

| ft | m | ft | m | ft | m | ft | m |
|----|-----|----|--------|----|--------|-----|--------|
| 1 | 0.304 8 | 26 | 7.924 8 | 51 | 15.545 | 76 | 23.165 |
| 2 | 0.609 6 | 27 | 8.229 6 | 52 | 15.850 | 77 | 23.470 |
| 3 | 0.914 40 | 28 | 8.534 4 | 53 | 16.154 | 78 | 23.774 |
| 4 | 1.219 2 | 29 | 8.839 2 | 54 | 16.459 | 79 | 24.079 |
| 5 | 1.524 0 | 30 | 9.144 0 | 55 | 16.764 | 80 | 24.384 |
| 6 | 1.828 8 | 31 | 9.448 8 | 56 | 17.069 | 81 | 24.689 |
| 7 | 2.133 6 | 32 | 9.753 6 | 57 | 17.374 | 82 | 24.994 |
| 8 | 2.438 4 | 33 | 10.058 | 58 | 17.678 | 83 | 25.298 |
| 9 | 2.743 2 | 34 | 10.363 | 59 | 17.983 | 84 | 25.603 |
| 10 | 3.048 0 | 35 | 10.668 | 60 | 18.288 | 85 | 25.908 |
| 11 | 3.352 8 | 36 | 10.973 | 61 | 18.593 | 86 | 26.213 |
| 12 | 3.657 6 | 37 | 11.278 | 62 | 18.898 | 87 | 26.518 |
| 13 | 3.962 4 | 38 | 11.582 | 63 | 19.202 | 88 | 26.822 |
| 14 | 4.267 2 | 39 | 11.887 | 64 | 19.507 | 89 | 27.127 |
| 15 | 4.572 0 | 40 | 12.192 | 65 | 19.812 | 90 | 27.432 |
| 16 | 4.876 8 | 41 | 12.497 | 66 | 20.117 | 91 | 27.737 |
| 17 | 5.181 6 | 42 | 12.802 | 67 | 20.422 | 92 | 28.042 |
| 18 | 5.486 4 | 43 | 13.106 | 68 | 20.726 | 93 | 28.346 |
| 19 | 5.791 2 | 44 | 13.411 | 69 | 21.031 | 94 | 28.651 |
| 20 | 6.096 0 | 45 | 13.716 | 70 | 21.336 | 95 | 28.956 |
| 21 | 6.400 8 | 46 | 14.021 | 71 | 21.641 | 96 | 29.261 |
| 22 | 6.705 6 | 47 | 14.326 | 72 | 21.946 | 97 | 29.566 |
| 23 | 7.010 4 | 48 | 14.630 | 73 | 22.250 | 98 | 29.870 |
| 24 | 7.315 2 | 49 | 14.935 | 74 | 22.555 | 99 | 30.175 |
| 25 | 7.620 0 | 50 | 15.240 | 75 | 22.860 | 100 | 30.480 |

The conversions in Table 22-3 are based on a conversion factor of 1 m = 3.280 8 ft. The conversions in Table 22-4 are based on a conversion factor of 1 ft = 0.304 8 m.

# Frequency to wavelength conversion for very high frequencies

At very high frequencies certain components, such as the elements of receiving or transmitting antennas, become small enough to be measured easily. Knowing the frequency at which such elements work and using data such as that contained in Table 22-5 makes it easy to convert frequency into lengths in meters.

In Table 22-5, $\lambda$ represents the wavelength in meters; MHz is the frequency in megahertz.

Table 22-5. Wavelength to frequency (VHF and UHF).

| m | MHz | m | MHz | m | MHz | m | MHz | m | MHz |
|---|---|---|---|---|---|---|---|---|---|
| 0.1 | 3 000 | 2.6 | 115 | 5.1 | 58.8 | 7.6 | 39.5 | 10.1 | 29.7 |
| 0.2 | 1 500 | 2.7 | 111 | 5.2 | 57.7 | 7.7 | 39.0 | 10.2 | 29.4 |
| 0.3 | 1 000 | 2.8 | 107 | 5.3 | 56.6 | 7.8 | 38.5 | 10.3 | 29.1 |
| 0.4 | 750 | 2.9 | 103 | 5.4 | 55.6 | 7.9 | 38.0 | 10.4 | 28.8 |
| 0.5 | 600 | 3.0 | 100 | 5.5 | 54.5 | 8.0 | 37.5 | 10.5 | 28.0 |
| 0.6 | 500 | 3.1 | 96.8 | 5.6 | 53.6 | 8.1 | 37.0 | 10.6 | 28.3 |
| 0.7 | 429 | 3.2 | 93.8 | 5.7 | 52.6 | 8.2 | 36.6 | 10.7 | 28.0 |
| 0.8 | 375 | 3.3 | 90.9 | 5.8 | 51.7 | 8.3 | 36.1 | 10.8 | 27.8 |
| 0.9 | 333 | 3.4 | 88.2 | 5.9 | 50.8 | 8.4 | 35.7 | 10.9 | 27.5 |
| 1.0 | 300 | 3.5 | 85.7 | 6.0 | 50.0 | 8.5 | 35.3 | 11.0 | 27.3 |
| 1.1 | 273 | 3.6 | 83.3 | 6.1 | 49.2 | 8.6 | 34.9 | 11.1 | 27.0 |
| 1.2 | 250 | 3.7 | 81.1 | 6.2 | 48.4 | 8.7 | 34.5 | 11.2 | 26.8 |
| 1.3 | 231 | 3.8 | 78.9 | 6.3 | 47.6 | 8.8 | 34.1 | 11.3 | 26.5 |
| 1.4 | 214 | 3.9 | 76.9 | 6.4 | 46.9 | 8.9 | 33.7 | 11.4 | 26.3 |
| 1.5 | 200 | 4.0 | 75.0 | 6.5 | 46.2 | 9.0 | 33.3 | 11.5 | 26.1 |
| 1.6 | 188 | 4.1 | 73.2 | 6.6 | 45.5 | 9.1 | 33.0 | 11.6 | 25.9 |
| 1.7 | 176 | 4.2 | 71.4 | 6.7 | 44.8 | 9.2 | 32.6 | 11.7 | 25.6 |
| 1.8 | 167 | 4.3 | 69.8 | 6.8 | 44.1 | 9.3 | 32.3 | 11.8 | 25.4 |
| 1.9 | 158 | 4.4 | 68.2 | 6.9 | 43.5 | 9.4 | 31.9 | 11.9 | 25.2 |
| 2.0 | 150 | 4.5 | 66.7 | 7.0 | 42.9 | 9.5 | 31.6 | 12.0 | 25.0 |
| 2.1 | 143 | 4.6 | 65.2 | 7.1 | 42.3 | 9.6 | 31.3 | | |
| 2.2 | 136 | 4.7 | 63.8 | 7.2 | 41.7 | 9.7 | 30.9 | | |
| 2.3 | 130 | 4.8 | 62.5 | 7.3 | 41.1 | 9.8 | 30.6 | | |
| 2.4 | 125 | 4.9 | 61.2 | 7.4 | 40.5 | 9.9 | 30.3 | | |
| 2.5 | 120 | 5.0 | 60.0 | 7.5 | 40.0 | 10.0 | 30.0 | | |

*Example* What is the frequency of a wave whose length is 10 cm?

*Solution* Using Table 22-5 find the number 10 under the heading of m. Divide by 100 to obtain 10 cm. Also multiply 30 MHz by 100 to obtain 3 000 MHz. An easier technique is to move to the right to find 0.1 m or one-tenth of a meter. Continue moving to the right and locate the answer in the MHz column —3 000 MHz.

*Example* What is the wavelength of a wave whose frequency is 25 MHz?

*Solution* In the MHz column you will see 25.0 in Table 22-5. Move to the left and you will see that the corresponding wavelength is 12 m.

# Inches to millimeters and millimeters to inches

Table 22-6 supplies a convenient way of converting decimal inches to millimeters. Use Table 22-7 for converting millimeters to decimal inches. The range of both

*Table 22-6. Decimal inches to millimeters.*

| in | mm | in | mm | in | mm | in | mm |
|-------|---------|-------|--------|-------|-------|-------|-------|
| 0.001 | 0.025 4 | 0.200 | 5.080 | 0.480 | 12.19 | 0.760 | 19.30 |
| 0.002 | 0.050 8 | 0.210 | 5.334 | 0.490 | 12.45 | 0.770 | 19.56 |
| 0.003 | 0.076 2 | 0.220 | 5.588 | 0.500 | 12.70 | 0.780 | 19.81 |
| 0.004 | 0.101 6 | 0.230 | 5.842 | 0.510 | 12.95 | 0.790 | 20.07 |
| 0.005 | 0.127 0 | 0.240 | 6.096 | 0.520 | 13.21 | 0.800 | 20.32 |
| 0.006 | 0.152 4 | 0.250 | 6.350 | 0.530 | 13.46 | 0.810 | 20.57 |
| 0.007 | 0.177 8 | 0.260 | 6.604 | 0.540 | 13.72 | 0.820 | 20.83 |
| 0.008 | 0.203 2 | 0.270 | 6.858 | 0.550 | 13.97 | 0.830 | 21.08 |
| 0.009 | 0.228 6 | 0.280 | 7.112 | 0.560 | 14.22 | 0.840 | 21.34 |
| 0.010 | 0.254 0 | 0.290 | 7.37 | 0.570 | 14.48 | 0.850 | 21.59 |
| 0.020 | 0.508 0 | 0.300 | 7.62 | 0.580 | 14.73 | 0.860 | 21.84 |
| 0.030 | 0.762 0 | 0.310 | 7.87 | 0.590 | 14.99 | 0.870 | 22.10 |
| 0.040 | 1.016 | 0.320 | 8.13 | 0.600 | 15.24 | 0.880 | 22.35 |
| 0.050 | 1.270 | 0.330 | 8.38 | 0.610 | 15.49 | 0.890 | 22.61 |
| 0.060 | 1.524 | 0.340 | 8.64 | 0.620 | 15.75 | 0.900 | 22.86 |
| 0.070 | 1.778 | 0.350 | 8.89 | 0.630 | 16.00 | 0.910 | 23.11 |
| 0.080 | 2.032 | 0.360 | 9.14 | 0.640 | 16.26 | 0.920 | 23.37 |
| 0.090 | 2.286 | 0.370 | 9.40 | 0.650 | 16.51 | 0.930 | 23.62 |
| 0.100 | 2.540 | 0.380 | 9.65 | 0.660 | 16.76 | 0.940 | 23.88 |
| 0.110 | 2.794 | 0.390 | 9.91 | 0.670 | 17.02 | 0.950 | 24.13 |
| 0.120 | 3.048 | 0.400 | 10.16 | 0.680 | 17.27 | 0.960 | 24.38 |
| 0.130 | 3.302 | 0.410 | 10.41 | 0.690 | 17.53 | 0.970 | 24.64 |
| 0.140 | 3.556 | 0.420 | 10.67 | 0.700 | 17.78 | 0.980 | 24.89 |
| 0.150 | 3.810 | 0.430 | 10.92 | 0.710 | 18.03 | 0.990 | 25.15 |
| 0.160 | 4.064 | 0.440 | 11.18 | 0.720 | 18.29 | 1.000 | 25.40 |
| 0.170 | 4.318 | 0.450 | 11.43 | 0.730 | 18.54 | . . . . | . . . . |
| 0.180 | 4.572 | 0.460 | 11.68 | 0.740 | 18.80 | . . . . | . . . . |
| 0.190 | 4.826 | 0.470 | 11.94 | 0.750 | 19.05 | . . . . | . . . . |

tables can be easily extended by moving the decimal point an equal number of places in the same direction, for both columns.

*Example* What is the length in millimeters of a wave whose length is 0.280 in?

*Solution* Locate 0.280 in the inches column in Table 22-6. The answer, 7.112 mm, is shown in the column immediately to the right.

*Example* A wave is approximately 33 mm long. What is its length in inches?

*Solution* Locate 3.302 in the millimeters column in Table 22-6. By moving the decimal point one place to the right, you will have 33.02 mm. The corresponding distance in inches is 0.130, but you must move the decimal point one place to the right. The answer is 1.30 in.

*Example* A wave has a length of 0.92 mm. What is the corresponding length in inches?

*Solution* Using Table 22-7 locate 0.92 in the millimeters column. The length in inches is 0.036 156.

*Table 22-7. Millimeters to decimal inches.*

| mm | in | mm | in | mm | in | mm | in |
|------|-------------|------|-----------|------|-----------|------|-----------|
| 0.01 | 0.000 393 7 | 0.26 | 0.010 218 | 0.51 | 0.020 043 | 0.76 | 0.029 868 |
| 0.02 | 0.000 786   | 0.27 | 0.010 611 | 0.52 | 0.020 436 | 0.77 | 0.030 261 |
| 0.03 | 0.001 179   | 0.28 | 0.011 004 | 0.53 | 0.020 829 | 0.78 | 0.030 654 |
| 0.04 | 0.001 572   | 0.29 | 0.011 397 | 0.54 | 0.021 222 | 0.79 | 0.031 047 |
| 0.05 | 0.001 965   | 0.30 | 0.011 790 | 0.55 | 0.021 615 | 0.80 | 0.031 440 |
| 0.06 | 0.002 358   | 0.31 | 0.012 183 | 0.56 | 0.022 008 | 0.81 | 0.031 833 |
| 0.07 | 0.002 751   | 0.32 | 0.012 576 | 0.57 | 0.022 401 | 0.82 | 0.032 226 |
| 0.08 | 0.003 144   | 0.33 | 0.012 969 | 0.58 | 0.022 794 | 0.83 | 0.032 619 |
| 0.09 | 0.003 537   | 0.34 | 0.013 362 | 0.59 | 0.023 187 | 0.84 | 0.033 012 |
| 0.10 | 0.003 937   | 0.35 | 0.013 755 | 0.60 | 0.023 580 | 0.85 | 0.033 405 |
| 0.11 | 0.004 323   | 0.36 | 0.014 148 | 0.61 | 0.023 973 | 0.86 | 0.033 798 |
| 0.12 | 0.004 716   | 0.37 | 0.014 541 | 0.62 | 0.024 366 | 0.87 | 0.034 191 |
| 0.13 | 0.005 109   | 0.38 | 0.014 934 | 0.63 | 0.024 759 | 0.88 | 0.034 584 |
| 0.14 | 0.005 502   | 0.39 | 0.015 327 | 0.64 | 0.025 152 | 0.89 | 0.034 977 |
| 0.15 | 0.005 895   | 0.40 | 0.015 720 | 0.65 | 0.025 450 | 0.90 | 0.035 370 |
| 0.16 | 0.006 288   | 0.41 | 0.016 113 | 0.66 | 0.025 938 | 0.91 | 0.035 763 |
| 0.17 | 0.006 681   | 0.42 | 0.016 506 | 0.67 | 0.026 331 | 0.92 | 0.036 156 |
| 0.18 | 0.007 074   | 0.43 | 0.016 899 | 0.68 | 0.026 724 | 0.93 | 0.036 549 |
| 0.19 | 0.007 467   | 0.44 | 0.017 292 | 0.69 | 0.027 117 | 0.94 | 0.036 942 |
| 0.20 | 0.007 860   | 0.45 | 0.017 685 | 0.70 | 0.027 510 | 0.95 | 0.037 335 |
| 0.21 | 0.008 253   | 0.46 | 0.018 078 | 0.71 | 0.027 903 | 0.96 | 0.039 370 |
| 0.22 | 0.008 646   | 0.47 | 0.018 471 | 0.72 | 0.028 296 | 0.97 | 0.037 728 |
| 0.23 | 0.009 039   | 0.48 | 0.018 864 | 0.73 | 0.028 689 | 0.98 | 0.038 121 |
| 0.24 | 0.009 432   | 0.49 | 0.019 257 | 0.74 | 0.029 082 | 0.99 | 0.038 514 |
| 0.25 | 0.009 825   | 0.50 | 0.019 650 | 0.75 | 0.029 475 | 1.00 | 0.038 907 |

# Electrical and physical values

Conversion factors for a variety of electrical and physical values are shown in Table 22-8. Its use does require some arithmetic, usually either division or multiplication. However, it does have a wide range.

## Table 22-8. Conversion of electrical and physical values.

| To convert from | To | Multiply by |
|---|---|---|
| Amperes | milliamperes | $10^3$ |
| | microamperes | $10^6$ |
| Å | cm | $10^{-8}$ |
| Atmospheres | mm of mercury at $0\,°C$ | 760 |
| | inches of mercury at $32\,°F$ | 29.921 |
| | feet of water at $39.1\,°F$ | 33.90 |
| | lb/ft$^2$ | 2 116.2 |
| | lb/in$^2$ | 14.696 |
| | kg/m$^2$ | $1.033 \times 10^4$ |
| | tons per square foot | 1.058 |
| Atomic mass unit | mass of electron | $5.488 \times 10^{-4}$ |
| | mass of proton | 1.007 6 |
| | mass of neutron | 1.009 0 |
| | mass of $\alpha$ particle | 4.002 9 |
| | $H_1$ atom mass | 1.008 1 |
| | MeV | 931.1 |
| | gram mass | $1.659\ 8 \times 10^{-24}$ |
| Bars | dynes per square centimeter | $1.0 \times 10^4$ |
| Btu | ft·lb | 778.0 |
| | horsepower-hours | $3.929 \times 10^{-4}$ |
| | J | 1 055 |
| | kilogram-meters | 107.6 |
| | kilogram-calories | 0.252 0 |
| Btu | kWh | $2.930 \times 10^{-4}$ |
| Btu/h | hp | $3.929 \times 10^{-4}$ |
| Btu/(h-ft$^2$-°F) | W/cm$^2$-°C | $5.68 \times 10^{-4}$ |
| | W/in$^2$-°F | $2.035 \times 10^{-3}$ |
| Btu (h-ft$^2$)/(°F/in) | (cal/s-cm$^2$)/(°C/cm) | $0.344 \times 10^{-3}$ |
| Btu/min | ft-lb/s | 12.97 |
| | hp | 0.023 58 |
| | kW | 0.017 58 |
| Btu/lb | gram-calorie per gram | 0.555 6 |
| Btu/lb-°F | gram-calorie per gram-°C | 1.0 |
| Btu/s | kilogram-meters/s | 107.6 |
| | kW | 1.054 9 |
| Btu/ft$^2$ | gram-calorie per square centimeter | 0.271 2 |
| cm Hg (0°C) | atmospheres | 0.013 158 |
| cm Hg (0°C) | lb/in$^2$ | 0.193 37 |
| | inches of water | 5.352 |
| cm/°C | in/°F | 0.218 7 |
| Centipoises | pound/foot-hours | 2.42 |
| ft$^3$/min | gal/s | 0.124 7 |
| | cm$^3$/s | 472.0 |
| ft$^3$/lb | cm$^3$g | 62.43 |

*Table 22-8. Continued.*

| To convert from | To | Multiply by |
|---|---|---|
| $ft^3/s$ | gal/min | 448.8 |
| $m^3$ | $ft^3$ | 35.31 |
| Curie | disintegration/s | $3.7 \times 10^{10}$ |
| Degrees | radians | 0.017 45 |
| Degrees per second | rev/min | 0.166 7 |
| Dynes | g | $1.019\ 7 \times 10^{-3}$ |
| | poundals | $7.233 \times 10^{-5}$ |
| Dynes per centimeter | lb/ft | $6.852 \times 10^{-5}$ |
| Dynes per square centimeter | atmospheres | $9.869 \times 10^{-7}$ |
| Electron volts | calories (15 °C) | $3.828 \times 10^{-20}$ |
| | J | $1.602 \times 10^{-19}$ |
| | calorie/mole | 23.05 |
| | $cm^{-1}$ | 8 065.9 |
| | kWh | $4.45 \times 10^{-26}$ |
| | microns wavelength | 1.240 |
| ergs | J | $1 \times 10^{-7}$ |
| Farads | microfarads | $10^6$ |
| | picofarads | $10^{12}$ |
| Feet of $H_2O$ at 39.2 °F | atmospheres | 0.029 50 |
| | inches of Hg at 32 °F | 0.882 7 |
| ft·lb | Btu (mean) | $1.284\ 9 \times 10^{-3}$ |
| | J (abs.) | 1.355 8 |
| | kWh | $3.766 \times 10^{-7}$ |
| ft·lb/min | kW | $2.260 \times 10^{-5}$ |
| ft·lb/s | Btu/min | 0.077 12 |
| | kW | $1.356 \times 10^{-3}$ |
| Gal (U.S. liquid) | $in^3$ | 231.0 |
| | $ft^3$ | 0.133 68 |
| | $cm^3$ | 3 785 |
| Gal/min (of water) | lb/h of water | 500.8 |
| Gram-calories | Btu | $3.968 \times 10^{-3}$ |
| | J | 4.186 |
| Gram-calorie per square centimeter | $Btu/ft^2$ | 3.687 |
| Gram-centimeter | Btu | $9.297 \times 10^{-8}$ |
| | ergs | 980.7 |
| Gram-mole gas | cubic centimeter gas (0 °C and 760 mm) | $2.24 \times 10^4$ |
| Grams | dynes | 980.665 |
| | lb | $2.204\ 6 \times 10^{-3}$ |
| Grams of matter | eV | $5.61 \times 10^{32}$ |
| Grams/$cm^3$ | lb/$in^3$ | 0.036 13 |
| | lb/$ft^3$ | 62.43 |
| Gram/$cm^2$ | centimeters of Hg | 0.073 56 |

*Table 22-8. Continued.*

| To convert from | To | Multiply by |
|---|---|---|
| | atmosphere | $9.678 \times 10^{-4}$ |
| | lb/ft$^2$ | 2.048 |
| H | mH | $10^3$ |
| hp (mech.) | ft·lb/s | 550 |
| | kilowatts | 0.745 7 |
| | watts | 745.7 |
| Horsepower hours | Btu | $2.545 \times 10^{-3}$ |
| | J (abs.) | $2.684 \times 10^4$ |
| | kg-cal | 641.3 |
| Inches of Hg at 32 °F | atmospheres | 0.033 42 |
| | lb/in$^2$ | 0.491 2 |
| Inches of H$_2$O at 39.2 °F (4 °C) | atmospheres | $2.458 \times 10^{-3}$ |
| | centimeters of Hg | 0.186 8 |
| | lb/ft$^2$ | 5.202 |
| J (abs.) | Btu (mean) | $9.480 \times 10^{-4}$ |
| | Wh | $2.778 \times 10^{-4}$ |
| | electron volts | $0.624\ 2 \times 10^{19}$ |
| | ergs | $1 \times 10^7$ |
| | kilogram-calories | $2.39 \times 10^{-4}$ |
| J/g-°C | Btu/lb-°F | 0.238 9 |
| Kilogram-calorie | horsepower-hours | $1.559 \times 10^{-3}$ |
| Kilogram-meter | Btu | $9.297 \times 10^{-3}$ |
| | ergs | $9.807 \times 10^7$ |
| kWh | average noon sunlight on 1 m$^2$ | $\cong 1$ |
| | Btu | $3.413 \times 10^3$ |
| | grams U$^{235}$ fissioned | $\cong 4.2 \times 10^{-5}$ |
| kWh | horsepower hours | 1.341 0 |
| | J (abs.) | $3.6 \times 10^6$ |
| | pounds coal consumed | $\cong 0.75$ |
| kW | Btu/min | 56.90 |
| | kilogram-calorie/minute | 14.33 |
| | W | $10^3$ |
| Liter-atmospheres | Btu | 0.096 07 |
| Liters/kilogram | ft$^3$/lb | 0.016 02 |
| Liters/minute | ft$^3$/s | $5.885 \times 10^{-4}$ |
| | gal/h | 15.851 |
| Lumens | W | 0.001 496 |
| N | dynes | $1 \times 10^5$ |
| Nm | J (abs.) | 1.000 |
| Ω | kΩ | $10^{-3}$ |
| | MΩ | $10^{-6}$ |
| Pound-Celsius (Centigrade) unit | Btu | 1.8 |

*Table 22-8. Continued.*

| To convert from | To | Multiply by |
|---|---|---|
| Pound-Fahrenheit unit | Btu | 1.0 |
| Pound-mol gas | cubic feet of gas (60 °F at 1 atm) | 379.4 |
| Pounds of $H_2O$ (4 °C) | gal | 0.119 8 |
| Pounds of $H_2O$ at 64 °F | $ft^3$ | 0.016 033 |
| lb/gal | $g/cm^3$ | 0.119 8 |
| $lb/in^2$ | atmospheres | 0.068 05 |
| | feet of $H_2O$ at 39.1 °F | 2.307 |
| | millimeters of Hg at 32 °F | 51.7 |
| Radian | degrees | 57.296 |
| Radian/s | revolutions/s | 0.159 2 |
| Roentgens | $esu/cm^3$ change in air | 1 (STP) |
| Volt-coulomb | J | 1.000 |
| Wh | Btu | 3.413 |
| Ws | J | 1.000 |
| W | Btu/min | 0.056 90 |
| | erg/s | $1 \times 10^7$ |
| | ft-lb/min | 44.25 |
| | hp | $1.341 \times 10^{-3}$ |
| | lumens | 668.5 |
| | mW | 1 000 |
| | kW | $10^{-3}$ |
| | MW | $10^{-6}$ |

# Frequency conversions

Three common measures of frequency are used in electronics. To convert from one to the other, see Table 22-9.

# Frequency allocations

Figure 22-1 shows frequency allocations from 15 to 806 MHz.

*Table 22-9.*
*Frequency conversion.*

| Given this value | Multiply by this value to get | | |
|---|---|---|---|
| | Hz | kHz | MHz |
| Hz | — | $10^{-3}$ | $10^{-6}$ |
| kHz | $10^3$ | — | $10^{-3}$ |
| MHz | $10^6$ | $10^3$ | — |

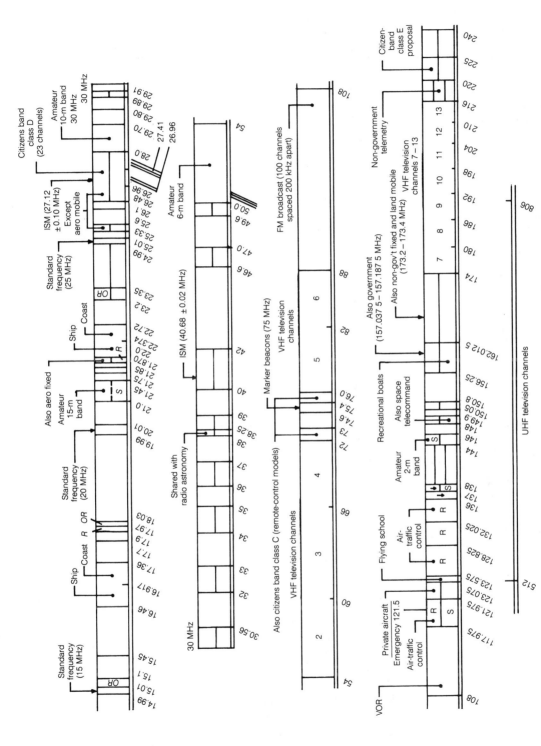

Fig. 22-1. *FCC (Federal Communications Commission) frequency allocations, 15 MHz to 806 MHz.*

# Measuring systems

Two measuring systems are in common use—the English and the metric. The advantage of the metric system is that it is decimal, and so it is easy to move from basic metric units to multiples or submultiples. The meter and the gram are bases for the metric system. Commonly used abbreviations are m for meters, mm for millimeters, cm for centimeters, dm for decimeters, dam for decameters, hm for

*Table 22-10. English units of measure and metric equivalents.*

### Distance

| | | |
|---|---|---|
| 1 micron | = | 0.001 mm |
| 1 micron | = | 0.000 001 m |
| 1 mm | = | 0.039 370 0 in |
| 1 mm | = | 0.003 28 ft |
| 1 cm | = | 10 mm |
| 1 cm | = | 0.393 700 in |
| 1 cm | = | 0.032 808 ft |
| 1 cm | = | 0.010 936 11 yd |
| 1 m | = | 39.370 0 in |
| 1 m | = | 3.280 833 333 ft |
| 1 m | = | 1.093 61 yd |
| 1 dm | = | 10 cm |
| 1 dm | = | 3.937 in |
| 1 m | = | 10 dm |
| 1 m | = | 100 cm |
| 1 m | = | 1 000 mm |
| 1 dkm | = | 10 m |
| 1 dkm | = | 393.7 in |
| 1 hm | = | 10 dkm |
| 1 hm | = | 328 ft, 1 in |
| 1 km | = | 10 hm |
| 1 km | = | 0.621 37 mi |
| 1 myriameter | = | 10 km |
| 1 myriameter | = | 6.213 7 mi |
| 1 in | = | 25.400 05 mm |
| 1 in | = | 2.540 005 cm |
| 1 in | = | 0.025 400 05 m |
| 1 ft | = | 304.800 6 mm |
| 1 ft | = | 30.480 06 cm |
| 1 ft | = | 0.304 800 6 m |
| 1 yd | = | 0.914 4 m |
| 1 mi | = | 1.609 km |

### Area

| | | |
|---|---|---|
| 1 in$^2$ | = | 6.452 cm$^2$ |
| 1 ft$^2$ | = | 0.092 9 m$^2$ |
| 1 yd$^2$ | = | 0.84 m$^2$ |
| 1 mi$^2$ | = | 2.589 km$^2$ |
| 1 mm$^2$ | = | 0.001 55 in$^2$ |
| 1 cm$^2$ | = | 0.155 0 in$^2$ |
| 1 m$^2$ | = | 1.196 yd$^2$ |
| 1 km$^2$ | = | 0.386 1 mi$^2$ |

### Volume

| | | |
|---|---|---|
| 1 in$^3$ | = | 16.387 2 cm$^3$ |
| 1 ft$^3$ | = | 0.028 32 m$^3$ |
| 1 yd$^3$ | = | 0.764 6 m$^3$ |
| 1 liquid qt | = | 0.946 4 L |
| 1 liquid gal | = | 3.785 L |
| 1 dry qt | = | 1.101 L |
| 1 cm$^3$ | = | 0.061 02 in$^3$ |
| 1 m$^3$ | = | 1.308 yd$^3$ |
| 1 mL | = | 0.033 81 liquid oz |
| 1 L | = | 1.057 liquid qt |
| 1 L | = | 0.908 1 dry qt |

### Weight

| | | |
|---|---|---|
| 1 grain | = | 0.064 8 g |
| 1 oz (avoir.) | = | 28.35 g |
| 1 oz (troy) | = | 31.10 grains |
| 1 lb (avoir.) | = | 0.453 6 kg |
| 1 lb (troy) | = | 0.373 2 kg |
| 1 ton (short) | = | 0.907 2 metric ton |
| 1 g | = | 15.43 grains (troy) |
| 1 g | = | 0.032 15 oz (troy) |
| 1 g | = | 0.035 27 oz (avoir.) |
| 1 kg | = | 2.205 lbs (avoir.) |
| 1 kg | = | 2.679 lbs (troy) |
| 1 metric ton | = | 1.102 tons (short) |

hectometers, km for kilometers, and mym for myriameters. The micron is represented by $\mu$ (Greek letter mu). Table 22-10 shows the relationships between English measure and metric equivalents.

Table 22-11 supplies data on length, volume, and mass in metric form.

*Table 22-11. Length, volume, and mass in metric form.*

| Prefix | Exponential form | Symbol | Length | Volume | Mass |
|--------|------------------|--------|--------|--------|------|
| tera | $10^{12}$ | T | terameter | teraliter | teragram |
| giga | $10^{9}$ | G | gigameter | gigaliter | gigagram |
| mega | $10^{6}$ | M | megameter | megaliter | megagram |
| kilo | $10^{3}$ | k | kilometer | kiloliter | kilogram |
| hecto | $10^{2}$ | h | hectometer | hectoliter | hectogram |
| deca | $10^{1}$ | da | decameter | decaliter | decagram |
| | $10^{0}$ | | meter | liter | gram |
| deci | $10^{-1}$ | d | decimeter | deciliter | decigram |
| centi | $10^{-2}$ | c | centimeter | centiliter | centigram |
| milli | $10^{-3}$ | m | millimeter | milliliter | milligram |
| micro | $10^{-6}$ | $\mu$ | micrometer | microliter | microgram |
| nano | $10^{-9}$ | n | nanometer | nanoliter | nanogram |
| pico | $10^{-12}$ | p | picometer | picoliter | picogram |
| femto | $10^{-15}$ | f | femtometer | femtoliter | femtogram |
| atto | $10^{-18}$ | a | attometer | attoliter | attogram |

Note: There are certain standards to observe in using these abbreviations. Thus, a microsecond could be abbreviated as $\mu$s. Conceivably, a thousandth of a microsecond could then be written as m$\mu$s or a millimicrosecond. The preferred form is ns or nanosecond.

# Proper (common) fractions and millimetric equivalents

Fractions are the usual form when measurements are made with a foot rule or yardstick. Table 22-12 shows the metric equivalents in millimeters and also decimal equivalents for fractions of an inch, from $1/64$ to 1 in, in $1/64$-in steps.

When the conversion involves a whole number plus a fraction, use a factor of 25.400 05 mm for each inch.

*Example*   Convert $2^{17}/_{64}$ in to millimeters.
*Solution*   Two inches = $2 \times 25.400\,05$ mm, or 50.800 1 mm. Table 22-12 shows that $^{17}/_{64}$ inch = 6.746 875 mm. 50.800 1 mm + 6.746 875 mm = 57.546 975 mm.

*Table 22-12. Fractional and decimal*
*inches and millimeter equivalents.*

**(1 in = 25.400 05 mm)**

| | in | | | in | |
|---|---|---|---|---|---|
| **Fraction** | **Decimal** | **mm** | **Fraction** | **Decimal** | **mm** |
| 1/64 | 0.015 625 | 0.396 875 | 33/64 | 0.515 625 | 13.096 875 |
| 1/32 | 0.031 250 | 0.793 750 | 17/32 | 0.531 250 | 13.493 750 |
| 3/64 | 0.046 875 | 1.190 625 | 35/64 | 0.546 875 | 13.890 625 |
| 1/16 | 0.062 500 | 1.587 500 | 9/16 | 0.562 500 | 14.287 500 |
| 5/64 | 0.078 125 | 1.984 375 | 37/64 | 0.578 125 | 14.684 375 |
| 3/32 | 0.093 750 | 2.381 250 | 19/32 | 0.593 750 | 15.081 250 |
| 7/64 | 0.109 375 | 2.778 125 | 39/64 | 0.609 375 | 15.478 125 |
| 1/8 | 0.125 000 | 3.175 000 | 5/8 | 0.625 000 | 15.875 000 |
| 9/64 | 0.140 625 | 3.571 875 | 41/64 | 0.640 625 | 16.271 875 |
| 5/32 | 0.156 250 | 3.968 750 | 21/32 | 0.656 250 | 16.668 750 |
| 11/64 | 0.171 875 | 4.365 625 | 43/64 | 0.671 875 | 17.065 625 |
| 3/16 | 0.187 500 | 4.762 500 | 11/16 | 0.687 500 | 17.462 500 |
| 13/64 | 0.203 125 | 5.159 375 | 45/64 | 0.703 125 | 17.859 375 |
| 7/32 | 0.218 750 | 5.556 250 | 23/32 | 0.718 750 | 18.256 250 |
| 15/64 | 0.234 375 | 5.953 125 | 47/64 | 0.734 375 | 18.653 125 |
| 1/4 | 0.250 000 | 6.350 000 | 3/4 | 0.750 000 | 19.050 000 |
| 17/64 | 0.265 625 | 6.746 875 | 49/64 | 0.765 625 | 19.446 875 |
| 9/32 | 0.281 250 | 7.143 750 | 25/32 | 0.781 250 | 19.843 750 |
| 19/64 | 0.296 875 | 7.540 625 | 51/64 | 0.796 875 | 20.240 625 |
| 5/16 | 0.312 500 | 7.937 500 | 13/16 | 0.812 500 | 20.637 500 |
| 21/64 | 0.328 125 | 8.334 375 | 53/64 | 0.828 125 | 21.034 375 |
| 11/32 | 0.343 750 | 8.731 250 | 27/32 | 0.843 750 | 21.431 250 |
| 23/64 | 0.359 375 | 9.128 125 | 55/64 | 0.859 375 | 21.828 125 |
| 3/8 | 0.375 000 | 9.525 000 | 7/8 | 0.875 000 | 22.225 000 |
| 25/64 | 0.390 625 | 9.921 875 | 57/64 | 0.890 625 | 22.621 875 |
| 13/32 | 0.406 250 | 10.318 750 | 29/32 | 0.906 250 | 23.018 750 |
| 27/64 | 0.421 875 | 10.715 625 | 59/64 | 0.921 875 | 23.415 625 |
| 7/16 | 0.437 500 | 11.112 500 | 15/16 | 0.937 500 | 23.812 500 |
| 29/64 | 0.453 125 | 11.509 375 | 61/64 | 0.953 125 | 24.209 375 |
| 15/32 | 0.468 750 | 11.906 250 | 31/32 | 0.968 750 | 24.606 250 |
| 31/64 | 0.484 375 | 12.303 125 | 63/64 | 0.984 375 | 25.003 125 |
| 1/2 | 0.500 000 | 12.700 000 | 1 | 1.000 000 | 25.400 050 |

# Temperature conversions

Components used in electronics, including wire, can have a positive temperature coefficient, with resistance increasing as temperature rises, or a negative temperature coefficient, with resistance varying inversely with temperature. Some components are specifically designed to have a zero temperature coefficient, with resistance not affected by temperature changes.

Although temperature is often specified in degrees Fahrenheit, the use of the Celsius temperature scale (formerly known as Centigrade) is becoming more acceptable in electrical and electronic applications (see Fig. 22-2). The following

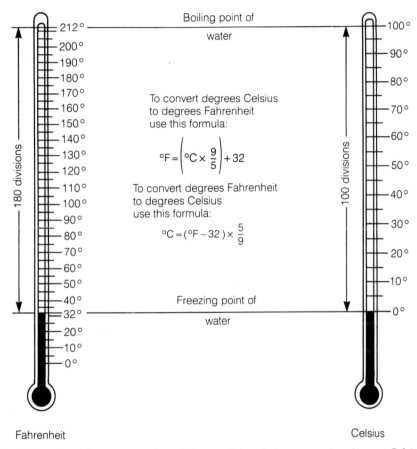

Fig. 22-2. *Graphic representation of degrees Fahrenheit compared to degrees Celsius.*

tables aren't complete for all possible values of Fahrenheit degrees and Celsius degrees but are intended for the more common values encountered in electrical and electronic applications.

Tables 22-13 and 22-14 are based on the formulas:

$$C=(F-32)\times5/9$$
$$F=(C\times9/5)+32$$

| Table 22-13. Degrees Celsius to degrees Fahrenheit. | | | | | |
|---|---|---|---|---|---|
| °C | °F | °C | °F | °C | °F |
| −100 | −148 | 5 | 41 | 105 | 221 |
| −95 | −139 | 10 | 50 | 110 | 230 |
| −90 | −130 | 15 | 59 | 115 | 239 |
| −85 | −121 | 20 | 68 | 120 | 248 |
| −80 | −112 | 25 | 77 | 125 | 257 |
| −75 | −103 | 30 | 86 | 130 | 266 |
| −70 | −94 | 35 | 93 | 135 | 275 |
| −65 | −85 | 40 | 104 | 140 | 284 |
| −60 | −76 | 45 | 113 | 145 | 293 |
| −55 | −67 | 50 | 122 | 150 | 302 |
| −50 | −58 | 55 | 131 | 155 | 311 |
| −45 | −49 | 60 | 140 | 160 | 320 |
| −40 | −40 | 65 | 149 | 165 | 329 |
| −35 | −31 | 70 | 158 | 170 | 338 |
| −30 | −22 | 75 | 167 | 175 | 347 |
| −25 | −13 | 80 | 176 | 180 | 356 |
| −20 | −4 | 85 | 185 | 185 | 365 |
| −15 | 5 | 90 | 194 | 190 | 374 |
| −10 | 14 | 95 | 203 | 195 | 383 |
| −5 | 23 | 100 | 212 | 200 | 392 |
| 0 | 32 | | | | |

| Table 22-14. Degrees Fahrenheit to degrees Celsius. | | | |
|---|---|---|---|
| °F | °C | °F | °C |
| −100 | −73 | 70 | 21 |
| −90 | −68 | 75 | 24 |
| −80 | −62 | 80 | 27 |
| −70 | −57 | 85 | 29 |
| −60 | −51 | 90 | 32 |
| −50 | −46 | 95 | 35 |
| −40 | −40 | 100 | 38 |
| −30 | −34 | 105 | 41 |
| −20 | −29 | 110 | 43 |
| −10 | −23 | 115 | 46 |
| −5 | −21 | 120 | 49 |
| 0 | −18 | 125 | 52 |
| 5 | −15 | 130 | 54 |
| 10 | −12 | 135 | 57 |
| 15 | −9 | 140 | 60 |
| 20 | −7 | 145 | 63 |
| 25 | −4 | 150 | 66 |
| 30 | −1 | 155 | 68 |
| 35 | 1.7 | 160 | 71 |
| 40 | 4 | 165 | 74 |
| 45 | 7 | 170 | 77 |
| 50 | 10 | 175 | 79 |
| 55 | 13 | 180 | 82 |
| 60 | 16 | 185 | 85 |
| 65 | 18 | 190 | 88 |
| | | 195 | 91 |
| | | 200 | 93 |

All Celsius and Fahrenheit temperatures in Table 22-14 are positive unless otherwise indicated.

# Electronic units

Formulas in electronics use basic units—units such as the volt, ohm, ampere, farad, and henry. However, problems invariably supply information in terms of multiples of the ohm or submultiples of the farad and the henry. Thus, before any solution can be attempted, it is often necessary to convert multiples or submultiples to basic units.

Quite frequently the numbers used in formulas will be very large whole numbers or long decimals. In either case, it is highly advantageous to be able to use powers of ten and to be familiar with the rules for dividing and multiplying numbers using exponents. Table 22-15 shows the method of expressing small and large numbers using powers of ten.

*Table 22-15. Powers of 10.*

| | |
|---|---|
| $10^0 = 1$ | $10^0 = 1$ |
| $10^1 = 10$ | $10^{-1} = 0.1$ |
| $10^2 = 100$ | $10^{-2} = 0.01$ |
| $10^3 = 1\ 000$ | $10^{-3} = 0.001$ |
| $10^4 = 10\ 000$ | $10^{-4} = 0.000\ 1$ |
| $10^5 = 100\ 000$ | $10^{-5} = 0.000\ 01$ |
| $10^6 = 1\ 000\ 000$ | $10^{-6} = 0.000\ 001$ |
| $10^7 = 10\ 000\ 000$ | $10^{-7} = 0.000\ 000\ 1$ |
| $10^8 = 100\ 000\ 000$ | $10^{-8} = 0.000\ 000\ 01$ |
| $10^9 = 1\ 000\ 000\ 000$ | $10^{-9} = 0.000\ 000\ 001$ |
| $10^{10} = 10\ 000\ 000\ 000$ | $10^{-10} = 0.000\ 000\ 000\ 1$ |
| $10^{11} = 100\ 000\ 000\ 000$ | $10^{-11} = 0.000\ 000\ 000\ 01$ |
| $10^{12} = 1\ 000\ 000\ 000\ 000$ | $10^{-12} = 0.000\ 000\ 000\ 001$ |

# Symbols and prefixes for powers of 10

Numbers have names. The prefixes given in Table 22-16 are helpful in identifying particular values of powers of 10. Thus, a gigahertz (prefix giga) corresponds to $10^9$.

*Table 22-16. Powers of 10, symbols, and prefixes.*

| Power of ten | Prefix | Symbol |
|---|---|---|
| $10^{12}$ | tera | T |
| $10^9$ | giga | G |
| $10^6$ | mega | M |
| $10^3$ | kilo | k |
| $10^2$ | hecto or hekto | h |
| 10 | deka | da |
| $10^{-1}$ | deci | d |
| $10^{-2}$ | centi | c |
| $10^{-3}$ | milli | m |
| $10^{-6}$ | micro | $\mu$ |
| $10^{-9}$ | nano | n |
| $10^{-12}$ | pico | p |
| $10^{-15}$ | femto | f |
| $10^{-18}$ | atto | a |

Table 22-17 supplies data on the conversion of electronic units from one form to another. Most of the conversion factors are supplied as powers of 10.

*Table 22-17. Electronic units.*

| Multiply | By | To convert to |
|---|---|---|
| ampere-turns | 1.257 | Gb |
| A | $10^{12}$ | pA |
| A | $10^6$ | $\mu$A |
| A | $10^3$ | mA |
| cm | 10 | mm |
| degrees | 60 | min |
| F | $10^{12}$ | pF |
| F | $10^6$ | $\mu$F |
| F | $10^3$ | mF |
| G | 1 | lines/cm$^2$ |
| G | 6.452 | lines/in$^2$ |
| H | $10^6$ | $\mu$H |
| H | $10^3$ | mH |
| Hz | $10^{-6}$ | MHz |
| Hz | $10^{-3}$ | kHz |
| Hz | $10^9$ | gHz |
| hp | 0.745 7 | kW |
| hp | 745.7 | W |
| J | 1 | W |
| J | 10 | |
| kHz | $10^3$ | Hz |
| kHz | $10^{-3}$ | MHz |
| kHz | $10^6$ | gHz |
| kVA | $10^3$ | VA |
| kV | $10^3$ | V |
| kWh | $3.6 \times 10^6$ | J |
| kW | $10^3$ | W |
| kW | 1.341 | hp |
| lines/cm$^2$ | 1 | G |
| lines/cm$^2$ | 6.452 | lines/in$^2$ |
| MHz | $10^6$ | Hz |
| MHz | $10^3$ | kHz |
| MW | $10^{-6}$ | W |
| M$\Omega$ | $10^6$ | $\Omega$ |
| m | $10^2$ | cm |
| m | 3.281 | ft |
| m | 39.37 | in |
| m | $10^3$ | mm |
| $\mu$A | $10^{-6}$ | A |
| $\mu$A | $10^{-3}$ | mA |
| $\mu$F | $10^{-6}$ | F |
| $\mu$F | $10^6$ | pF |
| $\mu$H | $10^{-6}$ | H |

*Table 22-17. Continued.*

| Multiply | By | To convert to |
|---|---|---|
| $\mu$H | $10^{-3}$ | mH |
| $\mu$S | $10^{-6}$ | S |
| $\mu$V | $10^{-3}$ | mV |
| $\mu$V | $10^{-6}$ | V |
| $\mu$W | $10^{-6}$ | W |
| $\mu$W | $10^{-3}$ | mW |
| mA | $10^{-3}$ | A |
| mA | $10^{3}$ | $\mu$A |
| mH | $10^{-3}$ | H |
| mm | $10^{-3}$ | m |
| mm | $10^{-1}$ | cm |
| mS | $10^{-3}$ | S |
| mV | $10^{-3}$ | V |
| mV | $10^{3}$ | $\mu$V |
| mV | $10^{-6}$ | kV |
| mW | $10^{-3}$ | W |
| mW | $10^{3}$ | $\mu$W |
| mW | $10^{-9}$ | MW |
| mils | $10^{-3}$ | in |
| min | 60 | s |
| min | $^1/_{60}$ | deg |
| $\Omega$ | $10^{-3}$ | k$\Omega$ |
| $\Omega$ | $10^{-6}$ | M$\Omega$ |
| pA | $10^{6}$ | $\mu$A |
| pF | $10^{-12}$ | F |
| p$\Omega$ | $10^{-12}$ | $\Omega$ |
| radians | 57.3 | deg |
| s | $^1/_{3600}$ | deg |
| s | $10^{3}$ | ms |
| s | $10^{6}$ | $\mu$s |
| s | $^1/_{60}$ | min |
| S | $10^{6}$ | $\mu$S |
| S | $10^{3}$ | mS |
| cm$^2$ | $1\,973 \times 10^2$ | circular mils |
| in$^2$ | $1\,273 \times 10^3$ | circular mils |
| square mils | 1.273 | circular mils |
| VA | $^1/_{1000}$ | kVA |
| V | $10^{6}$ | $\mu$V |
| V | $10^{3}$ | mV |
| V | $10^{-3}$ | kV |
| Wh | $36 \times 10^2$ | J |
| Ws | 1 | J |
| W | $10^{6}$ | $\mu$W |
| W | $10^{3}$ | mW |
| W | $10^{-3}$ | kW |
| W | $10^{-6}$ | MW |

*Table 22-18. Electronic unit multiples and submultiples.*

### Current

| Unit | Conversion |
|------|------------|
| $\mu$A | mA$\times$1 000 |
| $\mu$A | A$\times$1 000 000 |
| mA | A$\times$1 000 |
| mA | $\mu$A/1 000 |
| A | $\mu$A/1 000 000 |
| A | mA/1 000 |

### Resistance

| Unit | Conversion |
|------|------------|
| $\Omega$ | k$\Omega\times$1 000 |
| $\Omega$ | M$\Omega\times$1 000 000 |
| k$\Omega$ | $\Omega$/1 000 |
| k$\Omega$ | M$\Omega\times$1 000 |
| M$\Omega$ | $\Omega$/1 000 000 |
| M$\Omega$ | k$\Omega$/1 000 |

### Voltage

| Unit | Conversion |
|------|------------|
| V | kV$\times$1 000 |
| V | MV$\times$1 000 000 |
| kV | V/1 000 |
| kV | MV$\times$1 000 |
| MV | V/1 000 000 |
| MV | kV/1 000 |
| mV | V$\times$1 000 |
| mV | $\mu$V/1 000 |
| $\mu$V | V$\times$1 000 000 |
| $\mu$V | mV$\times$1 000 |

### Power

| Unit | Conversion |
|------|------------|
| W | mW/1 000 |
| W | $\mu$W/1 000 000 |
| W | kW$\times$1 000 |
| W | MW$\times$1 000 000 |
| kW | W/1 000 |

### Power

| Unit | Conversion |
|------|------------|
| kW | MW$\times$1 000 |
| MW | kW/1 000 |
| MW | W/1 000 000 |
| mW | W$\times$1 000 |
| mW | $\mu$W/1 000 |
| $\mu$W | mW$\times$1 000 |
| $\mu$W | W$\times$1 000 000 |

### Capacitance

| Unit | Conversion |
|------|------------|
| $\mu$F | F$\times$1 000 000 |
| pF | F$\times$1 000 000 000 000 |
| $\mu$F | pF/1 000 000 |
| pF | $\mu$F$\times$1 000 000 |
| F | $\mu$F/1 000 000 |
| F | pF/1 000 000 000 000 |

### Inductance

| Unit | Conversion |
|------|------------|
| $\mu$H | H$\times$1 000 000 |
| $\mu$H | mH$\times$1 000 |
| mH | H$\times$1 000 |
| mH | $\mu$H/1 000 |
| H | mH/1 000 |
| H | $\mu$H/1 000 000 |

### Frequency

| Unit | Conversion |
|------|------------|
| Hz | cycles per second (cps) |
| Hz | mHz$\times$1 000 |
| Hz | MHz$\times$1 000 000 |
| mHz | Hz/1 000 |
| mHz | MHz$\times$1 000 |
| MHz | Hz/1 000 000 |
| MHz | mHz/1 000 |

## Table 22-19. Conversions using powers of 10.

### Current

| Unit | Conversions | |
|------|------|------|
| $\mu A$ | $mA \times 10^3$ | $mA/10^{-3}$ |
| $\mu A$ | $A \times 10^6$ | $A/10^{-6}$ |
| $mA$ | $A \times 10^3$ | $A/10^{-3}$ |
| $mA$ | $\mu A \times 10^{-3}$ | $\mu A/10^3$ |
| $A$ | $\mu A \times 10^{-6}$ | $\mu A/10^6$ |
| $A$ | $mA \times 10^{-3}$ | $mA/10^3$ |

### Resistance

| Unit | Conversions | |
|------|------|------|
| $\Omega$ | $k\Omega \times 10^3$ | $k\Omega/10^{-3}$ |
| $\Omega$ | $M\Omega \times 10^6$ | $M\Omega/10^{-6}$ |
| $k\Omega$ | $\Omega/10^3$ | $\Omega \times 10^{-3}$ |
| $k\Omega$ | $M\Omega \times 10^3$ | $M\Omega/10^{-3}$ |
| $M\Omega$ | $\Omega/10^6$ | $\Omega \times 10^{-6}$ |
| $M\Omega$ | $k\Omega/10^3$ | $k\Omega \times 10^{-3}$ |

### Voltage

| Unit | Conversions | |
|------|------|------|
| $V$ | $kV \times 10^3$ | $kV/10^{-3}$ |
| $V$ | $MV \times 10^6$ | $MV/10^{-6}$ |
| $kV$ | $V/10^3$ | $V \times 10^{-3}$ |
| $kV$ | $MV \times 10^3$ | $MV/10^{-3}$ |
| $MV$ | $V/10^6$ | $V \times 10^{-6}$ |
| $MV$ | $kV/10^3$ | $kV \times 10^{-3}$ |
| $mV$ | $V \times 10^3$ | $V/10^{-3}$ |
| $mV$ | $\mu V/10^3$ | $\mu V \times 10^{-3}$ |
| $\mu V$ | $V \times 10^6$ | $V/10^{-6}$ |
| $\mu V$ | $mV \times 10^3$ | $mV/10^{-3}$ |

### Power

| Unit | Conversions | |
|------|------|------|
| $W$ | $mW/10^3$ | $mW \times 10^{-3}$ |
| $W$ | $\mu W/10^6$ | $\mu W \times 10^{-6}$ |

### Power

| Unit | Conversions | |
|------|------|------|
| $W$ | $kW \times 10^3$ | $kW/10^{-3}$ |
| $W$ | $MW \times 10^6$ | $MW/10^{-6}$ |
| $kW$ | $W/10^3$ | $W \times 10^{-3}$ |
| $kW$ | $MW \times 10^3$ | $MW/10^{-3}$ |
| $MW$ | $kW/10^3$ | $kW \times 10^{-3}$ |
| $MW$ | $W/10^6$ | $W \times 10^{-6}$ |
| $mW$ | $W \times 10^3$ | $W/10^{-3}$ |
| $mW$ | $\mu W \times 10^3$ | $\mu W \times 10^{-3}$ |
| $\mu W$ | $mW \times 10^3$ | $mW/10^{-3}$ |
| $\mu W$ | $\mu W \times 10^6$ | $W/10^{-6}$ |

### Capacitance

| Unit | Conversions | |
|------|------|------|
| $\mu F$ | $F \times 10^6$ | $F/10^{-6}$ |
| $pF$ | $F \times 10^{12}$ | $F/10^{-12}$ |
| $\mu F$ | $pF/10^6$ | $pF \times 10^{-6}$ |
| $pF$ | $\mu F \times 10^6$ | $\mu F/10^{-6}$ |
| $F$ | $\mu F/10^6$ | $\mu F \times 10^{-6}$ |
| $F$ | $pF/10^{12}$ | $pF \times 10^{-12}$ |

### Inductance

| Unit | Conversions | |
|------|------|------|
| $\mu H$ | $H \times 10^6$ | $H/10^{-6}$ |
| $\mu H$ | $mH \times 10^3$ | $mH/10^{-3}$ |
| $mH$ | $H \times 10^3$ | $H/10^{-3}$ |
| $mH$ | $\mu H/10^3$ | $\mu H \times 10^{-3}$ |
| $H$ | $mH/10^3$ | $mH \times 10^{-3}$ |
| $H$ | $\mu H/10^6$ | $\mu H \times 10^{-6}$ |

### Inches and Mils

| Unit | Conversions | |
|------|------|------|
| $mils$ | $in/10^3$ | $in \times 10^{-3}$ |
| $in$ | $mils \times 10^3$ | $mils/10^{-3}$ |

*Table 22-20. Conversion factors for electronic multiples and submultiples.*

| To convert these | To these, multiply by the figures below | | | | | | | | | | | | | |
|---|---|---|---|---|---|---|---|---|---|---|---|---|---|---|
| | Pico— | Nano— | Micro— | Milli— | Centi— | Deci— | Units | Deka— | Hekto— | Kilo— | Myria— | Mega— | Giga— | Tera— |
| Pico— | | $10^{-3}$ | $10^{-6}$ | $10^{-9}$ | $10^{-10}$ | $10^{-11}$ | $10^{-12}$ | $10^{-13}$ | $10^{-14}$ | $10^{-15}$ | $10^{-16}$ | $10^{-18}$ | $10^{-21}$ | $10^{-24}$ |
| Nano— | $10^{3}$ | | $10^{-3}$ | $10^{-6}$ | $10^{-7}$ | $10^{-8}$ | $10^{-9}$ | $10^{-10}$ | $10^{-11}$ | $10^{-12}$ | $10^{-13}$ | $10^{-15}$ | $10^{-18}$ | $10^{-21}$ |
| Micro— | $10^{6}$ | $10^{3}$ | | $10^{-3}$ | $10^{-4}$ | $10^{-5}$ | $10^{-6}$ | $10^{-7}$ | $10^{-8}$ | $10^{-9}$ | $10^{-10}$ | $10^{-12}$ | $10^{-15}$ | $10^{-16}$ |
| Milli— | $10^{9}$ | $10^{6}$ | $10^{3}$ | | $10^{-1}$ | $10^{-2}$ | $10^{-3}$ | $10^{-4}$ | $10^{-5}$ | $10^{-6}$ | $10^{-7}$ | $10^{-9}$ | $10^{-12}$ | $10^{-15}$ |
| Centi— | $10^{10}$ | $10^{7}$ | $10^{4}$ | $10^{1}$ | | $10^{-1}$ | $10^{-2}$ | $10^{-3}$ | $10^{-4}$ | $10^{-5}$ | $10^{-6}$ | $10^{-8}$ | $10^{-11}$ | $10^{-14}$ |
| Deci— | $10^{11}$ | $10^{8}$ | $10^{5}$ | $10^{2}$ | $10^{1}$ | | $10^{-1}$ | $10^{-2}$ | $10^{-3}$ | $10^{-4}$ | $10^{-5}$ | $10^{-7}$ | $10^{-10}$ | $10^{-13}$ |
| Units | $10^{12}$ | $10^{9}$ | $10^{6}$ | $10^{3}$ | $10^{2}$ | $10^{1}$ | | $10^{-1}$ | $10^{-2}$ | $10^{-3}$ | $10^{-4}$ | $10^{-6}$ | $10^{-9}$ | $10^{-12}$ |
| Deka— | $10^{13}$ | $10^{10}$ | $10^{7}$ | $10^{4}$ | $10^{3}$ | $10^{2}$ | $10^{1}$ | | $10^{-1}$ | $10^{-2}$ | $10^{-3}$ | $10^{-5}$ | $10^{-8}$ | $10^{-11}$ |
| Hekto— | $10^{14}$ | $10^{11}$ | $10^{8}$ | $10^{5}$ | $10^{4}$ | $10^{3}$ | $10^{2}$ | $10^{1}$ | | $10^{-1}$ | $10^{-2}$ | $10^{-4}$ | $10^{-7}$ | $10^{-10}$ |
| Kilo— | $10^{15}$ | $10^{12}$ | $10^{9}$ | $10^{6}$ | $10^{5}$ | $10^{4}$ | $10^{3}$ | $10^{2}$ | $10^{1}$ | | $10^{-1}$ | $10^{-3}$ | $10^{-6}$ | $10^{-9}$ |
| Myria— | $10^{16}$ | $10^{13}$ | $10^{10}$ | $10^{7}$ | $10^{6}$ | $10^{5}$ | $10^{4}$ | $10^{3}$ | $10^{2}$ | $10^{1}$ | | $10^{-2}$ | $10^{-5}$ | $10^{-8}$ |
| Mega— | $10^{18}$ | $10^{15}$ | $10^{12}$ | $10^{9}$ | $10^{8}$ | $10^{7}$ | $10^{6}$ | $10^{5}$ | $10^{4}$ | $10^{3}$ | $10^{2}$ | | $10^{-3}$ | $10^{-6}$ |
| Giga— | $10^{21}$ | $10^{18}$ | $10^{15}$ | $10^{12}$ | $10^{11}$ | $10^{10}$ | $10^{9}$ | $10^{8}$ | $10^{7}$ | $10^{6}$ | $10^{5}$ | $10^{3}$ | | $10^{-3}$ |
| Tera— | $10^{24}$ | $10^{21}$ | $10^{18}$ | $10^{15}$ | $10^{14}$ | $10^{13}$ | $10^{12}$ | $10^{11}$ | $10^{10}$ | $10^{9}$ | $10^{8}$ | $10^{6}$ | $10^{3}$ | |

*Table 22-21. Conversion factors
for units of measurements.*

| To convert from | To | Multiply by |
|---|---|---|
| **Length** | | |
| mils | mm | 0.025 4 |
| mils | in | 0.001 |
| mm | in | 0.039 37 |
| cm | in | 0.393 7 |
| cm | ft | 0.032 81 |
| in | m | 0.025 4 |
| ft | m | 0.304 8 |
| yd | m | 0.914 4 |
| km | mi | 0621 4 |
| | | |
| **Area** | | |
| cir. mils | $in^2$ | 0.000 000 785 4 |
| cir. mils | $mils^2$ | 0.785 4 |
| cir. mils | $mm^2$ | 0.000 506 6 |
| $mm^2$ | $in^2$ | 0.001 55 |
| $mil^2$ | $in^2$ | 0.000 001 |
| $cm^2$ | $in^2$ | 0.155 |
| $ft^2$ | $m^2$ | 0.092 9 |
| | | |
| **Volume** | | |
| $in^3$ | L | 0.016 39 |
| $in^3$ | gal | 0.004 329 |
| L | gal | 0.264 17 |
| $cm^3$ | $in^3$ | 0.061 02 |
| $cm^3$ | gal | 0.000 264 |
| | | |
| **Power** | | |
| ft• lb/min | hp | 0.000 030 3 |
| ft• lb/s | hp | 0.001 818 |

While it is more convenient to handle electronic conversions using powers of 10, such conversions can also be made using whole numbers. Table 22-18 shows the conversion relationships for current, resistance, voltage, power, capacitance, inductance, and frequency. Use Table 22-19 for conversions with powers of 10.

Table 22-20 supplies conversion factors for electronic multiples and sub-multiples. Table 22-21 gives conversion factors for units of measurement.

# 23
CHAPTER

# Decimal numbers

## Exponents

Formulas, charts, and tables are all involved with numbers. Often, to be able to get maximum advantage, you must manipulate the numbers or change them from one form to another.

One way of expressing a number is through the use of exponents. An exponent indicates the number of times a number is to be multiplied by itself. Thus, $9^2$ is an abbreviated way of writing $9 \times 9$. It is much easier and quicker to write $9^{18}$ than to write the digit 9 a total of 18 times including a long series of multiplication signs. Table 23-1 supplies numbers from 1 to 100 with exponents from 3 to 8.

*Example*   What is the value of $9^6$?
*Solution*   Locate 9 in the left-hand column in Table 23-1. Move to the right to the column headed by $n^6$. The value of $9^6$ is given as 531 441.

*Example*   What is the value of $23^7$?
*Solution*   Locate 23 in the left-hand column. Move to the right to the column headed by $n^7$. The number shown here is 3.404 825. However, this number must be multiplied by $10^9$ as indicated in the column heading. Hence, $23^7$ equals 3.404 825 $\times 10^9$. To get the final answer, move the decimal point 9 places to the right. 3.404 825 $\times 10^9$ equals, 3 404 825 000.

## Square roots of numbers

Table 23-2 lists the square roots of numbers from 1 to 100.

*Example*   What is the square root of 19? (Solution on page 504.)

## Table 23-1. Powers of numbers.

| n | $n^3$ | $n^4$ | $n^5$ | $n^6$ | $n^7$ | $n^8$ |
|---|---|---|---|---|---|---|
| 1 | 1 | 1 | 1 | 1 | 1 | 1 |
| 2 | 8 | 16 | 32 | 64 | 128 | 256 |
| 3 | 27 | 81 | 243 | 729 | 2 187 | 6 561 |
| 4 | 64 | 256 | 1 024 | 4 096 | 16 384 | 65 536 |
| 5 | 125 | 625 | 3 125 | 15 625 | 78 125 | 390 625 |
| 6 | 216 | 1 296 | 7 776 | 46 656 | 279 936 | 1 679 616 |
| 7 | 343 | 2 401 | 16 807 | 117 649 | 823 543 | 5 764 801 |
| 8 | 512 | 4 096 | 32 768 | 262 144 | 2 097 152 | 16 777 216 |
| 9 | 729 | 6 561 | 59 049 | 531 441 | 4 782 969 | 43 046 721 |
| | | | | | | $\times 10^8$ |
| 10 | 1 000 | 10 000 | 100 000 | 1 000 000 | 10 000 000 | 1.000 000 |
| 11 | 1 331 | 14 641 | 161 051 | 1 771 561 | 19 487 171 | 2.143 589 |
| 12 | 1 728 | 20 736 | 248 832 | 2 985 984 | 35 831 808 | 4.299 817 |
| 13 | 2 197 | 28 561 | 371 293 | 4 826 809 | 62 748 517 | 8.157 307 |
| 14 | 2 794 | 38 416 | 537 824 | 7 529 536 | 105 413 504 | 14.757 891 |
| 15 | 3 375 | 50 625 | 759 375 | 11 390 625 | 170 859 375 | 25.628 906 |
| 16 | 4 096 | 65 536 | 1 048 576 | 16 777 216 | 268 435 456 | 42.949 673 |
| 17 | 4 913 | 83 521 | 1 419 857 | 24 137 569 | 410 338 673 | 69.757 574 |
| 18 | 5 832 | 104 976 | 1 889 568 | 34 012 224 | 612 220 032 | 110.199 606 |
| 19 | 6 859 | 130 321 | 2 476 099 | 47 045 881 | 893 871 739 | 169.835 630 |
| | | | | | $\times 10^9$ | $\times 10^{10}$ |
| 20 | 8 000 | 160 000 | 3 200 000 | 64 000 000 | 1.280 000 | 2.560 000 |
| 21 | 9 261 | 194 481 | 4 084 101 | 85 766 121 | 1.801 089 | 3.782 286 |
| 22 | 10 648 | 234 256 | 5 153 632 | 113 379 904 | 2.494 358 | 5.487 587 |
| 23 | 12 167 | 279 841 | 6 436 343 | 148 035 889 | 3.404 825 | 7.831 099 |
| 24 | 13 824 | 331 776 | 7 962 624 | 191 102 976 | 4.586 471 | 11.007 531 |
| 25 | 15 625 | 390 625 | 9 765 625 | 244 140 625 | 6.103 516 | 15.258 789 |
| 26 | 17 576 | 456 976 | 11 881 376 | 308 915 776 | 8.031 810 | 20.882 706 |
| 27 | 19 683 | 531 441 | 14 348 907 | 387 420 489 | 10.460 353 | 28.242 954 |
| 28 | 21 952 | 614 656 | 17 210 368 | 481 890 304 | 13.492 929 | 37.780 200 |
| 29 | 24 389 | 707 281 | 20 511 149 | 594 823 321 | 17.249 876 | 50.024 641 |
| | | | | $\times 10^8$ | $\times 10^{10}$ | $\times 10^{11}$ |
| 30 | 27 000 | 810 000 | 24 300 000 | 7.290 000 | 2.187 000 | 6.561 000 |
| 31 | 29 791 | 923 521 | 28 629 151 | 8.875 037 | 2.751 261 | 8.528 910 |
| 32 | 32 768 | 1 048 576 | 33 554 432 | 10.737 418 | 3.435 974 | 10.995 116 |
| 33 | 35 937 | 1 185 921 | 39 135 393 | 12.914 680 | 4.261 844 | 14.064 086 |
| 34 | 39 304 | 1 336 336 | 45 435 424 | 15.448 044 | 5.252 335 | 17.857 939 |
| 35 | 42 875 | 1 500 625 | 52 521 875 | 18.382 656 | 6.433 930 | 22.518 754 |
| 36 | 46 656 | 1 679 616 | 60 466 176 | 21.767 823 | 7.836 416 | 28.211 099 |
| 37 | 50 653 | 1 874 161 | 69 343 957 | 25.657 264 | 9.493 188 | 35.124 795 |

Table 23-1. Continued.

| n | $n^3$ | $n^4$ | $n^5$ | $n^6$ | $n^7$ | $n^8$ |
|---|---|---|---|---|---|---|
| 38 | 54 872 | 2 085 136 | 79 235 168 | 30.109 364 | 11.441 558 | 43.477 921 |
| 39 | 59 319 | 2 313 441 | 90 224 199 | 35.187 438 | 13.723 101 | 53.520 093 |
| | | | | $\times 10^9$ | $\times 10^{10}$ | $\times 10^{12}$ |
| 40 | 64 000 | 2 560 000 | 102 400 000 | 4.096 000 | 16.384 000 | 6.553 600 |
| 41 | 68 921 | 2 825 761 | 115 856 201 | 4.750 104 | 19.475 427 | 7.984 925 |
| 42 | 74 088 | 3 111 696 | 130 691 232 | 5.489 032 | 23.053 933 | 9.682 652 |
| 43 | 79 507 | 3 418 801 | 147 008 443 | 6.321 363 | 27.181 861 | 11.688 200 |
| 44 | 85 184 | 3 748 096 | 164 916 224 | 7.256 314 | 31.927 781 | 14.048 224 |
| 45 | 91 125 | 4 100 625 | 184 528 125 | 8.303 766 | 37.366 945 | 16.815 125 |
| 46 | 97 336 | 4 477 456 | 205 962 976 | 9.474 297 | 43.581 766 | 20.047 612 |
| 47 | 103 823 | 4 879 681 | 229 345 007 | 10.779 215 | 50.662 312 | 23.811 287 |
| 48 | 110 592 | 5 308 416 | 254 803 968 | 12.230 590 | 58.706 834 | 28.179 280 |
| 49 | 117 649 | 5 764 801 | 282 475 249 | 13.841 287 | 67.822 307 | 33.232 931 |
| | | | | | $\times 10^{11}$ | $\times 10^{13}$ |
| 50 | 125 000 | 6 250 000 | 312 500 000 | 15.625 000 | 7.812 500 | 3.906 250 |
| 51 | 132 651 | 6 765 201 | 345 025 251 | 17.596 288 | 8.974 107 | 4.576 794 |
| 52 | 140 608 | 7 311 616 | 380 204 032 | 19.770 610 | 10.280 717 | 5.345 973 |
| 53 | 148 877 | 7 890 481 | 418 195 493 | 22.164 361 | 11.747 111 | 6.225 969 |
| 54 | 157 464 | 8 503 056 | 459 165 024 | 24.794 911 | 13.389 252 | 7.230 196 |
| 55 | 166 375 | 9 150 625 | 503 284 375 | 27.680 641 | 15.224 352 | 8.373 394 |
| 56 | 175 616 | 9 834 496 | 550 731 776 | 30.840 979 | 17.270 948 | 9.671 731 |
| 57 | 185 193 | 10 556 001 | 601 692 057 | 34.296 447 | 19.548 975 | 11.142 916 |
| 58 | 195 112 | 11 316 496 | 656 356 768 | 38.068 693 | 22.079 842 | 12.806 308 |
| 59 | 205 379 | 12 117 361 | 714 924 299 | 42.180 534 | 24.886 515 | 14.683 044 |
| | | | $\times 10^8$ | $\times 10^{10}$ | | $\times 10^{13}$ |
| 60 | 216 000 | 12 960 000 | 7.776 000 | 4.665 600 | 27.993 600 | 16.796 160 |
| 61 | 226 981 | 13 845 841 | 8.445 963 | 5.152 037 | 31.427 428 | 19.170 731 |
| 62 | 238 328 | 14 776 336 | 9.161 328 | 5.680 024 | 35.216 146 | 21.834 011 |
| 63 | 250 047 | 15 752 961 | 9.924 365 | 6.252 350 | 39.389 806 | 24.815 578 |
| 64 | 262 114 | 16 777 216 | 10.787 418 | 6.871 948 | 43.980 465 | 28.147 498 |
| 65 | 274 625 | 17 850 625 | 11.602 906 | 7.541 889 | 49.022 279 | 31.864 481 |
| 66 | 287 496 | 18 974 736 | 12.523 326 | 8.265 395 | 54.551 607 | 36.004 061 |
| 67 | 300 763 | 20 151 121 | 13.501 251 | 9.045 838 | 60.607 116 | 40.606 768 |
| 68 | 314 432 | 21 381 376 | 14.539 336 | 9.886 748 | 67.229 888 | 45.716 324 |
| 69 | 328 509 | 22 667 121 | 15.640 313 | 10.791 816 | 74.463 533 | 51.379 837 |
| | | | $\times 10^8$ | $\times 10^{10}$ | | $\times 10^{14}$ |
| 70 | 343 000 | 24 010 000 | 16.807 000 | 11.764 900 | 8.235 430 | 5.764 801 |
| 71 | 357 911 | 25 411 681 | 18.042 294 | 12.810 028 | 9.095 120 | 6.457 535 |
| 72 | 373 248 | 26 873 856 | 19.349 176 | 13.931 407 | 10.030 613 | 7.222 041 |

*Table 23-1. Continued.*

| n | n³ | n⁴ | n⁵ | n⁶ | n⁷ | n⁸ |
|---|-----|-----|-----|-----|-----|-----|
| 73 | 389 017 | 28 398 241 | 20.730 716 | 15.133 423 | 11.047 399 | 8.064 601 |
| 74 | 405 224 | 29 986 576 | 22.190 066 | 16.420 649 | 12.151 280 | 8.991 947 |
| 75 | 421 875 | 31 640 625 | 23.730 469 | 17.797 852 | 13.348 389 | 10.011 292 |
| 76 | 438 976 | 33 362 176 | 25.355 254 | 19.269 993 | 14.645 195 | 11.130 348 |
| 77 | 456 533 | 35 153 041 | 27.067 842 | 20.842 238 | 16.048 523 | 12.357 363 |
| 78 | 474 552 | 37 015 056 | 28.871 744 | 22.519 960 | 17.565 569 | 13.701 144 |
| 79 | 493 039 | 38 950 081 | 30.770 564 | 24.308 746 | 19.203 909 | 15.171 088 |
|    |         |            | $\times 10^8$ | $\times 10^{10}$ | $\times 10^{12}$ | $\times 10^{14}$ |
| 80 | 512 000 | 40 960 000 | 32.768 000 | 26.214 400 | 20.971 520 | 16.777 216 |
| 81 | 531 441 | 43 046 721 | 34.867 844 | 28.242 954 | 22.876 792 | 18.530 202 |
| 82 | 551 368 | 45 212 176 | 37.073 984 | 30.400 667 | 24.928 547 | 20.441 409 |
| 83 | 571 787 | 47 458 321 | 39.390 406 | 32.694 037 | 27.136 051 | 22.522 922 |
| 84 | 592 704 | 49 787 136 | 41.821 194 | 35.129 803 | 29.509 035 | 24.787 589 |
| 85 | 614 125 | 52 200 625 | 44.370 531 | 37.714 952 | 32.057 709 | 27.249 053 |
| 86 | 636 056 | 54 700 816 | 47.042 702 | 40.456 724 | 34.792 782 | 29.921 793 |
| 87 | 658 503 | 57 289 761 | 49.842 092 | 43.362 620 | 37.725 479 | 32.821 167 |
| 88 | 681 472 | 59 969 536 | 52.773 192 | 46.440 409 | 40.867 560 | 35.963 452 |
| 89 | 704 969 | 62 742 241 | 55.840 594 | 49.698 129 | 44.231 335 | 39.365 888 |
|    |         |            | $\times 10^9$ | $\times 10^{11}$ | $\times 10^{13}$ | $\times 10^{15}$ |
| 90 | 729 000 | 65 610 000 | 5.904 900 | 5.314 410 | 4.782 969 | 4.304 672 |
| 91 | 753 571 | 68 574 961 | 6.240 321 | 5.678 693 | 5.167 610 | 4.702 525 |
| 92 | 778 688 | 71 639 296 | 6.590 815 | 6.063 550 | 5.578 466 | 5.132 189 |
| 93 | 804 357 | 74 805 201 | 6.956 884 | 6.469 902 | 6.017 009 | 5.595 818 |
| 94 | 830 584 | 78 074 896 | 7.339 040 | 6.898 698 | 6.484 776 | 6.095 689 |
| 95 | 857 375 | 81 450 625 | 7.737 809 | 7.350 919 | 6.983 373 | 6.634 204 |
| 96 | 884 736 | 84 934 656 | 8.153 727 | 7.827 578 | 7.514 475 | 7.213 896 |
| 97 | 912 673 | 88 529 281 | 8.587 340 | 8.329 720 | 8.079 828 | 7.837 434 |
| 98 | 941 192 | 92 236 816 | 9.039 208 | 8.858 424 | 8.681 255 | 8.507 630 |
| 99 | 970 299 | 96 059 601 | 9.509 900 | 9.414 801 | 9.320 653 | 9.227 447 |
| 100 | 1 000 000 | 100 000 000 | 10.000 000 | 10.000 000 | 10.000 000 | 10.000 000 |

**Solution**  Use Table 23-2 to find the square roots of numbers. Find 19 in the column headed by the letter n (abbreviation for number). The square root of 19 appears immediately to the right in the column identified by $\sqrt{n}$. The square root of 19 is 4.358 9.

# Cube roots of numbers

Table 23-3 can be used to find the cube roots of numbers ranging from 1 to 10 000.

## Table 23-2. Square roots of numbers.

| n | $\sqrt{n}$ | n | $\sqrt{n}$ | n | $\sqrt{n}$ | n | $\sqrt{n}$ |
|---|---|---|---|---|---|---|---|
| 1 | 1.000 0 | 26 | 5.099 0 | 51 | 7.141 4 | 76 | 8.717 8 |
| 2 | 1.414 2 | 27 | 5.196 2 | 52 | 7.211 1 | 77 | 8.775 0 |
| 3 | 1.732 1 | 28 | 5.291 5 | 53 | 7.280 1 | 78 | 8.831 8 |
| 4 | 2.000 0 | 29 | 5.385 2 | 54 | 7.348 5 | 79 | 8.888 2 |
| 5 | 2.236 1 | 30 | 5.477 2 | 55 | 7.416 2 | 80 | 8.944 3 |
| 6 | 2.449 5 | 31 | 5.567 8 | 56 | 7.483 3 | 81 | 9.000 0 |
| 7 | 2.645 8 | 32 | 5.656 9 | 57 | 7.549 8 | 82 | 9.055 4 |
| 8 | 2.828 4 | 33 | 5.744 6 | 58 | 7.615 8 | 83 | 9.110 4 |
| 9 | 3.000 0 | 34 | 5.831 0 | 59 | 7.681 1 | 84 | 9.165 2 |
| 10 | 3.162 3 | 35 | 5.916 1 | 60 | 7.746 0 | 85 | 9.219 5 |
| 11 | 3.316 6 | 36 | 6.000 0 | 61 | 7.810 2 | 86 | 9.273 6 |
| 12 | 3.464 1 | 37 | 6.082 8 | 62 | 7.874 0 | 87 | 9.327 4 |
| 13 | 3.605 6 | 38 | 6.164 4 | 63 | 7.937 3 | 88 | 9.380 8 |
| 14 | 3.741 7 | 39 | 6.245 0 | 64 | 8.000 0 | 89 | 9.434 0 |
| 15 | 3.873 0 | 40 | 6.324 6 | 65 | 8.062 3 | 90 | 9.486 8 |
| 16 | 4.000 0 | 41 | 6.403 1 | 66 | 8.124 0 | 91 | 9.539 4 |
| 17 | 4.123 1 | 42 | 6.480 7 | 67 | 8.185 4 | 92 | 9.591 7 |
| 18 | 4.242 6 | 43 | 6.557 4 | 68 | 8.246 2 | 93 | 9.643 7 |
| 19 | 4.358 9 | 44 | 6.633 2 | 69 | 8.306 6 | 94 | 9.695 4 |
| 20 | 4.472 1 | 45 | 6.708 2 | 70 | 8.366 6 | 95 | 9.746 8 |
| 21 | 4.582 6 | 46 | 6.782 3 | 71 | 8.426 1 | 96 | 9.798 0 |
| 22 | 4.690 4 | 47 | 6.855 7 | 72 | 8.485 3 | 97 | 9.849 0 |
| 23 | 4.795 8 | 48 | 6.928 2 | 73 | 8.544 0 | 98 | 9.899 5 |
| 24 | 4.899 0 | 49 | 7.000 0 | 74 | 8.602 3 | 99 | 9.949 9 |
| 25 | 5.000 0 | 50 | 7.071 1 | 75 | 8.660 3 | 100 | 10.000 0 |

## Table 23-3. Cube roots of numbers.

| n | $\sqrt[3]{n}$ | $\sqrt[3]{10n}$ | $\sqrt[3]{100n}$ | n | $\sqrt[3]{n}$ | $\sqrt[3]{10n}$ | $\sqrt[3]{100n}$ |
|---|---|---|---|---|---|---|---|
| 1 | 1.000 000 | 2.154 435 | 4.641 589 | 13 | 2.351 335 | 5.065 797 | 10.913 93 |
| 2 | 1.259 921 | 2.714 418 | 5.848 035 | 14 | 2.410 142 | 5.192 494 | 11.186 89 |
| 3 | 1.442 250 | 3.107 233 | 6.694 330 | 15 | 2.466 212 | 5.313 293 | 11.447 14 |
| 4 | 1.587 401 | 3.419 952 | 7.368 063 | 16 | 2.519 842 | 5.428 835 | 11.696 07 |
| 5 | 1.709 976 | 3.684 031 | 7.937 005 | 17 | 2.571 282 | 5.539 658 | 11.934 83 |
| 6 | 1.817 121 | 3.914 868 | 8.434 327 | 18 | 2.620 741 | 5.646 216 | 12.164 40 |
| 7 | 1.912 931 | 4.121 285 | 8.879 040 | 19 | 2.668 402 | 5.748 897 | 12.385 62 |
| 8 | 2.000 000 | 4.308 869 | 9.283 178 | 20 | 2.714 418 | 5.848 035 | 12.599 21 |
| 9 | 2.080 084 | 4.481 405 | 9.654 894 | 21 | 2.758 924 | 5.943 922 | 12.805 79 |
| 10 | 2.154 435 | 4.641 589 | 10.000 00 | 22 | 2.802 039 | 6.036 811 | 13.005 91 |
| 11 | 2.223 980 | 4.791 420 | 10.322 80 | 23 | 2.843 867 | 6.126 926 | 13.200 06 |
| 12 | 2.289 428 | 4.932 424 | 10.626 59 | 24 | 2.884 499 | 6.214 465 | 13.388 66 |

Table 23-3. Continued.

| n | $\sqrt[3]{n}$ | $\sqrt[3]{10n}$ | $\sqrt[3]{100n}$ | n | $\sqrt[3]{n}$ | $\sqrt[3]{10n}$ | $\sqrt[3]{100n}$ |
|---|---|---|---|---|---|---|---|
| 25 | 2.924 018 | 6.299 605 | 13.572 09 | 63 | 3.979 057 | 8.572 619 | 18.469 15 |
| 26 | 2.962 496 | 6.382 504 | 13.750 69 | 64 | 4.000 000 | 8.617 739 | 18.566 36 |
| 27 | 3.000 000 | 6.463 304 | 13.924 77 | 65 | 4.020 726 | 8.662 391 | 18.662 56 |
| 28 | 3.036 589 | 6.542 133 | 14.094 60 | 66 | 4.041 240 | 8.706 588 | 18.757 77 |
| 29 | 3.072 317 | 6.619 106 | 14.260 43 | 67 | 4.061 548 | 8.750 340 | 18.852 04 |
| 30 | 3.107 233 | 6.694 330 | 14.422 50 | 68 | 4.081 655 | 8.793 659 | 18.945 36 |
| 31 | 3.141 381 | 6.767 899 | 14.581 00 | 69 | 4.101 566 | 8.836 556 | 19.037 78 |
| 32 | 3.174 802 | 6.839 904 | 14.736 13 | 70 | 4.121 285 | 8.879 040 | 19.129 31 |
| 33 | 3.207 534 | 6.910 423 | 14.888 06 | 71 | 4.140 818 | 8.921 121 | 19.219 97 |
| 34 | 3.239 612 | 6.979 532 | 15.036 95 | 72 | 4.160 168 | 8.962 809 | 19.309 79 |
| 35 | 3.271 066 | 7.047 299 | 15.182 94 | 73 | 4.179 339 | 9.004 113 | 19.398 77 |
| 36 | 3.301 927 | 7.113 787 | 15.326 19 | 74 | 4.198 336 | 9.045 042 | 19.486 95 |
| 37 | 3.332 222 | 7.179 054 | 15.466 80 | 75 | 4.217 163 | 9.085 603 | 19.574 34 |
| 38 | 3.361 975 | 7.243 156 | 15.604 91 | 76 | 4.235 824 | 9.125 805 | 19.660 95 |
| 39 | 3.391 211 | 7.306 144 | 15.740 61 | 77 | 4.254 321 | 9.165 656 | 19.746 81 |
| 40 | 3.419 952 | 7.368 063 | 15.874 01 | 78 | 4.272 659 | 9.205 164 | 19.831 92 |
| 41 | 3.448 217 | 7.428 959 | 16.005 21 | 79 | 4.290 840 | 9.244 335 | 19.916 32 |
| 42 | 3.476 027 | 7.488 872 | 16.134 29 | 80 | 4.308 869 | 9.283 178 | 20.000 00 |
| 43 | 3.503 398 | 7.547 842 | 16.261 33 | 81 | 4.326 749 | 9.321 698 | 20.082 99 |
| 44 | 3.530 348 | 7.605 905 | 16.386 43 | 82 | 4.344 481 | 9.359 902 | 20.165 30 |
| 45 | 3.556 893 | 7.663 094 | 16.509 64 | 83 | 4.362 071 | 9.397 796 | 20.246 94 |
| 46 | 3.583 048 | 7.719 443 | 16.631 03 | 84 | 4.379 519 | 9.435 388 | 20.327 93 |
| 47 | 3.608 826 | 7.774 980 | 16.750 69 | 85 | 4.396 830 | 9.472 682 | 20.408 28 |
| 48 | 3.634 241 | 7.829 735 | 16.868 65 | 86 | 4.414 005 | 9.509 685 | 20.488 00 |
| 49 | 3.659 306 | 7.883 735 | 16.984 99 | 87 | 4.431 048 | 9.546 403 | 20.567 10 |
| 50 | 3.684 031 | 7.937 005 | 17.099 76 | 88 | 4.447 960 | 9.582 840 | 20.645 60 |
| 51 | 3.708 430 | 7.989 570 | 17.213 01 | 89 | 4.464 745 | 9.619 002 | 20.723 51 |
| 52 | 3.732 511 | 8.041 452 | 17.324 78 | 90 | 4.481 405 | 9.654 894 | 20.800 84 |
| 53 | 3.756 286 | 8.092 672 | 17.435 13 | 91 | 4.497 941 | 9.690 521 | 20.877 59 |
| 54 | 3.779 763 | 8.143 253 | 17.544 11 | 92 | 4.514 357 | 9.725 888 | 20.953 79 |
| 55 | 3.802 952 | 8.193 213 | 17.651 74 | 93 | 4.530 655 | 9.761 000 | 21.029 44 |
| 56 | 3.825 862 | 8.242 571 | 17.758 08 | 94 | 4.546 836 | 9.795 861 | 21.104 54 |
| 57 | 3.848 501 | 8.291 344 | 17.863 16 | 95 | 4.562 903 | 9.830 476 | 21.179 12 |
| 58 | 3.870 877 | 8.339 551 | 17.967 02 | 96 | 4.578 857 | 9.864 848 | 21.253 17 |
| 59 | 3.892 996 | 8.387 207 | 18.069 69 | 97 | 4.594 701 | 9.898 983 | 21.326 71 |
| 60 | 3.914 868 | 8.434 327 | 18.171 21 | 98 | 4.610 436 | 9.932 884 | 21.399 75 |
| 61 | 3.936 497 | 8.480 926 | 18.271 60 | 99 | 4.626 065 | 9.966 555 | 21.472 29 |
| 62 | 3.957 892 | 8.527 019 | 18.370 91 | 100 | 4.641 589 | 10.000 00 | 21.544 35 |

*Example*   What is the cube root of 16?

*Solution*   Locate 16 in the column headed by the letter n. Move to the right to the column identified by $\sqrt[3]{n}$. The cube root of 16 is 2.519 842.

Two additional columns are marked $\sqrt[3]{10n}$ and $\sqrt[3]{100n}$. These are used for extending the values in the n column. Thus, if n is 18, then 10n is $10 \times 18 = 180$, and 100n is $100 \times 18 = 1\,800$. Consequently, 18 in the n column can represent 18, 180, or 1 800.

*Example*   What is the cube root of 3 100?

*Solution*   Locate 31 in the n column. 3 100, however, is $31 \times 100$, and so the answer will be found in the $\sqrt[3]{100n}$ column. Move horizontally to this column and the answer is given as 14.581 00. The last two zeros of this number do not contribute to its value and so can be omitted. $\sqrt[3]{3\,100}$ equals 14.581.

# Numbers and reciprocals

The reciprocal of a number is the inverse of that number, or the number divided into 1. The reciprocal of 5 is 1/5; the reciprocal of 87 is 1/87. Table 23-4 supplies reciprocals of numbers ranging from 0.1 to 100.

*Example*   What is the reciprocal of 27?

*Solution*   Locate 27 in the n column. Move to the right and in the column headed by 1/n the answer is given as 0.037 0.

Table 23-4 can be extended by moving the decimal point as required.

*Example*   What is the reciprocal of 160?

*Solution*   160 does not appear directly in the table. Instead, locate 16 in the n column and change it to 160. In the 1/n column, the corresponding reciprocal value for 16 is 0.062 5. Add another zero directly after the decimal point (equivalent to dividing the reciprocal by 10) and the answer becomes 0.006 25.

To find the reciprocal of 1 600, divide the answer in the 1/n column (corresponding to 16) by 100, or insert two zeros after the decimal point. Hence the reciprocal of 1 600 is 0.000 625.

Large numbers can be more simply written in exponential form as shown in Table 23-5. For whole numbers the exponent indicates the number of zeros following digit 1. Thus, $10^3$ indicates the digit 1 is to be followed by three zeros (1 000). For decimal values the exponent indicates the number of digits following the decimal point. Thus, $10^{-2}$ indicates 0.01.

### Table 23-4. Numbers and reciprocals.

| n | 1/n | n | 1/n | n | 1/n |
|---|---|---|---|---|---|
| 0.1 | 10.000 0 | 28 | 0.035 7 | 65 | 0.015 4 |
| 0.2 | 5.000 0 | 29 | 0.034 5 | 66 | 0.015 2 |
| 0.3 | 3.333 3 | 30 | 0.033 3 | 67 | 0.014 9 |
| 0.4 | 2.500 0 | 31 | 0.032 3 | 68 | 0.014 7 |
| 0.5 | 2.000 0 | 32 | 0.031 3 | 69 | 0.014 5 |
| 0.6 | 1.666 6 | 33 | 0.030 3 | 70 | 0.014 3 |
| 0.7 | 1.428 6 | 34 | 0.029 4 | 71 | 0.014 1 |
| 0.8 | 1.250 0 | 35 | 0.028 6 | 72 | 0.013 9 |
| 0.9 | 1.111 1 | 36 | 0.027 8 | 73 | 0.013 7 |
|  |  | 37 | 0.027 0 | 74 | 0.013 5 |
| 1 | 1.000 0 | 38 | 0.026 3 | 75 | 0.013 3 |
| 2 | 0.500 0 | 39 | 0.025 6 | 76 | 0.013 2 |
| 3 | 0.333 3 | 40 | 0.025 0 | 77 | 0.013 0 |
| 4 | 0.250 0 | 41 | 0.024 4 | 78 | 0.012 8 |
| 5 | 0.200 0 | 42 | 0.023 8 | 79 | 0.012 7 |
| 6 | 0.166 7 | 43 | 0.023 3 | 80 | 0.012 5 |
| 7 | 0.142 9 | 44 | 0.022 7 | 81 | 0.012 3 |
| 8 | 0.125 0 | 45 | 0.022 2 | 82 | 0.012 2 |
| 9 | 0.111 1 | 46 | 0.021 7 | 83 | 0.012 0 |
| 10 | 0.100 0 | 47 | 0.021 3 | 84 | 0.011 9 |
| 11 | 0.090 9 | 48 | 0.020 8 | 85 | 0.011 8 |
| 12 | 0.083 3 | 49 | 0.020 4 | 86 | 0.011 6 |
| 13 | 0.076 9 | 50 | 0.020 0 | 87 | 0.011 5 |
| 14 | 0.071 4 | 51 | 0.019 6 | 88 | 0.011 4 |
| 15 | 0.066 7 | 52 | 0.019 2 | 89 | 0.011 2 |
| 16 | 0.062 5 | 53 | 0.018 9 | 90 | 0.011 1 |
| 17 | 0.058 8 | 54 | 0.018 5 | 91 | 0.011 0 |
| 18 | 0.055 5 | 55 | 0.018 2 | 92 | 0.010 9 |
| 19 | 0.052 6 | 56 | 0.017 9 | 93 | 0.010 8 |
| 20 | 0.050 0 | 57 | 0.017 5 | 94 | 0.010 6 |
| 21 | 0.047 6 | 58 | 0.017 2 | 95 | 0.010 5 |
| 22 | 0.045 5 | 59 | 0.016 9 | 96 | 0.010 4 |
| 23 | 0.043 5 | 60 | 0.016 7 | 97 | 0.010 3 |
| 24 | 0.041 7 | 61 | 0.016 4 | 98 | 0.010 2 |
| 25 | 0.040 0 | 62 | 0.016 1 | 99 | 0.010 1 |
| 26 | 0.038 5 | 63 | 0.015 9 | 100 | 0.010 0 |
| 27 | 0.037 0 | 64 | 0.015 6 |  |  |

## Table 23-5. Powers of 10.

| Power | Decimal | Fractions |
|-------|---------|-----------|
| $10^{-6}$ | = 0.000 001 | = $\dfrac{1}{1\ 000\ 000} = \dfrac{1}{10^6}$ |
| $10^{-5}$ | = 0.000 01 | = $\dfrac{1}{100\ 000} = \dfrac{1}{10^5}$ |
| $10^{-4}$ | = 0.000 1 | = $\dfrac{1}{10\ 000} = \dfrac{1}{10^4}$ |
| $10^{-3}$ | = 0.001 | = $\dfrac{1}{1\ 000} = \dfrac{1}{10^3}$ |
| $10^{-2}$ | = 0.01 | = $\dfrac{1}{100} = \dfrac{1}{10^2}$ |
| $10^{-1}$ | = 0.1 | = $\dfrac{1}{10} = \dfrac{1}{10^1}$ |
| $10^{0}$ | = 1 | = $\dfrac{1}{1} = \dfrac{1}{10^0}$ |
| $10^{1}$ | = 10 | = $\dfrac{1}{10^{-1}}$ |
| $10^{2}$ | = 100 | = $\dfrac{1}{10^{-2}}$ |
| $10^{3}$ | = 1 000 | = $\dfrac{1}{10^{-3}}$ |
| $10^{4}$ | = 10 000 | = $\dfrac{1}{10^{-4}}$ |
| $10^{5}$ | = 100 000 | = $\dfrac{1}{10^{-5}}$ |
| $10^{6}$ | = 1 000 000 | = $\dfrac{1}{10^{-6}}$ |

# Significant figures in the decimal system

Some of the tables that have been presented consist of whole numbers only, and others are comprised of whole numbers followed by decimals, with the two—whole numbers and decimals—separated by a decimal point.

The numbers used in these tables are known as significant figures. A number, such as 35, in one of the tables, indicates an accuracy of two significant figures. The number 35 could have been written as 35, but the decimal point is generally omitted. This means that a decimal point, technically, should follow the two numbers to indicate the extent of accuracy.

If the number 35 is written as 35.0, from a numbers viewpoint it is the same as 35, but there is a difference. Thirty-five represents a two-digit accuracy, or the number has an accuracy of two significant figures. If it is written 35.0, the accuracy has improved for the number now has three significant figures. 35.0 implies an order of accuracy to a tenth.

If you now take the same number and write it as 35.00, it has four significant figures. Again, from a numerical value viewpoint, 35.00 is the same as 35.0 and that is the same as 35. But 35.00 tells you that this number is accurate to the hundredths. If 35 represents a voltage, 35 means 35 V. 35.0 means a more precise measurement, and 35.00 is still more exact.

When using a formula or a table, it isn't always desirable to have answers with a high degree of accuracy. It all depends if the component being used must have a value that is as precise as possible. For most applications, this isn't necessary.

Formulas and tables do not take the need for a high order of accuracy into consideration. If, in using a formula such as Ohm's law to determine the value of resistance, with the voltage at 65 V, and the current at 165 mA, the resistance is:

$$R = E/I = 65/0.165 = 393.939\ 3\ \Omega$$

This number has seven significant figures. For most applications a resistor having a value of 400 $\Omega$ would be satisfactory. If this resistor has a tolerance of plus or minus 5%, then its uppermost limit would be 400+20, or 420 $\Omega$. Its lowermost limit would be 400−20 or 380 $\Omega$.

# Absolute error

The difference between a calculated value using a formula and the actual value of a component is known as absolute error. In the example given above, the absolute error would be 400−393=7. Even here, the amount of absolute error is simply an approximation since the resistor could have an actual value, as indicated, of 380 to 420 $\Omega$. Absolute error is based on calculations rather than measurements.

# Relative error

If a table or the use of a formula supplies one answer, but the component you select has some other value, the ratio of the two is known as relative error.

Assume a table or a formula calls for a resistor having a value of 87 Ω, but the actual value of the resistor as measured, is 90 Ω. 90−87=3. If we now divide 3 by the exact value, 90 in this case, you will have 3/90=0.033 3. This can be converted to a percent by moving the decimal point two places to the right. The relative error in this example is 3.33%.

# Rounding numbers

Quite often values of components used in electronics are reasonable approximations. The greater the accuracy demanded of a component, the more expensive it is. So from a point of view of practical economics, there is no benefit in using a component having 1% precision when a tolerance of 20% will do. While some of the tables presented indicate results having four- or five-figure significance, that much precision might not be required.

As an example, the number 3.141 6 is often used in electronics problems. This number, having five significant figures, can be rounded off by reducing it to four significant figures, or even three.

The rule for rounding off is simple. If the last digit, the digit at the extreme right, has a value of less than 5, just discard it. However, if it has a value of 5 or more, discard it also, but make the digit to its immediate left larger by 1.

*Example*   Round off the numbers 712.1 and 921.19 by one digit.
*Solution*   712.1 has four significant figures. It can be rounded off by dropping the decimal. The rounded number then becomes 712.

921.19 has five significant figures. It can be rounded off by dropping the rightmost digit and then increasing the number to its left by 1, leaving 921.2.

*Example*   Round the following number to three significant figures.

612.375 9

*Solution*   The rightmost digit has a value of 9, discard it and increase the digit preceding it by 1.

612.376

The rightmost digit has a value greater than 5, drop it and increase the digit preceding it by 1.

612.38

The number has five significant digits. Again, drop the last number and increase the digit preceding it by 1.

$$612.4$$

Now drop the rightmost digit, and since its value is less than 5, no further number changes need be made. The final number, significant to three places is 612.

# Numerical prefixes

Numerical prefixes are used to indicate the value of a number. A hexagon is a six-sided figure and is recognizable as such because the prefix, *hex*, means six. Some of the prefixes as listed in Table 23-6 are seldom used.

*Table 23-6. Numerical prefixes.*

| Number | Greek prefix | Latin prefix |
|--------|--------------|--------------|
| 1/2 | hemi- | semi- |
| 1 | mono- or mon- | uni- |
| 1 1/2 | — | sesqui- |
| 2 | di- | bi-; duo- |
| 3 | tri- | tri- or ter- |
| 4 | tetra- or tetr- | quadri- or quadr- |
| 5 | penta- or pent- | quinque- or quinqu- |
| 6 | hexa- or hex- | sexi- or sex- |
| 7 | hepta- or hept- | septi- or sept- |
| 8 | octa- or oct- or octo- | octo- |
| 9 | ennea- or enne- | nona-; novem- |
| 10 | deca- or dec- | decem- |
| 11 | hendeca- or hendec- | undeca- or undec- |
| 12 | dodeca- or dodec- | doudec- |
| 13 | trideca- or tridec- | tredec- |
| 14 | tetradeca- or tetradec- | quatuordec- |
| 15 | pentadeca- or pentadec- | quindec- |
| 16 | hexadeca- or hexadec- | sextodec- |
| 17 | heptadeca- or heptadec- | septendec- |
| 18 | octadeca- or octadec- | octodec- |
| 19 | nonadeca- or nonadec- | novemdec- |
| 20 | eicosa- or eicos- | viginti- |
| 21 | heneicosa- or heneicos- | |
| 22 | docosa- or docos- | |
| 23 | tricosa- or tricos- | |
| 24 | tetracosa- or tetracos- | |
| 25 | pentacosa- or pentacos- | |
| 26 | hexacosa- or hexacos- | |
| 27 | heptacosa- or heptacos- | |

*Table 23-6. Continued.*

| Number | Greek prefix | Latin prefix |
|--------|--------------|--------------|
| 28 | octacosa- or octacos- | |
| 29 | nonacosa- or nonacos- | |
| 30 | triaconta- or triacont- | triginti- |
| 31 | hentriaconta- or hentriacont- | |
| 32 | dotriaconta- or dotriacont- | |
| 40 | tetraconta- or tetracont- | quadragin- |
| 50 | pentaconta- or pentacont- | quinquagin- |
| 60 | hexaconta- or hexacont- | sexagin- |

# Decimal equivalents

A proper fraction is any fraction that has a value of less than 1. Although fractions are commonly used, they are often inconvenient when applied to practical problems. Problem solving is often easier if the decimal equivalent to a fraction is used. These are listed in Table 23-7.

Table 23-8 lists some of the more commonly used values in electronic formulas and tables.

*Table 23-7. Decimal equivalents.*

| Fraction | Decimal | Fraction | Decimal |
|----------|---------|----------|---------|
| $1/64$ | 0.015 6 | $17/64$ | 0.265 6 |
| $1/32$ | 0.031 2 | $9/32$ | 0.281 2 |
| $3/64$ | 0.046 9 | $19/64$ | 0.296 9 |
| $1/16$ | 0.062 5 | $5/16$ | 0.312 5 |
| $5/64$ | 0.078 1 | $21/64$ | 0.328 1 |
| $3/32$ | 0.093 7 | $11/32$ | 0.343 7 |
| $7/64$ | 0.109 4 | $23/64$ | 0.359 4 |
| $1/8$ | 0.125 0 | $3/8$ | 0.375 0 |
| $9/64$ | 0.140 6 | $25/64$ | 0.390 6 |
| $5/32$ | 0.156 2 | $13/32$ | 0.406 2 |
| $11/64$ | 0.171 9 | $27/64$ | 0.421 9 |
| $3/16$ | 0.187 5 | $7/16$ | 0.437 5 |
| $13/64$ | 0.203 1 | $29/64$ | 0.453 1 |
| $7/32$ | 0.218 7 | $15/32$ | 0.468 7 |
| $15/16$ | 0.234 4 | $31/64$ | 0.484 4 |
| $1/4$ | 0.250 0 | $1/2$ | 0.500 0 |

## Table 23-8. Commonly used values in electronic formulas and tables.

| | |
|---|---|
| $\epsilon \doteq 2.718\ 281\ 83$ | $\log 8 \doteq 0.903\ 090$ |
| $1/\epsilon \doteq 0.367\ 879\ 44$ | $\log 9 \doteq 0.954\ 243$ |
| $\log \epsilon \doteq 0.434\ 294\ 48$ | $\log 10 = 1.000\ 000$ |
| $\pi \doteq 3.141\ 592\ 65$ | $1\ \text{radian} = 180°/\pi \doteq 57°17'\ 44.8''$ |
| $2\pi \doteq 6.283\ 185\ 30$ | $1° = \pi/180° \doteq 0.017\ 453\ 29\ \text{radian}$ |
| $1/\pi \doteq 0.318\ 309\ 89$ | $\sqrt{\ } = 1.000\ 0$ |
| $\pi^2 \doteq 9.869\ 044\ 0$ | $\sqrt{2} \doteq 1.414\ 2$ |
| $\sqrt{\pi} \doteq 1.772\ 453\ 85$ | $\sqrt{3} \doteq 1.732\ 1$ |
| $\log \pi \doteq 0.497\ 149\ 87$ | $\sqrt{4} = 2.000\ 0$ |
| $\log 1 = 0.000\ 000$ | $\sqrt{5} \doteq 2.236\ 1$ |
| $\log 2 \doteq 0.301\ 030$ | $\sqrt{6} \doteq 2.449\ 5$ |
| $\log 3 \doteq 0.477\ 121$ | $\sqrt{7} \doteq 2.645\ 8$ |
| $\log 4 \doteq 0.602\ 060$ | $\sqrt{8} \doteq 2.828\ 4$ |
| $\log 5 \doteq 0.698\ 970$ | $\sqrt{9} = 3.000\ 0$ |
| $\log 6 \doteq 0.778\ 151$ | $\sqrt{10} \doteq 3.162\ 3$ |
| $\log 7 \doteq 0.845\ 098$ | $j = \sqrt{-1}$ |

[logs are to base 10]

# 24
## CHAPTER

# Mathematics

## Square versus circular area

It is sometimes necessary to convert from square to circular area or from circular to square area. Square area is equal to circular area when the side of the square is equal to 0.886 23 multiplied by the diameter of the circle. Conversely, the areas are equal when the diameter of the circle equals 1.128 38 multiplied by the side of the square.

*Example*   A circle has a diameter of 2 in. What is the length of any side of a square such that the area of the square is the same as the area of the circle?

*Solution*   The length of any side of the square is equal to the diameter of the circle multiplied by 0.886 23. Therefore, $2 \times 0.886\ 23 = 1.772\ 46$ in. The area of the square is $1.772\ 46 \times 1.772\ 46$, or $3.141\ 6$ in$^2$. The area of the circle is $\pi r^2$. The radius equals one-half the diameter, $1/2 \times 2$, or 1. The area of the circle is equal to $\pi r^2 = 3.141\ 6 \times 1^2 = 3.141\ 6$. Thus, a circle with a diameter of 2 in has the same area as a square, each of whose sides has a length of 1.772 46 in.

Table 24-1 supplies the circumference and area of circles having diameters ranging from $1/32$ in to 100 in.

## Functions of angles

Table 24-2 supplies the trigonometric functions of the angle included between the base and the hypotenuse of a right-angle triangle. The sine (sin) of the angle is the ratio of the altitude to the hypotenuse; that is, the altitude divided by the length of the hypotenuse. The cosine (cos) is the base divided by the hypotenuse. The tangent (tan) is the altitude divided by the base, and the cotangent (cot) is the base divided by the altitude. The sec (secant) is the ratio of the hypotenuse to the base, and the csc (cosecant) is the ratio of the hypotenuse to the altitude.

*Table 24-1. Circumference and area of circles.*

| Diameter | Circumference ($\pi$d) | Area $\pi\left(\dfrac{d}{2}\right)^2$ | Diameter | Circumference ($\pi$d) | Area $\pi\left(\dfrac{d}{2}\right)^2$ |
|---|---|---|---|---|---|
| 1/32 | 0.098 17 | 0.000 8 | 10 | 31.416 | 78.539 8 |
| 1/16 | 0.196 35 | 0.003 1 | 11 | 34.558 | 95.033 2 |
| 3/32 | 0.294 52 | 0.006 9 | 12 | 37.699 | 113.097 |
| 1/8 | 0.392 70 | 0.012 3 | 13 | 40.841 | 132.732 |
| 5/32 | 0.490 87 | 0.019 2 | 14 | 43.982 | 153.938 |
| 3/16 | 0.589 05 | 0.027 6 | 15 | 47.124 | 176.715 |
| 7/32 | 0.687 22 | 0.037 6 | 16 | 50.265 | 201.062 |
| 1/4 | 0.785 40 | 0.049 1 | 17 | 53.407 | 226.980 |
| 9/32 | 0.883 57 | 0.062 1 | 18 | 56.549 | 254.469 |
| 5/16 | 0.981 75 | 0.076 7 | 19 | 59.690 | 283.529 |
| 11/32 | 1.079 92 | 0.092 8 | 20 | 62.832 | 314.159 |
| 3/8 | 1.178 10 | 0.110 4 | 21 | 65.973 | 346.361 |
| 13/32 | 1.276 27 | 0.129 6 | 22 | 69.115 | 380.133 |
| 7/16 | 1.374 45 | 0.150 3 | 23 | 72.257 | 415.476 |
| 15/32 | 1.472 62 | 0.172 6 | 24 | 75.398 | 452.389 |
| 1/2 | 1.570 80 | 0.196 3 | 25 | 78.540 | 490.874 |
| 17/32 | 1.668 97 | 0.221 7 | 26 | 81.681 | 530.929 |
| 9/16 | 1.767 14 | 0.248 5 | 27 | 84.823 | 572.555 |
| 19/32 | 1.865 32 | 0.276 9 | 28 | 87.965 | 615.752 |
| 5/8 | 1.963 49 | 0.306 8 | 29 | 91.106 | 660.520 |
| | | | 30 | 94.248 | 706.858 |
| 21/32 | 2.061 67 | 0.338 2 | 31 | 97.389 | 754.768 |
| 11/16 | 2.159 84 | 0.371 2 | 32 | 100.531 | 804.248 |
| 23/32 | 2.258 02 | 0.405 7 | 33 | 103.673 | 855.299 |
| 3/4 | 2.356 19 | 0.441 8 | 34 | 106.814 | 907.920 |
| 25/32 | 2.454 37 | 0.479 4 | 35 | 109.956 | 962.113 |
| 13/16 | 2.552 54 | 0.518 5 | 36 | 113.097 | 1 017.88 |
| 27/32 | 2.650 72 | 0.559 1 | 37 | 116.239 | 1 075.21 |
| 7/8 | 2.748 89 | 0.601 3 | 38 | 119.381 | 1 134.11 |
| 29/32 | 2.847 07 | 0.645 0 | 39 | 122.522 | 1 194.59 |
| 15/16 | 2.945 24 | 0.690 3 | 40 | 125.66 | 1 256.64 |
| 31/32 | 3.043 42 | 0.737 1 | 41 | 128.81 | 1 320.25 |
| 1 | 3.142 | 0.785 4 | 42 | 131.95 | 1 385.44 |
| 2 | 6.283 | 3.141 6 | 43 | 135.09 | 1 452.20 |
| 3 | 9.425 | 7.068 6 | 44 | 138.23 | 1 520.53 |
| 4 | 12.566 | 12.566 4 | 45 | 141.37 | 1 590.43 |
| 5 | 15.708 | 19.635 0 | 46 | 144.51 | 1 661.90 |
| 6 | 18.850 | 28.274 3 | 47 | 147.65 | 1 734.94 |
| 7 | 21.991 | 38.484 5 | 48 | 150.80 | 1 809.56 |
| 8 | 25.133 | 50.265 5 | 49 | 153.94 | 1 885.74 |
| 9 | 28.274 | 63.617 3 | 50 | 157.08 | 1 963.50 |

Table 24-1. Continued.

| Diameter | Circumference ($\pi$d) | Area $\pi\left(\dfrac{d}{2}\right)^2$ | Diameter | Circumference ($\pi$d) | Area $\pi\left(\dfrac{d}{2}\right)^2$ |
|---|---|---|---|---|---|
| 51 | 160.22 | 2 042.82 | 76 | 238.76 | 4 536.46 |
| 52 | 163.36 | 2 123.72 | 77 | 241.90 | 4 656.63 |
| 53 | 166.50 | 2 206.18 | 78 | 245.04 | 4 778.36 |
| 54 | 169.65 | 2 290.22 | 79 | 248.19 | 4 901.67 |
| 55 | 172.79 | 2 375.83 | 80 | 251.33 | 5 026.55 |
| 56 | 175.93 | 2 463.01 | 81 | 254.47 | 5 153.00 |
| 57 | 179.07 | 2 551.76 | 82 | 257.61 | 5 281.02 |
| 58 | 182.21 | 2 642.08 | 83 | 260.75 | 5 410.61 |
| 59 | 185.35 | 2 733.97 | 84 | 263.89 | 5 541.77 |
| 60 | 188.50 | 2 827.43 | 85 | 267.04 | 5 674.50 |
| 61 | 191.64 | 2 922.47 | 86 | 270.18 | 5 808.80 |
| 62 | 194.78 | 3 019.07 | 87 | 273.32 | 5 944.68 |
| 63 | 197.92 | 3 117.25 | 88 | 276.46 | 6 082.12 |
| 64 | 201.06 | 3 216.99 | 89 | 279.60 | 6 221.14 |
| 65 | 204.20 | 3 318.31 | 90 | 282.74 | 6 361.73 |
| 66 | 207.35 | 3 421.19 | 91 | 285.88 | 6 503.88 |
| 67 | 210.49 | 3 525.65 | 92 | 289.03 | 6 647.61 |
| 68 | 213.63 | 3 631.68 | 93 | 292.17 | 6 792.91 |
| 69 | 216.77 | 3 739.28 | 94 | 295.31 | 6 939.78 |
| 70 | 219.91 | 3 848.45 | 95 | 298.45 | 7 088.22 |
| 71 | 223.05 | 3 959.19 | 96 | 301.59 | 7 238.23 |
| 72 | 226.19 | 4 071.50 | 97 | 304.73 | 7 389.81 |
| 73 | 229.34 | 4 185.39 | 98 | 307.88 | 7 542.96 |
| 74 | 232.48 | 4 300.84 | 99 | 311.02 | 7 697.69 |
| 75 | 235.62 | 4 417.86 | 100 | 314.16 | 7 853.98 |

# Polarity of trigonometric functions

Table 24-3 supplies the polarity of the trigonometric functions in each of the four quadrants.

# Phase angle for the tangent

The ratio of the altitude of a right-angle triangle to its base is called the tangent. This ratio and the corresponding phase angle are supplied in Table 24-4.

# Oblique triangles

The included angles or the dimensions of the sides of an oblique triangle can be determined with the formulas supplied in Table 24-5.

The values of trigonometric functions are supplied in Table 24-6.

*Table 24-2. Angles and their functions.*

| Angle A | sin A | cos A | tan A | cot A | sec A | csc A |
|---------|-------|-------|-------|-------|-------|-------|
| 0° | 0 | 1 | 0 | ∞ | 1 | ∞ |
| 30° | $\dfrac{1}{2}$ | $\dfrac{\sqrt{3}}{2}$ | $\dfrac{\sqrt{3}}{3}$ | $\sqrt{3}$ | $\dfrac{2\sqrt{3}}{3}$ | 2 |
| 45° | $\sqrt{2}$ | $\dfrac{\sqrt{2}}{2}$ | 1 | 1 | $\sqrt{2}$ | $\sqrt{2}$ |
| 60° | $\dfrac{\sqrt{3}}{2}$ | $\dfrac{1}{2}$ | $\sqrt{3}$ | $\dfrac{\sqrt{3}}{3}$ | 2 | $\dfrac{2\sqrt{3}}{3}$ |
| 90° | 1 | 0 | ∞ | 0 | ∞ | 1 |
| 120° | $\dfrac{\sqrt{3}}{2}$ | $-\dfrac{1}{2}$ | $\sqrt{3}$ | $-\dfrac{\sqrt{3}}{3}$ | $-2$ | $\dfrac{2\sqrt{3}}{3}$ |
| 180° | 0 | $-1$ | 0 | ∞ | $-1$ | ∞ |
| 270° | $-1$ | 0 | ∞ | 0 | ∞ | $-1$ |
| 360° | 0 | 1 | 0 | ∞ | 1 | ∞ |

*Table 24-3. Polarity of trigonometric functions.*

| Quadrant | $\sin\theta$ | $\cos\theta$ | $\tan\theta$ | $\cot\theta$ | $\sec\theta$ | $\csc\theta$ |
|----------|------|------|------|------|------|------|
| I | + | + | + | + | + | + |
| II | + | − | − | − | − | + |
| III | − | − | + | + | − | − |
| IV | − | + | − | − | + | − |

*Table 24-4. The ratio tan $\theta$ = alt/base and corresponding values of phase angle, $\theta$.*

| Phase angle (°) | Ratio | Phase angle (°) | Ratio | Phase angle (°) | Ratio |
|-----------------|-------|-----------------|-------|-----------------|-------|
| 0 | 0.000 0 | 7 | 0.122 8 | 14 | 0.249 3 |
| 1 | 0.017 5 | 8 | 0.140 5 | 15 | 0.267 9 |
| 2 | 0.034 9 | 9 | 0.158 4 | 16 | 0.286 7 |
| 3 | 0.052 4 | 10 | 0.176 3 | 17 | 0.305 7 |
| 4 | 0.069 9 | 11 | 0.194 4 | 18 | 0.324 9 |
| 5 | 0.087 5 | 12 | 0.212 6 | 19 | 0.344 3 |
| 6 | 0.105 1 | 13 | 0.230 9 | 20 | 0.364 0 |

Table 24-4. Continued.

| Phase angle (°) | Ratio | Phase angle (°) | Ratio | Phase angle (°) | Ratio |
|---|---|---|---|---|---|
| 21 | 0.383 9 | 44 | 0.965 7 | 67 | 2.355 9 |
| 22 | 0.404 0 | 45 | 1.000 0 | 68 | 2.475 1 |
| 23 | 0.424 5 | 46 | 1.035 5 | 69 | 2.605 1 |
| 24 | 0.445 2 | 47 | 1.072 4 | 70 | 2.747 5 |
| 25 | 0.466 3 | 48 | 1.110 6 | 71 | 2.904 2 |
| 26 | 0.487 7 | 49 | 1.150 4 | 72 | 3.077 7 |
| 27 | 0.509 5 | 50 | 1.191 8 | 73 | 3.270 9 |
| 28 | 0.531 7 | 51 | 1.234 9 | 74 | 3.487 4 |
| 29 | 0.554 3 | 52 | 1.279 9 | 75 | 3.732 1 |
| 30 | 0.577 4 | 53 | 1.327 0 | 76 | 4.010 8 |
| 31 | 0.600 9 | 54 | 1.376 4 | 77 | 4.331 5 |
| 32 | 0.624 9 | 55 | 1.428 1 | 78 | 4.704 6 |
| 33 | 0.649 4 | 56 | 1.482 6 | 79 | 5.144 6 |
| 34 | 0.674 5 | 57 | 1.539 9 | 80 | 5.671 3 |
| 35 | 0.700 2 | 58 | 1.600 3 | 81 | 6.313 8 |
| 36 | 0.726 5 | 59 | 1.664 3 | 82 | 7.115 4 |
| 37 | 0.753 6 | 60 | 1.732 1 | 83 | 8.144 3 |
| 38 | 0.781 3 | 61 | 1.804 0 | 84 | 9.514 4 |
| 39 | 0.809 8 | 62 | 1.880 7 | 85 | 11.43 |
| 40 | 0.839 1 | 63 | 1.962 6 | 86 | 14.30 |
| 41 | 0.869 3 | 64 | 2.050 3 | 87 | 19.08 |
| 42 | 0.900 4 | 65 | 2.144 5 | 88 | 28.64 |
| 43 | 0.932 5 | 66 | 2.246 0 | 89 | 87.29 |

Table 24-5. Angular and linear
dimensions of oblique triangles.

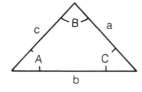

| Parts given | Parts to be found | Formula |
|---|---|---|
| a b c | A | $\cos A = \dfrac{b^2 + c^2 - a^2}{2bc}$ |
| a b A | B | $\sin B = \dfrac{b \times \sin A}{a}$ |
| a A B | C | $C = 180° - (A + B)$ |

Table 24-5. Continued.

| Parts given | Parts to be found | Formula |
|---|---|---|
| a A B | b | $b = \dfrac{a \times \sin B}{\sin A}$ |
| a A B | c | $c = \dfrac{a \sin C}{\sin A} = \dfrac{a \sin (180° - A - B)}{\sin A}$ |
| b A C | B | $B = 180° - (A + C)$ |

*Table 24-6. Trigonometric functions.*

| Degrees | Sine | Tangent | Cotangent | Cosine | Degrees |
|---|---|---|---|---|---|
| 0 | 0.000 0 | 0.000 0 | . . . . . | 1.000 0 | 90 |
| 1 | 0.017 5 | 0.017 5 | 57.29 | 0.999 8 | 89 |
| 2 | 0.034 9 | 0.034 9 | 28.636 | 0.999 4 | 88 |
| 3 | 0.052 3 | 0.052 4 | 19.081 | 9.998 6 | 87 |
| 4 | 0.069 8 | 0.069 9 | 14.301 | 0.997 6 | 86 |
| 5 | 0.087 2 | 0.087 5 | 11.430 | 0.996 2 | 85 |
| 6 | 0.104 5 | 0.105 1 | 9.514 4 | 0.994 5 | 84 |
| 7 | 0.121 9 | 0.122 8 | 8.144 3 | 0.992 5 | 83 |
| 8 | 0.139 2 | 0.140 5 | 7.115 4 | 0.990 3 | 82 |
| 9 | 0.156 4 | 0.158 4 | 6.313 8 | 0.987 7 | 81 |
| 10 | 0.173 6 | 0.176 3 | 5.671 3 | 0.984 8 | 80 |
| 11 | 0.190 8 | 0.194 4 | 5.144 6 | 0.981 6 | 79 |
| 12 | 0.207 9 | 0.212 6 | 4.704 6 | 0.978 1 | 78 |
| 13 | 0.225 0 | 0.230 9 | 4.331 5 | 0.974 4 | 77 |
| 14 | 0.241 9 | 0.249 3 | 4.010 8 | 0.970 3 | 76 |
| 15 | 0.258 8 | 0.267 9 | 3.732 1 | 0.965 9 | 75 |
| 16 | 0.275 6 | 0.286 7 | 3.487 4 | 0.961 3 | 74 |
| 17 | 0.292 4 | 0.305 7 | 3.270 9 | 0.956 3 | 73 |
| 18 | 0.309 0 | 0.324 9 | 3.077 7 | 0.951 1 | 72 |
| 19 | 0.325 6 | 0.344 3 | 2.904 2 | 0.945 5 | 71 |
| 20 | 0.342 0 | 0.364 0 | 2.747 5 | 0.939 7 | 70 |
| 21 | 0.358 4 | 0.383 9 | 2.605 1 | 0.933 6 | 69 |
| 22 | 0.374 6 | 0.404 0 | 2.475 1 | 0.927 2 | 68 |
| 23 | 0.390 7 | 0.424 5 | 2.355 9 | 0.920 5 | 67 |
| 24 | 0.406 7 | 0.445 2 | 2.246 0 | 0.913 5 | 66 |
| 25 | 0.422 6 | 0.466 3 | 2.144 5 | 0.906 3 | 65 |
| 26 | 0.438 4 | 0.487 7 | 2.050 3 | 0.898 8 | 64 |
| 27 | 0.454 0 | 0.509 5 | 1.962 6 | 0.891 0 | 63 |
| 28 | 0.469 5 | 0.531 7 | 1.880 7 | 0.882 9 | 62 |
| 29 | 0.484 8 | 0.554 3 | 1.804 0 | 0.874 6 | 61 |
| **Degrees** | **Cosine** | **Cotangent** | **Tangent** | **Sine** | **Degrees** |

*Table 24-6. Continued.*

| Degrees | Sine | Tangent | Cotangent | Cosine | Degrees |
|---------|------|---------|-----------|--------|---------|
| 30 | 0.500 0 | 0.577 4 | 1.732 1 | 0.866 0 | 60 |
| 31 | 0.515 0 | 0.600 9 | 1.664 3 | 0.857 2 | 59 |
| 32 | 0.529 9 | 0.624 9 | 1.600 3 | 0.848 0 | 58 |
| 33 | 0.544 6 | 0.649 4 | 1.539 9 | 0.838 7 | 57 |
| 34 | 0.559 2 | 0.674 5 | 1.482 6 | 0.829 0 | 56 |
| 35 | 0.573 6 | 0.700 2 | 1.428 1 | 0.819 2 | 55 |
| 36 | 0.587 8 | 0.726 5 | 1.376 4 | 0.809 0 | 54 |
| 37 | 0.601 8 | 0.753 6 | 1.327 0 | 0.798 6 | 53 |
| 38 | 0.615 7 | 0.781 3 | 1.279 9 | 0.788 0 | 52 |
| 39 | 0.629 3 | 0.809 8 | 1.234 9 | 0.777 1 | 51 |
| 40 | 0.642 8 | 0.839 1 | 1.191 8 | 0.766 0 | 50 |
| 41 | 0.656 1 | 0.869 3 | 1.150 4 | 0.754 7 | 49 |
| 42 | 0.669 1 | 0.900 4 | 1.110 6 | 0.743 1 | 48 |
| 43 | 0.682 0 | 0.932 5 | 1.072 4 | 0.731 4 | 47 |
| 44 | 0.694 7 | 0.965 7 | 1.035 5 | 0.719 3 | 46 |
| 45 | 0.707 1 | 1.000 0 | 1.000 0 | 0.707 1 | 45 |

# Values of $\pi$

Various values of $\pi$ appear regularly in calculations involving electronic problems. This constant is used for finding values of inductive reactance, capacitive reactance, frequency, and in problems involving resonance. Table 24-7 supplies coefficients, powers, roots, and logs of $\pi$.

*Table 24-7. Coefficients, powers, roots, and logs of $\pi$.*

| $\pi$ | formula | $\pi$ | formula |
|-------|---------|-------|---------|
| $1\,\pi$ | 3.141 592 65 | $\pi/2$ | 1.570 796 33 |
| $2\,\pi$ | 6.283 185 31 | $\pi/3$ | 1.047 197 55 |
| $3\,\pi$ | 9.424 777 96 | $\pi/4$ | 0.785 398 16 |
| $4\,\pi$ | 12.566 370 61 | $\pi/5$ | 0.628 318 53 |
| $5\,\pi$ | 15.707 963 27 | $\pi/6$ | 0.523 598 78 |
| $6\,\pi$ | 18.849 555 92 | $\pi/7$ | 0.448 498 95 |
| $7\,\pi$ | 21.991 148 57 | $\pi/8$ | 0.392 699 08 |
| $8\,\pi$ | 25.132 741 22 | $\pi/9$ | 0.349 065 85 |
| $9\,\pi$ | 28.274 333 87 | $\pi/10$ | 0.314 159 265 |
| $10\,\pi$ | 31.415 926 52 | $4\pi/3$ | 4.188 790 20 |
| $1/\pi$ | 0.318 309 89 | $3\pi$ | 0.954 929 666 |
| $1/2\pi$ | 0.159 154 94 | $4/\pi$ | 1.273 239 54 |
| $1/3\pi$ | 0.106 103 | $\pi^2$ | 9.869 604 40 |

*Table 24-7. Continued.*

| π formula | | π formula | |
|---|---|---|---|
| $\sqrt{\pi}$ | 1.772 453 85 | log π | 0.497 149 9 |
| $1/\pi^2$ | 0.101 321 18 | log 2π | 0.798 179 9 |
| $1/\sqrt{\pi}$ | 0.564 189 58 | log 4π | 1.099 209 9 |
| $\sqrt{3/\pi}$ | 0.977 205 02 | log $\pi^2$ | 0.994 299 7 |
| $\sqrt{4/\pi}$ | 1.128 379 17 | log $\sqrt{\pi}$ | 0.248 |
| $\sqrt[3]{\pi}$ | 1.464 491 89 | | |

# Mathematical symbols

Electronics is a science that relies on mathematics for the solution of problems. Consequently, the symbols used in mathematics have become an integral part of textbooks having electronics as the subject. Table 24-8 lists those mathematical symbols most often found associated with electronics.

*Table 24-8. Mathematical symbols.*

| Symbol | Name or meaning | Description |
|---|---|---|
| 0, 1, 2, 3, 4, etc. | Whole numbers | Cardinal (Arabic) numbers used in mathematics. Arithmetic numbers. |
| No.; no.; # | Abbreviation | Symbol that may precede a number. |
| i; ii; iii; iv; v<br>I; II; III; IV; V | Roman numerals | Have limited use—introductory pages of books; watches; building cornerstones, etc. |
| $2^{1}/_{2}$; $3^{1}/_{8}$; $4^{1}/_{9}$ | Mixed numbers | Consist of a whole number and a proper fraction. |
| $^{1}/_{2}$; $^{3}/_{4}$; $^{4}/_{7}$ | Proper fractions | Numerator is always smaller in value than the denominator. |
| $^{5}/_{2}$; $^{4}/_{3}$; $^{9}/_{5}$ | Improper fractions | Numerator is always larger in value than the denominator. |

*Table 24-8. Continued.*

| Symbol | Name or meaning | Description |
|---|---|---|
| + | Plus sign | Positive; plus; addition. Can be used to indicate direction. |
| − | Minus sign | Negative; minus; subtraction. Can be used to indicate direction. |
| ± | Plus-minus | Positive or negative; plus or minus; addition or subtraction. |
| ∓ | Minus-plus | Negative or positive; minus or plus; subtraction or addition. |
| a; x; y; z; etc. | Literal numbers | Letters used as a symbol for a number. |
| $25/100$; $4/10$; $6589/1000$ | Decimal fractions | Fractions having 10, 100, 1 000 etc., in the denominator. |
| 1.657; 0.354; 0.25 | Decimal point | Point used in decimal fractions when denominator is omitted. Also known as the radix point. The number to the right of the decimal point is the decimal fraction. In some countries a comma is used in place of the decimal point. In the U.S. the decimal point is put at the bottom; in England at the center or higher. In the binary number system, it is known as the binary point. In the hexadecimal system, it is called the hexadecimal point. The term *decimal point* refers only to the use of the radix point in the decimal system. |
| ∞ | Infinity | An indefinitely great number or amount. |
| × | Multiply by | Multiplication. A form of repeated addition. |

*Table 24-8. Continued.*

| Symbol | Name or meaning | Description |
|---|---|---|
| x·y | Dot between letters means multiplication | x is multiplied by y. |
| (x) (y) | Adjacent parentheses means multiplication | x is multiplied by y. |
| x ÷ y | Straight line with dot above and below means division | x is divided by y. |
| x/y | Slant line indicates division | x is divided by y. Also, ratio of x to y. |
| $\frac{x}{y}$ | Horizontal line means division | x is divided by y. Also, ratio of x to y. |
| x : y | Dots indicate division or ratio | x is divided by y; the ratio of x to y. |
| = or :: | Equals signs | Signs representing equality between two quantities; is equal to. |
| ≅ | Approximation sign | Quantities separated by this sign are approximately equal. Congruent. |
| ≠ | Inequality sign | Quantities separated by this sign are not equal. |
| < | Inequality | Is less than. |
| << | Inequality | Is much less than. |
| > | Inequality | Is greater than. |
| >> | Inequality | Is much greater than. |
| ≥ | Inequality or equality | Is equal to or is greater than. |
| ≤ | Inequality or equality | Is equal to or is less than. |
| ∴ | Therefore | Symbol used in geometry and logic. |
| → or lim | Approaches as a limit | Variable approaches a constant but never reaches it. |
| | Perpendicular | Lines that form right angles. |
| ‖ | Parallel | Lines that are parallel to each other. |
| ∡ | Angle; also positive angle | The angle formed by two lines. |

*Table 24-8. Continued.*

| Symbol | Name or meaning | Description |
|---|---|---|
| ∡ s | Angles | Two or more angles. |
| | Negative angle | An angle whose sine, tangent, cosecant, and contangent are negative. |
| Δ | Change, increase or decrease, increment | Capital Greek letter delta. A similar symbol is used to represent triangles in trigonometry. |
| √ | Square root | Quantity raised to the one-half power. |
| sin | Sine of included base angle | Ratio of altitude to hypotenuse in a right-angle triangle. |
| cos | Cosine of included base angle | Ratio of base to hypotenuse in a right-angle triangle. |
| tan | Tangent of included base angle | Ratio of altitude to the base in a right-angle triangle. |
| cot | Cotangent of included base angle | Ratio of base to altitude in a right-angle triangle. |
| sec | Secant of included base angle | Ratio of hypotenuse to base in a right-angle triangle. |
| csc | Cosecant of included base angle | Ratio of hypotenuse to altitude in a right-angle triangle. |

# Greek alphabet

Lowercase and capital letters of the Greek alphabet are used in electronics, especially in formulas. Table 24-9 lists these letters and their commonly used designations.

*Table 24-9. Greek alphabet.*

| Name | Capital | Lowercase | Commonly used to designate |
|---|---|---|---|
| Alpha | A | $\alpha$ | Angles, area, coefficients |
| Beta | B | $\beta$ | Angles, flux density, coefficients |
| Gamma | $\Gamma$ | $\gamma$ | Conductivity, specific gravity |
| Delta | $\Delta$ | $\delta$ | Variation, density |
| Epsilon | E | $\epsilon$ | Base of natural logarithms |
| Zeta | Z | $\zeta$ | Impedance, coefficients, coordinates |
| Eta | H | $\eta$ | Hysteresis coefficient, efficiency |
| Theta | $\Theta$ | $\theta$ | Temperature, phase angle |
| Iota | I | $\iota$ | Unit vector |
| Kappa | K | $\varkappa$ | Dielectric constant, susceptibility |
| Lambda | $\Lambda$ | $\lambda$ | Wave length |
| Mu | M | $\mu$ | Micro, amplification factor, permeability |
| Nu | N | $\nu$ | Reluctivity |
| Xi | $\Xi$ | $\xi$ | Coordinates |
| Omicron | O | $o$ | — |
| Pi | $\Pi$ | $\pi$ | 3.141 6 (ratio of circumference to diameter) |
| Rho | P | $\varrho$ | Resistivity |
| Sigma | $\Sigma$ | $\sigma$ | Sign of summation |
| Tau | T | $\tau$ | Time constant, time phase displacement |
| Upsilon | $\Upsilon$ | $\upsilon$ | — |
| Phi | $\Phi$ | $\varphi$ | Magnetic flux, angles |
| Chi | X | $\chi$ | Electric susceptibility, angles |
| Psi | $\Psi$ | $\psi$ | Dielectric flux, phase difference |
| Omega | $\Omega$ | $\omega$ | $\Omega$, ohms; $\omega$, angular velocity |

# 25
CHAPTER

# Analog and digital signal transmission

## Transmission space

The amount of frequency space available for signal transmission is limited. Signals radiated by two different services can occupy the same band of frequencies provided they are sufficiently separated geographically that they are unable to interfere with each other. The assignment of frequency space is governed by three factors: frequency, location, and power.

The utilization of the spectrum is governed by the ITU (International Telecommunications Union), which is composed of 154 nations that ratified and agreed to the Telecommunications Convention established in 1981. Any problems that involve the ITU are handled by a Frequency Registration board located in the offices of the ITU, Place des Nations, CH-1211, in Geneva, 20, Switzerland. The ITU has established world wide standards for electromagnetic communications.

The ITU has the assistance of adjunct organizations that specialize in various aspects of communications. They include the CCIR (International Radio Consultative Committee) and the CCITT (International Telegraph and Telephone Consultative Committee). In addition, the ITU works with a number of other groups that have an interest in the allocation and utilization of frequency space. These include the ICAO (International Civil Aviation Organization), the IMCO (International Maritime Consultative Organization), the WMO (World Meterological Organization), the IEC (International Electrotechnical Commission), and the IOC (International Oceanographic Commission).

The national members of the ITU have agreed to abide by the rules governing frequency, location, power, and time of transmitted signals. However, each nation has the right to set up its own laws for communications, provided that these come within the agreed-upon rules of the ITU.

The ITU has divided the geographic areas of the earth into three regions, identified as 1, 2, and 3, with the U.S. in region 2. The ITU has assigned groups of frequencies for various kinds of transmitting services. Each nation has its own governing authority for transmission.

In the U.S., this authority is the FCC (Federal Communications Commission). The FCC, following the rules set up by the ITU, issues licenses, describes operating procedures, assigns call letters, delineates the type or types of transmission to be used, the operating power, and controls most other frequency space issues. Violation of FCC rules and regulations can result in loss of an operating license, a fine and possible imprisonment. The power of the FCC is limited by Congress.

# Modulation

Modulation consists of a selected technique used for the transmission of information. There are various kinds of modulation, and they can be associated with direct or alternating currents. In the case of ac, they can be of various frequencies. Codes are associated with modulation and a number of different codes are available.

# The Morse code

The oldest and possibly the simplest code is the International Morse code, sometimes called the Continental code. It is still widely used in stock exchange ticker tape, for wired and wireless communications, for amateur radio, for police and fire alarm systems, and for fixed and mobile transmissions by the military. The modulation method consists of the spaced interruption of a direct current for wired systems. The current is switched on and off by a telegraph key that functions as a single-pole, single-throw switch. Because there are only two operating states, current on and current off, the system is binary. It is alphanumeric but can be used for the transmission of punctuation.

Table 25-1 lists the letters, numbers and punctuation used by the International Morse Code. Each dot (referred to as dit) and each dash (called a dash) represents the momentary flow of current. The dots and dashes indicate current on time; the spacing between them is current-off time. A dash has a length three times that of a dot.

# Q code

Two techniques are used for saving time when transmitting with a wired or wireless system. The first is to use abbreviations when possible; the second is to take

*Table 25-1. International Morse code.*

| Code | Value | Code | Value |
|---|---|---|---|
| •— | A | •———— | 1 |
| —••• | B | ••——— | 2 |
| —•—• | C | •••—— | 3 |
| —•• | D | ••••— | 4 |
| • | E | ••••• | 5 |
| ••—• | F | —•••• | 6 |
| ——• | G | ——••• | 7 |
| •••• | H | ———•• | 8 |
| •• | I | ————• | 9 |
| •——— | J | ————— | 0 (zero) |
| —•— | K | •—•—•— | Period (.) |
| •—•• | L | ——••—— | Comma (,) |
| —— | M | ••——•• | Question mark (?) |
| —• | N | ———••• | Colon (:) |
| ——— | O | —•—•—• | Semicolon (;) |
| •——• | P | —•——•— | Parenthesis ( ) |
| ——•— | Q | •—••• | Wait sign (AS) |
| •—• | R | —•••— | Double dash (break) |
| ••• | S | •••••••• | Error (erase sign) |
| — | T | —••—• | Fraction bar (/) |
| ••— | U | •—•—• | End of message (AR) |
| •••— | V | •••—•— | End of transmission (SK) |
| •—— | W | •••———••• | International distress signal (SOS) |
| —••— | X | | |
| —•—— | Y | | |
| ——•• | Z | | |

advantage of codes. In this case, a code consists of three letters that represent an instruction or a complete sentence. Table 25-2 lists some of these, generally referred to as the Q code. Some telegraphy transmissions consist entirely of 5- or 6-letter codes. At the receiving end, these codes are translated into English sentences, often with the help of a book containing a listing of such codes and their meanings.

# The carrier

With wired telegraphy, a carrier consists of a direct current made to move between various locations. The transmission of a message by dot and dash interruptions of that current is referred to as modulation. The Morse code can take advantage of a direct current for the wired transmission of information. It can also be used in connection with high-frequency ac, which is then broadcast. The ac is referred to as a continuous wave, abbreviated as *CW*.

*Table 25-2. Portion of the Q code.*

| Abbreviations | Question | Answer |
|---|---|---|
| QRA | What is the name of your station? | The name of my station is . . . |
| QRB | How far approximately are you from my station? | The approximate distance between our stations is . . . nautical miles (*or* . . . kilometers). |
| QRD | Where are you bound and where are you from? | I am bound for . . . from . . . |
| QRE | What is your estimated time of arrival at . . . (place)? | My estimated time of arrival at . . . (place) is . . . hrs. |
| QRF | Are you returning to . . . (place)? | I am returning to . . . place *or* Return to . . . (place). |
| QRG | Will you tell me my frequency (*or* that of . . . )? | Your frequency (*or* that of . . .) is . . . kHz (*or* MHz). |
| QRH | Does my frequency vary? . . | Your frequency varies. |
| QRI | How is the tone of my transmission? | The tone of your transmission is . . . (1) Good; (2) Variable; (3) Bad. |
| QRK | What is the readability of my signals (*or* those of . . .)? | The readability of your signals (*or* those of . . .) is . . . (1) Unreadable; (2) Readable now and then; (3) Readable, but with difficulty; (4) Readable; (5) Perfectly readable. |
| QRL | Are you busy? . . . . . . . . . . . | I am busy (*or* I am busy with . . .). Please do not interfere. |
| QRM | Are you being interfered with? . . . | I am having interference. |
| QRN | Are you troubled by static? . . | I am troubled by static. |
| QRO | Shall I increase power? . . . . | Increase power. |
| QRP | Shall I decrease power? . . . . | Decrease power. |

# Types of modulation

There are a number of types of modulation, including:

Amplitude modulation (AM)
Frequency modulation (FM)
Pulse code modulation (PCM)
Pulse amplitude modulation (PAM)
Pulse duration modulation (PDM)
Pulse position modulation (PPM)
Differential pulse code modulation (DPCM)
Delta modulation (DM)

# Broadcast telegraphy

High-frequency ac, as indicated in Fig. 25-1, is used when telegraphy is of the broadcast type. The technique is essentially the same as wired telegraphy. The car-

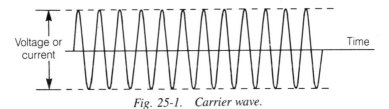

*Fig. 25-1.   Carrier wave.*

rier is interrupted by a key acting as a single-pole, single-throw switch. The illustration in Fig. 25-2 shows the transmission of the letter *C* (dash dot dash dot). In the absence of keying the carrier is unmodulated, and there is no transmission of data. Keying the carrier to produce an alphanumeric representation is one form of modulation.

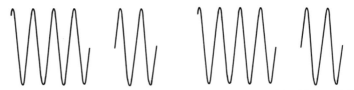

*Fig. 25-2.   Keyed carrier wave for radio transmission of the Morse Code.*

# Amplitude modulation

In AM (amplitude modulation), the carrier is high frequency ac. The frequency of the ac wave is established by the FCC (Fig. 25-3A). Modulation consists of superimposing an audio voltage (Fig. 25-3B) on the carrier with the resulting waveform shown in Fig. 25-3C. The effect of AM is to change the amplitude of the carrier.

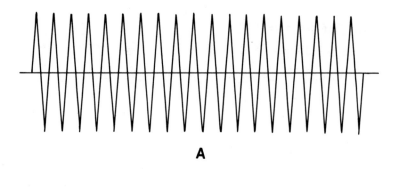

**A**

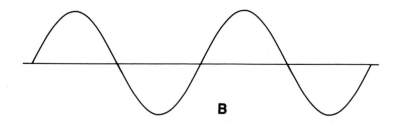

**B**

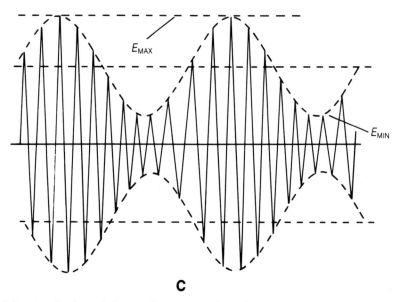

**C**

*Fig. 25-3. Amplitude modulation. Carrier signal (A); audio waveform (B); resultant modulated waveform (C).*

# Frequency modulation

With FM (frequency modulation), the amplitude of the carrier remains constant, but its frequency changes as indicated in Fig. 25-4. The change in carrier frequency is proportional to the amplitude of the modulating waveform. Both AM and FM are forms of analog modulation.

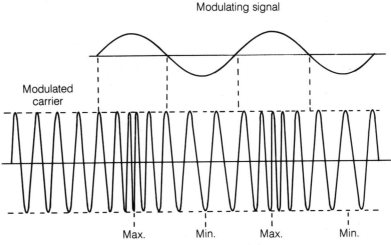

Fig. 25-4. *Frequency modulation.*

# Pulse code modulation

As in analog forms of modulation, pulse modulation can be nonbroadcast or broadcast. With the nonbroadcast type, the carrier wave is eliminated, as with compact discs.

Pulse code modulation (PCM) consists of rectangular pulses that represent the data. With pulse code modulation, the data is sampled at a predetermined rate. Each sample is quantitized and then is represented by a digital binary code. Quantitization consists of sampling a large number of instantaneous values and converting these values into binary numbers. The binary numbers are then represented by pulses.

# Pulse amplitude modulation

With pulse amplitude modulation (PAM), the time duration of each pulse remains constant, but the amplitude of each pulse is determined by the instantaneous amplitude of the original analog waveform (Fig. 25-5B).

# Pulse duration modulation

With pulse duration modulation (PDM) the amplitude of each pulse remains constant but each pulse has a different time duration (Fig. 25-5C).

# Pulse position modulation

Pulse position modulation (PPM) uses pulses that have constant amplitude and also constant time duration, but the amount of separation between the pulses can be used to represent the original analog data (Fig. 25-5D).

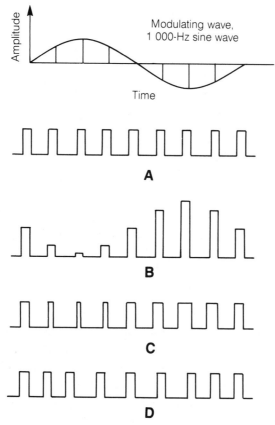

*Fig. 25-5. Pulse modulation. Pulse carrier (A); pulse amplitude modulation (B); pulse duration modulation (C); pulse position modulation (D).*

# Differential pulse code modulation

Abbreviated as DPCM, differential pulse code modulation depends on the difference between a predecessor pulse and the one following it.

# Delta modulation

Delta modulation is a type of pulse modulation. The pulses are produced by a clock circuit with the pulses having a constant amplitude and constant spacing. The pulses can be regarded as ac because they are positive or negative. When the modulating waveform is moving in a positive direction, positive pulses are generated. If that waveform moves negatively, negative pulses are produced. However, if the modulating waveform is constant—moving neither in a positive nor in a negative direction—the pulses are alternately positive and negative.

The Greek letter delta (Δ) is commonly used to indicate a change, from which this type of modulation derives its name.

# Codes

The Morse code is just one of a number of different codes. A code might be selected for security, for the ability to pack the maximum amount of data into the smallest possible code, or for other reasons. A code can use the dot-dash-space method as in the Morse code or a set of symbols that represent data. Other codes include ASCII, Baudot, Gray, Hollerith, and teletype. A code converter can be used for changing one code into another. Thus, the input at the transmitting end can be one code and converted at the receiving end into a different code. *Encoding* is the process of converting analog information into binary coded form. *Decoding* is the technique of extracting data in analog form from a binary encoded signal.

### ASCII

This is an abbreviation for American Standard Code for Information Exchange and is used extensively for the transmission of data. Numbers from 0 to 127 are represented by special characters as indicated in Table 25-3. The seven-bit code was developed by the American National Standards Institute in 1963. The code is alphanumeric and consists of 128 digits, letters, symbols, and special control codes. As an example, the number 5 is represented by 011 0101. A control code, start of text, is 000 0010. ASCII can represent both uppercase and lowercase letters.

As a first step in using the ASCII code, locate the number or symbol in the grid. The top row gives the first group of binary or hex numbers, the left column provides the second. For example, A would be hex 41 or binary 100 0001, and 5 would be hex 35 or binary 011 0101. ASCII also uses an error-checking signal known as a parity bit. The ASCII code uses hexadecimal-to-binary conversion (see chapter 15).

Table 25-3. U.S. ASCII seven-bit code.

| 2nd digits Hex | Binary | 0 000 | 1 001 | 2 010 | 3 011 | 4 100 | 5 101 | 6 110 | 7 111 | Code | Value |
|---|---|---|---|---|---|---|---|---|---|---|---|
| 0 | 0000 | NUL | DLE | SP | 0 | @ | P | ` | p | NUL | null/idle |
| 1 | 0001 | SOH | DC1 | ! | 1 | A | Q | a | q | SOH | start of heading |
| 2 | 0010 | STX | DC2 | " | 2 | B | R | b | r | STX | start of text |
| 3 | 0011 | ETX | DC3 | # | 3 | C | S | c | s | ETX | end of text |
| 4 | 0100 | EOT | DC4 | $ | 4 | D | T | d | t | EOT | end of transmission |
| 5 | 0101 | ENQ | NAK | % | 5 | E | U | e | u | ENQ | enquiry |
| 6 | 0110 | ACK | SYN | & | 6 | F | V | f | v | ACK | acknowledge |
| 7 | 0111 | BEL | ETB | ' | 7 | G | W | g | w | BEL | audible signal |
| 8 | 1000 | BS | CAN | ( | 8 | H | X | h | x | BS | back space |
| 9 | 1001 | HT | EM | ) | 9 | I | Y | i | y | HT | horizontal tab |
| A | 1010 | LF | SUB | * | : | J | Z | j | z | LF | line feed |
| B | 1011 | VT | ESC | + | ; | K | [ | k | { | VT | vertical tab |
| C | 1100 | FF | SS | , | < | L | \ | l | \| | FF | form feed |
| D | 1101 | CR | GS | - | = | M | ] | m | } | CR | carriage return |
| E | 1110 | SO | RS | • | > | N | ↑ | n | ~ | SO | shift-out |
| F | 1111 | SI | US | / | ? | O | ← | o | DEL | SI | shift-in |
| | | | | | | | | | | DLE | data link escape |

1st hex and binary digits

*Table 25-3. Continued.*

| 2nd digits | | 1st hex and binary digits | | | | | | | Code | Value |
| Hex Binary | 0 000 | 1 001 | 2 010 | 3 011 | 4 100 | 5 101 | 6 110 | 7 111 | | |
|---|---|---|---|---|---|---|---|---|---|---|
| | | | | | | | | | DC1 | device controls |
| | | | | | | | | | DC2 | |
| | | | | | | | | | DC3 | |
| | | | | | | | | | DC4 | device control stop |
| | | | | | | | | | NAK | negative acknowledge |
| | | | | | | | | | SYN | synchronous idle |
| | | | | | | | | | ETB | end of transmission block |
| | | | | | | | | | CAN | canceled |
| | | | | | | | | | EM | end of medium |
| | | | | | | | | | SUB | substitute character |
| | | | | | | | | | ESC | escape |
| | | | | | | | | | SS | start of special sequence |
| | | | | | | | | | GS | group separator |
| | | | | | | | | | RS | record separator |
| | | | | | | | | | US | unit separator |
| | | | | | | | | | DEL | delete |

## Baudot code

The Baudot code is a digital code consisting of five bits, each having an equal time duration. The code is illustrated in Table 25-4.

*Table 25-4. Baudot five-bit telegraphy code.*

| Start | 1 | 2 | 3 | 4 | 5 | Stop | Lowercase | Uppercase International telegraph alphabet #2 | U.S. teletype commercial keyboard |
|---|---|---|---|---|---|---|---|---|---|
| | • | • | | | | • | A | - | - |
| | • | | | • | • | • | B | ? | ? |
| | | • | • | • | | • | C | : | : |
| | • | | | • | | • | D | Who are you? | $ |
| | • | | | | | • | E | 3 | 3 |
| | • | | • | • | | • | F | (*) | ! |
| | | • | | • | • | • | G | (*) | & |
| | | | • | | • | • | H | (*) | # |
| | | • | • | | | • | I | 8 | 8 |
| | • | • | | • | | • | J | Bell | Bell |
| | • | • | • | • | | • | K | ( | ( |
| | | • | | | • | • | L | ) | ) |
| | | | • | • | • | • | M | . | . |
| | | | • | • | | • | N | , | , |
| | | | | • | • | • | O | 9 | 9 |
| | | • | • | | • | • | P | 0 | 0 |
| | • | • | • | | • | • | Q | 1 | 1 |
| | | • | | • | | • | R | 4 | 4 |
| | • | | • | | | • | S | ' | ' |
| | | | | | • | • | T | 5 | 5 |
| | • | • | • | | | • | U | 7 | 7 |
| | | • | • | • | • | • | V | = | ; |
| | • | • | | | • | • | W | 2 | 2 |
| | • | | • | • | • | • | X | / | / |

*Table 25-4. Continued.*

| Start | 1 | 2 | 3 | 4 | 5 | Stop | Lowercase | Uppercase International telegraph alphabet #2 | U.S. teletype commercial keyboard |
|:---:|:---:|:---:|:---:|:---:|:---:|:---:|:---:|:---:|:---:|
| | | | • | • | • | • | Y | 6 | 6 |
| | • | | | | • | • | Z | + | " |
| | | | | | | • | blank | | |
| | • | • | • | • | • | • | letters shift | | |
| | • | • | | • | • | • | figures shift | | |
| | | | • | | | • | space | | |
| | | | | • | | • | carriage return | | |
| | | • | | | | • | line feed | | |

(*) = not allocated; for each country's internal use.

• = positive current mark.

*Table 25-5. Speed rates for the Baudot code.*

| Baud rate | Length of pulse (ms) | wpm* |
|:---:|:---:|:---:|
| 45.45 | 22.0 | 60.6 |
| 45.45 | 22.0 | 61.3 |
| 45.45 | 22.0 | 65.0 |
| 50 | 20.0 | 66.7 |
| 56.92 | 17.6 | 75.9 |
| 56.92 | 17.6 | 76.7 |
| 74.20 | 13.5 | 99.0 |
| 74.20 | 13.5 | 100.0 |
| 100 | 10.0 | 133.3 |

*Words per minute.

The code is not only alphanumeric but also includes symbols and codes for control operations. The code has 31 possible combinations. It can represent the 26 letters of the alphabet in addition to five control signals. Only capital letters are transmitted. The start bit is a space, and the stop bit is a mark.

The baud rate is a determination of the speed of transmission for digital code. The baud is a unit of code per second. Commonly, Baudot data speed ranges from about 45.45 to 100 baud, as indicated in Table 25-5.

# Index

chips, microchips, 364, 367-370

integrated circuits (IC), 364, 367-370

printed circuit board (PCB) connections, 404

rectifiers, semiconductor rectifiers, 364-367

transistors (*see* transistors)

sones-to-phones conversion, 216

sound and acoustics (*see also* decibels; sine waves), 207-219

absorption of sound, 215-218

audio cables, 385, 389

audio frequencies, capacitive reactance, 105

bandwidth, 438

decibel volume (loudness), ordinary sounds, 213-214

decibels (*see* decibels)

fundamental frequency, 209-211

microphone cables, 385

musical instruments frequency range, 207-211, 212

octaves, 211

phon/sone conversion, 216

piano, frequency range, 211-212

pitch, 209-211

radio frequencies, capacitive reactance, 104

reverberation time, 217-219

sine waves (*see also* sine waves)

velocity of sound in air, 212

velocity of sound in liquids, 214

velocity of sound in solids, 214

wavelength, 207-208

sound pressure level, microphones, 186

speakers, 458

specific resistance, 437

square roots, 501, 505

square waves, 78-79

standing-wave ratio, antennas, 275

stub mast antennas, 261

susceptance, 425

switches, symbols, 459-460

switching-circuits, 330-333

0 or 1 state, 332

AND condition, 330

AND-OR condition, 331-332

gates, 332-333

OR condition, 330

symbols

electronic components, 441-463

Greek alphabet, 525, 526

mathematical symbols, 522-525, 526

# T

T-connectors, 397

T-section filters, 224

tangents, phase angle, 517, 518

tantalum-foil electrolytic capacitors, 99

taps, sliding tap resistor, 20

telegraphy, broadcast, 531

television (*see* video)

television interference (tvi), 289, 293-295

temperature coefficient of resistance, 437

temperature systems conversions, 491-492

temporary magnets, 170-171

test instrument connectors/leads, 397-398

thermistors, 366

thermocouples, 470

time constants

resistive-capacitive, 127-130

resistive-inductive, 130-133

series RC circuits, 414

series RL circuits, 414

time units, sine waves, 70

tip jacks, 405

tip plugs, 391-392

tolerance, resistors, 20-24

transformers (*see also* coils), 430-432

current, 431

eddy current loss, 432

hysteresis loss, 432

impedance and turns ratio, 157-161

maximum ac magnetizing force, 433

primary and secondary turns, 430-431

symbols, 457-458

turns ratio and impedance, 157-161, 430, 432

voltage, 431-432

transient response, microphones, 186

transistors

alpha and beta, 353-356

base-current amplification factor, 439

functions of transistors, 357-358

input resistance, 356

output resistance, 356-357

power, 356

power gain, 439

resistance gain, 439

symbols, 363-364

types of transistors, 358-363

voltage gain, 439

transmission line cable, 381

triacs, 366

triangles, oblique, 517, 519, 520

trigonometric functions, 520-521

polarity, 517, 518

trimmers, conductive plastic, 24

true power, 425-426

truth tables, 350-351

AND condition, 340-341

OR condition, 339-340

tuner sensitivity in decibels, 201

tunnel diodes, 366

turns ratio, transformer impedance, 157-161

turnstile antenna, 260

twin lead cables, 381

impedance, 390

shielded, 382

signal loss, 388

# U

unity coefficient of coupling, coils, 108-109

# V

V-type antennas, 265

varactors, 366

variable resistors, 20

varistors, 365-366

VCRs, 300-301

# Other Bestsellers of Related Interest

**BEGINNER'S GUIDE TO
READING SCHEMATICS—***Robert J. Traister*

Electronics diagrams can be as easy to read as a road map with the help of this how-to handbook. You'll learn what each symbol stands for and what the cryptic words and numbers with each one mean. Block diagrams show where sections of complicated circuits are located and how they relate to each other, and flowcharts show you what should be happening where and when. 140 pages, 123 illustrations. **Book No. 1536, $11.95 paperback, $14.95 hardcover**

**ELECTRONIC CONVERSIONS: Symbols and
Formulas—2nd Edition—***Rufus P. Turner and
Stan Gibilisco*

This revised and updated edition supplies all the formulas, symbols, tables, and conversion factors commonly used in electronics. Exceptionally easy to use, the material is organized by subject matter. Its format is ideal and you can save time by directly accessing specific information. Topics cover only the most-needed facts about the most-often-used conversions, symbols, formulas, and tables. 280 pages, 94 illustrations. **Book No. 2865, $14.95 paperback, $21.95 hardcover**

**HOW TO READ ELECTRONIC CIRCUIT
DIAGRAMS—2nd Edition—***Robert M. Brown,
Paul Lawrence, and James A Whitson*

In this updated edition of a classic handbook, the authors take an unhurried approach to the task, Basic electronic components and their schematic symbols are introduced early. More specialized components, such as transducers and indicating devices, follow—enabling you to learn how to use block diagrams and mechanical construction diagrams. Before you know it, you'll be able to identify schematics for amplifiers, oscillators, power supplies, radios, and televisions. 224 pages 213 illustrations. **Book No. 2880, $14.95 paperback, $20.95 hardcover**

**THE CET STUDY GUIDE—2nd Edition
—***Sam Wilson*

Written by the Director of CET Testing for ISCET (International Society of Certified Electronics Technicians), this completely up-to-date and practical guide gives you a comprehensive review of all topics covered in the Associate and Journeyman exams. Example questions help you pinpoint your own strengths and weaknesses. Most importantly, the author provides the answers to all the questions and offers valuable hints on how you can avoid careless errors when you take the actual CET exams. 336 pages, 179 illustrations. **Book No. 2941, $16.95 paperback, $23.95 hardcover**

**THE CET EXAM BOOK—2nd Edition
*Ron Crow and Dick Glass***

An excellent source for update or review, this book includes information on practical mathematics, capacitance and inductance, oscillators and demodulators, meters, dependency logic notation, understanding microprocessors, electronics troubleshooting, and more! Thoroughly practical, it is an essential handbook for preparing for the Associate CET test! 266 pages, 211 illustrations. **Book No. 2950, $13.95 paperback, $21.95 hardcover**

**ELECTRONIC DATABOOK—4th Edition
—***Rudolf F. Graf*

If it's electronic, it's here—detailed and comprehensive! Use this book to broaden your electronics information base. Revised and expanded to include all up-to-date information, this fourth edition makes any electronic job easier and less time-consuming. You'll find information that will aid in the design of local area networks, computer interfacing structure, and more! 528 pages, 131 illustrations. **Book No. 2958, $24.95 paperback, $34.95 hardcover**

**THE COMPLETE ELECTRONICS CAREER GUIDE**—*Joe Risse*

Here is all the information you need to get started in a satisfying career in electronics! This book presents electronics from a career prespective. It shows what jobs are out there, what it takes to get one, and whether a career in electronics is right for you. Just a few of the careers covered include broadcasting and communications, robotics, electronic assembly, sales engineering, aerospace and aviation electronics instrumentation, computer servicing, and others. 208 pages, 64 illustrations. **Book No. 3110, $12.95 paperback, $19.95 hardcover**

**THE BENCHTOP ELECTRONICS REFERENCE MANUAL**—2nd Edition —*Victor F. C. Veley*
Praise for the first edition:

*". . . a one-stop source of valuable information on a wide variety of topics . . . deserves a prominent place on your bookshelf."* —*Modern Electronics*

Veley has completely updated this edition and added new sections on mathematics and digital electronics. All of the most common electronics topics are covered: AC, DC, circuits, communications, microwave, and more. 784 pages, 389 illustrations. **Book No. 3414, $29.95 paperback, $39.95 hardcover**

**Prices Subject to Change Without Notice.**

## Look for These and Other TAB Books at Your Local Bookstore

## To Order Call Toll Free 1-800-822-8158

or write to TAB Books, Blue Ridge Summit, PA 17294-0840.

| Title | Product No. | Quantity | Price |
|---|---|---|---|
| | | | |
| | | | |
| | | | |
| | | | |

☐ Check or money order made payable to TAB Books

Charge my ☐ VISA ☐ MasterCard ☐ American Express

Acct. No. _____ Exp. _____

Signature: _____

Name: _____

Address: _____

City: _____

State: _____ Zip: _____

Subtotal $ _____

Postage and Handling
($3.00 in U.S., $5.00 outside U.S.) $ _____

Add applicable state and local
sales tax $ _____

TOTAL $ _____

TAB Books catalog free with purchase; otherwise send $1.00 in check or money order and receive $1.00 credit on your next purchase.

*Orders outside U.S. must pay with international money order in U.S. dollars.*

**TAB Guarantee: If for any reason you are not satisfied with the book(s) you order, simply return it (them) within 15 days and receive a full refund.**                    **BC**